# Dictionary of environmental science and technology

Second Edition

ANDREW PORTEOUS

D1324819

## JOHN WILEY & SONS

Chichester · New York · Brisbane · Toronto · Singapore

Second edition published 1996 by John Wiley & Sons Ltd,
Baffins Lane, Chichester,
West Sussex PO19 1UD, England
National    01243 779777
International (+44) 1243 779777

Reprinted June 1996

Revised edition published 1992
First edition published 1991 by Open University Press

*Other Wiley Editorial Offices*

John Wiley & Sons, Inc., 605 Third Avenue,
New York, NY 10158-0012, USA

Jacaranda Wiley Ltd, 33 Park Road, Milton
Queensland 4064, Australia

John Wiley & Sons (Canada) Ltd, 22 Worcester Road,
Rexdale, Ontario M9W 1L1, Canada

John Wiley & Sons (Asia) Pte Ltd, 2 Clementi Loop #02--1,
Jin Xing Distripark, Singapore 0512

***Library of Congress Cataloging-in-Publication Data***

Porteous, Andrew
    Dictionary of environmental science and technology / Andrew Porteous. —
2nd ed.
        p. cm.
    ISBN 0-471-96075-6 (alk. paper)
    1. Environmental sciences—Dictionaires.    2. Sanitary engineering —
Dictionaires.    I. Title.
GE10.P67    1996                                                    95–46374
628'.03—dc20                                                        CIP

***British Library Cataloguing in Publication Data***
A catalogue record for this book is available from the British Library

ISBN 0-471-96075-6

Typeset in 10/11 pt Times by Vision Typesetting, Manchester
Printed and bound in Great Britain by Biddles Limited, Guildford and King's Lynn

# Contents

# *Preface to Second Edition*

Time marches on as they say, and this second edition is my way of documenting advances in environmental practices, knowledge and perceptions.

The opportunity has been taken to greatly expand the contents. I hope it meets with your approval.

My grateful thanks to my colleagues and my secretary Mrs Morine Gordon for typing this second edition.

Balaam (Numbers, Chapter 24), 'I came to curse [this task] but stayed to bless it!'

Andrew Porteous
*Professor of Environmental Engineering*
*Faculty of Technology*
*Open University*

# Preface

This text springs from an earlier attempt when, along with my colleague Geoffrey Holister, we endeavoured to lay the foundations for widespread environmental literacy in *The Environment – a dictionary of the world around us* published in 1976. This new work is not quite so ambitious; it is focused on the science and technology of environmental protection and resource management, as this is where the environmental payoffs are greatest.

The text has been principally written in basic SI units of kg, m, s, but in Tables and Figures, the units may be sub-values of the basic SI units, e.g. $g/m^3$ or kg/h. The SI system is admirable for setting out theory and equations but can become cumbersome if numerical values have to be written frequently in terms of powers of ten. For example volume flows have been written as litre/s which is much more familiar than using $10^{-3}$ m/s. Equations are written for temperatures in K, but Tables and Figures are given in °C because this is the common unit of practical measurement. On occasions, concentrations have been given in non-SI units as these are often enshrined in current legislation or codes of practice.

I should like to put on record my thanks to my colleagues Judy Anderson and Rod Barratt, who have reviewed the text with diligence, and to Lesley Booth who had the trying task of typing it. My colleagues Keith Attenborough, David Cooke, David Yeoman, have kindly commented on specific entries. Caryl Hunter-Brown, OU liaison librarian, compiled the invaluable directory of environmental organizations.

Andrew Porteous
*Professor of Environmental Engineering*
*Faculty of Technology*
*Open University*

# Introduction

Stanley Clinton Davis in his 1987 Royal Society of Arts Lecture 'The European Year of the Environment' gave two reasons why the public wanted action by government on environmental issues now rather than later. These are given below.

> All governmental decisions tend to be taken under pressure from particular special interest groups. And it is a sad fact that the producers of pollution are on the whole better at exerting such pressure than those who have to live with its consequences. The manufacturers of nitrate fertilizers are well organized to ensure that their views are known in official circles. Those who worry about the fish that subsequently die are not. There are, of course, exceptions to this rule. There are well organized campaigns in which the view of the man and woman in the street is fully brought home to the man (and sometimes woman) in the Ministerial Office – I vividly recall the recent example of baby seals – but they are rare. I have direct explanations of why they should continue to do whatever it is they are doing. And I receive a steady stream of letters from the public inevitably tending to be more personal, and less well argued. I find that I have to be constantly on the alert not to let myself be misled by the different levels of presentation.

> The second reason why people are better than governments at spotting environmental needs goes still deeper. It is because environmental policy is fundamentally about the *future*, while governmental decision making is often too exclusively focused on the present.

These were brought home to me when I heard two senior British politicians respectively declaim on a television discussion programme 'PCBs cause toxic algae blooms in the North Sea' and 'Fitting catalytic converters to cars may reduce sulphur dioxide but will still produce gases that deplete the ozone layer'.

This book is written in the hope that it may contribute to environmental literacy, by providing basic definitions and data plus demonstrate the nature of the issues. Where appropriate, the technology and/or measures that are already available to prevent pollution and aid resource conservation are outlined. As our understanding of these becomes clearer, we may at least begin to appreciate a number of things that we must obviously *not* do and that is a start of sorts.

In writing this book I have had to select information from an almost infinite source of data, much of it conflicting. In making this selection, I have naturally had to apply my own value judgements as to which data are relevant and important and which are not. In doing this, I have attempted to be as objective as possible, but under such circumstances perfect objectivity is quite impossible. We cannot assess the value of data without passing a tacit judgement on the character of the source. Objectivity does not consist in giving equal weight to all statements.

Environmental problems are essentially multi-faceted and demand at least a nodding acquaintance with many previously separate specialisms – ecology, economics, sociology, technology, physics, chemistry, and so on. The world is an enormously complex system and it is in the nature of complex systems that the characteristics of the connections between the constituent parts are often more important than the nature of the separate parts themselves.

This book is designed for a multi-access approach on the part of the reader – it can be dipped into, as well as read straight through. The reason for this format is the obvious diversity of backgrounds and interests of the readers. Most of you who read this book will have some specialist knowledge of some aspect of our technological society, and are likely to be interested in one particular aspect of our environment more than another. The format of this book allows you to select those areas of interest, although because of the complex nature of environmental issues, you will find that *wherever* you start in this book, you will be led inexorably by the references to other areas of the problem that are probably new to you. But that is the nature of the environment; everything is related, in some way or another, to everything else.

As Chief Seattle of Suquamish Indians said of the earth over 130 years ago:

> . . . *The earth does not belong to man; man belongs to the earth . . . all things are connected . . .*
> *Man did not weave the web of life, he is merely a strand in it. Whatever he does to the web, he does to himself.*

# Acknowledgements

Grateful acknowledgement is made to the following sources for permission to reproduce material in this book:

*Figure 1*: based on material supplied courtesy of CEGB Research, with kind permission of the Editor.

*Figure 3*: London Scientific Services, *London Air Pollution Monitoring Network*, Second Report, 1987.

*Figures 6 and 7*: based on material supplied courtesy of Hamworthy Engineering, Poole, Dorset.

*Figures 8 and 41*: based on material in *Renewable Energy*, Issue 6, January 1989, Department of the Environment.

*Figures 12 and 13*: based on material supplied courtesy of Keep Britain Tidy Group Schools Research Project, Brighton Polytechnic, Science Unit 2, Glass.

*Figure 23*: taken from S. H. Schneider and R. D. Dennett, 'Climatic barriers to long-term energy growth', *Ambio*, vol. 4, no. 2, 1975.

*Figures 24–28*: based on material in 'Focus on Combined Heat & Power', *Energy Management*, Issue 7, October 1988.

*Figures 42, 51 and 52*: Open University, Course PT272, 'Environmental Control and Public Health'.

*Figure 43*: from the Royal Commission on Environmental Pollution, *Report No. 1*, HMSO, 1971.

*Figure 44*: from *Coal and Energy Quarterly*, no. 7, Winter 1975.

*Figure 47*: Ashman, R., 'Air Pollution Control' in A. Porteous (ed.), *Developments in Environmental Control and Public Health*, Applied Science Publishers, 1979.

*Figure 49*. from Pringle, L. *Ecology: Science of Survival*, Macmillan Co., 1971.

*Figure 64*: from *Energy Conservation*, HMSO, 1974.

*Figure 73*: based on material supplied courtesy of CEGB Research, September 1988, with kind permission of the Editor.

*Figure 80*: from Tucker, A., *The Toxic Metals*, Pan/Ballantine, 1972.

*Figure 83*: from Burns, W., *Noise and Man*, 2nd edition, John Murray, 1973.

*Figure 84*: data from Borgstrom, G., *Too Many*, Macmillan, 1967.

*Figure 87*: courtesy of Motherwell Bridge Engineering, PO Box 4, Logans Road, Motherwell, Scotland.

*Figures 88 and 89*: courtesy of Warwickshire County Council. Coping with Landfill Gas Report, no. 4/4, November 1988, County Surveyors Society.

*Figure 92*: Campbell, D. V., *Landfill – An environmentally acceptable method of waste disposal and an economic source of energy*. Paper C86/85, Proc. 1. Mech. E. Conference, Agricultural, Industrial and Municipal Waste Management, Warwick, April 1985.

*Figure 111*: from Warman, H. R., 'World energy prospects and North-Sea oil', *Coal and Energy Quarterly*, no. 6, Autumn 1975, p. 24.

*Figure 117*: Claerbout, J. 'Material and Energy Recycling of PVC: Case Studies', in Karl J. Thomé-Kozmiensky (ed.), *Recycling International*, vol. 3, EF-Verlag.

*Figure 138*: from *Land, Air and Sea – Research in Man's Environment*, NERC, 1975.

*Figure 151*: courtesy of Blue Circle Group.

*Figure 153*: Knox, K., Ch. 4, in A. Porteous (ed.), *Hazardous Wastes Management Handbook*, Applied Science Publishers, 1985.

*Figure 154*: from Metropolitan Water Division, Thames Water Authority, *A Brief Description of the Undertaking*, 1972.

# Abbreviations

| | |
|---|---|
| ADI | Acceptable daily intake |
| A/G ratio | Arithmetic–geometric ratio |
| ATU | Allythiourea |
| BATNEEC | Best Available Techniques Not Entailing Excessive Cost |
| BHC | Benzene hexachloride |
| BOD | Biochemical oxygen demand ($BOD_x$ signifies measurement over $x$ days) |
| BPEO | Best Practicable Environmental Option |
| BPM | Best Practicable Means |
| Ci | curie |
| CNL | Corrected noise level |
| COD | Chemical oxygen demand |
| c.o.p. | Coefficient(s) of performance |
| dB | decibel |
| dB(A) | decibels A-scale |
| DDE | 1,1-dichloro-2,2-bis (p-chlorophenyl)ethene |
| DDT | Dichlorodiphenyltrichloroethane |
| DDVP | Dichlorvos 2,2-dichlorovinyl dimethyl phosphate |
| DES | Diethylstilboestrol |
| DNA | Deoxyribonucleic acid |
| DO | Dissolved oxygen |
| DOA | Dioctyladipate |
| EDTA | Ethylenediamine tetraacetate disodium salt or ethylenediamine tetraacetic acid |
| EPA | Environmental Protection Act |
| EPA | Environmental Protection Agency (USA) |
| EPNdB | Effective perceived noise level |
| ERTS | Earth Resources Technology Satellite |
| FMC | Field moisture capacity |
| GNP | Gross national product |
| HMIP | Her Majesty's Inspectorate of Pollution |

| Hz | hertz |
| ICRP | International Commission for Radiological Protection |
| IPC | Integrated Pollution Control |
| IWC | International Whaling Commission |
| K | Kelvin |
| $L_{10}$ | 10 per cent level (road traffic noise index) |
| LAWDC | Local Authority waste disposal company |
| $LC_{50}$ | Lethal concentration (50 per cent survival) |
| $LD_{50}$ | Lethal dose (50 per cent survival) |
| LFG | Landfill gas |
| MAC | Maximum allowable concentration |
| MAFF | Ministry of Agriculture, Fisheries and Food |
| MHD | Magnetohydrodynamic generator |
| MLVSS | Mixed liquor volatile suspended solids |
| MPN | Most probable number |
| MSW | Municipal Solid Wastes |
| NNI | Noise and number index |
| OECD | Organization for Economic Co-operation and Development |
| PAN | Peroxyacetylnitrate |
| PCB | Polychlorinated biphenyl(s) |
| PNdB | Perceived noise decibels |
| pphm | parts per hundred million |
| ppm | parts per million |
| PTFE | Teflon |
| PVC | Polyvinyl chloride |
| RBE | Relative biological effectiveness |
| RCEP | Royal Commission on Environmental Protection |
| R/P ratio | Reserves–production ratio |
| SCP | Single-cell protein |
| SI | International System of Units |
| SMD | Soil moisture deficit |
| STP | Standard temperature and pressure |
| TCDD | Tetrachlorodibenzo-*p*-dioxin |
| TLV | Threshold limiting value |
| TNI | Traffic noise index |
| UDC | Underdeveloped country |
| VCM | Vinyl chloride monomer |
| VOC | Volatile organic compounds |
| WHO | World Health Organization |
| WLM | Working level month |

# A

**Abatement.** Reduction or lessening (of pollution) or doing away with a nuisance, by legislative or technical means, or both.

**Absolute temperature scale.** ⇨ TEMPERATURE.

**Absorption**[1]. The taking up, usually, of a liquid or gas into the body of another material (the absorbent). Thus, for instance, an air pollutant may be removed by absorption in a suitable solvent. Not to be confused with ADSORPTION.

**Absorption**[2]. The taking up of radiant energy by a material it encounters or passes through. This can be the basis of measuring the concentration of a substance or identifying a number of substances.

**Absorption chillers.** District heating (DH) can also be used as the energy source for absorption chillers. The benefits of encouraging potential consumers to install absorption chillers for their chilled water and air conditioning requirements are two-fold. Firstly absorption systems use no CFCs or HFCs, so are environmentally friendly; secondly they have very few moving parts and are consequently relatively easy to maintain.

The chilling effect is produced in the absorption cycle by the evaporation of a fluid – the refrigerant – in the lower pressure side of a closed system. The refrigerant vapour is recovered by the absorbing action of a second fluid – the absorbent. In the higher pressure side of the system the absorbent containing the dissolved refrigerant is boiled, the refrigerant vapour and the absorbent being returned to the lower pressure side for further use.

Indirect fired machines can use medium temperature hot water or steam in the range 104 °C to 130 °C, making them ideal for connection to the DH network.

Absorption chillers also produce a large benefit to the DH operator, in that they increase the summer demand for heat, thus providing a balanced loading over as much of the year as possible, to utilize the available energy efficiently. If a building's winter heating and summer cooling requirements can be met using the district heating network, it will greatly enhance the efficiency of the scheme and increase annual revenues.

**Abstraction.** The permanent or temporary withdrawal of water from any source of supply, so that it is no longer part of the resources of the locality.

**Acceptable daily intake (ADI).** The acceptable daily intake of a chemical is the daily intake which, during an entire lifetime, appears to be without risk on the basis of all known facts at the time. It is expressed in milligrams of chemical per kilogram of body weight (mg/kg). ADIs may be unconditional, conditional, or temporary.

**Accuracy.** The nearness of a measurement to the true value of the thing being measured. cf. PRECISION.

**Acid.** A substance which, in solution in water, splits up to a greater or lesser extent into ions, including hydrogen IONS. Strong acids are those which split up to a large degree; they are usually corrosive. Acids neutralize ALKALIS with the formation of salts.

**Acid dewpoint.** As commonly used, this term applies to the temperature at which dilute acid (e.g. sulphuric acid) appears as a condensate (liquid droplets) when a flue gas containing sulphur trioxide and water vapour is cooled below saturation temperature. This value is related to the moisture and sulphur trioxide content of the flue gas. Therefore, gases must be released to atmosphere well above this temperature otherwise substantial corrosion will take place in the stack. The corrosion of many combustion installations especially if fired on heavy fuel oil is due to the flue gases being cooled below the acid dewpoint.

**Acid mine drainage.** Many mining operations, particularly those that work sulphide ores, e.g. nickel, copper or coal mining where pyrites (iron sulphide) are present, can, through a combination of air and moisture, form acidic and metal-bearing solutions. This combination of acids and metals can have severe local effects on the ecology of streams and rivers, and the metals can enter food chains and further affect life. If, as is likely, iron sulphates are formed, this will result in brown coating on rocks and stream beds, further despoiling the appearance of the area.

In the UK the problems of acid mine drainage are not severe, but in nickel and copper mining areas (e.g. Sudbury, Ontario) the devastation has to be seen to be believed.

The only way of controlling or preventing acid mine drainage other than not working the ores is to control the pH by raising it with lime and neutralizing the drainage water. The tips can then be grassed and the subsequent exclusion of air will prevent the oxidizing of the sulphides. The successful application of revegetation techniques requires the initial neutralization of the acidic effluents. ($\Rightarrow$ pH)

**Acid rain.** Most rainfall is slightly acidic due to carbonic acid from the carbon dioxide content of the atmosphere, but 'acid rain' in the pollution sense is produced by the conversion of the primary pollutants sulphur dioxide and nitrogen oxides to sulphuric acid and nitric acid respectively. These processes are complex, depending on the physical dispersion processes and the rates of the chemical conversions. The basic cycle is shown in Figure 1.

However, the term 'acid rain' is misleading because the phenomenon and associated effects are broader than it suggests. Pure water has a neutral pH value of 7. Carbon dioxide in the atmosphere dissolves in rain reducing its pH to 5.6 and naturally occurring oxides of sulphur and nitrogen are responsible for unpolluted rain having a pH of about 5.0. Lower values of pH may result from strong acids produced from fuel use. An alternative unit of acidity is the 'microequivalent of hydrogen ion per litre', written as $\mu$eq $H^+$/l, as defined by

$$pH = -\log_{10}(\mu eq \ H^+/l \times 10^{-6})$$

Hence, a value of $10 \mu$eq $H^+$/l equates to a pH $= -\log_{10}(10^{-5}) = 5$.

The following definitions are in use:

*Acid precipitation:* rainfall or snow with acidity greater than $10 \mu$eq $H^+$/l or pH lower than 5.

*Acid mists:* fog, mist or low cloud in which the water has an acidity greater than $10 \mu$eq $H^+$/l or pH lower than 5.

*Acid deposition:* total deposition of acid (hydrogen ions) or acid-forming compounds e.g. $SO_2$ and $NOX_x$ by both wet and dry deposition.

*Acid rain:* Precipitation and other deposition pathways which are more acidic than pH 5.0.

*Source:* 'The down-side of cleaner air', *Financial Times*, 28 Feb. 1995.

The solutions to acid rain are as a consequence not simple. DESULPHURIZATION of flue gases alone will not necessarily

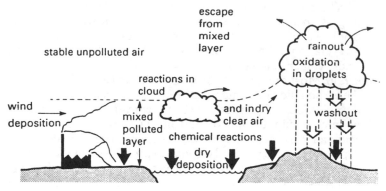

**Figure 1**   Acid rain cycle.

effect a cure. The high levels of NITROGEN OXIDE and UNBURNT HYDROCARBONS from motor vehicles are also implicated.

Damage to trees is usually worst on hills where mists collect and deposition is greater. The soils are often poor and acidic to begin with, hence the additional stresses imposed by acid rain alone or in conjunction with ozone can prove too much as has been found in the German mountain forests and increasingly so in the UK as well.

To cause soil acidification it is necessary to have:

(a) A source of hydrogen ions to exchange for the exchangeable BASES.

(b) Some means of removing the displaced base CATIONS.

Acid precipitation supplies both. The source of hydrogen IONS is externally applied ACID (sulphuric or nitric acid). Cations are removed by leachating of the displaced base cation by the mobile sulphate or nitrate anion.

(It has been estimated that about 75 per cent of the UK now receives under 20 kg of atmospheric sulphur per hectare per year which is less than half the amount needed to produce a full crop of oilseed rape, insufficient for either silage grass or cereals and barely enough for potatoes or sugar beet.)

A short-term remedial measure for water sources and lakes consists of dosing with lime to restore the pH balance. (⇨ DEPOSITION PROCESSES; FLUE GAS DESULPHURIZATION; pH)

**Actinides.** The group of elements, with ATOMIC NUMBERS from 89 to 103, including URANIUM and PLUTONIUM. All members of this group are produced in nuclear reactions or by radioactive decay. Many of the more long lived are RADIATION emitters.

**Action level.** The CONCENTRATION level of a substance at which there is negligible environmental impact or which triggers remedial/ameliorative action to safeguard public health. This is based on current knowledge and can be revised (usually downwards) (⇨ DIOXINS, NITROGEN OXIDES, RADON)

**Activated carbon.** Carbon obtained from vegetable or animal matter by roasting in a vacuum furnace. Its porous nature gives it a very high surface area per unit mass – as much as 1000 square metres per gram, which is 10 million times the surface area of one gram of water in an open container. For this reason, the substance is a very good adsorbent for aromatic and unsaturated aliphatic compounds. It is extensively used for odour control and air-freshening applications, and can adsorb large quantities of gases. For effluent treatment purposes, three specific modes of action can be identified:

1. *Adsorption* – Activated carbon adsorbs substances which are not, or only slightly, adsorbed by the activated sludge process.
2. *Catalysis* – Activated carbon accelerates the biological decomposition of adsorbed substances by increasing the reaction rate, even at lower temperatures.
3. *Adsorption/Desorption* – Activated carbon adsorbs substances temporarily, desorbing them into the sludge system at a later stage. This gives important benefits in the treatment of waste water.

Powdered activated carbon, suspended in water, can adsorb large quantities of oxygen. In an activated sludge effluent treatment system, this allows a higher aeration efficiency. The adsorbed oxygen is desorbed into the sludge once the oxygen concentration around the carbon particles is lowered by microbiological activity. The addition of powdered carbon increases the rate of sedimentation of the sludge by a nucleation process giving denser, more compact floccules.

This allows greater concentrations of sludge to be maintained in the aeration basin. The sludge itself drains better, allowing discharge transport costs to be reduced. It can also adsorb organic and inorganic substances that are not, or only partly, decomposed by the activated sludge system. This applies to most toxic compounds.

When granular activated carbon is used, it is commonly

regenerated, with the method chosen dependent on the nature of the adsorbed impurity and the ease with which it can be removed. The three most commonly used methods are:

*Steam* – This can be used if the adsorbed products are volatile, i.e. if they can be steam distilled.

*Chemical Regeneration* – It is sometimes the case that the adsorbed impurity can be desorbed from the surface by treatment with a chemical. Caustic soda has been used successfully with certain organic acid purification systems.

*Thermal Regeneration* – This is the most widely used method particularly in the large tonnage outlets in the sugar, glucose and water industries. This involves the removal of the carbon from the column and burning off the impurities under controlled conditions in a regeneration kiln.

**Activated sludge.** The sludge removed from the activated sludge sewage treatment process. It consists of BACTERIA and PROTOZOA which can live and multiply on the sewage. Because of this multiplication, the excess organisms require continuous removal. Part of the still active sludge is returned to the raw sewage (hence the term 'activated sludge'), and part (approximately 90 per cent) is sent for disposal to land, sea or incineration. (⇨ SEWAGE TREATMENT)

**Activated sludge process.** ⇨ SEWAGE TREATMENT.

**Activation product.** Material made radioactive as a result of irradiation particularly by neutrons in a nuclear reactor.

**Acute.** Of a disease or medical condition, coming to a rapid crisis, the opposite to CHRONIC.

**Adiabatic.** Without loss or gain of heat. When air rises, air pressure decreases and it expands adiabatically in the atmosphere; since it can neither gain nor lose heat its temperature will fall as it expands to fill a larger volume. This gives rise to the theoretical adiabatic temperature profile or lapse rate, which is used as a basis to comparison for actual temperature profiles (from ground level) and hence predictions of stack gas dispersion characteristics.

**Adipates.** ⇨ PHTHALATES.

**Adsorption.** A phenomenon in which molecules of a substance (the adsorbate) are taken up and held on the surface of a material (the adsorbent) (compare ABSORPTION). (⇨ ACTIVATED CARBON)

**Aeration zone.** The area above the SATURATION ZONE where WATER/LEACHATE infiltration percolates partially filling PORE spaces and voids. The aeration zone can be split into three: (i) the soil water zone, (ii) the intermediate zone and (iii) the capillary fringe.

**Aerobic processes.** Many bio-technology production and effluent treatment processes are dependent on micro-organisms which require oxygen for their metabolism. For example, water in an aerobic stream contains DISSOLVED OXYGEN and therefore organisms utilizing this can oxidize organic wastes to simple compounds as below:

$$\text{organic materials} + O_2 \xrightarrow{\text{micro-organisms}} CO_2 + H_2O + NH_3 \\ + \text{micro-organisms}$$

Ammonia ($NH_3$) can then be further oxidized to a compound containing the nitrate ion ($NO_3^-$) by bacteria and the organic waste such as sewage is thereby totally decomposed. (⇨ SEWAGE TREATMENT)

**Aerosol.** Minute liquid droplets or solids of particle size up to 100 $\mu$m suspended in a gaseous flow or the atmosphere. Due to their small size, they can be readily dispersed. There are two types: *condensation aerosols* formed when moisture-laden gases are cooled and *dispersion aerosols* formed from the break up of solids or atomization of liquids.

**Aerosol propellant.** Colloquially, the pressurized system used to disperse hair spray, deodorant, etc. More accurately, an inert liquid with a low boiling point, from the CHLOROFLUOROCAR-BONS or HYDROCARBONS, which vaporizes instantaneously at room temperatures on release of pressure. When the pressure in the aerosol canister is released, the vapour carries the AEROSOL of the desired substance to its target. The propellant then disperses into the atmosphere. Compounds containing chlorine are a hazard to the earth's OZONE SHIELD because of the free CHLORINE liberated in the upper atmosphere as a result of ultra-violet radiation.

The use of fluorocarbon propellants is intended to be virtually terminated by the early 1990s, except for medical purposes.

Other propellants are available, such as butane, which is highly flammable and therefore only suitable for spraying water-based products. CARBON DIOXIDE, nitrous oxide and nitrogen are possibly suitable for most spray purposes and are inert, but there is no commercial incentive for their adoption, and, as yet, no legal requirement that this be done to protect the environment.

**After-burner.** In COMBUSTION equipment, an after-burner may be fitted following the main zone of combustion in order to obtain virtually complete combustion, thus decreasing emissions of certain air pollutants.

After-burners may also be used in non-combustion plants to control pollutant emissions.

**Agent Orange.** ⇨ DIOXINS.

**Agricultural economics.** The application of traditional economic criteria to the production of food. It is unfortunate that government policies are usually designed by specialists whose experiences are essentially urban and industrial, and who have little or no understanding of the complex ecological balances required for a stable agricultural system.

Agriculture cannot be regarded simply as an industry. The majority of our industries are concerned with the conversion of raw materials – which are often regarded as virtually inexhaustible – into goods. The raw material of agriculture is the soil, which should be husbanded – not mined.

**Agriculture, energy and efficiency aspects.** The effectiveness of modern agriculture is invariably measured in traditional economic terms – percentage return on invested capital, yields per acre, yields per man-year, etc. It is in these terms that claims for the high efficiency of modern industrialized agricultural techniques are made. At first sight, these claims are persuasive: in Britain, average wheat yields are up from 0.95 tonne per acre in 1946 to 1.4 tonnes in 1968–9; barley up from 0.9 tonne per acre in 1946 to 1.4 tonne per acre in 1968–9; and with similar increases in most other field crops. Cereal yields approaching 3 tonnes per acre have been obtained in the late 1980s. (⇨ AGRICULTURAL ECONOMICS)

Unfortunately, close investigation shows that when such figures are apportioned and averaged per agricultural worker they misrepresent the true facts because of the nature of the oversimplifications that are made. For instance, figures showing increased yields per farm worker ignore all of those workers who have moved from farms into agriculture-related industries, and are engaged in the production of the machines, fuel and chemicals which give the remaining farm workers an artificially greater efficiency.

A more realistic measure of agricultural efficiency is given by the relative inputs and outputs of energy to the total system. This calculation is difficult when applied to a modern industrial-type agricultural system; nevertheless, the total energy budgets for British agriculture have been calculated (see Leach and Slesser, 1973) and compared with the efficiency of other agricultural systems (see Leach, 1974). The general conclusions were that, where a fossil-fuel subsidy *does not exist*, human energy can be used very efficiently. For example, maize growers in Yucatan produce 13 to 29 units of food energy for each unit of (predominantly human) energy expended; primitive gardeners in Tsembaga, New Guinea, produce 20 units of energy for each unit expended. British wheat growers, on the other hand, produce 2.2 energy units for each unit expended; potato growers gain 1.1 units per unit expended; sugar beet, by the time it has been refined to white sugar, represents 0.49 units gained per unit expended; a battery egg, including the food value of the hen carcase at the end of her laying life, represents 0.16 units for each unit expended; and a broiler chicken, 0.11 units gained.

*Source:* G. Leach and M. Slesser, *Energy Equivalent of Network Inputs to Food Producing Processors*, Strathclyde University, 1973.
G. Leach, *The Man–Food Equation*, Academic Press, 1974.

**Air.** The mixture of gases which constitutes the earth's ATMOSPHERE. The approximate composition of dry air is, by volume at sea level, nitrogen 78.0 per cent, oxygen 20.95 per cent, argon 0.93 per cent, and CARBON DIOXIDE 0.03 per cent, together with very small amounts of numerous other constituents. The water vapour content is highly variable and depends on atmospheric conditions. Air is said to be pure when none of the minor constituents is present in sufficient concentration to be injurious to the health of human beings or animals, to damage vegetation, or to cause loss of amenity (e.g. through the presence of dust, dirt, or odours or by diminution of sunshine).

**Air, primary.** ⇨ COMBUSTION.

**Air, secondary.** ⇨ COMBUSTION.

**Air classifier (or air separator).** Dry separation device used to separate shredded domestic refuse by density difference methods. It can be used in conjunction with ferrous-magnet and revolving screens to sort municipal waste or commercial wastes for recycling and WASTE-DERIVED FUEL production.

Figure 2 shows a vertical air classifier. The shredded refuse is fed through the air lock into a counter-current stream of air

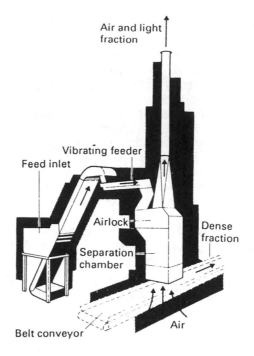

Air and light fraction

Vibrating feeder

Feed inlet

Airlock

Dense fraction

Separation chamber

Belt conveyor

Air

**Figure 2**   Air separator.

which entrains the light fractions (paper and plastics) while allowing the dense fraction (glass, tin cans, stones) to drop down onto the conveyor belt. The light fraction can then be further processed to a pelletised fuel or fed loose to a furnace as a fuel supplement.

**Air pollution index.** An arbitrary function of the concentration of one or more pollutants used to scale the severity of air pollution. For example, the following index has been used in the USA: 10 times the $SO_2$ concentration plus the CO concentration (both in ppm by volume) plus twice the coefficient of HAZE. The 'alarm' level on this scale is 50 (or more). The average value is 12. Measures of this nature are quite empirical and great caution is required in interpretation.

**Air quality.** Air quality usually refers to the CONCENTRATION in air of one or more pollutants. For many pollutants, air quality is expressed as an average concentration over a certain period of time, e.g. $mg/m^3$ over an 8 hour mean.

**Air Quality Act.** An act passed by the US Government in 1967 requiring the states to establish regional AIR QUALITY standards (and setting up a timetable for their doing so) and to control emissions in accordance with national criteria where these existed. It was amended by the Clean Air Act of 1970. (⇨ CLEAN AIR ACT (USA))

**Air quality criteria.** Describe the effects that may be expected to occur whenever and wherever the ambient air concentration of a pollutant reaches or exceeds certain values for specific time periods. Air quality criteria are descriptive.

**Air quality management.** The key principles of the UK Government air quality management strategy are:

- the setting of national standards and reduction targets for all main pollutants;
- supplementing national policies with new systems for local air management, focused on designated areas at risk;
- integrating air quality considerations with planning transport and other policies;
- promoting a balanced approach to emission control designated to secure the most cost-effective improvement process; and including maintaining control of domestic emissions, pressing the continuing improvement of industrial emissions on the basis of BATNEEC and securing early and substantial improvement in VEHICLE EMISSIONS.

*Source:* Department of the Environment, Press Release 14, 19 January 1995.

**Air quality standards.** These prescribe the concentrations of air pollutants that cannot legally be exceeded during a given time period in a specified location. Some examples of standards and their application to London conditions follow:

*Carbon monoxide.* There are no EC environmental standards for carbon monoxide. The WHO recommended guidelines are:

$$30\,mg/m^3, \text{1-hour mean}$$
$$10\,mg/m^3, \text{8-hour mean}$$

Recommended guidelines for shorter-term exposure are $60\,mg/m^3$ for 30 minutes, and $100\,mg/m^3$ for 15 minutes. These guidelines are designed to prevent carboxyhaemoglobin levels in the blood exceeding 2.5–3 per cent in non-smoking populations. The data below from London Scientific Services (LSS) give highest 8-hour average 9 a.m. to 5 p.m. carbon monoxide concentrations measured in the London area in 1986 and 1987, and the number of days 10 mg/m$^3$ was exceeded.

| | West London residential near airport | | Central London urban background | | Central London busy roadside | | East London industrial | |
|---|---|---|---|---|---|---|---|---|
| | 86 | 87 | 86 | 87 | 86 | 87 | 86 | 87 |
| Maximum 8-hour average (mg/m$^3$) | 8.1 | 12.3 | 3.7 | 9.6 | 12.6 | 14 | 5.9 | 6.4 |
| Number of days 10 mg/m$^3$ exceeded | 0 | 2 | 0 | 0 | 2 | 4 | 0 | 0 |

London Scientific Services, *London Air Pollution Monitoring Network*, Second Report, 1987.

*Nitrogen oxides.* The two main oxides of interest are nitric oxide (NO) and nitrogen dioxide ($NO_2$). The concentration of both together is quoted as $NO_x$. Nitric oxide is generally not considered harmful to health and is measured primarily to help understand atmospheric processes affecting levels of nitrogen dioxide and ozone.

The WHO recommended guidelines for $NO_x$ are:

$$400 \, \mu g/m^3, \text{ 1-hour mean}$$
$$150 \, \mu g/m^3, \text{ 24-hour mean}$$

The 1-hour guideline is designed to provide a margin of protection for asthmatics, and the 24-hour guideline to protect against chronic exposure. See Figure 3.

An EC air quality limit value has been set to protect human beings against the effects of nitrogen dioxide in the environment. To satisfy this limit value, 98 per cent of mean hourly nitrogen dioxide values recorded throughout the year must not exceed 200 $\mu g/m^3$. The EC have also specified guide values to improve the protection of human health and contribute to the long-term protection of the environment. The guide values are 50 $\mu g/m^3$

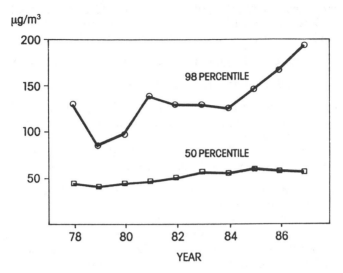

**Figure 3**　Nitrogen dioxide – annual 50 and 98 percentiles of hourly mean values at the central London background station from 1978 to 1987.

and 135 μg/m³ expressed as 50 and 98 percentiles of hourly mean values recorded throughout the year.

*Sulphur dioxide.* The WHO recommends the following guidelines for the protection of public health.

$$500 \, \mu g/m^3, \text{ 10-minute mean}$$
$$350 \, \mu g/m^3, \text{ 1-hour mean}$$

The 10-minute guideline incorporates a protection factor of 2, and the 1-hour guideline is a derived equivalent figure. Account should be taken of local circumstances if the frequency distribution in concentrations is likely to give a different equivalent 1-hour MEAN.

The EC has specified air quality limit values for sulphur dioxide for the protection human health in particular. To satisfy these limit values, 98 per cent of daily mean sulphur dioxide levels recorded over the year should not exceed 350 μg/m³, and 50 per cent of daily mean values should not exceed 120 μg/m³. These limits apply when smoke levels are less than 150 μg/m³ expressed as a 98 percentile, and less than 40 $\mu g/m^3$ expressed as a 50 percentile. More stringent limits apply at higher smoke levels but these are not generally relevant to London.

*Airborne lead.* The guideline recommended by the WHO is

0.5–1.0 $\mu g/m^3$ long-term mean (e.g. 1 year)

This incorporates a protection factor of 2, and is based on the assumption that 98 per cent of the population will be maintained below a blood level of 20 $\mu g/dl$.

To satisfy the EC air quality limit value for lead in air, the annual mean level should not exceed 2 $\mu g/m^3$.

*Airborne particulate matter.* There are no air quality criteria for judging airborne particulate matter as measured gravimetrically at the LSS network stations. The WHO has, however, recommended guidelines for airborne particulate matter, measured by the SMOKE SHADE method and by the HIGH VOLUME SAMPLING method, equal to annual means of 50 $\mu g/m^3$ and 120 $\mu g/m^3$ respectively. Even so, it is difficult to judge the significance of reported airborne particulate matter in terms of health because it is likely to depend critically on the relative proportions of organic and non-organic material.

Both the LSS 1987 report and data show the care and precision required in air pollution monitoring.

**Airborne particulate matter.** ⇨ DUST.

**Aircraft noise.** Aircraft noise consists of a build-up to a peak level and then a fall-off, occurring at intervals, as opposed to the

| NNI | Typical air-traffic conditions |
| --- | --- |
| 60 | This occurs only close to airports where there are many over-flights at low altitude. Noise levels can interfere with sleep and conversation in ordinary houses and may also interfere even within sound-insulated houses. |
| 45 | This occurs mostly near busy routes from airports. Many aircraft are heard at noise levels which can interfere with conversation in ordinary houses. |
| 35 | The overflying is typically irregular at noise levels which are noticeable and occasionally will be intrusive within ordinary houses. |

(⇨ NOISE; SOUND; NOISE INDICES; HEARING; ROAD-TRAFFIC NOISE; INDUSTRIAL NOISE MEASUREMENT)

continuous but fluctuating noise from heavy road traffic. In social surveys around Heathrow Airport, the annoyance caused by noise from air traffic was found to depend on the peak perceived noise levels and on the number of aircraft heard in a given period. Hence an index was derived termed NOISE AND NUMBER INDEX (NNI) which combines the two quantities according to a formula for a given period during the day. Some values of NNI and an indication of the conditions associated with them are listed in the table.

**Albedo.** The ratio of light reflected from a particle, planet or satellite to that falling on it. Therefore, it always has a value less than or equal to 1. It is also used in nuclear physics.

The albedo of the earth plays an important part in the earth's radiation balance and influences the MEAN ANNUAL TEMPERATURE, and therefore the CLIMATE, on both a local and global scale.

**Aldehydes.** Organic compounds containing the group –CHO attached to a hydrocarbon. As air pollutants, a number of them have an unpleasant smell, e.g. in diesel exhaust, and can be an irritant to nose and eyes; many can be poisonous. (⇨ AUTOMOBILE EMISSIONS)

**Aldrin.** An agricultural insecticide which, together with other CHLORINATED HYDROCARBONS such as endrin, DDT, DIELDRIN, and BENZENE HEXACHLORIDE, are major and serious pollutants. They have all been found in significant quantities in the milk of human mothers in the United States (in the USA 99.5 per cent of human tissues taken during post-mortems in a 1971 study contained an average of 0.29 parts per million (ppm) of dieldrin). Aldrin is converted to dieldrin in the environment.

In October 1974 both aldrin and dieldrin were banned by the US government because of strong evidence that they are powerful CARCINOGENS. The use of both aldrin and dieldrin has been restricted since 1960. The UK Ministry of Agriculture, Fisheries and Food has allowed the use of aldrin for bulb crop protection against narcissus fly, but on 18 May 1989 ordered an immediate ban because of 'massive' concentrations of dieldrin in eels in the Newlyn River in Penzance. Rising concentrations had been noticed for several years and it took strong conservation group protests to have the ban implemented. However, there is concern that stockpiles have been accumulated and its use may continue in some areas unless these are completely recalled. (A court case in June 1990 revealed that a gamekeeper and fruit grower respectively had stockpiled a toxic agricultural chemical, endrin, since it was banned in 1983.) (⇨ RED LIST)

**Alert level.** A concentration of gaseous pollutants that has been defined by a competent authority as indicating an approaching, or constituting a potential or actual, hazard to health. Several different alert levels may be defined, ranging from a concentration at which a preliminary warning is issued to one that necessitates emergency action.

**Algae.** Extremely simple unicellular or multicellular plants which utilize the process of PHOTOSYNTHESIS for life. Most of them thrive in a wet environment (freshwater or marine) such as lakes, rivers and damp walls. They can cause problems in lakes and reservoirs if there is an excess of NUTRIENTS, as their excessive multiplication results in an algal 'bloom' and when they die the decay process is most unpleasant. Lakes that are rich in nutrients and as a consequence are highly productive in algae and other organic matter are said to be *eutrophic*.

A novel proposal is that algal blooms resulting from EUTROPHICATION should be harvested and fed to organisms higher up the FOOD CHAIN that have economic value, such as oysters. Others have suggested that algae should be cultivated in suitable nutrient liquids to produce vegetable protein.

Algae also form one of the constituent symbiotic partners in LICHENS. (⇨ SYMBIOSIS)

**Algal bloom.** ⇨ ALGAE.

**Algicide.** A chemical used for destroying algal growths.

**Alkali.** A substance which, on dissolving in water, splits up to a greater or lesser extent into IONS and results in an excess of hydroxyl ions over hydrogen ions. Alkalis neutralize ACIDS to form salts. The adjective is 'alkaline'.

**Alkali and Clean Air Inspectorate.** Until 1987, the Inspectorate (formerly known as the Alkali Inspectorate) was responsible for enforcement of the Alkali Works Regulation Act 1906, and the Health and Safety at Work etc. Act 1974 in England and Wales. The Inspectorate now forms part of Her Majesty's Inspectorate of Pollution. (⇨ BEST PRACTICABLE MEANS; BEST AVAILABLE TECHNOLOGY)

**Alkalinity.** The capacity of a water to neutralize acids due to the bicarbonate, carbonate or hydroxide content. It is usually expressed in milligrams per litre of calcium carbonate equivalent.

**Alpha particle (α-particle).** An alpha particle has a positive charge, consists of two protons and two neutrons (in effect the nucleus of a helium atom) and is emitted from the nucleus of an atom. A nucleus can spontaneously emit an alpha particle when its mass is greater than the combined masses of the product or daughter nucleus and the alpha particle. Spontaneous alpha emission takes place with many of the nuclei heavier than lead.

Alpha particles cause high ionization and are large in comparison with other radiating particles. They therefore lose energy very quickly. They have little penetrating force, 0.001–0.007 centimetres in soft tissue, but once inside the body, either by inhalation or through a wound, they are biologically very damaging.

**Alpha radiation.** A stream of fast-moving alpha particles emitted from the nuclei of radioactive elements. They are easily absorbed by matter. (⇨ RADIONUCLIDE)

**Aluminium.** Light metal produced by electrolysis of bauxite ore.

**Aluminium sulphate ($Al_2(SO_4)_3$).** A water additive used as a coagulant to remove particles that discolour water. It may be implicated in excess aluminium levels in humans, although aluminium cooking vessels, food containers, etc. may be alternative routes. In addition, ACID RAIN allows aluminium to be leached from soils in water catchment areas and hence into drinking water.

The EC maximum admissible concentration level for aluminium in drinking water is $200\,\mu g/1$. It is estimated (1989) that more than 2 million people in the UK drink water with aluminium levels in excess of this limit.

Attention was focused on this topic when a 20 tonne load of 8 per cent solution of aluminium sulphate accidentally entered the water supply of the CAMELFORD area of North Cornwall on 6 July 1988. A population of ca. 20 000 was put at risk, with some people and animals becoming violently ill plus the death of an estimated 60 000 fish in the rivers Camel and Allen. Exposure levels of up to 2000 times the EC limit were postulated for the initial period of this event. The after-affects of the chemicals dosed into the water supply to neutralize the aluminium sulphate reacted with water pipes which subsequently produced high levels of copper, lead and zinc in the drinking water. The ill effects *attributed* to this incident are skin troubles, arthritis, nausea and possibly kidney complaints. This is a classic case of (i) solving a problem in isolation and (ii) how easy it is for a seemingly minor act of carelessness to grossly pollute and consequently endanger a large group of people. (⇨ CHERNOBYL)

**Amino acid.** Essential component of PROTEINS, consisting of amino ($-NH_2$) and acidic carboxyl ($-COOH$) groups. There are some 20 different amino acids normally present in proteins and a balanced diet must consist of the right intakes of these acids. There are eight essential amino acids for humans that can only be obtained from the environment by heterotrophic means. The other non-essential amino acids can be manufactured by the human organism usually from the essential ones. 'First class' protein is the name given to protein that contains these eight essential amino acids and man's daily needs is estimated at 30–70 grams, some of which should preferably come from animal sources.

**Ammonia.** Compound $NH_3$. In effluent treatment terms, it is the equilibrium mixture of $NH_3$, $NH_4 OH$ and $NH_4^+$. This can cause both toxicity to fish and an oxygen demand in receiving waters. A high chlorine demand is also created if the water is to be used for drinking water. The presence of AMMONIACAL NITROGEN is taken as an indicator of sewage pollution in a stream.

Ammonia is present in very small amounts in exhaust gases from engines without catalytic converters and there is also evidence that emissions from vehicles fitted with three-way catalytic converters are very much higher than from non-catalyst vehicles. If all petrol cars were fitted with catalysts, they could contribute as much as 10 per cent of total UK emissions of ammonia.

Ammonia injection can also be used as a means of NITROGEN OXIDE control in combustion processes producing nitrogen and water vapour as the products.

*Source:* Royal Commission on Environmental Pollution 18th Report, published 1994.

**Ammonia slip.** The excess AMMONIA ($NH_3$) discharged in STACK GASES from SELECTIVE NON-CATALYTIC REDUCTION (SNCR) systems used for NITROGEN OXIDE ($NO_x$) control which make use of ammonia or UREA injection to achieve $NO_x$ reduction.

A balance has to be struck between $NO_x$ emissions and $NH_3$ slip. See Figure 4.

**Ammoniacal nitrogen.** Nitrogen present as ammonia and ammonium ion in liquid effluents. It has a high CHLORINE requirement in water treatment and is toxic to fish.

**Ammonium nitrate.** Formula $NH_4 NO_3$, made by reacting ammonia and nitric acid. Used as an explosive and (mainly) as a booster

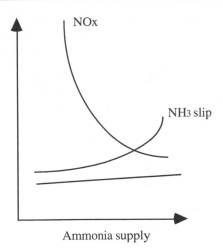

**Figure 4** $NO_x$ emissions in relation to the ammonia slip in a SNCR system based either on ammonia or urea.

fertilizer, plants can absorb nitrogen as ammonium ions or as nitrate ions when they are reduced to ammonia.

**Anaerobic digestion.** Although AEROBIC PROCESSES are almost always used for the reduction of BIOCHEMICAL OXYGEN DEMAND (BOD) in organic effluents, interest in ANAEROBIC PROCESSES (in the absence of free oxygen) is growing, particularly for high BOD/COD wastes from industry. It is also used as a sludge treatment process in conventional sewage treatment plants and for animal effluent on farms.

The overall process of anaerobic digestion occurs through the combined action of a consortium of four different types of microorganism. Hydrolytic and fermentative microorganisms including common food spoilage bacteria break down complex wastes into their component sub-units and then ferment them to short-chain fatty acids and carbon dioxide and hydrogen gases. Bacteria convert the complex mixture of short-chain fatty acids to acetic acid with release of more carbon dioxide and hydrogen gases to provide the main substrates for the methane bacterias.

Methane bacteria produce large quantities of methane and carbon dioxide from this acetic acid and also combine all of the available hydrogen gas with carbon dioxide to produce more methane. Sulphate-reducing bacteria are also present. They reduce sulphates and other sulphur compounds to hydrogen sulphide, most of which reacts with iron and other heavy metal salts to form insoluble sulphides, but there will always be some hydrogen sulphide remaining in the biogas.

The ultimate yield of biogas depends on the composition and biodegradability of the waste feedstock but its rate of production will depend on the population of bacteria, their growth conditions and the temperature of the fermentation. Microbial growth and natural biogas production is very slow at ambient temperatures. As a waste treatment process, the rate of anaerobic digestion is greatly increased by operating in the mesophilic temperature range (35–40 °C).

As this form of processing is expected to grow, several applications are summarized below to show the potential of this method.

*Application* 1. On waste containing a BOD of 12 000 milligrams per litre resulting from a starch and sugar content (*Process Engineering*, May 1975, p.6). The process is shown in Figure 5 and uses a digester which must be kept heated to around 35°C to allow sufficient bacterial activity. The result is methane, which is a source of light and heat. Between 85 and 95 per cent BOD is removed in this way, which means that the effluent from the digester would normally require aerobic treatment before final discharge.

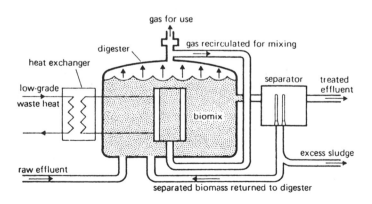

**Figure 5**　Anerobic effluent treatment.

The advantages that are claimed for the anaerobic process are that the plant is much smaller than that for an aerobic process, and the volume of sludge produced can be one-tenth of that from aerobic plants; it is said to be less objectionable to handle. It is extremely suitable for high BOD effluents which could overwhelm conventional sewage effluent plants. (⇨ EFFLUENT, PHYSICO-CHEMICAL EFFLUENT TREATMENT).

*Application 2.* Animal excreta. The Bethlehem Abbey of the Cistercian monastery in Co. Antrim, Ireland is heated completely by the biogas produced by the anaerobic digestion of a daily input of 12 tonnes of cow slurry and manure from its 300 head beef fattening farm and 22 000 broiler hens, respectively. The biogas saves the Abbey an estimated £1000 per month in fuel costs plus the fibre from the digested slurry is composted for 3 weeks at 60°C and bagged into 25 litre sacks which sell for £3.85. The compost is marketed under the brand name of 'Dungstead'.

*Application 3.* Whey treatment (Hamworthy Engineering Report, *Whey Processing for Profit*, Poole, Dorset, 1987).

WHEY is notoriously difficult to treat as $1 m^3$ is equivalent in BOD to sewage effluent from 600 people. A typical comparison of before and after treatment is:

| | Before | After | Percentage reduction |
|---|---|---|---|
| Total solids (mg/l) | 66 000 | 9400 | 85 |
| Volatile solids (mg/l) | 65 000 | 6500 | 90 |
| Chemical oxygen demand (mg/l) | 65 000 | 6500 | 90 |
| Biochemical oxygen demand (mg/l) | 45 000 | 2250 | 95 |
| Total nitrogen (mg/l) | 1500 | 1000 | 33.3 |

A waste of this strength would be highly uneconomic to treat by conventional aerobic means due to the high power requirement for aeration. Hence anaerobic treatment can be chosen. The individual components of whey, lactose, protein and fat are broken down by the digestion process to form BIOGAS. One cubic metre of whey yields $38 m^3$ of biogas at NORMAL TEMPERATURE AND PRESSURE (NTP).

The effluent from the digester has 90 per cent reductions in the COD/BOD and can, in some instances, be sent direct to sewer. If it is to be discharged to a watercourse, further treatment will be necessary to reduce the BOD, and total nitrogen and ammonia contents with the aid of an aerobic nitrification/denitrification system to meet ROYAL COMMISSION STANDARDS.

The schematic diagram for the South Caernarfon Creameries anaerobic digestion plant is shown in Figure 6. This plant treats 40–100 m$^3$ whey/day plus 10 m$^3$ of aerobic sludge from the creameries' existing biological filtration plant. Following the digester, the major pollutant remaining in the waste is nitrogen, the total nitrogen and ammonia nitrogen contents being 1000 mg/l and 750 mg/l respectively. The nitrogen and BOD remaining in the digester effluent are broken down in an aerobic nitrification/ denitrification system, to produce a Royal Commission effluent of 20 mg/l BOD, 30 mg/l SS and 10 mg/l ammonia (NH$_3$). The post-digestion flowsheet is given in Figure 7.

The biogas produced varies from 1560 m$^3$/day at low whey production to 4000 m$^3$/day at maximum production. An overall BOD reduction of 99.95 per cent is achieved.

On a maximum feed of 100 m$^3$ whey/24 hours and 38 m$^3$ biogas/m$^3$ whey and with the biogas having a nett CALORIFIC VALUE of 21.08 MJ/m$^3$, 2560 litres of heavy fuel oil boiler fuel can be saved every 24 hours, or alternatively, 250 kW of electrical power and a further 480 kW of hot water at 85.9°C if WASTE HEAT is recovered from the engine cooling water and exhaust.

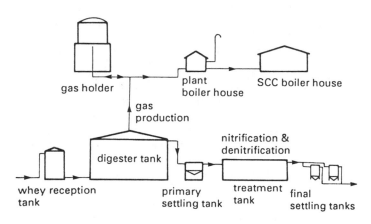

**Figure 6**   Schematic diagram of South Caernarvon Creameries digester and after treatment plant.

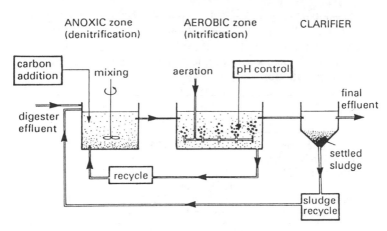

**Figure 7**   Post-anaerobic digestion flow sheet for creamery waste.

The addition of anaerobic treatment of the whey has led to substantial savings in disposal costs for South Caernarfon Creameries, plus all its steam demands are met and there is electrical power for sale. In addition, the plant is no longer in a position to pollute the local stream thus saving on fines as well as being a good environmental neighbour.

For high solids feedstocks such as MUNICIPAL SOLID WASTES (MSW) efficiency can be further increased by operation at thermophilic (50–60°C) temperatures.

The claimed advantages of thermophilic over mesophilic digestion are:

- higher gas production rates for similar retention times to mesophilic digestion;
- better separation of water from digestate;
- higher reduction of pathogens (disease-producing organisms).

The disadvantages are:

- higher energy consumption;
- higher fatty acids concentration in effluent;
- higher total solids (mg/l);
- more sensitivity to changes in process conditions and toxic substances.

The bacteria require mineral nutrients; a ratio of C:N of 10–16 is typical and N:P should be 7. Also, wastepaper can only be digested in low concentrations due to its effect on the C: N ratio.

Typical digester residence times are 25–35 days with biogas yields in the range 100–150 m³/tonne of input waste.

The table below gives typical yields and calorific values and steam production.

| | | Mixed MSW | Vegetable wastes |
|---|---|---|---|
| Gas production | (Nm³/ton) | 105 | 95 |
| Net calorific value | | | |
| (55–60% CH₄) | (MJ/Nm³) | 19.7–21.5 | 19.7–21.5 |
| | (MJ/tonne) | 2069–2258 | 1872–2043 |
| Density | (kg/Nm³) | 1.284–1.221 | 1.284–1.221 |
| Steam production | (ton/ton) | 0.66–0.72 | 0.60–0.65 |

*Sources: Biogas from Municipal Solid Waste: Overview of Systems and Markets for Anaerobic Digestion of MSW*, IEA Bioenergy Brochure, published by Ministry of Energy/Danish Energy Agency, Copenhagen, Denmark, July 1994. *Feasibility Study for Anaerobic Digestion of Solid Wastes in Leicestershire*, ETSU B/EW/00389/12/REP, Contractor Tebodin PLT, Slough, prepared by A. Snuverink, J. van Seeters and E. C. Griffiths, 1994.

**Anaerobic process.** Any process (usually chemical or biological) that is carried out without the presence of air or oxygen, e.g. in a watercourse that is heavily polluted with no dissolved oxygen present. The anaerobic decay processes produce METHANE (CH₄) and HYDROGEN SULPHIDE (H₂S) which is an evil-smelling toxic gas.

Some anaerobic organisms, e.g. the denitrifying bacteria, are poisoned by the presence of oxygen. (⟹ AEROBIC PROCESS)

**Anchovy fisheries.** The anchovy is a small fish of the herring family which can be easily converted to fish-meal. The major source of anchovy is the Peruvian fishing industry, which became the world's largest producer in the early 1960s. Approximately 96 per cent of the anchoveta catch from the Peruvian (Humboldt) Current was converted into fish-meal. It represents the world's largest ocean harvest of a single species, and its export effectively deprives the South American continent of 50 per cent more protein than it is producing as meat, including Argentinian deliveries to Europe and North America. (⟹ MAXIMUM SUSTAINABLE YIELD; WHALE HARVESTS)

**Angiosarcoma.** A rare form of cancer commonly associated with the liver. (⟹ POLYVINYL CHLORIDE)

**Anion.** Negatively charged ION.

**Anode.** An electrode which is at a positive potential.

**Antagonism.** A state in which the presence of two or more substances diminishes or decreases the toxic effects of the substances acting independently. It is the opposite of SYNERGISM.

**Antibiotic.** A chemical substance, produced by micro-organisms which has the capacity, in dilute solution, to inhibit the growth of or destroy bacteria. Originally derived from moulds or bacteria but now produced synthetically as well, antibiotics are used to treat and control infectious diseases in man, animals and food crops, to stimulate the growth of animals, and to preserve foods.

With the large-scale use of antibiotics, particularly in farming, the emergence of resistant bacteria has become an increasing problem. As resistance develops, an antibiotic becomes less effective and dosages must be increased. Particularly disturbing is the mechanism of transferable drug resistance, by means of which one species of bacteria previously susceptible to a certain antibiotic suddenly becomes resistant to it by virtue of cell-to-cell contact with a species of bacteria which is already resistant to the antibiotic. This resistance can be acquired to several drugs simultaneously.

**Anticyclonic blocking.** A winter occurrence when high pressure periodically builds over Russia and Scandinavia which diverts unsettled and rainy systems into the western Mediterranean. When it does not occur, persistent low pressure systems develop across northern Europe, especially Scandinavia, leaving high pressure across southern Europe and an absence of the rainfall commonly associated with low pressure.

The Andalucia area of Spain has been without significant rainfall during 1991–95 due to this phenomenon.

*Source: Daily Telegraph, 1 May 1995.*

**AOX.** AOX is defined as organic halogens subject to absorption. This is a measure of the amount of chlorine (and other halogens) combined with organic compounds.

**Aquifer.** An underground water-bearing layer of porous rock, e.g. sandstone, in which water can be stored and through which it can flow, after it has infiltrated from either the surface or another underground source. The London Chalk Basin is a typical example of such a aquifer; it supplies three million cubic metres of water per day. Water may also be extracted from the chalk under the Cotswolds to meet the expected demand from London by the end of the century. (⇨ INFILTRATION)

Polluted surface water can enter the SATURATION ZONE of an aquifer and lead to its contamination. Typical pollutants can be LEACHATE, NITRATE, SOLVENTS which can permeate via fissures, fractures and boreholes.

Aquifer management and RIVER REGULATION are now practised conjunctively in the UK to maximize the amount of water available for consumption from rainfall over a particular catchment area, e.g. the December 1988 rainfall in S.E. England was 17 mm against an average of 76 mm. Hence the water extracted from the chalk aquifer in early 1989 has been cut back substantially to conserve this source of supply.

**Arithmetic–geometric ratio (A/G ratio).** A principle used by geologists for estimating the abundance of certain types of ore deposits, the idea being that as the grade of the ore decreases arithmetically its abundance increases geometrically until the average abundance in the earth's crust is reached. This has led certain economists to assert that the problem of non-renewable mineral resources is a non-problem, since as demand increases mining will simply move to poorer and poorer ores which are assumed to be progressively more and more abundant.

As one would expect, the facts are against such a simplistically reassuring view of the situation. The A/G ratio is applicable only to a very limited number of ores, and only within certain limits. In addition, although the economic costs of working lower and lower grade ores might be absorbed in some cases, the energy costs could not. (⇨ ENERGY DEMAND)

**Artificial recharge.** The artificial replenishment of an AQUIFER by flooding the surface, ditches, excavations, wells, etc. This augments the natural infiltration and precipitation contribution. The UNSATURATED ZONE often acts as a giant free treatment plant, hence slightly polluted water may be used for recharge purposes.

**Artificial reef.** Reefs are rock chains near water-level which are often very useful for shore or wild-life protection. Plans are now afoot for the redundant steel jacket to the Odin gas production platform to become an artificial reef.

Esso, Norway, believe this could prove a model for future abandonments on the Norwegian continental shelf and have submitted a plan to the Norwegian Ministry of Industry and

Energy. Under the plan, the platform's deck and module will be taken onshore for scrapping or recycling while pipelines will be left in place.

The company believes an artificial reef offers the most cost-effective solution and is environmentally acceptable. (Deep sea dumping of the jacket or onshore scrapping are more expensive.)

Fishermen have generally been opposed to oil companies leaving what they consider to be 'debris' on the sea bed. The Esso spokesman said the 'artificial reef would provide an important area of study'.

*Source: Lloyd's List*, 5 April 1995.

**Asbestos.** A collective term for a group of magnesium silicate materials ($MgSiO_4$). It is a fibrous mineral of variable length with excellent resistance to fire, heat and chemical attack. It is widely used in building products, gaskets, brake linings, and roofing materials.

There are three main forms of asbestos: *blue asbestos* or *crocidolite*, which is so dangerous that its use is virtually banned now in the UK. A survey of male asbestos workers by the TUC Centenary Institute of Occupational Health suggests that 30 years after first exposure about one in 200 workers will be found to have died from mesothelioma (malignant tumours) associated with blue asbestos. Blue asbestos is still present in old buildings, boiler plant, ships, etc., and its presence is a substantial threat to any demolition workers.

*White asbestos*, or *chrysotile*, can be readily spun or woven into tape.

The third important variety is *amosite (brown asbestos)*. Other varieties of minor commercial importance are *tremolite*, *actinolite* and *anthophyllite*.

The dangers associated with asbestos are grave. Asbestosis, a scarring of the lung, cancers of the bronchii, pleura and peritoneum may result from breathing in the minute fibres. Asbestosis may result from exposures as short as six weeks in heavy dust concentrations. Brief exposure to blue asbestos can manifest itself later in life as mesothelioma, a specific and invariably fatal form of cancer.

In the UK, the 1931 Asbestos Regulations put the onus squarely on the employers to ensure that asbestos dust was removed at source, respirators being regarded as only secondary protection.

Astonishingly, these regulations did not address certain important uses of asbestos, and were apparently disregarded by employers, with a resulting high record of sickness and death in the industry. The Factory Inspectorate, which had the responsibility for ensuring that the factory acts relating to asbestos were properly enforced, appears to have been completely ineffective, with a total record of three prosecutions in 30 years.

The current legislation is under Control of Asbestos Work Regulations 1987. Its recommendations are:

1. Substitute material to be used where possible.
2. Exhaust ventilation to reduce atmospheric contamination to below the CONTROL LIMITS. Monitoring is mandatory and enforced by the Health and Safety Executive.

The current control limits for asbestos dust are: crocidolite, amosite and tremolite – 0.2 fibres per millilitre of air; chrysotile – 0.5 fibres per millilitre of air, averaged over any continuous period of 4 hours, where the fibres measured are five millionths of a metre in length or greater. The sample is to be taken by a prescribed membrane filter which is then scanned by a microscope at 400–450 magnification.

It would be optimistic in the extreme to assume that these stringent precautions are obeyed everywhere, particularly in the demolition and insulation industries. The Control of Asbestos in the Air Regulations 1990 (S.I. 1990 No. 556, in force since 5 April 1990) impose an emission limit of 0.1 mg/m³ for asbestos emission to the air by industrial installations.

Respiratory problems caused by asbestos include:

*Asbestosis.* This causes scarring and shrinkage of the lungs, and is thought to increase the risk of lung cancer by five times.
*Mesothelioma.* A cancer of the inner lining of the chest wall, associated primarily with exposure to asbestos (in the US it is called 'asbestos cancer'). There is no cure and patients usually die within a year of diagnosis.

Suspicions of a link between asbestos and lung disease were aired in the 1920s; the term 'asbestos' appeared in the *British Medical Journal* in 1927. The first full study of the effects of asbestos began the following year and the first legislation regulating its use was introduced in the UK in 1931.

Figures compiled in 1995 put asbestos-related deaths at an estimated 3000 a year in the UK. One in 40 men in their fifties who have been exposed to asbestos are expected to die of mesothelioma. Projected deaths from mesothelioma are expected to continue to rise until at least 2010, possibly 2025, and will peak at 5000 to 10000 deaths a year in the UK (Health and Safety Executive data).

*This material is a major hazard to health and no concentration of asbestos dust may be presumed safe.*

*Source: Lancet,* March 1995.

**Askarel.** An insulating non-flammable liquid used in electrical applications such as transformers, capacitors and special electrical power cables. Most well known are the polychlorinated biphenyls (PCBs).

**Aspergillus niger (AN).** A fungus capable of breaking down vegetable matter (e.g. carob pods, which come from the locust tree which is extensively grown in many developing countries) to produce SINGLE CELL PROTEIN (SCP). The strain used is called Ml and is capable of doubling its weight in 5 to 10 hours, provided inorganic nutrients are added.

A village-level technology based on this process is being developed to use carobs and other starchy vegetable matter as a source of protein for pigs and cattle.

**Asphyxiating pollutants.** Asphyxiation is deprivation of oxygen, either through obstruction of the air passages or, as in the case of CARBON MONOXIDE, the inability of the blood to carry oxygen. HYDROGEN SULPHIDE ($H_2S$) is an irritant at very low concentrations, but it can also paralyse the respiratory system at slightly higher concentrations causing death by asphyxiation.

*Source:* G. L. Waldbott, *Health Effects of Environmental Pollutants,* 2nd edition, C. V. Mosby, St. Louis, 1978.

**Aswan Dam.** ⇨ DAM PROJECTS.

**Atmosphere[1].** The envelope of air around the earth. The composition of our present atmosphere is the result of the CARBON CYCLE and the NITROGEN CYCLE and the atmosphere is renewed and maintained by these processes.

The atmosphere, like any other natural resource, is finite and 99 per cent of its mass is within 20 miles of the earth's surface. It is in contact at its inner edge with land and water and changes in the atmosphere can induce changes in land or water such as the amount of rainfall on a particular area. The changes can be direct or they can be complex chain reactions.

The role of the atmosphere in the earth's radiation balance is unique. The incoming solar radiation is absorbed by the earth and reradiated into outer space as long-wave radiation, but the two processes are not in balance except over, say, a year when the total incoming radiation may balance the total outgoing. A very small change in either the outgoing or incoming radiation can have very large effects. Hence, the concern for the CARBON DIOXIDE released by combustion of fossil fuels, the DUST particles released by combustion processes, the effects of AEROSOL PRO-PELLANTS on the ozone layer, or supersonic aircraft (⟹ ALBEDO; CLIMATE; CONCORDE; GREENHOUSE EFFECT; THERMAL POL-LUTION).

The atmosphere has two broad bands. The lower part, the TROPOSPHERE, decreases in temperature and density as altitude increases. The other band is the upper atmosphere, the STRATO-SPHERE, where temperature increases with altitude.

Because of its greater density, the lower troposphere contains roughly 80 per cent of the mass of the atmosphere. The depth of the troposphere varies widely, depending on latitude; it has an average depth of 8 km over the poles and 16 km over the equator. In the troposphere there is considerable mixing as the warmer air rises and the cooler air falls.

The stratosphere has the opposite temperature characteristic: the temperature rises as altitude increases, and so there is very little mixing.

NITROGEN OXIDES ($NO_x$) released by jet engines in the strato-sphere destroy ozone, but in the troposphere they contribute to a build-up of ozone, where it has a greater GREENHOUSE EFFECT. The residence time of $NO_x$ is 1–4 days in the troposphere but over 1 year in the stratosphere. Hence, the emissions from jet engines at high altitudes may have a disproportionate contribu-tion to global warming compared to the same mass of emissions at low levels (below 5 km).

**Atmosphere².** A unit of pressure equating to 760 mm of mercury. 1 normal atmosphere = $101\,325\,N/m^2$ or $14.72\,lb/in^2$. (⟹ STAN-DARD TEMPERATURE AND PRESSURE)

**Atom.** The smallest part of an ELEMENT that can take part in a chemical reaction. It consists of a positively-charged core, called the NUCLEUS, surrounded by negatively-charged ELECTRONS.

**Atomic mass (relative).** A number that gives the mass of an atom of an element relative to that of an isotope of carbon $^{12}C$, which has an assigned atomic mass of 12.

**Atomic mass unit (AMU).** One twelfth of the mass of an ATOM of carbon-12.

**Atomic number.** The number of ELECTRONS around the NUCLEUS of an ATOM. This determines the chemical behaviour of an atom. The atomic number is constant for each ELEMENT and its ISO-TOPES.

**Attenuation.** General term for a reduction in magnitude/intensity/concentration of a substance dispersed in a gaseous or liquid medium, e.g. LEACHATE from a LANDFILL SITE may be attenuated in its passage through underlying strata by the action of AB-SORPTION, ADSORPTION, and biological processes in any UN-SATURATED ZONE. Sound is attenuated by atmospheric absorp-tion and by interaction with absorbing surfaces.

**Automobile emissions.** Generic name for the emissions from car exhausts, the principal components of which are LEAD, NITRO-GEN OXIDES ($NO_x$), UNBURNT HYDROCARBONS, water vapour, CARBON MONOXIDE (CO), ALDEHYDES, and CARBON DIOXIDE ($CO_2$).

Carbon monoxide, lead and nitrogen oxides are major pollu-tants in our cities, due to increased automobile emissions resulting from an increase in the number of cars on the road and the use of high-powered engines. They have significant health effects, par-ticularly under SMOG and PHOTOCHEMICAL SMOG conditions. In busy streets the carbon monoxide concentration can rise to 15–20 parts per million (ppm); over 100 ppm have been measured in London. The background concentration of carbon monoxide in the atmosphere is less than 0.1 ppm.

The high compression ratios required for high-performance engines need fuels with 'anti-knock' properties. This has been achieved since the 1920s by adding tetraethyl- and tetramethyl-lead to the fuel, but now is being limited by legal and fiscal incentives. A large part of the lead – which is toxic – is emitted from the car exhaust in a fine particulate form that can be readily inhaled. Lead also poisons the catalytic reactors that are used in car exhaust systems to minimize emissions.

Photochemical smog is triggered by the action of sunlight on exhaust emissions of nitrogen oxides, and the amounts emitted from cars have increased as compression ratios, and hence engine temperatures, have increased, particularly in urban areas. It is estimated that in the United States from 1946 to 1968 total emissions of nitrogen oxides increased sevenfold, while they doubled in the UK between 1945 and 1980.

In California legislation has been enacted to effect reductions of 87 per cent for carbon monoxide, 95 per cent for hydrocarbons, and 75 per cent for oxides of nitrogen. This can be achieved by means of a CATALYTIC REACTOR in the car exhaust system which converts the unburnt hydrocarbons to carbon dioxide and water, and the carbon monoxide to carbon dioxide. The nitrogen oxides can also be reduced. Recirculation of exhaust and crankcase vapours back into the combustion chamber can also be carried out to further reduce emissions. The use of lead in petrol is also to be severely curtailed. EC legislation requires that by the end of 1992 almost all new cars sold will have to be fitted with catalytic reactors to achieve emission levels comparable to the US ones. (⇨ LEAN COMBUSTION)

**Autonomous house;** also known as the *Ecohouse*. The self-contained dwelling where man simulates the ways of ECOSYSTEMS, i.e. his wastes are converted to fuel by anaerobic digestion for METHANE production and the residues from the digestion are used for growing food. The food residues are composted and/or used for methane production. Solar energy is trapped by the GREENHOUSE EFFECT and used for house, crop and water heating. A windmill would be used for electricity. Thus, given sufficient space, sunshine, rainfall and wind, the autonomous house is in theory a self-contained system recycling its own wastes and using the sun as its energy input.

Much research and development needs to be done before this can be achieved. The practitioners still require an initial energy input to build the house, provide the raw materials such as glass, plastics, paint, etc., but the concept can lead to much more energy-efficient housing.

**Autotrophic organism.** An organism that does not require organic material for food from the environment but can manufacture food from inorganic chemicals. For example, chlorophyll-containing plants or algae can manufacture their food from water, carbon dioxide and nitrates using the sunlight for an energy source. HETEROTROPHIC ORGANISMS depend on the activities of the autotrophic ones for food. (⇨ PHOTOSYNTHESIS)

**Availability factor.** Availability factor is a power plant performance indicator widely used in the United States and Japan. It stipulates the amount to time over a given period during which the unit was available, whether synchronized to the grid or not.

$$\text{Availability factor} = \frac{\text{Hours operated or operable}}{\text{Total hours in period}}$$

(⇨ CAPACITY FACTOR)

**Azodrin.** ⇨ ORGANOPHOSPHORUS COMPOUNDS.

# B

**Background concentration of pollutants.** If the atmosphere in a particular area is polluted by some substance from a particular local source, then the background level of pollution is that concentration which would exist without the local source being present. Measurement would then be required to detect how much pollution the local source is responsible for.

Sometimes the word 'background' is used to mean the concentration of the substance some distance from the particular source and therefore largely uninfluenced by it.

The term is also used in radiation work to mean the background level of radiation from natural sources or from sources other than that being measured and is commonly used as a reference datum against which emissions are scaled for public information purposes. This is not quite the complete story, as background radiation sources are not the same as those from the nuclear power industry and hence attempts to equate both could be misleading. (⇨ IONIZING RADIATION, EFFECTS)

**Backwashing.** Cleaning of a filter or ion-exchange column by reversing the fluid flow through the bed so that the filter medium or ion-exchange resin is reclassified and dirt removed.

**Bacteria.** Class of organisms which do not possess CHLOROPHYLL. They usually multiply rapidly, by division. They are usually unicellular rods or rounded cells ca. 1 $\mu$m in diameter.

Bacteria occur virtually everywhere in the BIOSPHERE and are responsible, for example, for the souring of milk or the decay of dead animals. They participate in the CARBON CYCLE, can fix nitrogen in the NITROGEN CYCLE, and decompose sewage. As DECOMPOSERS they are responsible for the decomposition of organic matter into simple substances which can then be reincorporated into the biological cycles. They can also cause and transmit diseases, e.g. TYPHOID, tuberculosis, etc. (⇨ SEWAGE TREATMENT)

**Bacteria, use in mining.** ⇨ SOLUTION MINING.

**Baffle chamber.** A settling chamber containing a system of baffles, which enables coarse particulate matter (e.g. large particles of fly ash) to be removed from stack gases. This is done by changes of its direction and reduction of its velocity.

**Bag filter.** A closely woven bag for removing DUST from dust-laden gas streams. The fabric allows passage of the gas with retention of the dust. Separation efficiencies greater than 99 per cent can be obtained on DUSTS > 1 > 1 $\mu$m in diameter. Bag filters are cleaned either by mechanical shaking or reverse pressure by means of reversed gas jet. The former method requires that the plant be built in sections so that a section is isolated in sequence for cleaning. The latter method allows continuous operation.

**Bag house.** An installation with the function of abating the particulate content of gas streams by filtering these streams through large fabric bags.

**Ballast voyage.** The journey an oil tanker or other vessel makes with sea water carried in the oil-tanks as ballast to provide essential stability on the empty leg of its journey. The ballast is usually discharged at sea along with any residual oil from the tanks and consequently the practice is a major contributor to oil pollution. It is estimated that 20 million tonnes of oil were dumped on the world's seas in 1973 as a result of ballast-blowing. World oil consumption is about 3000 million tonnes per year, so less than 1 per cent is dumped at sea, but this is still an awful lot of oil which could cause marine pollution on a large scale. (⇨ OIL SLICK)

**Band screen.** An endless moving wire mesh band used for solids removal from liquid effluents.

**Bar.** Unit of pressure equal to $10^5$ N/m$^2$ or 750.7 mm of mercury. Used for high steam pressure measurements. 1 millibar = 100 N/m$^2$ which is used for barometric measurements. Standard atmospheric pressure = 1.01325 bar.

**Bar screen.** A set of regularly spaced bars used for removing large solids from sewage effluent. The bars may be mechanically raked for automatic solids removal.

**Barnes' formula.** Empirical relationship first proposed by A. G. Barnes (1916) for calculating the flow velocity in a sewer, i.e.

$$V = 107 R^{0.7} S^{0.5}$$

where $V$ = flow velocity (m/s), $R$ = hydraulic mean depth (m) and $S$ is slope of sewer.

**Figure 8** Location of preferred barrage line, showing tidal ranges in the Severn Estuary.

**Barrage.** A controlling dam or barrier placed in a stream or tidal flow. For renewable ENERGY purposes, a tidal barrage would have reversible turbines built into it so that both the ebbing and flowing tides can generate power.

One of the favoured UK sites for a tidal barrage is the Severn Estuary from Cardiff to Weston-Super-Mare as shown in Figure 8 which details the tidal ranges available. An ENVIRONMENTAL ASSESSMENT of such an undertaking must be made before implementations, as there could well be major changes in sediments, water quality, ECOSYSTEMS, effects on bird life, etc.

The Severn Estuary Scheme has been intensively studied and it is concluded that an installed turbine capacity of 8400 MW could produce a total annual energy output of 17 GWh at a 1989 cost of 3.7 p/kWh which is comparable to that of coal-fired power stations.

Other UK tidal power sites have been identified where similar generating costs could be anticipated. These include the Mersey Estuary which has a mean spring tide range of 8.4 m, possible installed capacity 600 MW; also the Conway Estuary, installed capacity 35 MW.

**Barrel.** Standard volumetric measure of crude oil production. Equivalent to 42 US gallons or 35 Imperial gallons or 159 litres.

**Base.** A substance which when dissolved in water generates hydroxide (OH⁻) IONS or is capable of reacting with an ACID to form a SALT. (⇨ ALKALI)

**Base exchange.** An ion-exchange water softening process in which calcium and magnesium ions are exchanged for sodium ions.

**Base process.** A process in which raw materials are fed into a plant in discrete batches rather than continuously. If air pollutants are produced by such a process they are usually rather more difficult to deal with than those produced by a continuous process. This was the main problem posed by early batch-fed refuse incinerators resulting in substantial combustion temperature reductions, incomplete combustion, DIOXIN emissions and gross air pollution. (⇨ INCINERATION)

**Becquerel (Bq).** A measure of the intensity (activity) of a radioactive material which has replaced the CURIE (Ci):

1 Bq is a rate of one disintegration per second. Hence

$$1\,Bq = 2.7 \times 10^{-11}\,Ci$$

or

$$1\,Ci = 3.7 \times 10^{10}\,Bq$$

*Note:* Both the intensity of the source (Bq) and the isotope must be known, e.g. STRONTIUM-90 decays by emission of a single $\beta$ 'particle'. Hence a 1 Bq source will emit one particle per second. However, COBALT-60 emits a $\beta$ particle and 2 $\gamma$ rays per disintegration; such a source of strength 1 Bq will therefore emit 3 'particles' per second.

Furthermore, should the source disintegrate to produce an unstable DAUGHTER PRODUCT, the overall intensity of the source at any time is the sum of the disintegration of the parent and daughter products.

**Benefit/cost.** The benefit/cost ratio seeks to compare the benefits of a particular action with its cost. In the case of anti-pollution expenditure the benefits of the effect of reduced pollution have to be assessed. A benefit/cost ratio of 10:1 implies that for every £1 spent on improving the pollution situation, the benefits can be valued at £10. The problem often arises in quantifying the benefits. If the polluter is divorced from the effects of the pollution, agreement may be hard to obtain.

**Bentonite.** A clay which can contain a large proportion of the mineral montmorillonite. This has the property of swelling when it comes in contact with water and thus if laid in sufficient thickness can make a very good LANDFILL SITE liner. The swelling bentonites are also used for drilling mud and dam sealing purposes.

**Benzene.** A clear, colourless, volatile, flammable liquid; an aromatic hydrocarbon, formula $C_6H_6$. Benzene is widely used in the chemical industries and is a minor constituent of petrol. It is highly toxic, and carcinogenic and is strongly implicated in leukaemia.

One estimate (US Govt.) is that there are 10 excess deaths per 1000 employees exposed to a working lifetime of 1 part per million benzene (US occupational exposure standard). The US EPA estimates leukaemia death RISK as $7 \times 10^{-6}$ over a lifetime exposure to $1 \mu g/m^3$ (0.3 ppb).

Wolffe has stated:

> Certainly, although very high (occupational) levels of benzene are linked with leukaemia any risks of cancer at lower levels of benzene appears to disappear into the background of 'spontaneous' leukaemia incidence [see Figures 9 and 10]. The existence of the background rate, and the shape of the graph in the second figure is used as an important justification for the concept of the 'threshold' dose which, translated into biological terms, argues for the presence of detoxification and DNA repair/maintenance mechanisms which operate efficiently at lower concentrations of carcinogen but which can be saturated at higher doses. However, such an interpretation

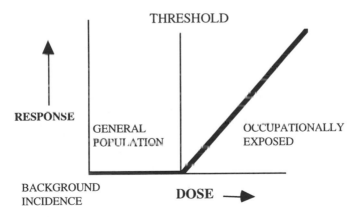

**Figure 9** Dose–response curve for relationship between benzene exposure and leukaemia assuming that background incidence of leukaemia is 'spontaneous' and unrelated to non-occupational exposure.

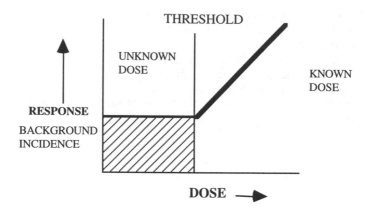

**Figure 10**   Revised dose–response curve for the relationship between benzene exposure and leukaemia suggesting that background incidence is related to non-occupational benzene exposure and that threshold dose is an arbitrary dividing line between known and unknown doses.

depends very much on how one views the revised dose response graph. The hypothesis that the background 'spontaneous' leukaemia incidence may also be related to non occupational benzene exposure casts the threshold dose in a new light. In this instance, the threshold becomes the artificial dividing line between the high (and fairly well-defined) levels of occupational exposure, to which thousands of adults are exposed, and the less-defined and generally lower levels to which millions of people are exposed in the course of their day-to-day activities. This background level of exposure to benzene is suggested to contribute to the background level of leukaemia and lymphoma in the non-occupationally exposed population.

Unleaded petrol contains more benzene than leaded petrol. It is estimated (*Royal Commission on Environmental Pollution*, 18th Report) that 78 per cent of atmospheric benzene is from petrol engine exhausts, 10 per cent from DIESEL engine exhausts and 10 per cent from fuel evaporation and oil refineries. CATALYTIC CONVERTERS, if used properly, can reduce benzene by up to 90 per cent.

Wolffe, has commented that:

detailed measurements of benzene and hydrocarbon exposure would, however, allow support for, or refutation of the hypothesis that internal combustion-powered vehicles cause cancer. There is a pressing need for biological and epidemiological investigation of non-occupational benzene exposure.

The EC maximum allowable concentration in drinking water is 10 parts per billion.

*Source:* Wolffe, S.P. Correlation between car ownership and leukaemia: Is non-occupational exposure to benzene from petrol and motor vehicle exhaust a causative factor in the leukaemia and lymphoma?, 1991.

**Benzene hexachloride (BHC).** The former name for the insecticide HCH (hexachlorocyclohexane). (⇨ CHLORINATED HYDRO-CARBONS, LINDANE)

**Benzo [alpha] pyrene.** A polynuclear aromatic HYDROCARBON that occurs in coal tar, SOOT and tobacco smoke. It has been shown to be carcinogenic in laboratory animals when administered by a number of different routes.

**Berylliosis.** ⇨ BERYLLIUM.

**Beryllium (Be).** A silver-white metal with very good resistance to corrosion, heat and stress. Its lightness and hardness find a wide variety of applications, especially in rocketry, aircraft, nuclear reactors. The main environmental sources are beryllium refining, alloying and fabricating operations and in the burning of coal.

Acute poisoning can result from exposure to airborne concentrations as low as 20 micrograms per cubic metre for less than 50 days. The lung tissue is inflamed and fevers, chills and shortness of breath can become chronic from such short exposures. Exposure leads to berylliosis.

In the UK the maximum allowable concentration at discharge has now been set at 0.1 micrograms per cubic metre. It has been classified as a HAZARDOUS POLLUTANT in the USA.

**Best Available Techniques Not Entailing Excessive Costs (BAT-NEEC).** This is essentially an updating and refinement of BEST PRACTICABLE MEANS (BPM) whereby a greater degree of control over emissions to land, air and water may be exercised. The techniques cover not only the hardware but also the way in which a particular process is operated, in fact, as operational management and staff training.

'Best available' is taken to mean demonstrably the most effective techniques for an operation at the appropriate scale and commercial availability even if they are not in general use or have not been used in the UK – this is a most important development from the previous BPM philosophy. 'Not entailing excessive cost' means that the benefits gained by using the best available techniques should be at a reasonable relationship to the cost of obtaining them, otherwise it may not be justifiable to obtain very small reductions in emissions at very high costs unless emissions are very toxic, in which case very high specifications may be imposed, e.g. as in hazardous waste incineration.

Judgement about what constitutes BATNEEC can differ between new and existing plants. For example the cost of adopting given techniques at an existing plant can depend upon its technical characteristics and could be much higher than those at a new plant which is purpose designed to incorporate these techniques from the beginning.

The benefits from adopting these techniques could also be less at an existing plant, depending on the remaining life and likely utilization, in other words a degree of flexibility is allowed to the Pollution Inspectorate.

**Best Practicable Environmental Option (BPEO).** This term was first used by the *Royal Commission on Environmental Pollution* 5th Report (1976) in order to take account of the total pollution from a process and the technical possibilities for dealing with it. It is seen as a successor to the BEST PRACTICABLE MEANS which was primarily aimed at emissions to air. BPEO in addition to controlling atmospheric emissions, includes

> appropriate measures to deal with any harmful discharges to water and for the treatment or disposal of other solid and liquid wastes to land. A BPEO should take into account the risk of transfer of pollutants from one medium to another.
>
> *RCEP*, 12th Report, BPEO, 1988.

Thus, BPEO specifically requires CROSS-MEDIA considerations which were often lacking under BPM. It does not imply that the best technology is to be used irrespective of cost. Quite the contrary, the costs and the benefits of action to deal with each problem will still be assessed. It establishes the option which provides the lease damage to the environment as a whole at an acceptable cost.

*Summary of steps in selection of a BPEO*

Step 1: *Define the objective.* State the objective of the project or proposal at the outset, in terms which do not prejudge the means by which that objective is to be achieved.

Step 2: *Generate options.* Identify all feasible options for achieving the objective: the aim is to find those which are both practicable and environmentally acceptable.

Step 3: *Evaluate the options.* Analyse these options, particularly to expose advantages and disadvantages for the environment. Use quantitative methods when these are appropriate. Qualitative evaluation will also be needed.

Step 4: *Summarize and present the evaluation.* Present the results of the evaluation concisely and objectively, and in a format which can highlight the advantages and disadvantages of each option. Do not combine the results of different measurements and forecasts if this would obscure information which is important to the decision.

Step 5: *Select the preferred option.* Select the BPEO from the feasible options. The choice will depend on the weight given to environmental impacts and associated risks, and to the costs involved. Decision-makers should be able to demonstrate that the preferred option does not involve unacceptable consequences for the environment.

Step 6: *Review the preferred option.* Scrutinize closely the proposed detailed design and the operating procedures to ensure that no pollution risks or hazards have been overlooked. It is good practice to have the scrutiny done by individuals who are independent of the original team.

Step 7: *Implement and monitor.* Monitor the achieved performance against the desired targets especially those for environmental quality. Do this to establish whether the assumptions in the design are correct and to provide feedback for future developments of proposals and designs.

Throughout Steps 1 to 7: *Maintain an audit trail.* Record the basis for any choices or decisions through all of these stages, i.e. the assumptions used, details of evaluation procedures, the reliability and origins of the data, the affiliations of those involved in the analytical work and a record of those who have taken the decisions.

*Note:* the boundaries between each of the steps will not always be clear-cut: some may proceed in parallel or may need to be repeated.

*Source: Royal Commission on Environmental Pollution,* 12th Report, 1988.

**Best Practicable Environmental Option (BPEO) Index.** A numerical estimate of the overall environmental input of a range of feasible pollution control options across the media of air, land, and water. The options can then be compared.

**Best Practicable Means (BPM).** An essentially pragmatic philosophy for the control of emissions from scheduled processes. The term was first used in the 1906 Alkali Act and was implemented by the now superseded UK ALKALI AND CLEAN AIR INSPECTORATE who were responsible for controlling emissions from industrial premises (oil refineries, chemical works, cement plant, etc.) listed in the schedule to the Act.

The ideology behind best practicable means caused much controversy as it was considered to be a loophole which allowed industry to emit noxious or injurious substances in greater amounts than would have been allowed had absolute standards been imposed. Basically, it implied that, while better emission standards may be obtainable, industry must not be unduly penalized in its operations as the costs of pollution are offset by the social and economic benefits of a thriving industrial sector. Thus, the effective emission standards may have surprisingly wide tolerances depending on local environmental factors and in practice often lagged behind those that were technologically quite feasible.

**Beta particles ($\beta$ particles).** Beta particles are electrons and therefore negatively charged. They are sparsely ionizing and have little penetrating ability, although more so than an ALPHA PARTICLE. Many atoms heavier than lead decay spontaneously by the emission of beta particles; however, a few emit positive particles (positrons) and these are also included under the term 'beta particle'. Positrons are equal in mass and charge to an electron.

In negative beta-particle emission, a neutron in the original nucleus breaks up into a proton and an electron; the latter is emitted. This decay will take place in nuclei which have too many neutrons for complete stability. Conversely, in positive beta-particle emission there are too many original protons and these break up into neutrons and positrons.

Also emitted in beta radiation are neutrinos, which have a zero charge and virtually zero mass.

**Beta radiation.** A stream of beta particles, i.e. electrons or positrons emitted from radioactive nuclei, possessing greater penetrating power than ALPHA RADIATION. ($\Rightarrow$ RADIONUCLIDE)

**Beverage cans.** A container for beverages or juices, usually steel or aluminium.

**Bilharziasis (bilharzia).** ⇨ SCHISTOSOMIASIS.

**Bi-metal container.** Container made out to two metals, with the body usually steel and the lid aluminium.

**Bioaccumulation.** ⇨ BIOLOGICAL CONCENTRATION.

**Biochemical oxygen demand (BOD).** A standard water-treatment test for the presence of organic pollutants. This biochemical test depends on the activities of bacteria and other microscopic organisms which in the presence of oxygen feed upon organic matter. The result of the test indicates the amount of dissolved oxygen in grams per cubic metre used up by the sample when incubated in darkness at 20°C for five days. (Some 'difficult' effluents may be incubated over 7 days. This is denoted by $BOD_7$.) This in turn gives a measure of the quantity of organic material present.

The factors which influence the test are:

1. The presence of toxic substances which may inhibit the growth of micro-organisms during the test.
2. The chemical uptake of oxygen by substances such as ammonia or nitrites.
3. A departure from the specified incubation temperature of 20°C.
4. The availability of dissolved oxygen and nutrients.

It is standard practice to suppress nitrification by the addition of Allythiourea (ATU). The BOD test may be supplemented by chemical oxygen demand (COD) tests, but these do not measure the biodegradable matter which is of central interest. Both tests are commonly run when a 'new' effluent is encountered. (⇨ DISSOLVED OXYGEN)

Typical BOD and COD values exerted by organic wastes are

| Type of waste | BOD (5 days, $g\,m^{-3}$) | COD ($g\,m^{-3}$) |
|---|---|---|
| Abattoir | 2600 | 4150 |
| Brewery | 550 | — |
| Distillery | 7000 | 10 000 |
| Domestic sewage | 350 | 300 |
| Pulpmill | 25 000 | 76 000 |
| Petroleum refinery | 850 | 1500 |
| Tannery | 2300 | 5100 |

shown in the table above. The COD column is the chemical oxygen demand, which uses potassium dichromate to oxidize the organic matter completely and so includes the oxygen demand from materials that cannot be treated biologically.
Both the BOD and COD tests are important for effluent monitoring. The BOD test requires five days where as the COD test can take only two hours.

Thus, once an effluent BOD:COD ratio has been determined, the monitoring of the effluent by the COD test automatically establishes the BOD, as BOD:COD ratios are relatively constant for particular effluents, but of course vary from one waste to another.

Typical BOD levels are given in the table below:

| Type of waste | BOD (mg per litre) |
| --- | --- |
| Treated domestic sewage | 20–60 |
| Raw domestic sewage | 300–400 |
| Liquid sewage silage | 10 000–20 000 |
| Pig slurry | 20 000–30 000 |
| Sludge effluent | 30 000–80 000 |
| Milk | 140 000 |

**Biodegradable materials.** Material which is capable of being broken down, usually by micro-organisms, into basic elements. Most ORGANIC wastes, such as food and paper, are biodegradable.

**Biodegradation.** The ability of natural decay processes to break down man-made and natural compounds to their constituent elements and compounds, for assimilation in, and by, the biological renewal cycles, e.g. wood is decomposed to carbon dioxide and water. Similarly, man-made paper can be decomposed. The problem is that many man-made materials such as plastics are not biodegradable and their quantity has increased substantially.

Since the 1950s the chemicals industry worldwide has increased its yearly production of bulk chemicals (from plastics to antifreeze) from 2 million tonnes to nearly 100 million tonnes in 1989. By their very nature, most of those chemicals are still around, in some shape or form, lingering and polluting long after their original job is done.

Action needs to be taken in both developing biodegradable environment-friendly alternatives and also restricting the use of those that are not strictly necessary for essential human or animal requirements.

**Biodiesel.** A DIESEL fuel substitute manufactured from oil seed rape (*Brassica napus*), 1 tonne of which will yield 300 kg oil which can be readily processed to RME (rape methyl ester). This is the main source in Europe, while soya oil is used in the United States and canola oil in Canada. This is an excellent (but costly) fuel for inner city vehicles due to the absence of SULPHUR OXIDES and low particulate emissions.

Currently, most rape seed oil is used in foodstuffs with 10 per cent going to chemical manufacture.

Reduced $CO_2$, negligible sulphur, reduced particulates and a product that biodegrades within a few days (making it particularly suitable for use on inland waterways) are the major benefits.

**Biofuel.** A fuel which is biological in origin, such as METHANE, BIOGAS and BIODIESEL, all of which can contribute to SUSTAINABILITY.

The smell of rancid cooking oil or vinegar from the exhaust fumes of buses running on biofuels are, at least for some people, a negative feature of these alternatives to petroleum-derived fuels (although to a committed user, they may be akin to silage odours to a farmer).

*Source: Pollution: Nuisance or Nemesis, HMSO, pp. 36–37.*

**Biogas.** Gas formed by ANAEROBIC DIGESTION of organic materials, e.g. whey or sewage sludge. Typical composition of 62 per cent methane, 38 per cent carbon dioxide. Can be used for heat and power purposes as spark ignition engines can be modified to use it as a fuel.

The table on page 46 gives the 1990 economics of processing slurry from 120 head equivalent dairy cattle (110 milking cows plus 50 followers), with a gravity/manual loading system (values are pounds sterling except where otherwise indicated).

The potential pollution reduction by biogas production from cattle, pig and poultry is highlighted by their respective population equivalent (in BOD terms) of 10.2, 3.5 and 0.1 times that of humans, respectively.

**Biological concentration.** The mechanism whereby filter feeders such as limpets, oysters and other shellfish concentrate HEAVY METALS or other stable compounds present in dilute concentrations in sea or fresh water. One extreme example is that of limpets collected from rocks in the Severn Estuary which

| | |
|---|---|
| 5 lid digester, fitted control house, stirring, heating, PLC controller, blower and separator, loading pump | 42000 |
| Civil engineering at £88 per cubic metre of tank | 8800 |
| Separator | 5200 |
| Separator tower and electrical work | 2800 |
| Rainwater control | 200 |
| Electrical engineering | 500 |
| Gas bags, $25\,m^3 \times 2$ | 3000 |
| Total | 62500 |
| | |
| Department of Agriculture grant 50% of digester, gas regulator bag and slurry separator but not gas use equipment | 31250 |
| Remainder to find for digester | 31250 |
| Domestic boilers to burn gas at home | 900 |
| Total cost to farmer | 32150 |

## INCOME

| | |
|---|---|
| Fibre in 250 days/year during housing of cattle | 150 tonnes |
| Plus imported slurries | 50 tonnes |
| Total for the year | 200 tonnes |

Gas $100\,m^3$ gas per day with a digester heating requirement of $34\,m^3$ biogas per day, the available gas is valued at $(100 - 34) \times 6.25\,kWh/m^3 \times £0.022/kWh \times 365$ days = £3312 per year

## SUMMARY

| | |
|---|---|
| Fibre income per year (200 t) | 6000 at £30/t |
| Gas income per year | 3300 |
| Total | 9300 |

£400 per year maintenance = net input of £8900 per year.
System cost to farmer £32000 if all work is done by outside contractors.
Payback time 3.5 years at zero inflation, less with inflation.
Other benefits include an estimated 0.5% continuous annual improvement in grass production, odour control, weed seed destruction, pollution reduction by 85%, improved pasture land and ability to graze pastures soon after slurry spreading, the possibility of organic pastures fed with clover nitrogen and large scale organic farming.

*Source:* Reproduced by permission of Green Land Systems Ltd.

contained 550 parts per million of CADMIUM of which the ingestion of 50 milligrams could be lethal. The limpets also contained 900 ppm zinc. The Severn Estuary is near a large zinc smelter and it is not improbable that the high concentrations resulted from industrial activity.

Another example is the 'Irish Seabird Wreck 1969' in which 10 000 sea birds were affected by PCBs and pesticides. Analyses showed the following PCB concentrations: the PCB level in the Firth of Clyde, 23–30 October 1989, was 0.01 micrograms per litre yet the concentration in mussels (less shell) ranged from 200–800 micrograms per kilogram, a concentration of up to 80 000 times. The pesticide concentration of the Firth of Clyde in the same period (DIELDRIN and DDT) was ca. 0.001 to 0.005 micrograms per litre. The corresponding concentration range in shellfish was 100–300 micrograms per kilogram, a concentration factor of up to 300 000 times. (⇨ PCBs)

This illustrates how minute traces of toxic compounds can be concentrated biologically and enter the food chain of human beings or in this case, be strongly implicated in the Irish Seabird Wreck, Autumn 1969 (NERC Series C No. 4, 1971). (⇨ MINAMATA DISEASE)

*Source:* B. Silcok, 'Limpets reveal metal poison danger', *The Sunday Times*, 6 June 1971.

**Biological filter.** A bed of inert granular material, e.g. plastic rings, clinker, etc., on which settled sewage effluent is distributed and allowed to percolate. In doing so, the effluent is aerated thus allowing aerobic BACTERIA and FUNGI to reduce its BIOCHEMICAL OXYGEN DEMAND. It is in effect an AEROBIC PROCESS and not a filter *per se*.

**Biological indicator.** The use of living organisms of plants to detect environmental changes. (⇨ BIOTIC INDEX; GLADIOLI)

**Biological systems.** ⇨ FOOD CHAIN; CARBON CYCLE; NITROGEN CYCLE.

**Biomass.** The mass of living organisms forming a prescribed population in a given area of earth's surface. It is usually expressed in grams per square metre ($g/m^2$).

**Biome.** A large, well-defined biological community, e.g. tundra, grassland, coral atoll. The biome has a particular form of vegetation and associated animals which have become adapted to the local conditions. In other words, it is a balanced ecological community.

**Biometallurgy.** The application of microbiological processes to mineral treatment especially for the extraction of non-ferrous metals from sulphide ores. (⟹ SOLUTION MINING)

**Biophotolysis.** The storage and use of electrons produced in the first stages of PHOTOSYNTHESIS which can then be used to make free hydrogen. This is another research route which investigates the use of energy from the sun to produce electricity or fuels such as hydrogen, in this case directly. Experimental work on algal systems has shown that this can be done in the laboratory.

**Bioreactor (aerobic).** Many industrial processes rely on the use of gas–liquid bioreactors which require the intimate contact of oxygen with the biological broth or effluent as the case may be.

As the characteristics of biological broths are different from those of liquid effluents, two designs of aerobic bioreactors have evolved, namely, the external loop and the internal loop configurations, respectively. The gas injected region is termed the *riser* and the relatively gas free region, being heavier, is the *downcomer*. In the external loop, the riser and the downcomer are usually two separate tubes linked by connecting zones near the top and the bottom of the reactor. The internal loop is often of 'split cylinder' construction with a vertical baffle which allows internal circulation.

The performance characteristics of the two designs are quite different. The external loop has low oxygen transfer rates and high turbulence which suits effluent treatment whereas biological broths are better processed by the high oxygen transfer rates and low shear of the internal loop which is used extensively for mammalian cell culture. (⟹ AEROBIC PROCESSES; DEEPSHAFT TREATMENT)

*Source:* Y. Christi, 'Airlift reactors design and diversity', *Chemical Engineering*, February 1989, pp. 41–43.

**Bioremediation.** The use of biological methods to remediate/restore CONTAMINATED LAND. Typical methods make use of tailored MICROBES and break down PHENOLS which are major contaminants in many sites, especially old town gasworks. (⟹ MICROBES, CONTAMINATED LAND)

**Biosphere.** The region of the earth and its atmosphere in which life exists. It is an envelope extending from up to 6000 metres above to 10000 metres below sea level and embraces Alpine plant life and the ocean deeps. The special conditions which exist in the biosphere to support life are a supply of water; a supply of usable energy; the existence of interfaces, i.e. areas where the liquid, solid and gaseous states meet; the presence of nitrogen, phosphorus, potassium and other essential nutrients and trace elements; a suitable temperature range; and a supply of air (although there are anaerobic forms of life).

**Biospheric cycles.** To maintain the biosphere and thus life, essential materials must be recycled so that after use they are returned in a re-usable form. In order to sustain this recycling a supply of energy is required – solar energy.

The principal elements in the biosphere are nitrogen, oxygen, hydrogen and carbon. As these elements constitute 99 per cent of all living things, including plants, they are obviously of prime importance. Other elements are also important, e.g. phosphorus, potassium, sulphur and calcium, and all have their own renewal cycles, so that they become available for reuse.

The major biospheric cycles are the CARBON CYCLE, the OXYGEN CYCLE, the NITROGEN CYCLE, and the water or HYDROLOGICAL CYCLE. They are all interlinked and operate simultaneously and maintain all life on earth.

**Bituminous coal.** The most common form of British coal. It produces or contains a relatively high proportion of volatile HYDROCARBONS (over 13 per cent) and, burned in the traditional domestic grate, it will give rise to SMOKE.

**Black List.** Actually List 1 in EC *Framework* DIRECTIVE (76/464/ EEC) on pollution caused by certain dangerous substances discharged into the aquatic environment. It lists 129 substances considered to be sufficiently toxic, persistent or bioaccumulative that steps should be taken to eliminate pollution by them. This is done by setting of limit values in subsequent *Daughter* Directives from the Black List. The UK government has decided on an initial list of 23 substances to be included on a RED LIST where substances on this list will be subject to a requirement, to be introduced progressively for existing plant, to use BATNEEC.

**Blanket.** 'Fertile' material (usually depleted URANIUM) placed round a fast reactor core to capture neutrons and create more fissile material (usually plutonium).

**Blasting.** A means of fragmentation to uncover coal, rock, etc. This is now strictly controlled. Figure 11 shows Northern Coal's 1993 blast monitoring results, using a fully computerized system to monitor properties located on the boundaries of their site operations.

A total of 3220 recordings have been made, not one of which has exceeded the authorized limits for their operations.

*Source: RJB Mining, Gosforth, Newcastle upon Tyne.*

**Bloom.** The sudden upsurge in numbers of ALGAE in a body of water, usually due to an excess of NUTRIENTS.

**Blowdown[1].** Rapid depressurization, e.g. of a PWR primary coolant circuit.

**Blowdown[2].** HYDROCARBONS purged during refinery start-up or shut-down.

**Blowdown[3].** In boiler operation, a process for reducing the concentration of dissolved solids in the boiler water, or for removal of sludge from a boiler.

**Blue-green algae.** Algae which, in addition to chlorophyll, possess red and blue pigments. Some are capable of fixing nitrogen directly and so make it available for plants and animals; others convert nitrates to nitrites. They are, therefore, involved in the NITROGEN CYCLE. Many are capable of producing toxins which have caused death to animals, birds, and fish in many countries. These toxins have been linked to human illness through skin contact and ingestion.

**BMPW Score.** The Biological Monitoring Working Party Score. This is arrived at by first obtaining a sample of the macro-invertebrates (creatures longer than 1mm) in a river. The macroinvertebrates are then classified into different taxia. Each taxon has a score between 1 and 10, with the species more sensitive to pollution having a higher score. The BMWP score is the sum of the scores for each taxa found. A BMWP score greater than 100 generally indicates good water quality.

**Body burden.** The amount of radioactive or toxic material in a human or animal body at any time.

**Boreholes.** ⇨ WELLS AND BOREHOLES.

**Bottle.** Glass or plastic container, which can be returnable or non-returnable. The ratio of non-returnable to returnable bottles in the UK is ca. 25:1, whereas in Norway, all glass bottles are returnable.

Bottles come in different colours, shapes and sizes. Since conservationists argue strongly against the throw-away philosophy, with some justification on ENERGY consumed, a strong case can be made for bottle standardization and the enforced use of returnables so that both energy and materials are conserved and waste minimized as non-returnable bottles

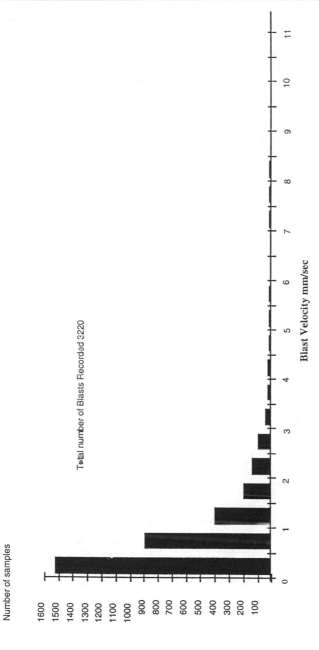

**Figure 11** Total blasts recorded 1993.

account for around 8 per cent by weight of UK municipal wastes.

Figure 12 gives the energy required $(kWh_e)$ to make a 0.3 litre refillable beer bottle and a non-refillable one but the former can last for many trips, requiring only the energy for transport, washing and refilling. Thus, as shown in Figure 13 the refillable bottle pays for itself in energy terms after two trips. The glass industry is promoting BOTTLE BANKS to recover bottles for remelting. However, the real pay-off in energy terms comes from the use of returnable refillable bottles.

A 1994 report for the US Department of Commerce concluded:

Recycling of glass containers saves some energy but not a significant quantity compared to reuse. The energy saved is about 13% of the energy required to make glass containers from virgin raw materials. This estimate includes energy required for the entire product life cycle, starting with raw materials in the ground and ending with either final waste disposition in a landfill or recycled material collection, processing, and return to the primary manufacturing process. The actual savings depend on local factors, including

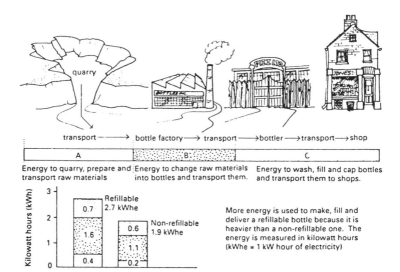

**Figure 12** Comparison of energy required to manufacture half pint refillable and non-refillable bottles.

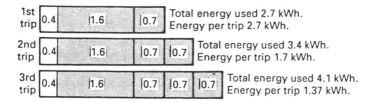

**Figure 13**   Energy saved by the use of refillable bottles.

population density; locations of landfills, recovery facilities and glass plants; and process efficiencies at the specific facilities available. The savings increase if wastes must be transported long distances to a landfill or if the containers are made in an inefficient furnace. They decrease if there is no local Municipal Recycling Facility (MRP) or glass plant, or if material losses in the recycling loop are high.

The options for disposition of used glass containers can be compared on the basis of the important decision factors, including energy saved, landfill space required, and emissions. The table on page 54 gives a qualitative comparison of the options for the recycling or disposal of glass containers. In order of energy saved, the options for disposition of glass containers are reuse, recycling to the same product, recycling to a lower value product, and landfill.

The options with minimum energy use generally have the lowest other impacts as well. No real trade-off exists. The energy savings are small, and the balance can be altered by local or regional conditions. Although the analysis was performed for US conditions, they are not that dissimilar to conditions in the UK. (⟳ ENERGY ANALYSIS)

*Source:* Argonne National Laboratory, *Energy Implications of Glass Container Recycling* report DE94000288, February 1994.

**Bottle bank.** A scheme promoted by the glass industry for the collection of glass bottles for CULLET. The percentage of domestic glass containers returned for remelting via the bottle banks in the UK is ca. 27.56 per cent (1994), cf. rates of up to 60 per cent of container glass in Holland. The use of cullet effects an energy saving compared with that required for melting the basic raw materials. (No account is taken of excess fuel consumption by the public in visiting bottle banks.)

Comparison of options for disposal of glass containers

| Alternative | Energy impact | Environmental impact | Material sent to landfill | Comments |
|---|---|---|---|---|
| Reuse | Saves most of production energy | Avoids production emissions; possible impacts from water treatment | None | Refillable bottles currently have a low market share in the United States |
| Recycle to containers | Small reduction in production energy | Small reduction in emissions[a] | Cullet processing losses | Bottles now made from 20% old bottles; colour separation required |
| Recycle to fibreglass | Small reduction in production energy | Small reduction in emissions[a] | Cullet processing losses | Inefficient furnaces used |
| Recycle to reflective beads | Small reduction in production energy | Small reduction in emissions[a] | Cullet processing losses | Can use clean, mixed cullet |
| Recycle to glassphalt | Saves no energy | Full production emissions | Cullet processing losses | Can use clean, mixed cullet |
| Landfill | Saves no energy | Full production emissions | Maximum | Economic cost, no return |

[a]Dust and $SO_2$ reduced, but $NO_x$ increased

**Bottle Bill.** A legislative (US) measure to encourage container recycling. 'Bottle Bills' are now in force in 12 states (including Massachusetts where a multi-million dollar industry-financed campaign was waged against such legislation). One such attempt is the Vermont Container Law 1973: it should be noted that no container is banned, but a deposit refund (in Vermont's case; other states require a plain refund) must be paid by the retail outlet on presentation of the empty container. By this means, a high degree of control over packaging entering the waste stream can be achieved and refillable bottles' use encouraged as it is cheaper for the bottler to refill empty bottles. The Vermont Deposit–Redemption Cycle is shown in Figure 14.

A 5 cent minimum deposit must be charged on a container sold at a retail (**3**). It can be initiated at (**1**), (**2**) or (**3**). Glass containers destined for refilling usually have the deposit initiated at (**1**). Most other deposits are initiated at (**2**). Containers are redeemed for the refund value at any store or redemption centres (**5**) which sells that kind, size and brand. Retailers and redemption centres (**5**) are reimbursed the refund value by the Distributor (**6**) and also are paid a handling fee (20 per cent of the refund value, or a minimum of 2 cents). Distributors (**6**) normally pick up containers from (**5**) on regular schedules or upon request. Glass containers are typically returned to bottlers for refilling. Other container materials are sold for recycling (**7**) along with the packaging materials such as corrugated cartons, etc.

Vermont (and other US states) have made great environmental gains through the use of Bottle Bills. California has just

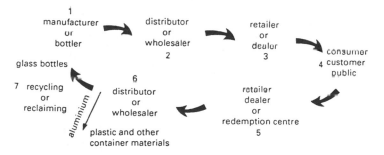

**Figure 14**  The deposit–redemption cycle – Vermont Container Law.

followed suit (1988) with a novel twist in that redemption centres will repay the deposit, which can be increased, until satisfactory return rates are achieved. The California 'Bottle Bill' of 10 October 1987:

- Requires the cost of beverage containers to the consumer to include an initial charge of one cent per container.
- Enables consumers to receive a refund, a bonus and the scrap value of the container at redemption centres, not at retail establishments that sell beverage containers.
- Enables recycling operators to have a more active role in container redemption than in other states.
- Establishes a state agency to administer the Act, including conducting audits of recovery rates. If these fall below 65 per cent, the deposit will be increased.

Enlightened European practice regarding containers is, perhaps, typified by that of Denmark where their Act No. 297 of June 8 1978, Reuse of Paper and Drink Packagings and the Reduction of Waste, has made significant contributions to resource conservation. The Danish National Agency for Environmental Protection has stated that without the present reuse of bottles and the ban on cans, the amounts of glass and metal waste would be much larger. They estimate that the percentage increase in glass in Danish municipal solid waste would rise by 46 per cent, were it not for these measures.

Metal beverage cans have now been phased out from a level of 43.5 million cans in 1976 to zero in 1982. There are no mandatory arrangements for wines and spirits (beer and soft drinks are of course covered) but there is an excise duty of 0.97 DKr on all new wines and spirit bottles and a 50 per cent increase in the excise duty was followed by a more than 50 per cent increase in the collection and cleaning of wine and spirit bottles, which in 1982 reached a record high of 77 million bottles thus recycled. Container deposit legislation has the power to ensure much more effective recycling than BOTTLE BANKS. (⇨ RECYCLING)

**Brackish water.** Water (non-potable) containing between 100 and 10 000 ppm of total dissolved solids. (⇨ BRINE, DESALINATION)

**Breakpoint chlorination.** The chlorination (disinfection) of drinking water until the chlorine demand of any organic impurities present is satisfied. Any further chlorine addition is then available as free chlorine.

**Breeder reactor.** ⇨ NUCLEAR REACTOR DESIGNS.

**Brine.** Water containing more than 100000 ppm of total dissolved solids (NaCl) which can yield salt after evaporation. Also the reject concentrate from DESALINATION plants. (⇨ CHLOR-ALKALI PROCESS)

**Btu (British Thermal Unit).** A unit of energy approximately equal to 1055 joules or 0.252 kilocalories. It is the amount of heat required to raise the temperature of 1 lb of water by 1 Fahrenheit Degree (60–61°F). 3413 Btu = 1 kilowatt-hour, the common unit in which electricity is measured. Although Btu is superseded by the joule, it remains in very common UK and US usage. *Note:* 100000 Btu = 1 therm.

**Bubble.** A concept of US origin which involves drawing a notional 'bubble' or dome-shaped boundary around a plant area, or even a state, and putting an upper limit on the total amount of a pollutant allowed to pass into the 'bubble'. Used in setting a limit or standard on total emissions. Hence an increase in pollution in one plant can be traded off against a decrease from one or more plants.

**Buffer capacity.** The capacity of certain solutions to oppose a change in properties, especially when ACID or ALKALI is added.

**Buoyancy.** The upward force that acts on a body that is totally immersed in a fluid. It is equal to the weight of the fluid displaced by the body, so if the density of the immersed body is less than that of the fluid, there will be a net upward force on the body, e.g. hot gases discharged from a chimney into cooler air will rise due to buoyancy until they obtain the temperature of the surroundings. The effect of buoyancy of hot flue gases can be to give an EFFECTIVE CHIMNEY HEIGHT which can be two or three times greater than the actual chimney height, thus greatly aiding dispersion of any pollutants or ODOURS.

**Burn out[1].** A measure of the efficiency of INCINERATION which relies on determinations of fixed carbon and putrescible matter in the residue after combustion.

**Burn out[2].** The failure of a nuclear reactor heat-exchange surface which can result in the loss of heat-exchange fluids and may, in very extreme circumstances, result in the release of radio-active products. It is equivalent to MELT-DOWN in US terminology.

**'Buy-back' programmes.** Programmes to purchase recyclable materials from the public.

**By-product.** An additional product which can be produced as an ancillary to the principal product, e.g. animal by-products from the offal remaining from meat processing or compost from the fines of a waste derived fuel plant.

**Byssinosis.** ⇨ FIBROSIS.

# C

**Cadmium (Cd).** A soft silvery HEAVY METAL, atomic weight 112.4. It is used in semiconductors, control rods for nuclear reactors, electroplating bases, POLYVINYL CHLORIDE (PVC) manufacture, and batteries. World production is around 18 000 tonnes per year produced in conjunction with zinc smelting.

Cadmium serves no biological functions, is toxic to almost all systems, and is absorbed into the human organism without regard to the amount stored. Very small doses can cause severe vomiting, diarrhoea and colitis; pneumonia can develop as a consequence. Poisoning has developed from sources such as home-made punch stored in a cadmium-plated bowl. Changes in the heartbeat rate of persons in atmospheres containing as little as 0.002 micrograms per cubic metre have been reported (maximum allowable concentration is 100 micrograms per cubic metre). Hypertension is a common complaint of those who have ingested the metal at doses well below those regarded as toxic. Long exposure in lead and zinc smelting areas in Japan led to the so-called Ouch-ouch (Itai-itai) disease where the sufferers had severe joint pains and eventual immobility as a result of skeletal collapse. Cadmium leads to bone porosity and inhibition of bone-repair mechanisms.

Figure 15 shows how close cooperation by industry can significantly lower Cd levels in sewage SLUDGE. This enables it to be spread on land (not always the case in the UK due to high Cd levels).

**Caesium-137.** A radioactive isotope of the metal caesium. Reported caesium-137 levels (Friends of the Earth, February 1989) of up to 4400 becquerels per kilogram in silt samples taken from the River Esk which is near the Ravenglass Estuary, the site of the Sellafield (Windscale) nuclear fuel reprocessing plant should be compared to the National Radiological Protection Board's generalized DERIVED WORKING LIMIT of 900.

**Calcium deficiency.** DDT and other CHLORINATED HYDROCARBONS

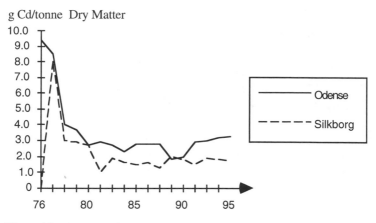

**Figure 15** *Source:* J.A. Hansen, 'A new biowaste agenda', *ISWA Times,* issue number 2, 1995.

such as polychlorinated biphenyls severely interfere with a bird's ability to metabolize calcium. The result is the production of egg shells which are so thin that they are crushed by the weight of the nesting birds.

**Calibration.** All the operations for the purpose of determining the values of errors of a measuring instrument (and if necessary to determine other metrological properties). Calibration of an instrument results in a correction factor or a series of correction factors that can subsequently be applied to readings given by the instrument.

**Calorie (cal).** Physicists define the calorie as the heat required to raise 1 gram of water 1°C. This is a minute quantity of heat in practical terms. The term used on diet sheets refers to the kilocalorie (Cal), which is 1000 times larger.

**Calorific value, gross.** The number of heat units measured as being liberated when unit mass of fuel is burned in oxygen saturated with water vapour under standardized conditions. (The remainder being gaseous oxygen, CARBON DIOXIDE, sulphur dioxide, and nitrogen, water and ash.) Used as a reference for fuel evaluation purposes.

**Calorific value, net.** The gross calorific value less the heat of evaporation of the water originally contained in the fuel *and* that formed during its combustion.

**Camelford.** Area of N. Cornwall, UK where a 20 tonne load of ALUMINIUM SULPHATE solution was accidentally tipped into public drinking water supplies on 6 July 1988.

Various ailments were attributed to this incident and out-of-court settlements have been accepted by most claimants (*Guardian* 19 May 1994).

**CAMP.** An acronym for Continuous Air Monitoring Program (USA).

**Can[1].** The metal container in which nuclear fuel is sealed to prevent contact with the coolant or escape of radioactive fission products, etc. It may be made of Magnox, Zircaloy or stainless steel which is stripped off in nuclear fuel reprocessing to retrieve the spent fuel.

**Can[2].** Tin plate, sheet aluminium or bimetallic (tin plate and aluminium) container for beverages and foodstuffs. (⇨ RE-CYCLING).

Figure 16 shows the can consumption in the UK for (a) 1994 and (b) 1980–1994.

**Can banks.** Containers used by public deposit for the collection of beverage and other cans.

**Cancer.** A disordered growth of cells which can invade and destroy body organs and thereby cause incapacity or death. It can be initiated by CARCINOGENS. The many causes of cancer are subjects of major medical research. (⇨ ASBESTOS; HEAVY METALS; IONIZING RADIATION)

**due to air pollution.** ⇨ CARCINOGEN.

**due to asbestos.** ⇨ ASBESTOS.

**due to chloroprene.** ⇨ CHLOROPRENE.

**due to chrome wastes.** ⇨ CHROME WASTES.

**due to DES.** ⇨ DES (DIETHYLSTILBOESTROL); FEED-CONVER-SION RATIO.

**due to hair dyes.** ⇨ MUTAGEN.

**due to organophosphates.** ⇨ ORGANOPHOSPHORUS COMPOUNDS.

**due to pesticides.** ⇨ ALDRIN.

**due to PVC.** ⇨ PVC (POLYVINYL CHLORIDE).

**Candle.** Formerly used in air pollution measurement or studies of sulphur dioxide monitoring. It consisted of a porcelain cylinder covered with linen impregnated with lead dioxide. The amount of lead sulphate formed over, say, one month was measured and average concentrations of sulphur dioxide determined.

**Capacity factor.** Capacity factor is a good indicator of the service

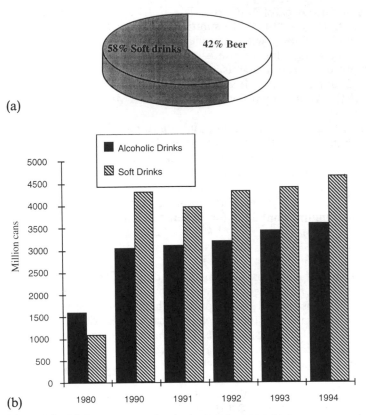

**Figure 16** (a) Can consumption in the UK for 1994. (b) Can consumption in the UK for 1980–1994.
*Source:* Canmaker Report 1995.

supplied to the power grid by a power generating unit or group of units if operating on base load, i.e.

$$\text{Capacity factor} = \frac{\text{Output for period}}{(\text{Rated capacity})(\text{Period duration})}$$

($\Rightarrow$ AVAILABILITY FACTOR)

**Carbohydrates.** Group of organic compounds composed of CARBON, HYDROGEN and OXYGEN. Contains the sugars, i.e. monosaccharides and disaccharides and the polysaccharides (starch and CELLULOSE). General formula: $C_x(H_2O)_y$.

**Carbon.** A non-metallic element, ATOMIC NUMBER 6, relative atomic mass 12.011, symbol C. Carbon compounds make up all living matter and, consequently, fuels such as coal and petroleum.

**Carbonation.** The process by which atmospheric CARBON DIOXIDE penetrates the pores of concrete in humid conditions and depassivates (neutralizes) the natural ALKALINITY of concrete by reacting with calcium hydroxide to give calcium carbonate. As the pH falls below 9, moisture can start to corrode the reinforcement steel. As this corrodes it expands (up to seven fold) and causes cracking and eventual destruction of the concrete. When SUSTAINABILITY is desired, measures need to be built in to ensure longevity of concrete structures.

**Carbon, activated.** ⇨ ACTIVATED CARBON.

**Carbon black.** Finely divided carbon, usually produced from the combustion of gaseous and liquid HYDROCARBONS with restricted air supply. Since it is not easy to remove the last traces of the product from gases leaving the stack, the production of carbon black is often a source of intense air pollution.

**Carbon cycle.** The atmosphere is a reservoir of gaseous CARBON DIOXIDE but to be of use to life this must be converted into suitable organic compounds, i.e. fixed, as in the production of plant stems. This is done by the process of PHOTOSYNTHESIS. The productivity of an area of vegetation is measured by the rate of carbon fixation. For example, tropical rain forest growth will fix 1–2 kilograms per square metre per year, whereas barren areas such as tundra will fix less than 1 per cent of this amount. Herbivores eat the plant material and carnivores eat them. The CARBON fixed by photosynthesis is eventually returned to the atmosphere as plants and animals die and the dead organic matter is consumed by the decomposer organisms. Thus the global carbon cycle on land (there is a complementary oceanic one as well) is as pictured in Figure 17.

In Figure 17, the atmosphere contains 700 thousand million ($700 \times 10^9$) tonnes of carbon as carbon dioxide. The fixed amount of carbon in the biomass (plants and animals except the decomposers) is $550 \times 10^9$ tonnes and the dead organic matter reservoir is $1200 \times 10^9$ tonnes. The land-based system is in a steady state, i.e. the inputs and outputs balance. The input has been estimated as $110 \times 10^9$ tonnes carbon per year from the ATMOSPHERE as carbon dioxide, of which $60 \times 10^9$ tonnes per

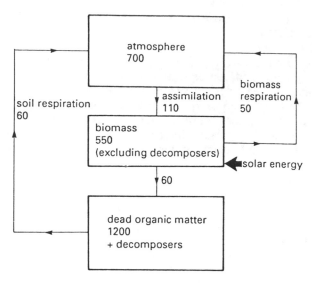

**Figure 17**   Carbon cycle on land. The quantities in Figures 17–19 are in thousand million ($10^9$) tonnes. The figures in the boxes show the amount of carbon either fixed or available for fixation. The figures with the arrows show the flow rates of carbon which is assimilated or respired per year.

year are fixed by photosynthesis and $50 \times 10^9$ tonnes per year are returned to the atmosphere as carbon dioxide from respiration processes after being used as an energy source to keep the plants alive. The dead organic matter reservoir has an input of $60 \times 10^9$ tonnes per year as FIXED CARBON which is decomposed to $60 \times 10^9$ tonnes per year as carbon dioxide. The system is in balance, powered by solar energy. In the process, oxygen is also produced so that the carbon and OXYGEN CYCLES interlink.

Complementing the CARBON CYCLE on land is the equivalent at sea. This differs from the land cycle in that carbon dioxide is soluble in sea water and is thus available for the phytoplankton. These are single-celled plants that take up carbon dioxide and, by photosynthesis, use it to produce carbohydrates and oxygen which also dissolves in the water, so the oceanic system is virtually closed. Further up the FOOD CHAIN, zooplankton and fish consume the CARBON fixed by the phytoplankton-centre > centre > 2 and in turn die and decompose, as do the

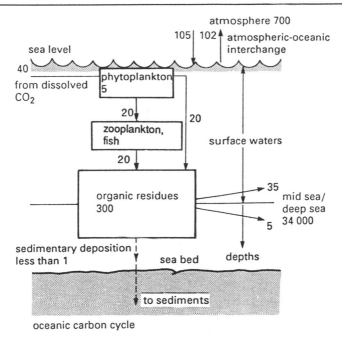

oceanic carbon cycle

**Figure 18** Oceanic carbon cycle.

uneaten phytoplankton, so that the carbon dioxide is replaced. Less than $1 \times 10^9$ tonnes carbon per year are deposited to sediments. Thus the oceanic system is also in a quasi steady state as shown in Figure 18. The inputs to the organic residues reservoir are $20 \times 10^9$ tonnes per year from fish and zooplankton and $20 \times 10^9$ tonnes per year from phytoplankton. This is balanced by an output of $35 \times 10^9$ tonnes per year as carbon dioxide to the surface layers and $5 \times 10^9$ tonnes per year carbon dioxide to the depths. Now, the mixing time for exchange between surface layers and the depths is roughly 1000 years and it is this mixing time which controls the balance between carbon dioxide in the atmosphere and the oceans. Thus any extra carbon dioxide from man's fossil fuel combustion activities will increase the atmospheric carbon dioxide concentration and, since the natural system is a closed cycle, the assimilation of excess carbon dioxide will take 1000 years before equilibrium is attained through atmospheric–oceanic interchange. As we

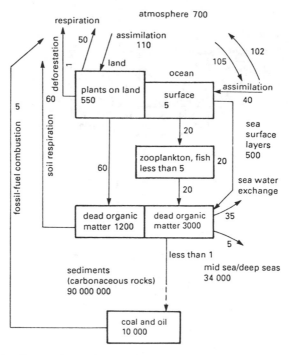

**Figure 19**   Composite carbon cycle.

cannot control the carbon (or any other) cycle, we should be careful not to disturb natural systems too far.

Man's fossil fuel COMBUSTION activities account for $5 \times 10^9$ tonnes per year, plus perhaps $1 \times 10^9$ tonnes per year from deforestation, of carbon as carbon dioxide which is not wholly assimilated by the oceanic interchange mechanism. ($\Rightarrow$ CARBON DIOXIDE)

The composite carbon cycle in the BIOSPHERE is shown in Figure 19 and includes the additional fossil fuel input and atmospheric–oceanic interchange fluxes which amount to an uptake of $105 \times 10^9$ tonnes of carbon per year and a release of $102 \times 10^9$ tonnes carbon per year, i.e. a net absorption of $3 \times 10^9$ tonnes of carbon per year which does not match the fossil fuel combustion and deforestation $CO_2$ contributions.

A controlling factor in the absorption of $CO_2$ in sea water is its ability to dissolve $CO_2$. This is influenced by sea water temperature. The greater the temperature, the less is the ability. Thus, the GREENHOUSE EFFECT could accelerate the increase of the $CO_2$ content of the atmosphere by raising sea-water temperatures. This atmosphere–ocean interaction is the subject of detailed scientific scrutiny.

Attention is also being focused on the role of plankton as they absorb $CO_2$ during photosynthesis, most of which is returned by respiration in a very short time scale. However, some sink to the depths where the carbon may then be permanently locked in the sediments thus removing some $CO_2$ from the atmosphere.

*Source:* Oceans and the Global Carbon Cycle. *Biochemical Ocean Flux Study*, BOFS Office, Plymouth Marine Laboratory, May 1987.

**Carbon dioxide (CO₂).** Gas produced by the complete combustion of carbonaceous materials, by decay organisms such as aerobic DECOMPOSERS, by FERMENTATION, and by the action of acid on limestone. It is exhaled by plants and animals and utilized in PHOTOSYNTHESIS in the CARBON CYCLE.

The COMBUSTION of fossil fuels, slash and burn agriculture and forest clearance has raised the atmospheric carbon dioxide concentration from 290 ppm in the last century to 350 ppm (1988). Most of the increase in 1988 was particularly high at 2.5 ppm as opposed to the average increase of 1.5 ppm over the last ten years. Estimates of the doubling of atmosphere $CO_2$ concentrations from the pre-industrial level of 280 ppm range from 30 years, if the oceans warm up and less $CO_2$ is dissolved, to ca. 50 years if current growth rates are maintained.

CARBON DIOXIDE is partly responsible for the atmospheric GREENHOUSE EFFECT – estimates vary, but 50 per cent may be attributable to $CO_2$. This has led to the claim that nuclear power can reduce $CO_2$ emissions. The Rocky Mountains Institute has shown that constructing a nuclear plant every 3 days at a cost of $80 billion/year over 37 years would not reduce global $CO_2$ emissions.

The base measures that would help steady the atmospheric $CO_2$ concentration are:

1. The use of national and international programmes on energy conservation to reduce fossil fuel combustion, plus a tax on fossil fuel based energy.
2. Phasing out CFCs and other substances which are depleting the ozone layer, by 1995.

3. Reversing the current level of deforestation, through the protection of the rainforests and large tree planting programmes.
4. More public transport and swingeing private vehicle taxes.
5. A switch to renewable sources of energy such as SOLAR ENERGY, WATER and TIDAL POWER.

The contribution of the other GREENHOUSE GASES should not be overlooked. Methane has a 'greenhouse factor' 30 times that of $CO_2$, hence measures should be directed towards its control, e.g. the reduction of LANDFILL GAS releases from the decomposition of DOMESTIC REFUSE in LANDFILL SITES could substantially reduce the greenhouse gas emissions from developed countries. Similarly, natural gas pipeline leakage should also be controlled. Most attention is, however, focused on $CO_2$ with coal fired power stations and automobiles as the main developed country targets. ($\Rightarrow$ CLIMATE; GREENHOUSE EFFECT; GREENHOUSE GASES)

**Carbon, fixed.** Not a precise constituent of a fuel, but in the PROXIMATE ANALYSIS that proportion determined by deducting the sum of moisture, volatile matter and ash from 100 per cent.

**Carbon monoxide (CO).** A colourless odourless gas, lighter than air, formed as a result of incomplete combustion. It is a chemical poison when inhaled, as it can penetrate tissues and is absorbed into the bloodstream where it combines with the haemoglobin of blood cells, 300 times faster than OXYGEN and thus deprives the brain and heart tissues of oxygen. ($\Rightarrow$ AUTOMOBILE EMISSIONS; SMOG; ASPHYXIATING POLLUTANTS; AIR QUALITY CRITERIA)

**Carbon tax.** $\Rightarrow$ GREENHOUSE GASES.

**Carbonization.** The treatment of coal or other solid fuels by heat in a closed vessel usually with the object of producing coke or gas. High-temperature carbonization is used for the production of coal gas and (at 1300°C) blast furnace coke. Low-temperature carbonization (600°C) is used to produce 'smokeless fuels'. Carbonization also gives many by-products (e.g. coal tar) of great importance to the chemical industries.

**Carcinogen.** Any compound or element which will induce CANCER in man or other animals. It is now accepted that a large proportion of human cancers are directly associated with environmental agents, in particular chemical and physical agents, e.g. some chlorinated hydrocarbon PESTICIDES and

ASBESTOS fibres. Carcinogens can be inhaled, e.g. tobacco smoke and asbestos fibres; ingested; absorbed through the skin (insecticides, hair dyes).

A no-effect level cannot be assumed for carcinogens as the development of cancer is the result of an accumulation of irreversible cellular damage. In contrast, exposure to toxic substances can in many cases result in little or no damage below certain levels and they are eventually excreted from the human system.

*Health Hazards of the Human Environment* (World Health Organization, 1972) lists the following classes of environmental carcinogenic agents:

1. *Polynuclear compounds.* These occur in tobacco and coal smoke, exhaust gases; a major compound is BENZO (ALPHA) PYRENE (BP) found in all kinds of soot and smoke.
2. *Aromatic amines.* These are mainly found in industrial environments; they can also contaminate foodstuffs and plastics.
3. *Chlorinated hydrocarbons.* A major group which embraces industrial solvents, DDT, ALDRIN, DIELDRIN, and other pesticides. Some members of this family are notorious for their ubiquity, persistence and ability to accumulate in living systems. (⇨ CHLORINATED HYDROCARBONS; RED LIST)
4. *N-nitroso compounds.* These can be found in industrial solvents of chemicals. They may be formed in the human intestine through bacterial action following the ingestion of nitrites.
5. *Inorganic substances.* Some toxic metals are included in this class, particularly beryllium, selenium, cadmium.
6. *Naturally occurring agents.* These are the so-called toxins that occur in spoiled food
7. *Hormonal carcinogens.* Hormonal imbalances can induce cancer. The growth-promoting chemical DES (DIETHYL-STILBOESTROL) has been implicated as a carcinogen.

The known range is very extensive and any substance that acts as a MUTAGEN must be considered a potential carcinogen until proved otherwise.

**Carnot efficiency.** This is a direct outcome of the Second Law of Thermodynamics and places the upper limit on the maximum possible conversion of heat (thermal energy) to work in a heat engine. It is usually derived as:

$$\text{Carnot efficiency} = \frac{T_{\text{source}} - T_{\text{sink}}}{T_{\text{source}}}$$

where the TEMPERATURES are on the ABSOLUTE TEMPERATURE SCALE. Thus, comparing a heat engine supplied with steam at 600°C (873 K) and the 'choice' of a sink in the Arctic at 5°C (278 K) or a UK river at 30°C (303 K) the respective Carnot efficiencies are:

$$\eta_{\text{Arctic}} = \frac{(873 - 278)}{873} \times 100 = 68\%$$

$$\eta_{\text{UK}} = \frac{(873 - 303)}{873} \times 100 = 65\%$$

This shows the influence on efficiency that sink temperatures can have, and why they should be as low as practicable in order to maximize efficiency. Conversely source temperatures should be as high as possible.

If a COMBINED HEAT AND POWER system is being considered, $T_{\text{sink}}$ could typically be 130°C in order to provide hot water at a useful temperature. In this case the Carnot efficiency becomes 54 per cent. This loss in conversion is of course made up by the use of the rejected heat in the associated heat network, thus leading to a higher overall efficiency.

In practice, the actual efficiencies are lower than the Carnot efficiency due to the effects of friction and flow losses. Nevertheless, the Carnot efficiency is a useful basis for comparison. (⇨ LAWS OF THERMODYNAMICS; DIRECT ENERGY CONVERSION)

**Car tyre.** Consumer product which presents a serious recycling problem. It is mainly composed of vulcanized rubbers, steel wire and fabric cord. The rubbers are chemically crosslinked with sulphur and form the bulk of the product comprising the carcass and tread. The tread is patterned with deep groves to assist water removal when driving in the wet and with use the tread is worn down in a more or less uniform manner. Current legislation (Jan 1992) imposes a minimum groove depth of 1.6 mm over the complete circumference, so a tyre with grooves less than this is illegal. Tyre replacement gives rise to a recycling problem, although the carcass may still be roadworthy. A small proportion may be retreaded, a practice more widely prevalent with truck tyres owing to their greater cost and complexity. The large proportion, however, are not fit for further use and are currently dumped as landfill or elsewhere. Since it is a hollow product, such tyre dumps easily catch fire, creating a serious air pollution problem (particularly from the CARBON BLACK soot and SULPHUR DIOXIDE gas emitted) as well as potential water run-off. A small proportion of waste tyres are incinerated to reclaim some heat energy in specially designed furnaces. Some *buffed crumb* is produced in retreading and can be recycled into sports surfaces, where the particles are bonded together with polyurethane adhesive.

**Catalysis.** The acceleration (or retardation) of chemical or biochemical reactions by a relatively small amount of a substance (the CATALYST), which itself undergoes no permanent chemical change, and which may be recovered when the reaction has finished.

**Catalyst.** A substance or compound that speeds up the rate of chemical or biochemical reactions. Catalysts used by the chemical industry are usually metals, such as vanadium which speeds up synthesis reactions. ENZYMES are organic catalysts.

One very interesting application of catalysis is the combustion of natural gas without the presence of flames. The catalyst is a platinum-coated ceramic tube which radiates heat (thermal energy) and because combustion is at a lower temperature than normal, could be instrumental in reducing the production of NITROGEN OXIDES.

**Catalytic converter.** ⇨ CATALYTIC REACTOR.

**Catalytic reactor** (colloquially catalytic converter). An attachment fitted to the exhaust pipes of unleaded fuel internal combustion engines which converts UNBURNT HYDROCARBONS and CARBON MONOXIDE to carbon dioxide and water vapour by means of a platinum mesh CATALYST. Three-way catalytic reactors will in addition convert NITROGEN OXIDES to nitrogen using platinum, palladium and rhodium catalysts, supported on a ceramic shell to withstand and sustain the high temperatures required for the reactions to take place. Should the vehicle cover only a small mileage (less than 15 miles in winter), the reactor may not reach an adequate working temperature and therefore be rendered ineffective. European market for three-way catalytic reactors after 1992 is estimated at 20 million units per year (*Financial Times*, 5 February 1990) worth an estimated £1 billion a year for the catalytic cores alone. This shows that pollution control is big business.

Diesel engines can also be fitted with a catalytic converter which is electrically heated for start-up purposes so that unburnt hydrocarbon and soot emissions can be reduced.

**Catchment area.** The natural drainage area for PRECIPITATION, the collection area for water supplies, or a river system. The area is defined by the notional line, or watershed, on surrounding high land.

**Cathode.** An electrode which is at a negative potential.

**Cathodic protection.** Cathodic protection works by applying a d.c. power source to reverse the natural flow of electrical current caused by galvanic corrosion. This stops the steel reinforcement in a structure from rusting. See Figure 20.

For it to work satisfactorily, there needs to be a structural integrity, sound concrete, continuity of reinforcement, an ANODE system, a power source and monitoring and control systems. Should there be any gaps in the steel in a severely corroded structure, these need to be bridged first.

An electrode is applied to concrete adjacent to the reinforcement and connected to a positive source of direct current, around 100 mV, and the reinforcement is connected to the negative side of the current supply. This flow of current acts in direct opposition to that naturally occurring as a result of corrosion and provides the protection.

Anodes come in many shapes and forms and can be made from a variety of conductive materials, including meshes, plates, wires and sprays, made of zinc, titanium, graphite and conductive polymers.

**Cation.** A positively charged ION.

**Caustic soda (NaOH).** Sodium hydroxide, an alkali used, for example, in the manufacture of wood-free PAPER and PULP.

**Cellulose.** A carbohydrate polymer constructed from the glucose building blocks formed by the action of PHOTOSYNTHESIS. It is the chief structural element and major constituent of the cell walls of trees and other higher plants where it is bound up with a glue (lignin) to form plant stems. The fixation of carbon in the carbon cycle is such that about 100000 million tonnes of cellulose are produced annually. Many important derivatives stem from it, e.g. cellulose nitrate and cellulose acetate.

A major use is in paper manufacture when it has been freed from its matrix of lignin and other organic matter by treatment with sodium sulphite or alkali to yield a pulp feedstock of cellulose fibres.

Cellulose is the principal chemical constituent of DOMESTIC

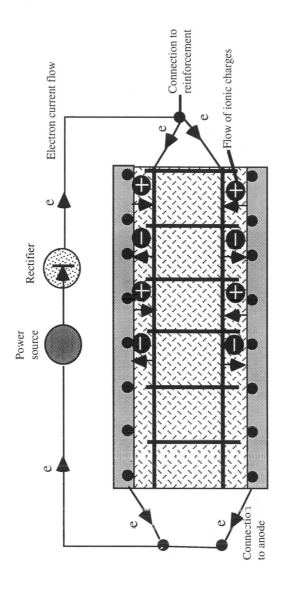

**Figure 20** Cathodic protection.
*Source:* 'Cathodic protection', *Surveyor*, January 1994.

REFUSE and can be processed to yield a variety of products such as glucose, SINGLE-CELL PROTEIN and ethyl alcohol. (⇨ HYDROLYSIS; PYROLYSIS; COMPOSTING; RECYCLING)

**Cellulose economy.** The use of the sun's energy through the process of PHOTOSYNTHESIS and the subsequent production of CELLULOSE which is then processed, by enzymatic means or by chemical HYDROLYSIS, to yield glucose which can then be fermented to yield many industrial products, protein or fuels.

The cellulose economy could use strains of plants which have a high energy conversion efficiency, e.g. sugar-cane and sorghum in tropical areas which can fix 3 per cent of the incoming solar energy. These plants are then processed by the chemical or biochemical means already mentioned. This technique of energy cropping would also use marginal lands, swamps, etc., as productive areas. Species would be used that are capable of continuous cropping, such as fast-growing deciduous trees which re-sprout from stump when cut and avoid the need for replanting. In the UK and USA, willows and hybrid poplars can be so cropped. One strain of hybrid poplar has been planted at 3700 trees per acre and produces eight harvests per planting with unit energy costs within an order of magnitude of those from fossil fuels.

Cellulose processing as a means of living off the sun's income is a fast-growing area of research and development especially for the production of SINGLE-CELL PROTEIN. (⇨ ENZYMES)

**Cement kiln – waste combustion.** Cement is made by heating a mixture of clay or shale plus chalk or lime in a ROTARY KILN up to 250 m long × 8 m diameter rotating at 1 rpm. The process can be wet, semi-dry or dry and the fuel can be pulverized coal, oil or gas. As the coal ash is similar in composition to the clay or shale, it can stay in the cement clinker.

As one of the kiln operator's major costs is fuel, and even a modest sized kiln can consume 8–10 tons of coal per hour, the cement kiln could, therefore, solve a disposal problem and also benefit the cement manufacturer by reducing fuel costs.

France used MSW and industrial waste as fuels in 12 cement kilns in 1986, and has subsequently increased capacity both to utilize the energy content in the waste and destroy hazardous wastes. This approach is also borne out by Dutch experience in that most inorganic materials bind very efficiently to cement clinker. Fuels with an 'appreciable' amount of cadmium, mercury or thallium should not be used. Regular analysis of the supplementary fuels and the binding efficiency is required. However, cement kilns do have potential as waste combustion outlets for PACKAGING wastes.

This development has not passed unnoticed by hazardous waste (merchant) incinerator operators one of whom has claimed that this 'will lead to unnecessary safety risks and POLLUTION levels in the vicinity of the operation . . .'. This is a space to watch as the formal authorization procedures to be followed for permanent use in the UK will set the parameters for any similar solvent-based fuel combustion. (The House of Commons Environment Committee, Second Report, *The Burning of Secondary Liquid Fuel in Cement Kilns*, 7 June 1995)

In response to considerable public pressure, The House of Commons Environment Committee considered the matter and recommended:

- That the government formally classify Secondary Liquid Fuel (SLF) as waste (if necessary, by amending the law) thus applying the Duty of Care to all shipments of SLF to kilns and formally bringing the burning of SLF in kilns within the scope of the EU Hazardous Waste Incineration Directive.
- If the burning of SLF is to continue, Her Majesty's Inspectorate of Pollution (HMIP) must enforce their own standards more effectively, and not be put off by commercial arguments such as the possible termination of contracts. For example, trials should be limited in duration and should not be allowed to continue unless the conditions of the HMIP Protocol can be complied with.
- That the adaptation of kilns to burn SLF should be designated a 'substantial change' under Part I of the Environmental Protection Act 1990, thus requiring a new authorization.
- That as a minimum HMIP apply additional controls to ensure that the emissions from burning coal plus SLF are not more polluting than those from burning coal alone. This would mean that:
  - the specification for SLF should be tightened up so that it is a cleaner fuel than coal, in terms of metals content, etc., and/or
  - effective abatement technology should be fitted, in order to trap the emissions which currently escape during kiln upsets.

- That HMIP carry out a more detailed analysis of the precipitator dust from kilns which burn SLF and its potential to contaminate soils and groundwater. If it is found to be an environmental hazard, it should be reclassified as Special Waste, which would bring it under a strict regulatory regime. It is also recommended that HMIP take immediate action to tackle the more general pollution problems associated with kilns.

- That the Government carry out a thorough survey of human and animal health in the vicinity of the kilns (both those burning SLF and those burning only coal) in order to establish or disprove the alleged links between kiln pollution and health effects.

As an example of how diverse wastes are being used as fuels, Rugby Cement has received the go-ahead from HMIP for a six-month trial burning of SECONDARY LIQUID FUEL in its Southam cement kilns in Warwickshire. The company will be permitted to substitute up to 40 per cent of the fuel for the kilns with wastes from nylon manufacture, produced on Teesside by DuPont. HMIP says the wastes contain only trace amounts of chlorine and sulphur. Provided stringent environmental requirements are met (especially with respect to *dioxin*), this area is set to grow.

**CEMFUEL.** ⇨ SECONDARY LIQUID FUEL.

**Ceramic filter.** Ceramics are inert, inorganic, brittle materials that are heat bonded. Ceramic filtration media offer potential benefits: they can withstand high temperatures and corrosive atmospheres, so no pretreatment of the gas stream is necessary.

Particles are collected on the surface of the element and as filtration continues the layer of deposited particles becomes thicker forming a cake which contributes to the pressure drop across the filter. At some point the cake must be removed for efficient filtration, usually by reverse pulse cleansing.

Figure 21 shows a typical schematic ceramic filter system.

**Cetane number.** A number on a conventional scale, indicating the ignition quality of a diesel fuel under the standardized conditions. It is expressed as the percentage by volume of cetane (i.e. hexadecane) in a reference mixture having the same ignition delay as the fuel for analysis. The higher the cetane number, the shorter the ignition delay. The lower the number, the longer the delay and the greater the noise levels produced in combustion.

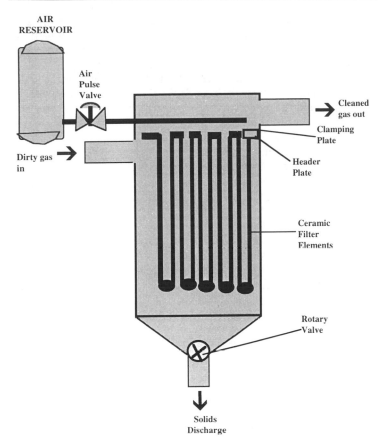

AIR
RESERVOIR

Air
Pulse
Valve

Cleaned
gas out

Clamping
Plate

Dirty gas
in

Header
Plate

Ceramic
Filter
Elements

Rotary
Valve

Solids
Discharge

**Figure 21** A typical ceramic filter.
*Source:* 'Filter tips', *The Chemical Engineer*, 6 April 1995.

**Charcoal.** A form of CARBON produced by the DESTRUCTIVE DISTILLATION of wood.

**Chelating agents.** A group of organic compounds which can incorporate metal ions into their structure and so obtain a soluble, stable and readily excretable substance (known chemically as a chelate complex). Thus toxic metals such as lead (normally poorly excreted) can be removed from blood and bones of the human organism and excreted. The use of chelating agents is a standard treatment in such cases of poisoning.

Chelating agents are, however, unable to penetrate the body's cellular material and to overcome this one method is to encapsulate them in substances known as liposomes which can penetrate the cell walls. This is an important application as the liver, for example, is one of the vital organs where PLUTONIUM and other HEAVY METALS accumulate. Thus intracellular deposits may now be purged from the human organism.

The medical use of chelating agents must be very carefully controlled as the sudden release of the burden of heavy metals can damage the liver and kidneys.

Another extremely useful application of chelating agents is the use of disodium ethylenediamine-tetraacetate (EDTA) to remove (sequester) minute quantities of cooper, iron, etc. in foods, thereby preventing discoloration. EDTA is also a preservative as it can inhibit the bacterial growth responsible for food poisoning.

**Chemical oxygen demand (COD).** The amount of oxygen consumed in the complete oxidation of carbonaceous matter in an effluent sample. This is done in a standard test which uses potassium dichromate as the oxidizing agent. (⇨ DISSOLVED OXYGEN; BIOCHEMICAL OXYGEN DEMAND)

**Chemical symbol.** Either a single capital letter or a combination of a capital and lower case one which is used to represent an atom of a chemical element, e.g. Na for sodium (from the Latin name, *natrium*) and S for sulphur (from the English).

**Chemosphere.** Part of the lower section of the atmosphere in which chemical processes such as molecular dissociation and recombination take place during the day and night respectively, under the influence of ultraviolet radiation. (⇨ IONOSPHERE)

**Chernobyl.** On 26 April 1986 the unit IV NUCLEAR REACTOR at the Chernobyl Nuclear Power Station in the Soviet Union caught fire and exploded. An estimated 100 million Ci (⇨ CURIES) of radioactivity were emitted with up to half deposited within 30 kilometres of the plant. An official Soviet report estimated that the average external exposure to radiation in 1986 to Soviet citizens in the fallout path was doubled. The fallout will continue to contribute to background levels for many decades. Calculations indicate an increase in the death rate from cancer of up to 0.05 per cent, i.e. adding perhaps 5000 to the 9.5 million people who would normally be expected to die of cancer over the next 70 years.

However, there are also internal radiation aspects through the ingestion of contaminated materials. The two principal

ISOTOPES are iodine-131 which has a HALF-LIFE of 8 days and CAESIUM-137, half-life 30 years. The iodine-131 uptake in plants is dependent on soil type. Poor soils allow much greater uptakes and it is for this reason and the prevailing deposition pattern that hill farmers' lambs in some parts of Wales and the Lake District are still (1989) contaminated with caesium-137 which in turn means that they cannot be sold. Indeed, it is now posited that the UK effects of the Chernobyl caesium-137 fallout will be felt for up to 150 years. In 1990, 400 000 acres of UK hill farming land were under sheep sale restrictions due to the effects of fallout.

Before the explosion, the annual incidence of childhood thyroid cancer in the Gomel region of Belarus was one in 2 million, about the same as in Britain. But by 1994 this had risen to one in 10 000 – a 200-fold increase. In some pockets of Belarus the incidence is ten times higher.

Over the same period, there was more than a 100-fold increase in the incidence of thyroid cancer in the most northerly parts of Ukraine. In the Bryansk and Kaluga regions of Russia, which lie to the north of Chernobyl, there has been an eight-fold increase since 1986.

The cause of the Chernobyl incident was an experimental reduction in power from 3 200 MW to ca. 1000 MW. However, the reactor could not be stabilized quickly enough and power dropped to 30 MW. Neutrons absorbing xenon in the core prevented the raising of the reactor power level to above 200 MW. To counteract this, control rods were withdrawn and the reactor became unstable and a slow nuclear chain reaction ensued which blew the top off the reactor with the reactor's graphite moderator subsequently catching fire. This was eventually controlled by dropping 5000 tonnes of boron, limestone and sand into the reactor from helicopters.

Several points stand out from the story of Chernobyl:

(a) Human error and miscalculations can never be eliminated.
(b) The long-term effects of its radiation release, notably caesium-137, may be around for decades.
(c) A major nuclear reactor explosion was tamed and the core sealed off by the extraordinary bravery of the Soviet firefighters, some of whom knowingly and willingly gave their lives in the process.
(d) Chernobyl has made many people ponder on the need for nuclear power and question whether it is strictly necessary.

*Souce:* USSR State Committee on the Utilization of Atomic Energy, *The Accident at Chernobyl Nuclear Power Plant and Its Consequences* (August 1986), p. 15.

The UK nuclear industry has stated (1992):

> The Chernobyl accident will cause a radiation dose of six millisieverts over the next 50 years to each of the 75 million people living nearest to it. But some British residents receive that each and every year from background radiation in their homes. The Chernobyl accident has rendered nearly 3,000 square kilometres of land sterile and useless. But major hydro schemes 'drown' areas up to three times as large all over the world.
>
> Each source of energy poses risks. Yet none of them pose risks which are unique or out of proportion to others. It is reckoned that the chance of another Chernobyl-scale accident anywhere in the world should be well below ten per cent during the next century, because of advances in plant technology and operation. At the same time, the world population is increasing and energy demand is likely to double by 2025. If the world is to meet this demand it will need to continue burning fossil fuel at the greatest rate consistent with a prudent attitude towards the threat of global climate change.
>
> It will need to develop all the practicable hydro schemes in the world.
>
> It will need nuclear.
>
> It cannot afford to abandon any of these options without jeopardising the ability to meet the demand. One might argue that the risks are trivial compared with the risks of having to do without any one of them.

From an industrial point of view, the above statements are not unreasonable. The looming energy crisis will need to be tackled on all fronts and the Chernobyl incident should not be used as a reason to impugn today's Western Nuclear Reactor Designs or plant operational procedures which are of the highest order.

Among the legacies of the Chernobyl disaster has been the proliferation of absurd claims. These achieved a surreal level at the time of the 1995 anniversary when Ukrainian officials reportedly claimed that 125 000 people had died as a result of the radiation released. It turned out that this was the total of all deaths in the affected area since the accident. But it was not the first time that the real victims of Chernobyl had been exploited to attract the sympathy of the West. *Network First* (ITV programme) told the tale of Igor Pavlovets, a child from Belarus born deformed after the accident. His case is a tragic one, but what evidence to link it to Chernobyl? Publicity for the programme asserts that 'over one million children' are deformed like Igor as a result of the accident. A sample survey of 500 children in Minsk, it says, has found only one to be completely healthy. But this may be a reflection of the poor health of Russian children in general and unconnected to the disaster at Chernobyl.

It may seem harsh to insist upon proper evidence in the face of misery, but only by doing so can the real causes of ill-health be properly addressed.

In an issue of *Nature*, Dr Valerie Beral of the Imperial Cancer Research Fund and a team from two Ukrainian scientific institutes examine the data on cancer of the thyroid, by far the most likely prompt response to a nuclear accident. Iodine-131 is a major part of the radioactive releases, contaminates grass, is eaten by cows and finds its way into milk. Since it is concentrated in the thyroid, that is where it is likely to cause damage, and children are at greater risk because their thyroids are smaller and they drink more milk.

The figures do indeed show an increase in cases. In the whole of the Ukraine, there were eight cases of thyroid cancer in children under 15 in 1986, 11 in 1989, and 42 in 1993. The data show that the risks are greatest among those who were living in the plant. There is no evidence of increased risk in children born after 1986, which is to be expected because iodine-131 decays relatively rapidly. Neither is there any evidence, Dr Beral says, of increases in any other forms of cancer.

Figures like these are no argument for complacency, but they do provide perspective. The actual evidence so far cannot possibly justify the wilder claims that have been made. Even a disaster needs to be kept in perspective.

*Source:* Nature Forum, June 1992.

Since 1977, Chernobyl has suffered a serious accident about once every 5 years. These events have left the site with only two of the original four units still in service today.

- In 1982, five years after first going critical, Chernobyl-1 experienced a core melt when coolant flow to a single fuel channel was inadvertently cut off. To this day, fuel fragments of the damaged sub-assemblies still remain inside the reactor.
- In 1986, an extreme power excursion completely destroyed Chernobyl-4. Lacking a containment, the reaction released

great quantities of radioactive materials making it the world's worst nuclear accident.

- Most recently, in 1991, an extensive turbine fire brought about the closure of Chernobyl-2. Serious core damage was narrowly averted after the fire disabled systems essential for core cooling.

*Source:* USA Department of Energy, *Most Dangerous Reactors Report,* August 1995.

**Chimney Height Memorandum.** A UK guidance document, not binding in law, giving a sample method of determining the height appropriate for certain new chimneys serving combustion plant, in order to avoid unacceptable pollution ($SO_2$ or other gaseous pollutants). The 3rd edition, 1981, takes account of very low sulphur fuels. The document takes into account the type of district, the capacity of the plant and other buildings in the vicinity.

**Chloracne.** A sub-lethal effect of exposure to DIOXIN and other chlorinated chemicals, which manifests itself as a painful skin disorder.

**Chlor-alkali process.** A process used for the manufacture of CHLORINE by ELECTROLYSIS of brine. This process was formerly a source of mercurial discharge to estuaries. ($\Rightarrow$ MINAMATA DISEASE)

**Chlorinated hydrocarbons.** One of three major groups of synthetic insecticides, others being ORGANOPHOSPHORUS COMPOUNDS and synthetic pyrethrins. The group includes DDT, ALDRIN, ENDRIN, BENZENE HEXACHLORIDE, DIELDRIN, and many others including ENDOSULFAN which is highly toxic to fish. An endosulfan spillage in the River Rhine in 1968 caused the death of millions of fish plus the cessation of drinking-water abstraction until the chemical contamination was thoroughly cleared.

In insects and other animals these compounds act primarily on the central nervous system. They also become concentrated in the fats of organisms and thus tend to produce fatty infiltration of the heart and fatty degeneration of the liver in vertebrates. In fishes they have the effect of preventing oxygen uptake, causing suffocation. They are also known to slow the rate of PHOTOSYNTHESIS in plants.

Their danger to the ECOSYSTEM resides in their great stability and the fact that they are broad-spectrum poisons which are very mobile because of their propensity to stick to dust particles and evaporate *with* water into the atmosphere. ($\Rightarrow$ PESTICIDES, ORGANOCHLORINES)

**Chlorine (Cl).** An element of the halogen group. A gas with an irritant smell which has severe effects on the lungs and respiratory system. Produced mainly by the CHLOR-ALKALI PROCESS, chlorine is widely used in the manufacture of organic chemicals, for example, carbon tetrachloride, HALOGENATED FLUOROCARBONS, AEROSOL PROPELLANTS, chloroform, PESTICIDES, PVC, and solvents, plastics and bleaching agents.

Euro Chlor estimates that about 12 tonnes of mercury a year are emitted through chlorine electrolysis. That compares with around 20000 tonnes a year occurring naturally, says Barrie Gilliat, the chairman of Euro Chlors technical committee. By contrast, it would cost about DM10bn to replace mercury electrolysis with membrane technology. 'Does it make any sense to spend DM10bn to eliminate 12 tonnes of mercury when nature and the rest of the world make 19988 tonnes?' he asks.

Chlorine is often used as a disinfectant in water treatment to protect public health and indeed, for developing countries, there is no practicable substitute. Its use as a disinfectant has materially improved sanitary conditions throughout the world. ($\Rightarrow$ WATER SUPPLY)

*Source:* 'Economy v Ecology', *Financial Times*, 15 February 1995.

**Chlorofluorocarbons (CFCs).** A class of chemical compounds commonly used as SOLVENTS, AEROSOL PROPELLANTS, REFRIGERANTS, and in foam productions, including CFC-11 ($CCl_3F$ or trichlorofluoromethane) or CFC-12 ($CCl_2F_2$ or dichlorodifluoromethane). The gases are chemically inert, but after release and entry into the STRATOSPHERE, where absorption of short-wave radiation takes place, each chlorofluorocarbon molecule decomposes with the release of the radical atomic chlorine which cannot exist independently and thus attacks OZONE ($O_3$) through catalytic chain reactions which lead to ozone depletion.

The decomposition of the TRICHLOROFLUOROMETHANE ($CFCl_3$) under ultraviolet radiation is given by equation (1):

$$CCl_3F \xrightarrow{ultraviolet} CCl_2F + Cl \qquad (1)$$

The chlorine atoms are extremely reactive and each chlorine atom can destroy up to $10^5$ molecules of ozone by converting ozone ($O_3$) molecules to ordinary oxygen ($O_2$) molecules before

**Table 1**  (DTI 1989)

| Number | Formula | Application | | | | Ozone depletion potential |
| | | Propellant | Refrigerant | Cleaning | Blowing | |
|---|---|---|---|---|---|---|
| CFC-11 | $CFCl_3F$ | Yes | Yes | Yes | Yes | 1.0 |
| CFC-12 | $CCl_2F_2$ | Yes | Yes | — | Yes | 1.0 |
| CFC-113 | $CCl_2FCCl_2$ | — | Yes | Yes | Yes | 0.8 |
| CFC-114 | $CClF_2CF_2C$ | Yes | Yes | — | Yes | 1.0 |
| CFC-115 | $CClF_2CF_3$ | — | Yes | — | Yes | 0.6 |
| Halons | Br compounds | Fire fighting | | | | 3.0 to 10.0 |
| 1211 | | | | | | |
| 1301 | | | | | | |
| 2402 | | | | | | |

Total  1985 world production = 800 000 tonnes

**Table 2**  (DTI 1989)

| Number | Formula | Application | | | | Flammability | Toxicity | Ozone depletion potential |
| | | Propellant | Refrigerant | Cleaning | Blowing | | | |
|---|---|---|---|---|---|---|---|---|
| CFC-22 | $CHF_2Cl$ | — | Yes | — | Yes | No | Low | 0.05 |
| CFC-142b | $CH_3CClF_2$ | Yes | — | — | — | Yes | Low | <0.05 |
| CFC-152a | $CH_3CHF_2$ | Yes | Yes | — | — | Yes | Low | 0 |

**Table 3** (DTI 1989)

| Number | Formula | Application | | | | Flammability | Toxicity | Ozone depletion potential |
| | | Propellant | Refrigerant | Cleaning | Blowing | | | |
|---|---|---|---|---|---|---|---|---|
| CFC-123 | $CHCl_2CF_3$ | — | Yes | — | Yes | No | Low | <0.05 |
| CFC-124 | $CHClFCF_3$ | — | Yes | — | Yes | No | Low | <0.05 |
| CFC-125 | $CHF_2CF_3$ | — | Yes | — | — | No | Unknown | 0 |
| CFC-134a | $CH_2=CF_3$ | — | Yes | — | — | No | Low/incomplete | 0 |
| CFC-141b | $CH_3CCl_2F$ | Yes | — | — | Yes | Yes | Low/incomplete | <0.05 |
| CFC-143a | $CH_3CF_3$ | — | Yes | — | — | Yes | Low/incomplete | 0 |

removal from the atmosphere by reaction with methane to form hydrochloric acid which is precipitated as given by equation (2):

$$Cl + CH_4 \rightarrow HCl + CH_3 \qquad (2)$$

The amount of chlorine in the atmosphere is estimated as 3 ppb. The target is to get it down to 2 ppb.

At the Antarctic, a different chemistry applies and inorganic forms of chlorine such as HCl and chlorine nitrate ($ClONO_2$) can be converted to the active forms of chlorine Cl and ClO respectively which then attack the ozone ($O_3$), as shown in equations (3), (4) and (5):

$$HCl + ClONO_2 \xrightarrow{\text{ice clouds}} Cl_2 + HNO_3 \qquad (3)$$

$$Cl_2 \xrightarrow{\text{ultraviolet}} Cl + Cl \qquad (4)$$

$$Cl + O_3 \rightarrow ClO + O_2 \qquad (5)$$

The virtual isolation of the polar atmosphere at certain times of the year is also a contributory factor in the ozone depletion process.

The gases are characterized by a long atmospheric lifetime because of their chemical and biological inertness. Their relative insolubility in water prevents rapid removal by cloud formation and precipitation.

The long atmospheric lifetime coupled with the delay period before the chlorofluorocarbons diffuse upwards and encounter the short-wave ultra-violet radiation mean that changes in the ozone layer will be observable through most of the next century even if production were held at current rates. Once again, the need for environmental acceptability tests for the products of the synthetic chemicals industry is demonstrated. ($\Rightarrow$ BUTANE; GREENHOUSE EFFECT)

Table 1 shows the present generation of chlorofluorocarbons and their associated OZONE DEPLETION POTENTIAL. Table 2 gives a listing of existing alternative HALOCARBONS and those under consideration are shown in Table 3. Figure 22 (National Engineering Laboratory, *Assessing Alternatives to CFCs*, 1989) shows how the properties of CFCs are determined by their chemical composition, e.g. R-12 is fully halogenated, R134a has no chlorine in its make-up. (*Note:* for $R_2$ on figure read CFC). As CFCs are phased out, many buildings using chilled water for air conditioning, where the refrigerant is currently a CFC, will have to use new plants or refrigerators. One environmentally

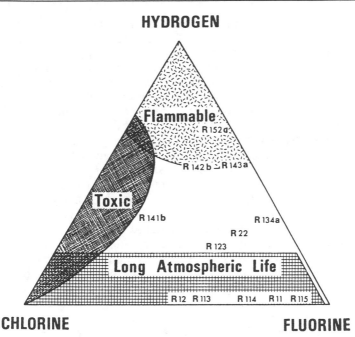

**Figure 22**  Assessing alternatives to CFCs.

*Source:* S. Pekbett, 'The changing atmosphere – ozone holes and greenhouses', *The Chemical Engineer*, August 1989, 14–18.

friendly alternative is provided by Citigen (London) which uses a COMBINED HEAT AND POWER (CHP) plant to supply electricity and both hot and cold water to key London buildings. The centralized system offers increased reliability and substantially lower running costs allied with better utilization of the CHP plant which supplies chilled water in summertime when the heat demand is at its lowest, thus balancing the load.

**Chlorophenols.** Major group of CHLORINATED HYDROCARBONS, PESTICIDES and biocides which in the UK account for over 85 per cent of the non-agricultural pesticide use, such as anti-rotting agents in non-woollen textiles and wood preservatives.

The chlorophenols act as biocides by inhibiting the respiration and energy-conversion processes of the micro-organisms. They are toxic to man above 40 parts per million (ppm), to fish above 1 ppm, whilst concentrations as low a 1 part per thousand million can taint water. Little is known about their ecological effects, which is not to imply that they are harmless in this respect.

**Chlorophyll.** A combination of green and yellow pigments, present in all 'green' plants, which captures light energy and enables the plants to form carbohydrate material from CARBON DIOXIDE and water in the process known as PHOTOSYNTHESIS. It is found in all ALGAE, phytoplankton, and almost all higher plants.

**Chloroprene.** A chemical used for the production of synthetic rubber, Neoprene, which had been identified as a possible CARCINOGEN. Workers in a Russian Neoprene factory showed higher than normal levels of lung and skin cancer.

**Chlorosis.** A condition of reduced CHLOROPHYLL in green plants marked by yellowing or whitening which can be caused by a variety of diseases, pollutants and ACID RAIN.

**Cholera.** Bacterial infection spread by contamination of drinking-water supplies by sewage effluent and infected food. It is prevalent in Eastern countries and those areas where the growth in tourism has not been matched by the development of safe water supplies and adequate sewage disposal facilities. Tourist camping sites and recreation centres near lakes and rivers are particularly sensitive areas where precautions must be taken.

Prevention of infection is by the use of separate piped water and sewage systems and chlorination of water supplies before use. (⇨ CHLORINE; WATER SUPPLY; SEWAGE TREATMENT)

**Chromatography.** An analytical technique used principally for separating and identifying the components of a sample by distributing them between a stationary and a moving phase. The stationary phase may be a solid, a liquid or a gas and may take the form of a column, a layer, or a film; the moving phase may be either a liquid or a gas. In gas chromatography (GC) the moving phase is a gas and the stationary phase is either a liquid (gas–liquid chromatography, GLC) or a solid (gas–solid chromatography, GSC). Gas chromatography, particularly with electron capture detection, can be used for the determination of certain CHLORINATED HYDROCARBON air pollutants; with flame ionization detection it can be used for the determination of CARBON MONOXIDE and other compounds.

**Chrome wastes.** Chromates and chromic acid are common wastes from certain industries, such as chromium plating and leather tanning. Chromates are soluble in water and are toxic to sewage treatment processes. (The hexavalent ion is the most toxic – it is present in chromic acid – and must be reduced to the trivalent state to form an insoluble product before being released into the environment.) Chromates act as irritants to eyes, nose and throat, and may cause dermatitis. On prolonged exposure there is liver and/or kidney damage and possible carcinogenic effects.

Hexavalent chromium may also cause abnormalities in chromosomes.

**Chromium (Cr).** A hard white metal. It is used as a steel-alloying element and in plating. The metal is stable and non-toxic, but there are water-soluble compounds which are extreme irritants and highly toxic. Chromium is a trace element essential for fat and sugar metabolism. Chromium aerosols can affect health in concentrations above 2.5 micrograms per cubic metre. Its presence as chromates can cause dermatitis at very low concentrations to individuals whose work has sensitized them, by contact with the compounds; for example, cement contains sufficient chromium (0.03 to 7.8 micrograms per gram) to cause dermatitis in sensitive people).

*Source:* G. L. Waldbott, *Health Effects of Environmental Pollutants*, 2nd edn, C. V. Mosby, St. Louis, 1978.

**Chronic.** Medical term used in relation to the effects of certain pollutants, such as ASBESTOS, to describe a response which develops as a result of long and continuous exposure to low concentrations. It is contrasted with ACUTE. (⇨ LEAD)

**Cladding.**

1. The protective layer covering the fuel in a nuclear fuel element.
2. The stainless-steel lining of a PWR coolant circuit.

**Clarifier.** A mechanical device for removing solids from water. It is used for treating the effluent from paper mills to remove fibrous tissues and fillers. It is also used in water treatment for domestic or industrial use. (⇨ DE-INKING; PULP; PAPER; WATER SUPPLY)

**Clay.** Fine-grained sedimentary rock of low PERMEABILITY which is capable of being shaped when moist. Consists of fine grains less than 4 $\mu$m in diameter. (⇨ MUD; SILT)

**Clean Air Act (UK).** The name given to two Acts passed by the United Kingdom Government. The 1956 Act dealt with the control of smoke from industrial and domestic sources. It was extended by the Act of 1968, particularly to control gas cleaning and heights of stacks of installations in which fuels are burned and also to deal with smoke from industrial open bonfires. Subsequent legislation has amended the Acts.

**Clean Air Act (USA).** The name given to two Acts passed by the US Government. The Act of 1963 affirmed the authority of the Federal Government in dealing with interstate pollution situations, although it recognized air pollution to be primarily a state and local problem. The Act of 1970 empowered the Federal Government to set national air pollution standards (in place of state standards) and required the states to meet those standards (but to develop their own ways of doing so). It also set up air quality control regions immediately.

**Clean coal technology.** There is no real alternative to coal based power generation in the long term other than nuclear power, and this may not necessarily be acceptable to many communities on grounds of both cost, perceived risk and the problems of NUCLEAR REACTOR WASTES.

Coal is commonly perceived as a dirty fuel because of its emissions of carbon dioxide and relatively low thermal efficiencies as well. The way round this is to utilize clean coal technology which relies on gasification and the use of combined cycle gas and steam turbines. This lifts the efficiency from around 35 per cent to over 55 per cent with the prospect of 60 per cent in the near future. The gasification process is a generic technology which can generate clean combustible gas from biomass and oil as well as coal.

**Climate.** The earth's climate has gradually evolved as the atmosphere has stabilized. The controlling factor is latitude which governs the intensity of incident solar radiation. Atmospheric currents and oceans iron out large inequalities in temperature. Of the incoming solar energy, an average of 30 per cent is reflected by the atmosphere back to space, 50 per cent is absorbed at the earth's surface, and the remainder is absorbed by the atmosphere, which in turn sets up atmospheric circulation patterns. Figure 23 shows on the left the incoming short-wave radiant energy and on the right the outgoing long-wave radiant energy. The input of solar energy (100) equals the output of short-wave (28) plus long-wave (72) radiation. Thus the system is balanced. (⇨ CARBON CYCLE)

A fraction of 1 per cent of the incoming solar energy is fixed

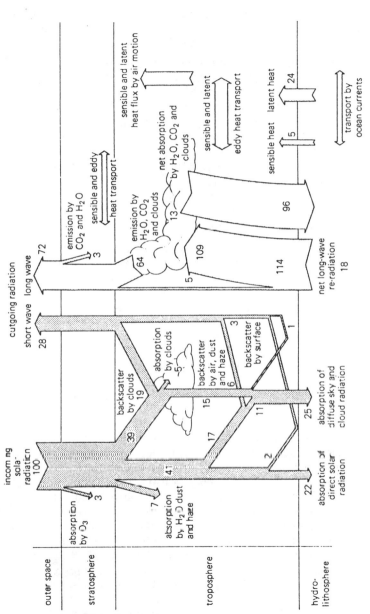

**Figure 23** The earth's radiation balance between incoming (solar) radiant energy (on the left) and outgoing (terrestrial) radiant energy (on the right). The figure also shows the distribution of energy in the global system.

by PHOTOSYNTHESIS to form trees and plants, and this is the total energy fixation on which life on earth depends. The main man-induced influences on global climatic patterns are:

1.  Increased CARBON DIOXIDE content (⇨ GREENHOUSE EFFECT; ENERGY, effect of conversion on CLIMATE).
2.  A change in the ALBEDO (which can affect the MEAN ANNUAL TEMPERATURE) by clearing jungles, creating reservoirs, and laying vast areas under concrete and asphalt.
3.  DUST/AEROSOL emission, which can reflect ionizing solar radiation.
4.  Stratospheric properties may also be affected by SUPERSONIC FLIGHT (⇨ CONCORDE) or possibly by CHLOROFLUOROCAR-BON propellant dissociation attacking the OZONE layer (⇨ AEROSOL PROPELLANT).
5.  OZONE SHIELD depletion.

Although these processes have some potential for changing the climate, it is unlikely that there will be drastic modification, but one caveat must be entered. *If the demand for fossil-fuel derived energy continues to grow, the heat balance of the earth could be upset as all energy used is ultimately degraded to heat* (thermal energy) which must be re-radiated into space to maintain the earth's radiation balance. This is linked with and to the GREENHOUSE EFFECT. The introduction and use of 'industrialized' energy sources such as fossil fuels and nuclear power mean that more energy must be re-radiated to space and the earth's radiation balance could be upset if the growth continues. The calculations are such that if the 'industrialized' energy input reaches 1 per cent of the solar input, then major alterations in the earth's climate can be expected as atmospheric circulation patterns will be disturbed, with unknown consequences. Currently, we are not near this figure, but the exponential growth of energy consumption shows it is not entirely inconceivable that this could be reached. Thus, the earth's climate patterns place an ultimate restriction on long-term energy growth. While we are nowhere near the postulated limit, it is important to remember that future energy production may well be in very concentrated locations so that very high-energy densities will take place locally and could conceivably cause major regional climate upsets. The atmospheric system is finely tuned and there are limits to what it, or any other natural system, will tolerate. The current global warming trend shows mankind cannot really control atmospheric processes and should exercise much more care. (⇨ EXPONENTIAL CURVE; THERMAL POLLUTION)

*Source:* S. H. Schneider and R. D. Dennett, 'Climate barriers to long-term energy growth', *Ambio*, vol. **4**, no. 2, 1975.

**Clouds.** A mass of water droplets formed in the atmosphere by condensation of water vapour around nuclei such as salt, dust and soil particles. Condensation commonly occurs when there is a drop in air temperature which cools the moist air mass to below its dewpoint (i.e. the temperature at which precipitation occurs in a water-vapour-laden gas stream).

**Club of Rome, The.** A body set up in 1968 by an international group of economists, scientists, technologists, politicians and others. The Club's object is the study of the interactions of economic, scientific, biological and social components of the present human situation, in the hope of eventually being able to predict, with some degree of certainty, the results of present policies and to formulate alternative policies where it is deemed necessary on environmental and survival grounds.

**Coal.** The world's coal and lignite (a low-grade coal, often brown in colour, with a relatively low heat value) reserves are greater than those of oil and gas. Two-thirds of these reserves are thought to be in Asia, but they are yet to be proven. Approximately 48 per cent of the world's known coal reserves are in the USSR, Eastern Europe, and China, 9 per cent in Western Europe, 6 per cent in Africa, 26 per cent in North America and 1 per cent in Latin America, 9 per cent in Australia and Asia.

World resources of reasonably accessible coal are sufficient to meet projected world demands for the next 300–500 years, although this estimate would have to be severely reduced if significant quantities are used to make synthetic liquid fuels as a replacement for oil.

The environmental costs of mining coal and lignite are, unfortunately, high. Much of the cheapest can be extracted only by surface mining on a large scale while the social costs of conventional mining are well known. Furthermore, when used as a fuel, coal is the dirtiest of the fossil fuels. It is possible that enlightened programmes of restoration, or remote mining, and of rigidly enforced controls on air pollution, may minimize such environmental impacts to acceptable levels, but the costs involved can be high. (⇨ TAR SANDS; OIL SHALES)

*Source:* B.P. Statistical Review of World Energy, July 1989.

**Coal equivalent (ce).** The amount of (standard) coal that would on combustion produce the same quantity of heat as a given mass of a given fuel. Coal equivalents can vary from one country to another.

**Coal gasification.** ⇨ IGCC.

**Cobalt-60.** Radioactive ISOTOPE of cobalt. Cobalt-60 emits high-energy gamma radiation and has a HALF-LIFE of over five years, thus rendering it liable to accumulate in areas of discharge. It is particularly damaging to biological systems because of the RADIATION hazard, and therefore its discharge is subject to very strict licensing procedures.

Cobalt-60 is an inevitable waste product from nuclear reactors, which use high cobalt steel fuel cans which are, of course, irradiated in the fission process. (⇨ NUCLEAR REACTOR DESIGNS; IONIZING RADIATION, EFFECTS)

**Coefficient of Performance (COP).** A measure of refrigeration plant or heat pump efficiency (sometimes termed reversed CARNOT) given by

$$COP = \frac{T_e}{T_e - T_c}$$

$T_e$ = evaporation temperature (K).
$T_c$ = condensing temperature (K).

A typical temperature range for a heat pump $(T_e - T_c)$ is 55K (55°C), i.e. $T_c$ = 278K (5°C), $T_e$ = 333K (60°C). Hence

$$COP = \frac{333}{55} = 6.05.$$

This means that 6 times the quantity of electrical energy supplied is discharged as 'heat' or thermal energy. In practice, the ideal (reversed carnot) COP cannot be achieved. Perhaps 60 per cent can be. HEAT PUMPS can make useful contributions to low grade heating application. (⇨ CARNOT EFFICIENCY)

**Co-generation.** ⇨ COMBINED HEAT AND POWER.

**Coliforms.** A group of bacteria whose absence from drinking water is regarded as a guarantee of freedom from pathogenic bacteria.

Great concern has been raised about sewage contamination of bathing waters. One well known Cornwall beach had, in

January 1993, levels of 2.6 million units per 100 ml water (legal level 10 000, and 500 for EC Blue Flag beaches).

**Coliform count.** A water-purity test: the number of presumptive coliform bacteria present in 100 millilitres of water. The organism *Escherichia coli* is used to indicate the presence of faecal matter which can spread enteric diseases. *Escherichia coli* is the main species of bacteria present in human excreta and while only 1 in 1 million of the bacteria may be pathogenic to man, the presence of any *E. coli* is an indication of polluted water. It cannot, therefore, be distributed for consumption without proper sterilization.

**Collection tax.** This is a form of the polluter-must-pay philosophy. For example, the non-returnable BOTTLE is subsidized by the community who pay for the cost of refuse collection and disposal. A collection tax would make the purveyor/manufacturer of goods in non-returnable bottles pay for their collection and disposal and, of course, may influence an increase in the use of returnable bottles. (⇨ RECYCLING, FINANCIAL INCENTIVES FOR)

**Combined cycle power station.** This type of plant is flexible in response and can be built in the 100–600 MW capacity range. It produces electrical power from both a gas turbine (ca. 1300°C gas inlet temperature), fuelled by natural gas or oil plus a steam turbine supplied with the steam generated by the 500°C exhaust gases from the gas turbine. The THERMAL EFFICIENCY of these stations is ca. 50 per cent compared with a maximum of 40 per cent from steam turbine coal-fired power stations.

This type of plant can be built in two years compared with six years for a coal-fired station and 10–15 years for nuclear. The proposed fuel in the UK is natural gas.

**Combined heat and power (CHP).** Combined heat and power or co-generation are the terms used to describe the joint production of heat (often in the form of steam) and power (usually in the form of electricity).

Large combined-cycle systems (with two ways of driving the same electric generator) are able to offer electrical efficiencies of 58 per cent, with 60 per cent in prospect within the next few years. Medium-sized combined-cycle plants are now capable of electrical efficiencies extended above 50 per cent. Such power stations, connected to industrial premises or district-heating networks, can achieve total efficiencies as high as 90 per cent. Merwedekanaal station in Utrecht, when operating in district-heating mode, has an 87 per cent efficiency.

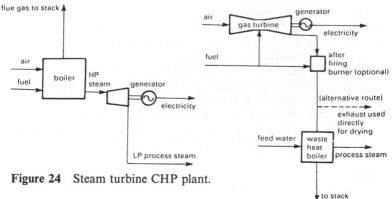

**Figure 24**  Steam turbine CHP plant.

**Figure 25**  Gas turbine CHP plant.

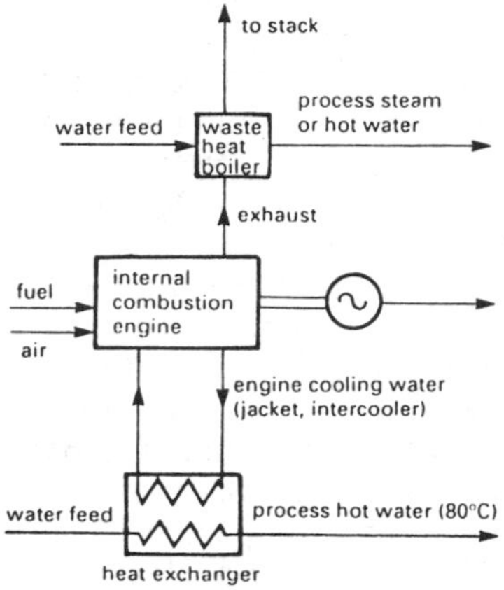

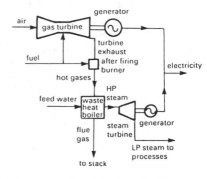

**Figure 26**  Internal combustion engine CHP plant.

**Figure 27**  Combined cycle CHP plant.

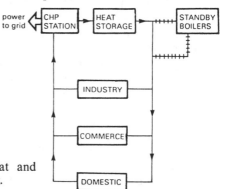

**Figure 28**  Combined heat and power with district heating.

Combined-cycle plants burning natural gas can be the least polluting thermal generators of electric power since natural gas produces virtually no $SO_2$ and relatively little $CO_2$ and unburnt hydrocarbons during combustion.

Another environmental advantage is the reduced cooling-water requirement, amounting to one-third less than for a conventional thermal power station. This is because only a minor part of the power in a combined-cycle plant is generated by the steam turbine, which largely determines the cooling requirement. As a result, the cooling water system can be smaller and less water has to be treated before returning it to the environment.

Combined-cycle plants can burn a wide range of fuels at high efficiency. These include all types of natural gas, oils from diesel to crude and heavy oils, and also coal and residual oils in various states. ($\Rightarrow$ IGCC).

In the UK some 3 per cent of electricity is produced from CHP plants. Approximately 95 per cent is generated from steam condensing turbines with an overall efficiency of 32–38 per cent. Modern gas turbines achieve much higher electrical efficiencies especially when operating in the combined-cycle mode with the exhaust heat from the gas turbines used to raise steam and drive a steam turbine. Combined cycles convert 45–50 per cent of fuel into electricity.

CHP plants can either be steam driven (Figure 24) where the output of electricity to steam is in the range of 1:2 to 1:6 or non-steam such as gas turbines where the high temperature products of combustion are used to produce heat in the form of steam or hot water (Figure 25), spark ignition engines or diesel engines (Figure 26). Non-steam plant can be combined with steam turbines to form a combined cycle and the residual heat used for the process applications (Figure 27).

In the case of a CHP plant the electricity produced can be used on site, sold privately to other consumers or alternatively fed into the national grid. The heat produced can be used on site for process (industrial CHP) or for space and water heating (micro-CHP) or distributed through a mains network (district heating) (Figure 28).

The CHP market can be broadly sub-divided into three main categories of industrial CHP, micro-CHP in buildings and large-scale CHP/district heating (DH). The common factor is a higher energy conversion efficiency and consequent fuel savings. ($\Rightarrow$ DISTRICT HEATING)

DH is widely adopted in many Western European cities and most Scandinavian towns. The UK lags badly in the DH field but there is widespread adoption of CHP in the process industries where a base heat load is guaranteed. Sewage SLUDGE digestion can also be assisted in a cost-effective manner by the adoption of CHP using the METHANE gas from the digestion as a fuel for modified internal combustion engines. The hot engine cooling jacket water is used to heat the digesters to enable optimum digestion to take place. Electrical power is also produced by the engines. Such a scheme cost the Northumbrian Water Authority £42800 in 1985 at their Cramplington treatment works. This generated £9000 per year in savings with a resulting PAY BACK of five years thus paving the way for the consideration of similar set-ups where populations in excess of 5000 are served. ($\Rightarrow$ ANAEROBIC DIGESTION)

The state-of-the-art is exemplified by the (1995) 'Mitte' new combined cycle power plant being built in the centre of Berlin which will, when completed, supply electricity and heat rated at 380 MW and 380 MJ/s, respectively, to the civic centre, government offices and other major complexes. The plant is designed for an electrical efficiency of 47.4 per cent and a thermal efficiency of almost 90 per cent. The plant's central location made its architecture and environmental compatibility important planning issues. ABB is the lead contractor.

*Source: Professional Engineering*, 6 April 1994.

The performance parameters of the Berlin Mitte district heating power plant (at the design point) and the main water–steam parameters are given in the tables on page 99.

**Combustion.** A chemical reaction in which a fuel combines with oxygen with the evolution of heat: 'burning'. The combustion of fuels containing carbon and hydrogen is said to be *complete* when these two elements are all oxidized to carbon dioxide and water, e.g. the combustion of carbon $C + O_2 = CO_2$, or hydrogen $2H_2 + O_2 = H_2O$. Incomplete combustion may lead to (1) appreciable amounts of carbon remaining in the ash; (2) emission of some of the carbon as carbon monoxide; and (3) reaction of the fuel molecules to give a range of products which are emitted as SMOKE. The combustion of volatile products can be rendered more complete if secondary air is admitted over or beyond the fuel bed for coal, or flame of oil and gas (air supplied through the fuel bed is termed 'primary air'). Air in excess of the

*Performance parameters*

| Integrated plant | |
|---|---|
| Ambient temperature (°C) | 0 |
| Gross electrical power (MW) | 380 |
| Heat output, normal/maximum (MJ/s) | 340/380 |
| Thermal efficiency (%) | 89.2 |
| Electrical efficiency (%) | 47.4 |
| Boiler/steam system | |
| Pressure levels | 2 |
| HP steam: Mass flow (t/h) | 210 |
| Exit pressure (bar) | 76.9 |
| Temperature (°C) | 525 |
| Feedwater temperature (°C) | 110 |
| LP steam: Mass flow (t/h) | 48 |
| Exit pressure (bar) | 5.3 |
| Temperature (°C) | 203 |
| District heating water preheater – heat recovered (MW) | 22.5 |
| District heating | |
| Medium | Hot water |
| Temperature out (°C) | 105 |
| Temperature in (°C) | 55 |
| Mass flow (t/h) | 5838 |

*Water–steam parameters*

| Water/steam section | HP |
|---|---|
| Steam generation rate$_{max}$ (kg/s) | 67.49 |
| Operating pressure max. (bar) | 90 |
| Steam drum pressure max. (bar) | 83.0 |
| Exit pressure max. (bar) | 82.0 |
| Exit temperature max. (°C) | 525 |
| Feedwater temperature, gas/oil (°C) | 110/130 |

*Source: ABB Review*, January 1995.

amount theoretically required for complete combustion ('excess air') is kept to the minimum compatible with complete combustion (and absence of smoke) in order to avoid undue loss of heat in the stack gases, poor heat transfer, and oxidation of sulphur dioxide to sulphur trioxide.

The proportions of the pollutants will vary according to the fuel analysis, e.g. chlorine is mainly present in coal, municipal solid waste and various spent solvents (which may be used as

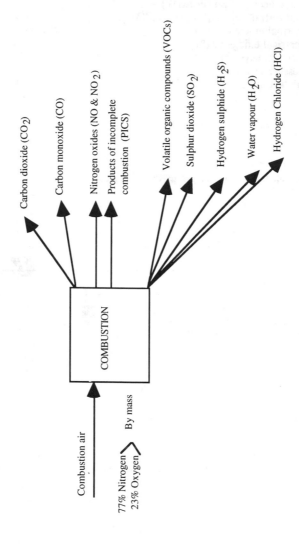

**Figure 29**  Possible pollutants emitted from a fuel containing carbon, hydrogen, sulphur and chlorine.

fuels) but not in petrol or DERV. A basic example of combustion air requirements is given below.

A brown COAL containing 70 per cent carbon and the remainder ash requires sufficient air to provide oxygen to oxidize the carbon to carbon dioxide. The oxygen requirement is dictated by chemistry and is such that 1 kg C requires 2.67 kg $O_2$ for complete combustion (and produces 3.67 kg $CO_2$ in the process). Hence, 0.7 kg C requires 1.87 kg $O_2$ for complete combustion providing 2.57 kg $CO_2$ in the process. However, excess air is always provided to ensure complete combustion (as far as possible). If 50 per cent excess air is provided, the actual amount of air provided to burn 1 kg brown coal is:

$$\left(\frac{\text{amount of oxygen required}}{\%\ \text{oxygen in air}}\right) \times 1.5$$

$$= \frac{1.87}{0.23} \times 1.5 = 12.2\ \text{kg air}$$

The nitrogen in the air provided is 77 per cent by mass. Hence, in burning 1 kg brown coal, 9.39 kg nitrogen is contained in the 12.2 kg air supplied. Thus, some nitrogen oxide formation is a certainty. The only question is how much and this depends on the technology employed. As we do not know this, we can compute the total mass of flue gases. Thus, the mass products of combustion of 1 kg brown coal are (at 50 per cent excess air):

| | |
|---|---|
| $CO_2$ | 2.57 kg |
| $N_2$ | 9.39 kg |
| $O_2$ | 0.94 kg |
| Total | 12.9 kg gaseous products of combustion or flue gas. |

It should be noted that 0.7 kg coal have been consumed in combustion, hence 0.7 kg carbon plus 12.2 kg air supplied a total of 12.9 kg; this is known as a MASS BALANCE and accords with the 12.9 mass of flue gases discharged.

Also, as the volume of gas occupied by a KILOGRAM MOLE of gas at STANDARD TEMPERATURE AND PRESSURE (STP) is $22.4\,\text{m}^3$, we can calculate the volume of flue gases emitted at STP from the combustion of 1 kg brown coal as follows, given that 1 kg mole $CO_2 = 44$ kg, 1 kg mole $N_2 = 28$ kg, 1 kg mole $O_2 = 32$.

Then from the combustion products table, the volume of flue gases discharged at STP from the combustion of 1 kg brown coal containing 70 per cent carbon in 50 per cent excess air is as follows:

$$CO_2 \text{ volume} = \frac{22.4 \times 2.57}{44} = 1.3 \, m^3$$

$$N_2 \text{ volume} = \frac{22.4 \times 9.39}{28} = 7.5 \, m^3$$

$$O_2 \text{ volume} = \frac{22.4 \times 0.94}{32} = 0.65 \, m^3$$

$$\text{Total} = 9.45 \, m^3$$

In addition, 0.3 kg of ash and particulates will be discharged as solids.

This simple calculation shows that very large quantities of combustion air are required to burn fuels and that commensurably large volumes of flue gases are emitted.

Theoretically, a 50 $MW_T$ (thermal) output boiler plant will consume

$$\frac{50}{\text{calorific value of coal (MJ/kg)}} \text{ kg coal per second.}$$

As the CALORIFIC VALUE of 70 per cent brown coal is 27 MJ/kg, this means our 50 $MW_T$ plant consumes

$$\frac{50}{27} \times \frac{60 \, (s)}{1000 \, (kg)} \times 60 \, (min) \times 24 \, (h) \text{ tonnes per day}$$

or 160 tonnes of coal per day (in practice, a greater amount will be consumed because of combustion losses). It emits (at 50 per cent excess air) 1 513 600 $m^3$ of stack gases at STP per day, which contain 411.2 tonnes of $CO_2$, i.e. 411.2 tonnes of $CO_2$ are emitted per day from the 50 $MW_T$ boiler plant.

*Note:* If the combustion efficiency is 90 per cent the plant will consume 11 per cent more fuel and the output will be increased accordingly. If 80 per cent efficient it will consume 25 per cent more fuel with correspondingly increased emissions. This is a point to note well where sustainability is concerned – efficiency is important.

**Commercial wastes.** ⇨ WASTES.

**Common Inheritance, Government White Paper.** A UK Government White Paper which looks at all levels of environmental concern and describes what the Government has done and proposes to do. It starts from general principles and objectives and it discusses the government's approach to the environmental problems affecting Britain, Europe and the world.

The main objectives are:

- protecting the physical environment through the planning system and other controls and incentives,
- using resources prudently, including increasing energy efficiency and recycling, and reducing waste,
- controlling pollution through effective inspectorates and clear standards, and
- encouraging greater public involvement, and making information available.

**Commons, The.** The concept of The Commons is one of common ownership of resources of value to the community. Originally the term referred to common land used for pasture, but it has recently been used to describe man's common environmental resources: land, air, water and so on. The relevance of the analogy has been pointed out by Garrett Hardin, an American biologist, who has shown that the more an individual (or corporation) exploits The Commons, the greater the harm to the community as a whole. Hence, protective measures are required to control man's greed otherwise we could all act as free enterprises putting little or nothing back for future generations.

It is sometimes said that market forces will take care of these problems. However, the price of a commodity depends on the difference between rate of supply and rate of demand. If supply slows down gradually, a pricing mechanism may evolve but too often the point of extinction is virtually reached before this happens. For example, fish can now be caught in much greater quantities by means of new technology and at a lower price because of this, yet this is a prime example of how to deplete a resource. The actual disappearance of blue whales, then the fin whales, etc., demonstrates this quite clearly. (Or the extinction of the dodo, the US passenger pigeon and virtually the buffalo.) In other words, market forces can lead to greater depletion rates; witness Saudi Arabia or Iran increasing oil production rates when revenues drop thus leading to faster resource depletion. (⇨ WHALE HARVESTS)

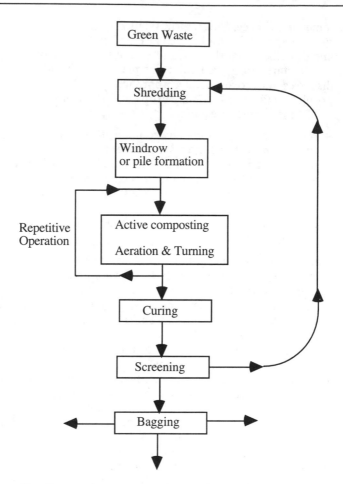

**Figure 30**   Composting system and operations. *Source:* P.A. Wheeler
and R.P. Bardos, *Preliminary Assessment of Heavy Metals in Anerobic
Digestion of Municipal Solid Waste*, ETSU Report, Harwell Laboratory,
Oxon., 1991.

*Source:* G. Hardin, 'The tragedy of The Commons', *Science* (*N Y*), vol. **162**, 1968, pp. 1243–8.

**Community liaison group.** A means of community involvement for large scale industrial operations (e.g. INCINERATION). A community liaison group could comprise representatives of the development company (or of the owners and operators as appropriate), planning authority representatives and a cross-section of local community representatives. A third party facilitator may be appropriate. The frequency of the meetings and their remit should be agreed by all parties on a basis which is relevant to each site.

**Compactor.** Equipment for the compression and volume reduction of waste materials usually into containers for onward transport.

**Composting.** A process of controlled decay which enables aerobic bacteria and other micro-organisms to decompose organic matter such as leaves, grasses, paper, and produce a stable end product suitable for soil dressing, or for landfill of DOMESTIC REFUSE in locations where crude refuse cannot be tipped. See Figure 30. Sales of compost are governed by the quality, distance to be hauled and state of the market. Compost from domestic refuse is rarely sold as it may be crudely decomposed and contain glass splinters, PLASTICS and HEAVY METALS. (⇨ AEROBIC PROCESSES). The table shows heavy metals in composts made from Municipal Solid Waste Units.

|                                       | Cadmium | Copper | Lead | Zinc |
|---------------------------------------|---------|--------|------|------|
| Little pre-sorting[1]                 | 4.8     | 370    | 610  | 830  |
| Wet drum pre-sorting[2]               | 8.9     | 190    | 260  | 550  |
| Pre-sorting for energy recovery[3]    | 8.5     | 190    | 280  | 590  |
| Source segregated[4]                  | 1.2     | 88     | 160  | 380  |

[1]Average value for composts where little pre-sorting was reported.
[2]Simulation value for composts produced after sorting using wet pulverization as a stand-alone process (most of the paper remaining).
[3]Simulation value for composts produced from trommel screen undersize as part of an integrated compost/energy recovery plant (most of the paper removed).
[4]Average values for composts produced from source-segregated materials.

**Concentrate and contain.** A means of waste management which relies on the wastes being concentrated and contained and kept separate from other environmental media. This is the opposite of DILUTE AND DISPERSE.

**Concentration, units for.** Many concentrations are expressed as parts per million (ppm) or parts per hundred million (pphm), where parts per million is based on proportional parts by volume. However, the International System of Units (SI) uses micrograms per cubic metre ($\mu g/m^3$) for air pollution, that is, weight per unit volume, and milligrams per litre (mg/l) or grams per cubic metre ($g/m^3$) for water pollution (*Note:* 'gramme' is the Continental spelling.) The difference in scales for air and water is due to the density difference of a cubic metre of water and a cubic metre of air.

Now in this text, as in many books on the subject, micrograms per cubic metre values for air pollution are not used exclusively; data are often in parts per million (as in US and many UK texts). To convert from parts per million to micrograms per cubic metre, we make use of the fact that 1 mole (i.e. a mass of gas equivalent to its molecular weight in grams) of any gas at standard temperature and pressure occupies a volume of 0.0224 cubic metres (22.4 litres). Thus,

Concentration in micrograms per cubic metre

$$= \frac{(\text{concentration in ppm}) \times (\text{molecular weight of substance})}{0.0224}$$

For example, sulphur dioxide ($SO_2$) has a molecular weight of

$$S(32) + O_2(2 \times 16) = 64$$

Therefore, the concentration in micrograms per cubic metre under standard conditions of 1 part per million sulphur dioxide is

$$\frac{1 \times 64}{0.0224} = 2857 \text{ micrograms per cubic metre}$$

Note: for water, parts per million by volume is the same as grams per cubic metre, as 1 gram of water occupies 1 cubic centimetre, and 1 cubic metre equals $10^6$ (one million) cubic centimetres. This relationship applies for dilute aqueous solutions. Where possible the units of parts per million should be avoided.

*Note:* The use of shorthand expressions of concentration such as ppb or ppm can be very misleading and should be used carefully. For example, the term ppb for parts per billion may mean one part in $10^9$ or $10^{12}$ depending on whether one uses the American or European billion. Usually ppb refers to 1 in $10^9$. Likewise the term ppt for parts per trillion is usually interpreted as 1 in $10^{12}$, although as we have seen this can be expressed as ppb. A further confusion is that the term ppt has been used to express salinity, and in this context marine scientists refer to 1 part in $10^3$ (i.e. 1 part per thousand). A further complication is that units such as ppb are ratios, but could be by weight to volume. This must be specified. Clearly a concentration in weight to volume cannot directly be expressed as ppm because the ratio must have both units the same. In such cases it is necessary to convert to common units. In the particular case of vapour concentrations in the air, ppm and similar units are inadvisable because vapours do not always behave exactly as gases, and so where concentration conversions are necessary this is not always straightforward.

Further caution is necessary when, for example, concentrations in animals tissues are quoted. While this is not always done, to be meaningful the concentration must specify whether it is in terms of wet weight, dry weight, ash weight or some other basis. Unfortunately, UK dust concentrations may still be given in grains per cubic foot $(gr/ft^3)$ $1\,gr/ft^3 = 2.288\,g/m^3$ at STP conditions.

**Concorde.** A SUPERSONIC (faster than sound) aircraft which, in order to achieve the very large thrust needed for its flight speed uses turbo-jet engines which are inherently noisier than the turbo-fan types.

There are two types of noise problem associated with Concorde and other supersonic aircraft: the sonic boom and the extremely large noise 'footprint' (i.e. the area which receives a certain specific noise level). The sonic boom characterizes all supersonic transport (SST) and is produced as the result of a build-up of air pressure in front of the aircraft. From American studies, the pattern of the Concorde 100 EPNdB footprint ranges 54 square miles, some 41 times larger than the DC-10-30 footprint. EPNdB – effective perceived noise level – is the unit used in noise certification based upon perceived noise decibels (PNdB) corrected for particular pure tone characteristics of jet noise that produce extra annoyance over and above that measured by PNdB ($\Rightarrow$ DECIBEL; NOISE CONTROL; PNdB). Any substantial reduction will require major engine redesign.

Concorde's early flights frequently broke Heathrow noise rules (measured in PNdB) on 26 take-offs out of 37. It has repeatedly created noise levels above the threshold of pain (133 decibels) (*Observer*, 19 October 1975). Also, the low-frequency sound emitted by Concorde is five times as intense as that from a 707. This type of sound can cause structural damage to buildings.

In addition to noise problems, there are fears that the exhaust emissions in the stratosphere could have far-reaching and destructive effects on climatic conditions. The stratospheric ozone layer may be decreased, causing an increase in the amount of ultra-violet solar radiation reaching the earth's surface. This aspect has yet to be proven and it is argued by the makers that the very much greater number of supersonic flights undertaken by military aircraft do not appear to have damaged the stratospheric ozone layer. The point really is, do we need to fly at this speed, given that quite acceptable and much more energy-efficient alternatives exist?

**Conservation of land.** The maintenance of areas of the countryside for leisure and the preservation of threatened species of animals and plants is one of the main objectives of a conservation policy. The existence of two bodies in the United Kingdom, the Nature Conservancy and the National Trust, indicates the British commitment to conservation. Conservation is essentially the preservation of man's environment in a condition to fulfil his needs for a healthy and satisfactory life as the pace of living and pressures of industrialized society increase. This is one aspect of THE COMMONS that requires external vigilance.

**Construction (Design & Management) Regulations 1994.** The UK Construction (Design & Management) (CDM) Regulations 1994 came into force on 31 March 1995.

The CDM Regulations seek to improve the overall management of health, safety and welfare and to ensure co-ordination between all the parties in both construction and throughout the life of a structure.

The CDM Regulations apply to:

● *construction work*
  – which is likely to last more that 30 days, or
  – which will involve more than 500 person days.
● *non-notifiable work* which involves 5 or more people on site at any one time.
● *any demolition work.*

**Contaminated land.** Land despoiled by contaminated uses such as gasworks, old landfill sites, foundries or tanneries, where high levels of heavy metals, phenols, acids or alkalis may be found.

Initially (UK) a contaminated land register was proposed but political pressures have seen it off. That does not mean the problem will go away, only that more research has to be done by would-be purchasers. *Caveat emptor.* (⇨ LIABILITY)

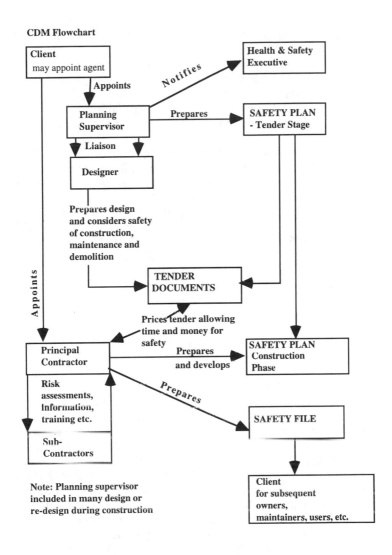

**Figure 31** Construction design and management flowchart (Institution of Civil Engineers, 1995).

A suitable-for-use approach is now adopted for contaminated land management.

At the heart of the policy is the recognition that not all contaminated land needs to be treated with the same degree of urgency.

Instead, it focuses the force of statute law on requiring action in cases where 'the contamination poses unacceptable actual or potential risks to health or the environment; and there are appropriate and cost-effective means available [to carry out remedial action] taking into account the actual or intended use of the site'.

The trend is towards a risk-based approach to contaminated land, finally laying to rest the idea of *carte blanche* contaminated land registers and confirming 'suitable for use' as a guiding principle (*CBI News*, April 1995).

Clause 54 in Part II of The Environment Act amends the Environmental Protection Act 1990, adding a new Part IIA comprising 16 sections which replace in relation to contaminated land, the wider and more general definitions and procedures for statutory nuisance to be found in Part III of the Act. 'Contamination land' is defined as follows: '. . . any land which appears to the local authority in whose area it is situated to be in such a condition, by reason of substances in, on or under the land, that (a) harm, or (b) pollution of controlled waters, is being, or is likely to be, caused.' A local authority (essentially a unitary authority or a district council) is obliged to have regard to guidance forthcoming from the ENVIRONMENT AGENCY in exercising this function. Each authority is obliged to identify contaminated land, closed landfill sites (defined as land in respect of which it appears that there is no waste disposal or waste management licence in force but where controlled waste has been deposited lawfully or unlawfully) and closed landfill sites which appear to be suitable for designation as 'special sites' by the Secretary of State. A 'special site' is a site in such a condition, by reason of substances in, on or under the land that (a) serious harm, or (b) serious pollution of controlled waters, is being, or is likely to be caused. It is the duty of the agency (in respect of a 'special site') and a local authority (in respect of a closed landfill site other than a special site) to prepare and publish a remediation statement. Such a statement must be served by the agency (in respect of a 'special site') or a local authority (in respect of any contaminated land other than a special site) or an 'appropriate person' who should be able thereby to ascertain what is required by way of remediation as well as the period for compliance. (The appropriate person is the person who caused or knowingly permitted the offending substances to be in, on or under the contaminated land.) This may not be the end of the matter but the problem of managing the effects of 200 years of heavy industry on a small island cannot be solved overnight.

As an example, there are an estimated 3000–4000 old town gas sites (where gas was manufactured using coal, before natural gas became available). These may be contaminated with coal-tar, phenols and sulphur compounds and asbestos.

**Continuous sampling.** Uninterrupted sampling of, say, air, usually at a fixed rate. Where the sample can be analysed continuously, the stream of gas may be passed through the measuring instrument continuously. Otherwise the sample is collected in an uninterrupted fashion for a given period and the total sample is finally analysed to give the mean composition of the air over the whole period.

**Continuously Regenerating Trap (CRT).** A device for particulate removal from DIESEL engined vehicle exhausts which uses NITROGEN DIOXIDE ($NO_2$) to oxidize particulates (at temperatures of more than 250°C). The fumes from the engine pass through a catalyst designed to maximize $NO_2$ levels, into the trap system which is made from extruded ceramic containing millions of microscopic 'honeycomb' channels. The fumes react with the $NO_2$ to form oxides of nitrogen and carbon dioxide. There is a small increase in CARBON DIOXIDE emissions of around one to two per cent.

Low sulphur diesel fuels are necessary as $SO_2$ absorbs water which condenses on the filter and lowers its performance.

The CRT features a combination of catalyst and filter technology to remove particulates as well as carbon monoxide, hydrocarbons and the characteristic diesel smell. It is a significant advance on other technologies that have been used for cleaning up diesel emissions because it is able continuously to remove and eliminate soot at the vehicle's normal operating temperature, i.e. the trap continuously regenerates itself.

An oxidation catalyst and filter are combined in the CRT and, inside a stainless steel skin, can be engineered to provide sufficient sound attenuation to replace an ordinary silencer. The CRT is mounted on the vehicle in place of a conventional silencer, and provides a novel combination of filter and catalyst technology.

Exhaust gases containing carbon monoxide and hydrocarbons pass through the catalyst where they react with oxygen in the gas stream and are converted into carbon dioxide and water vapour. In addition, the specially developed catalyst increases the proportion of nitrogen dioxide to nitrogen monoxide, a key feature of the soot removal process. The exhaust stream then passes into the filter, a porous ceramic with cells about four times as large as those in the catalyst and in which channels are blocked at either end alternately, in a chequer-board style.

As the exhaust gas enters a channel it is forced through the wall into an adjacent channel to exit the filter. In so doing, the soot carried in the exhaust stream is deposited on the surface of the filter. Here, the soot (or carbon) is oxidized by nitrogen dioxide in the exhaust stream. The soot particles are gradually oxidized away and disappear completely as the exhaust gases continue to flow through the filter.

The CRT requires low sulphur fuel and at present this means below 100 ppm. Europe is moving towards a standard 500 ppm in 1996 and it is expected that catalysts compatible with this level will be developed.

The continuous removal of the soot regenerates the filter so that it is working whenever the vehicle is in use. Importantly, the soot is removed at temperatures above 250°C, compared with 550–600°C necessary in earlier systems.

(New engines will have limited impact initially, because the durability of diesel engines means that trucks and buses remain in service for a long time. They typically have high mileages throughout their service lives. There will therefore be a large fleet of long-lived polluting diesel vehicles in our cities for ten years or more.)

*Source:* Johnson Matthey, Catalyst Systems Division, February 1995.

**Contraries.** Materials which have a detrimental effect upon reprocessing or recycling.

**Control limits.** Control limits are occupational exposure limits relating to personal exposure. They are exposure limits contained in Regulations, Approved Codes of Practice, in European Community Directives or have been adopted by the Health and Safety Commission, and should not normally be exceeded.

They have been set following detailed consideration of the available scientific and medical evidence to be 'reasonably practicable' for the whole spectrum of work activities in Great Britain. Control limits apply to substances such as ASBESTOS, isocyanates, LEAD and chloroethylene. (⇨ VINYL CHLORIDE)

**Control limits, occupational.** A means of controlling the DOSE received by a member of a workforce of a hazardous substance. There are two limits, the LONG-TERM EXPOSURE LIMIT and the SHORT-TERM EXPOSURE LIMIT. The former is a time weighted average over 8 hours and the latter over 10 minutes.

As an example, the revised control limits for NICKEL and its inorganic compounds (except nickel carbonyl) adopted in the UK (*UK Health and Safety Commission Newsletter*, April 1989) from 1 June 1989, are that occupational exposure to these substances must be controlled so as not to exceed:

- 0.5 milligrams per cubic metre ($mg/m^3$) for nickel and water-insoluble inorganic nickel compounds
- 0.1 $mg/m^3$ for water-soluble inorganic nickel compounds

Both of these limits are expressed as 8-hour time-weighted averages. As no specific short-term limits have been set, exposure during any 10-minute period should not exceed three times the 8-hour limit. (For the purpose of these limits a water-soluble nickel compound should be regarded as any single nickel salt or nickel complex which has a solubility greater than 10 per cent by weight in water at 20°C.)

Because of its widespread use many thousands of employees may be exposed in processes such as engineering, electroplating and nickel manufacture.

A combination of methods may be necessary to control exposure to these substances, ranging from total enclosure of the process and the use of automatic handling techniques, to partial enclosure, local exhaust ventilation and the use of respiratory protective equipment (RPE). Engineering control methods can generally ensure that the occupational exposure limits are not exceeded. However, where it is not possible to use such methods, or the required degree of control is not achieved, then suitable RPE will be necessary. Such activities may include furnace cleaning and spray drying operations.

**Control of Pollution Act 1974 (CoPA).** CoPA 1974 covers a wide range of legislation on pollution issues: Part I – waste disposal;

Part II – water pollution; Part III – noise; Part IV – atmospheric pollution; Parts V and VI – miscellaneous.

**Control rod.** A rod of neutron-absorbing material (e.g. cadmium, boron, hafnium) moved in or out of the reactor core to control the number of neutrons available for fission, and hence the power level of the reactor.

**Controlled landfill.** ⇨ LANDFILL SITE.

**Controlled tipping.** The most common method of DOMESTIC REFUSE disposal where the refuse is tipped in layers, compacted and covered at the end of every working day with an inert layer of suitable material to form a seal. Controlled tipping avoids refuse being blown from the site, or tipping in static or running water, and ensures the general orderly appearance and running of the tip. The opposite of this practice is open tips which constitute public nuisances and can threaten public health by provided breeding grounds for rodents and flies. In practice, many sites are less than perfect and illegal tipping of hospital hazardous wastes can take place.

The management of landfill sites in the UK is undergoing major changes as higher standards are imposed to prevent the migration of LANDFILL GAS and to ensure the integrity of HAZARDOUS WASTES that may be co-disposed with the domestic refuse. (⇨ LEACHATE)

**Controlled waste.** Waste controlled under the Control of Pollution Act, 1974. (⇨ WASTES)

**Coolant.** Gas, water, or liquid metal is circulated through a reactor core (primary coolant), to carry heat generated in it by fission and radioactive decay to boilers or heat exchangers where water (secondary coolant) is turned into steam for the turbines.

**Cooling tower precipitation.** The drizzle that occurs around water cooling towers that are not fitted with spray eliminators. If the circulating water within the tower contains salt, sewage, or other dissolved materials, such precipitation can be a source of pollution, and a breeding ground for the bacteria that cause Legionnaires' disease. In addition, certain wind conditions that produce a down-draught inside the tower, or high tangential velocities, may result in water droplets being blown out of the base of the tower. Cooling-tower precipitation is also referred to as 'drift' and 'carry-over'.

**Copolymerization.** ⇨ POLYMERIZATION.

**Copper (Cu).** A metal, commonly used for heat-transfer applications because of its high thermal conductivity. It is a very good electrical conductor and can be readily drawn or extruded into tubes and wire.

Copper is an essential constituent of living systems. However, copper ions ($Cu^{2+}$) are toxic to most forms of life, 0.5 parts per million being lethal to many algae. Most fish succumb to a few parts per million. In higher animals, brain damage is a characteristic feature of copper poisoning.

Copper pollution may arise from many sources. Soils receive high levels as a result of mining activities, intensive use of 'copper pellets' in pig rearing, or the application or copper fungicide. Effluents from factories and mines can cause serious water pollution.

**Corrected noise level (CNL).** Index of industrial noise. The level of noise emitted from industrial premises in dB(A), corrected for tonal character, intermittency and duration in accordance with the appropriate British Standard. The CNL measure is required because of the intermittent nature of many industrial noise sources. ($\Rightarrow$ DECIBEL A-SCALE; INDUSTRIAL NOISE MEASUREMENT)

**Cost-benefit analysis.** A technique which purports to evaluate the social costs and social benefits of investment projects in order to help decide whether or not such projects should be undertaken.

Cost-benefit methods have been offered to support the nuclear industry's claim that zero radiation release is too expensive. By using cost-benefit analysis, the minimum radiation levels that society can 'afford' to tolerate may be calculated. This approach is embodied in Figure 32, the thesis being that the dose that is 'as low as is readily achievable, economic considerations being taken into account', is thereby obtained.

One major objection to this approach is that exposure to radiation is an involuntary risk and is not one that an individual can opt out of. For example, a person driving a car takes a voluntary risk of an accident happening as opposed to completing the journey (benefit) and the costs and benefit apply to the same individual. In the case of nuclear power, costs and benefits do not apply to the same individuals and risks cannot be eliminated by voluntary action.

The questions which cost-benefit analyses have failed to answer in this and other areas are:

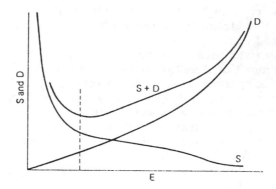

**Figure 32** Differential cost-benefit analysis reveals the radiation level from nuclear operations, according to the considerations being taken into account. The two curves S and D represent the full cost of achieving a standard of safety resulting in the collective dose (E), and the cost of the detriment resulting from that dose. Combining these two curves gives the U-shaped curve representing the total cost of the collective dose. The minimum of the S + D curve is the dose that is 'as low as readily achievable'.

1. How do you put a financial cost on a risk when there is no mechanism for people to buy themselves out of the risk?
2. How can 'emotive issues' be taken out of cost-benefit analysis when ultimately all value is derived from emotion?
3. How can one evaluate either costs or benefits to non-direct participants in a system which does not yet exist?
4. On what evidence is the same discount rate applied to costs and benefits? In particular, why apply a discount rate to children? Most people seem to apply an interest rate to their children.

Among economists themselves, perhaps the most devastating indictment of cost-benefit analysis has come from Schumacher:

> It can therefore never serve to clarify the situation and lead to an enlightened decision. All it can do is lead to self-deception or the deception of others; for to undertake to measure the immeasurable is absurd and constitutes but an elaborate method of moving from preconceived notions to foregone conclusions; all one has to do to obtain the desired results is to impute suitable values to the immeasurable costs and benefits. The logical absurdity, however, is not the greatest fault of the undertaking: what is

worse, and destructive of civilization, is the pretence that everything has a price or, in other words, that money is the highest of all values.

E. F. Schumacher, *Small is Beautiful*, Blond & Briggs, 1973.

The controversy over cost-benefit analysis is central to the study of the environment.

**Cost function.** The relationship between the degree to which a pollutant emission is reduced and the cost of attaining the reduction.

**Crank-case blow-by.** A mixture of air, petrol/diesel vapour, and exhaust gases that escapes from the cylinders of an internal combustion engine by blowing past the piston rings into the crank case. If this mixture is not returned to the cylinders to be burned, it can account for some 20 per cent of the total emission of hydrocarbons from the engine. Positive crank-case ventilation, in which the mixture is returned to the cylinders by means of suction from the intake manifold, is now obligatory in many countries.

**Critical group.** A method of monitoring pollution emission which relies on the identification of the group of individuals most at risk to a particular discharge. If they are receiving doses of exposures below those recommended, it is assumed that the population at large is therefore not exposed or subjected to the same discharge. This technique has been used for the monitoring of IONIZING RADIATION emissions from nuclear reactors. It depends on the success with which the group most at risk can be identified and an accurate assessment of their exposure to the discharge.

It is also proposed as a method of controlling discharges of HEAVY METALS to estuaries. For example, crab meat (wet weight) from South Devon has been found to contain up to 21 parts per million of CADMIUM, and if the group of individuals were identified as, say, 'large' crab-meat eaters, and if their daily intake was below the Food and Agricultural Organization/World Health Organization limit of 60 micrograms for a 70-kilogram man, then the population at large could be assumed to be not at risk.

Children and old people are two distinct critical groups that require separate consideration. Children take in approximately twice as much food as an adult in relation to body weight. Thus their exposure to risk is greater. Old people generally have weaker internal organs, and this again may militate against bodily defence mechanisms functioning correctly.

**Critical pathway.** In planning a nuclear installation, the environmental pathway is found for every nuclide that is to be released on discharge through food chains or otherwise to any member of the population. Some member(s) of the population is thereby found to constitute the critical case of exposure, i.e. CRITICAL GROUP. If it is reasonably certain that the dose in the critical case is less than the International Commission for Radiological Protection (ICRP) limit, then no other persons will be at any greater risk. However, ICRP recommends that every exposure should be reduced as low as is readily achievable within the economic and social framework. Efforts are also made to exploit any opportunity to reduce the dose still further (and indeed discharges for Sellafield (Windscale) were drastically reduced in the 1980s as evidence accumulated that there were worrying local concentrations of radionuclides in estuary silt and parts of local beaches. (⇨ CAESIUM-137; DERIVED WORKING LIMIT; IONIZING RADIATION, MAXIMUM PERMISSIBLE DOSE)

**Cross-media.** Term used in pollution control to mean POLLUTION CONVERSION, e.g. a hazardous waste incinerator discharging toxic dust from its dust collectors which requires further disposal would be assessed for cross-media effects.

**Cryogenic scrap recovery.** A form of scrap recovery which uses liquid nitrogen to freeze (at $-196°C$) the pot-pourri of metals and contaminants from used cars. The very low temperature causes mild steel to fracture like glass when put through an impactor. The steel can then be easily removed magnetically from its trapped contaminants and the result is a very high-grade scrap suitable for high-quality steel manufacture. The contaminants, which include copper and zinc, may be further processed for high-value scrap recovery. This method of recovery is being tested on tin cans in DOMESTIC REFUSE in north-east England. It allows excellent metals separation and 6000 tonnes per year of scrap steel are currently being recovered.

**Cullet.** Trade name for colour-graded glass fragments, offcuts, etc. suitable for remelting. Sources are mainly the glassworks themselves and bottling plants. DOMESTIC REFUSE is another source but the RECYCLING cost is often prohibitive when compared with the cost of the raw materials themselves as collection and transport costs outweigh the revenue from the sales.

**Curie (Ci).** Long established unit of radioactivity (now replaced by the BECQUEREL) defined as 37000 million ($3.7 \times 10^{10}$)

disintegrations per second which is approximately equal to the activity of 1 gram of radium-226. Thus an amount of radioactivity was expressed effectively by stating the number of grams of radium that could provide the number of disintegrations per second. In practice it was far too large and the picocurie was used, i.e. 1 million millionth of a curie ($10^{-12}$).

**Cyanide.** Generic name for a salt of hydrocyanic acid (HCN), e.g. potassium cyanide (KCN) used for dissolving gold from crushed rock. The solution is then reduced and filtered. Considerable water and soil pollution can result from improperly conducted gold dissolution operations.

Sodium cyanide (NaCN) is extensively used in the specialist steels industry. Like the other cyanides, it is extremely toxic if swallowed, breathed in as a dust or gas, or absorbed via eyes or mouth.

Cyanides form a highly toxic gas (hydrogen cyanide) when mixed with acids, so they must be kept apart. This illustrates the risks of indiscriminate industrial waste disposal and highly toxic substance such as cyanides must be disposed of by approved methods such as oxidation by OZONE or HYDROGEN PEROXIDE or CHLORINE. The trouble is that most cyanide wastes also contain metals such as cobalt or NICKEL and hence oxidize very slowly. Again the need for specialist professional waste treatment is highlighted in handling dangerous materials.

**Cycle.** A process or series of operations performed by a system in which conditions at the end of the process are the same as the original state. Thus in the CARBON CYCLE the carbon dioxide in the air is fixed by photosynthesis into plant life which in turn decomposes or is consumed and eventually ends up as carbon dioxide again. In the steam power station cycle, water is heated in the boiler (energy supplied) to steam which expands in the turbine to produce power (workout). The exhaust steam is condensed in the condenser (energy rejection). The resulting condensate (water) is pumped back through the boiler for the cycle to recommence.

**Cyclone dust separator.** A device for removing dust particles from air. The principle is shown in Figure 33. The dust-laden gas enters tangentially or axially and is spun in a helical path down the conical collector. The particles are flung to the wall by centrifugal force where they drop into the dust hopper while the clean gas leaves through a central 'core' tube at the top.

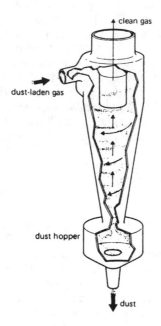

**Figure 33**  Cyclonic dust separation.

Cyclones can remove 75–95 per cent of particulates in their range of applicability; for smaller particle sizes ELECTROSTATIC PRECIPITATORS, BAG FILTERS or other devices are often used, e.g. to remove FLY ASH.

# D

**Damage function.** The relationship between the degree to which a pollutant emission is reduced and the cost of the damage prevented by the reduction.

**Dam projects.** Recent major dam projects have provided examples of high technology applied with the most benevolent of motives, but with inadequate consideration of any possible ecological side-effects. The most obvious example is the ASWAN DAM. Designed to produce power and to store water for a permanent irrigation system, the effect of the irrigation ditches on the spread of a serious disease, SCHISTOSOMIASIS (bilharzia) was not considered. Similar unexpected side-effects occurred with the Kariba Dam, which helped spread a fly-borne disease which has disrupted the agriculture of people living along the river banks. Furthermore, for most water projects of this type, silting of the reservoir will eventually eliminate all the temporary gains.

In the case of the ASWAN DAM, the changes produced in the flow of the Nile have had detrimental effects on the fisheries of the eastern Mediterranean. It will also probably have a deleterious effect on the soil fertility in the Nile Delta, since nutrients that were previously deposited annually by the flooding of the Nile will now be absent.

**Daughter product.** The NUCLIDE which results from the radioactive disintegration of a 'parent' nuclide. Daughter products can be radioactive as well.

**DDE (dichlorodiphenyldichloroethylene).** A METABOLITE of DDT, it is formed by the action of some soil micro-organisms on DDT (in water DDT is inert). It is more inert and persistent than DDT and large quantities of both products could be accumulating in the biosphere. In the presence of ultra-violet light (i.e. sunlight) it can photodegrade to about nine other products, including several chlorinated biphenyls which can also have severe biological effects.

DDE is one of the most abundant organochlorine compounds in the biosphere.

**DDT (*d*ichloro*d*iphenyl*t*richloroethane).** An organic PESTICIDE developed during the Second World War as a delousing agent and later used to combat the insect carriers of malaria, yellow fever and typhus, thereby saving many lives. However, in the early 1960s certain immune strains of mosquitoes and other disease carriers developed. At the same time a decline in the reproduction rate of fish and bird life occurred.

The key problem are the persistence of DDT and its high biological activity. It is estimated that there are several million tonnes in circulation in the biosphere, spread by air and water currents, and most of this ends up in the ocean. Small concentrations (0.01 parts per million) reduce PHOTOSYNTHESIS in marine plankton by 20 per cent, and can be lethal to fish as they concentrate it. Such a low concentration would not be lethal to man, but the use of DDT is now virtually banned world-wide because of its persistence. (⇨ BIOACCUMULATION)

**DDVP (Dichlorvos; 2,2-dichlorovinyl dimethyl phosphate).** An insecticide often used in pest strips sold for domestic use. The US Environmental Protection Agency has cautioned against its use in rooms where food is prepared, or where infants or aged person are confined – for example, in hospitals. Dietary deficiencies of protein, minerals, or vitamins may increase any ill effects. (⇨ PESTICIDES; ORGANOPHOSPHORUS COMPOUNDS)

**Decay.** The disappearance of a pollutant as a result of absorption or chemical reaction at the earth's surface, or to removal by rain, or to transformation into some other substance. (⇨ HALF-LIFE)

**Decay heat.** The heat produced by radioactive decay, especially of the fission products in irradiated fuel elements. This continues to be produced even after the reactor is shut down and the fuel removed.

**Decibel (dB).** A logarithmic measure used to compare the sound level of interest with a reference level. If we are concerned with sound power then reference is made to the smallest sound power that can be heard by someone with normal hearing at 1000 Hz. This reference power is $10^{-12}$ W. As an example we can deduce the sound power level of a jet aircraft on take-off. Now the noise of a jet aircraft at take-off (100 metres) is approximately 1 W which is $10^{12}$ as powerful as the reference power. Therefore, it is said to differ from the reference sound by

$$\log\left(\frac{power_2}{power_1}\right) = \log 10^{12} = 12\,bels$$

But bels are too large for convenience and so decibels are used instead, i.e. a factor of 10 is introduced. Hence, the jet aircraft at take-off has a sound power level of 120 dB with reference to a power of $10^{-12}$ W.

Sometimes it is necessary to compare sound pressures. Power is proportional to the mean square pressure under reflection-free conditions

$$dB = 10\log\left(\frac{power_2}{reference\ power_1}\right) = 20\log\left(\frac{pressure_2}{reference\ pressure_1}\right)$$

Reference pressure is $2 \times 10^{-5}$ N/m$^2$ (1 N/m$^2$ is referred to as a pascal (Pa)).

Thus, an increase of 3 dB in the sound power level corresponds to a doubling of the sound power which corresponds to an increase of 6 dB in the sound *pressure* level.

Decibels are also used in telecommunications work as a measure of the system response (e.g. signal-to-noise ratio).

The specification of the scale is important. For most purposes loudness is quoted in DECIBELS A-SCALE (dB(A)) although other scales are used as well. ($\Rightarrow$ HEARING; LOGARITHMS; SOUND)

**Decibels A-scale (dB(A)).** A frequency-weighted noise unit widely used for traffic and industrial noise measurement. The decibels A-scale corresponds approximately to the frequency response of the ear and thus correlates well with loudness. Other noise scales are used as well. ($\Rightarrow$ HEARING; LOGARITHMS; SOUND PNdB). For example, the TIME-WEIGHTED AVERAGE industrial exposure for 8 hours is 90 dB(A). An increase in 3 dB doubles the dose and the exposure scale becomes:

> 90 dB(A) for 8 hours
> 93 dB(A) for 4 hours
> 96 dB(A) for 2 hours
> 99 dB(A) for 1 hour

*Note:* If someone is exposed to 99 dB(A) for 1 hour, he/she must be returned to very low noise levels for the remainder of the 8 hour period. ($\Rightarrow$ DECIBEL; SOUND; HEARING; PNdB)

**Decommissioning.** The final closing down and putting into a state of safety of a nuclear reactor or other industrial plant or device when it has come to the end of its useful life.

**Decomposers.** Organisms, usually BACTERIA or FUNGI, which use dead plants or animals as sources of food. They break down this material, obtaining the energy needed for life and releasing minerals and nutrients back into the environment to be assimilated by other plant and animal life. They are an essential part of natural cycles and enable the components for life to be recycled. (⇨ CARBON CYCLE)

**Deepshaft treatment.** ⇨ SEWAGE TREATMENT.

**Defoliants.** ⇨ HERBICIDES.

**Degree-day.** The difference, expressed in degrees, between the mean temperature of a given day and a reference temperature used as a predictor for fuel consumption or demand. One heating degree-day is counted for each degree that the daily mean temperature is lower than a base temperature, usually 15.5°C in the UK. The total number of degree-days for the heating season is the sum of the degree-days for the different days of the season. Usually published monthly by area and used as a check on fuel consumption for heating purposes by large energy users.

**De-inking.** A process for removing ink from printed paper used in RECYCLING. De-inking takes place in four stages, as shown in Figure 34:

1. Pulping of the reclaimed paper with soda or other chemicals.
2. Centrifugal cleaning to remove paper clips, staples and dirt.
3. Screening to remove the freed fibres.
4. Washing to remove the printing ink solids either by mechanical or flotation methods.

The yield is ca. 75 per cent of the input paper.

PULP recovered from paper made from groundwood can be used for newsprint, magazines, and as a base for coated papers. Pulp recovered from non-groundwood paper can be used for the manufacture of writing/printing papers and tissues because of its longer fibre length.

The effluent from de-inking plants has a high BIOCHEMICAL OXYGEN DEMAND and requires considerable treatment before discharge.

**Demand.** The amount of a commodity that people are prepared to buy. If the price of a commodity varies, it would be reasonable to suppose that, at higher prices, less of the commodity will be

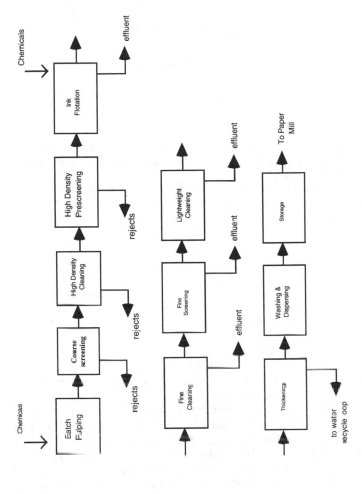

**Figure 34** Recycled fibre mill – flow diagram.

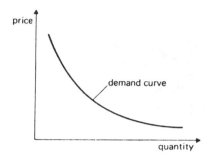

**Figure 35**   The demand curve.

bought. The fact that people still want the commodity does not mean that they demand it. If they cannot afford it, they exert no demand for it.

A graph showing the quantities of a commodity demanded at various prices is called a *demand curve* (see Figure 35). This shows, for each possible price, the total intended purchase of all buyers.

**Denitrifying bacteria.** ⇨ NITROGEN CYCLE.

**Density.** The weight of mass of a substance per unit volume.

**Density separation.** The separation by flotation in air or liquid to separate fractions of varying densities of certain types of material, e.g. paper and plastics from metals and other heavy contaminants in the waste stream for REFUSE DERIVED FUEL production.

**Deposition processes.** Air pollutants are removed from the atmosphere by two possible mechanisms – dry deposition and wet deposition. Dry deposition involves the pollutant impacting on to soil, water or vegetation at the earth's surface. Wet deposition involves the absorption of the pollutant into droplets either within clouds or below clouds, followed by removal by precipitation. Most pollutants are removed from the atmosphere by both processes. In dry climates, dry deposition will be the dominant process whereas in wet climates, wet deposition will dominate. In the UK, the amount of any pollutant removed by dry deposition approximately equals the amount removed by wet deposition averaged over a year.

Figure 36 shows the various deposition and transportation processes for the plumes from a power station stack. Typically, a

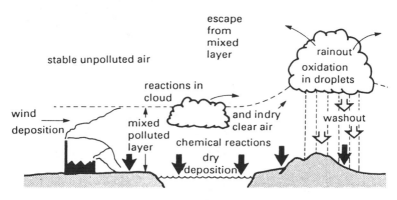

**Figure 36**   Acid rain deposition mechanisms.

plume from a 220 m stack first contacts the ground ca. 20 km downwind, by which time typical gas dilutions are 10000 times. Dry deposition takes place from the point of plume impact and is assumed to be proportional to SULPHUR DIOXIDE concentration. Wet deposition rates can be much greater and are also influenced by the concentration of photochemical oxidants in the atmosphere. The oxidants are also mainly man-made pollutants thus ACID RAIN needs to be tackled by several routes and not just DESULPHURIZATION.

**Derived working limit (DWL).** This is a specific figure for each radioisotope released into the environment which is derived by determining both the CRITICAL PATHWAY for the radiation release and the group of people (the CRITICAL GROUP) who are most likely to be affected, i.e. given a radiation dose which is much larger than the rest of the population. A derived working limit is then determined which limits the rate of discharge of the particular radioisotope to ensure that the ICRP dose limit would not be exceeded by the CRITICAL GROUP. A safety factor is always incorporated to guard against any uncertainties.

It should be noted that the determination of DWLs depends on a knowledge of the pathways, of the distribution of radioactive materials in the environment and of their ability to accumulate in food chains. The critical pathway may change with time and so can the critical group. Hence, continuous monitoring is necessary, e.g. for the discharge of ruthenium-106 from Sellafield (named Windscale at the time), the critical group used were laver bread (fried seaweed) eaters who ate the seaweed

| Exposure pathway | Critical material | Critical group | Daily consumption rate or annual exposure | Critical organ | Radionuclides contributing to exposure | Derived working level | ICRP recommended dose limit (mrem/year) | Typical exposure of population group involved (mrem/year) |
|---|---|---|---|---|---|---|---|---|
| Internal | Seaweed | Laver bread consumers (100 persons) | 160 g laver bread | Gastro-intestinal tract | $^{106}$Ru $^{144}$Ce | 300 pCi/g in the seaweed | 1500 | 600 |
| External | Estuarine silt | Fishermen (10 persons) | 350 h | Whole body | $^{95}$Zr $^{95}$Nb $^{106}$Ru | 1.4 mrem/h | 500 | 50 |
| Internal | Fish | Fishermen (100 persons) | 25 g | Gastro-intestinal tract, whole body | $^{106}$Ru $^{137}$Cs | 900 pCi/g 1800 pCi/g | 1500 500 | 2 0.5 |
| External | Fishing gear | Fishermen (100 persons) | 500 h | Hands | $^{106}$Ru $^{144}$Ce | 15 mrem/h | 7500 | 20 |

*Porphyra*, collected near Sellafield. However, these collectors retired and the *Porphyra* was collected elsewhere. The new critical group became salmon fisherman on the Ravenglass Estuary who received external exposure from radioactivity deposited on the shore.

The data opposite give an illustration of the 1970s exposure pathways for discharges from the Sellafield reprocessing plant to show the critical groups, the basis for estimation of the dose and the levels of exposure theoretically permitted for each critical group.

*Source:* A. W. Kenny and N. T. Mitchell, 'United Kingdom waste-management policy', *Management of Low- and Intermediate-Level Radioactive Wastes*, Proc. Symposium, Aix-en-Provence 1970, IAEA, Vienna, 1970.

(⇨ CAESIUM-137, IONIZING RADIATION EFFECTS; CRITICAL GROUP; CRITICAL PATHWAY; CURIE; REM)

**DERV.** Diesel-engined road vehicle. Also applied to a special grade of gas-oil with an extra-low sulphur content, which these vehicles use. (⇨ PETROLEUM)

**DES (diethylstilboestrol).** A growth-promoting hormone used in the USA in the cattle industry to cut down on feed required to produce a given weight gain. It has been found to be carcinogenic in animals and is therefore a cause for concern in connection with humans. (⇨ CARCINOGEN)

**Desalination.** The partial or complete removal of the dissolved solids in sea or saline water to make it suitable for domestic, agricultural or industrial purposes. The main techniques are ELECTRODIALYSIS, REVERSE OSMOSIS, freezing, and distillation. The processes and their applications are shown in Figure 37.

*Electrodialysis.* An electric current is passed through brackish or low-salinity water in a chamber in which many closely-spaced ion-selective membranes are placed, thus dividing the chamber into compartments. The electric current causes the salts to be concentrated in alternate compartments, with reduced salt content in the remainder, thus producing a reduced salt product. A principal disadvantage of electrodialysis is that power consumption is proportional to total dissolved solids or salinity.

*Reverse osmosis:* The reverse application of osmotic pressure. When salt water and fresh water are separated by a semi-permeable membrane, osmotic pressure causes the fresh water

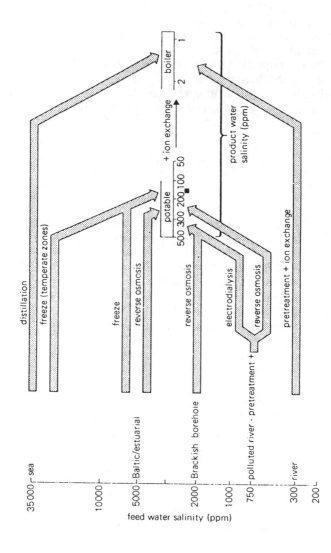

**Figure 37**  Desalination process application.

to flow through the membrane to dilute the saline water until osmotic equilibrium is established. Now applying this in reverse, if a greater pressure is applied to the salt-water side of the membrane, then relatively pure water will pass through it leaving a concentrated brine to be disposed of. The principal application may well be the treatment of industrial and sewage effluents or polluted rivers in the 750–1000 parts per million range either for re-use or to obtain an acceptable discharge. As with electrodialysis, power consumption is proportional to total dissolved solids. The likely development is a progression towards sea-water treatment but as yet no membrane has been developed which can sustain better than 99 per cent salt rejection which must be achieved before a potable water can be obtained from sea water with a total dissolved solids content of 35 000 ppm. The membranes are susceptible to bacterial fouling when applied to certain classes of effluent.

*Freezing:* The freezing of a salt solution causes crystals of pure water to nucleate and grow, leaving a brine concentrate behind. One commonly proposed freezing method is the use of a secondary refrigerant in which butane is evaporated in direct contact with sea water, resulting in the formation of ice crystals. The ice is separated, melted by the compressed butane vapour, the two liquids are decanted and the product water obtained and the butane recycled. There are substantial problems in constructing large-scale freezing plants. This is an example of a process that looked good at laboratory scale but which proved impracticable at large scale.

*Distillation:* The boiling or evaporation of sea water to form water vapour which is then condensed to yield a salt-free stream. Energy requirements are virtually independent of the feed-water salinity, and product purity of less than 50 parts per million can readily be achieved.

There are two main classes of distillation: heat consuming and power (mechanical energy) consuming. A heat-consuming process has lower energy input costs compared with the power-consuming process, as dictated by the conversion of heat to power. Over 85 per cent of the world's installed desalting capacity is accomplished by heat-consuming distillation processes.

*Source:* A. Porteous (ed.) *Desalination Technology*, Applied Science Publishers, 1983.

**Desertification.** Loss of productive land use from human action such as deforestation, SOIL exhaustion, salinized irrigated fields, depleted groundwater resources, over-grazing, drought, wind and water erosion. The table gives an example of the Aral Sea basin desertification causes.

| | Classes of land degradation | | | |
|---|---|---|---|---|
| *Types of land degradation* | *Slight* | *Moderate* | *Severe* | *Total* |
| Degradation of the vegetation cover | 53.5 | 21.9 | 1.6 | 77.0 |
| Wind erosion | 1.0 | 0.2 | 0.3 | 1.5 |
| Water erosion | 3.8 | 2.1 | — | 5.9 |
| Salinization of irrigated lands | 0.9 | 7.4 | 0.8 | 9.1 |
| Aral Sea level | 0.4 | 7.4 | 0.8 | 9.1 |
| Technogenic desertification | — | 0.4 | 1.0 | 2.4 |
| Waterlogging of desert rangelands | — | 0.4 | 0.1 | 0.5 |
| TOTAL (per cent) | 59.4 | 33.7 | 6.7 | 100.0 |

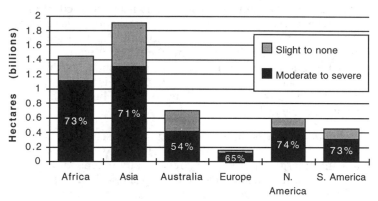

**Figure 38**   Drylands by continent (UNEP, revised 1992).

Pimentel *et al.* have stated:

Of the world's agricultural land, about one-third is devoted to crops and the remaining two-thirds is devoted to pastures for livestock grazing. About 80% of the world's agricultural land suffers moderate to severe erosion, and 10% suffers slight to moderate erosion. Croplands are the most susceptible to erosion because their soil is repeatedly tilled and left without a protective cover of vegetation.

However, soil erosion rates may exceed 100 tons ha$^{-1}$ year$^{-1}$ in severely overgrazed pastures. More than half of the world's pasturelands are overgrazed and subject to erosive degradation.

Soil erosion rates are highest in Asia, Africa and South America, averaging 30 to 40 tons ha$^{-1}$ year$^{-1}$, and lowest in the United States and Europe, averaging about 17 tons ha$^{-1}$ year$^{-1}$. The relatively low rates in the United States and Europe, however, greatly exceed the average rate of soil formation of about 1 ton ha$^{-1}$ year$^{-1}$. Erosion rates in the undisturbed forests range from only 0.004 to 0.05 ton ha$^{-1}$ year$^{-1}$.

*Sources: Our Planet*, Vol. 6, No. 5, 1994.
Pimentel *et al.*, 'Environmental and economic costs of soil erosion and conservation benefits' *Science*, Vol. 267, p. 1117, 24 February 1995.

**Destructive distillation.** The heating of solid substances in closed retorts in the absence of air, and condensation of the ensuing volatile gases. The process has been mooted for the utilization of DOMESTIC REFUSE for the production of METHANOL. It is akin to PYROLYSIS which is often used synonymously for the destructive distillation of refuse.

**Desulphurization.** ⇨ FLUE GAS DESULPHURIZATION.

**Detergents.** Cleaning agents which include, as part of their chemical make-up, petrochemical or other synthetically derived wetting agents. They are made of up three main parts:

1. *Surfactant* – a wetting agent which permits water to penetrate fabric more. The surfactant molecules provide a link between the dirt molecules and the water molecules.
2. *Builder* – a sequestering agent. These tie up hard-water ions to form large water-soluble ions. The water becomes alkaline, which is necessary for removal of dirt.
3. *Miscellaneous* – brighteners, perfumes, anti-redeposition agents and enzymes.

A major drawback in the use of 'hard' detergents is that they are non-biodegradable. There is now a move to incorporate only biodegradable substances in detergents – although the use of hard detergents is considered necessary in certain industries such as wool scouring. The decomposition or breakdown of some detergents leads to phosphorus becoming available in aquatic systems and where this is the limiting nutrient it can cause EUTROPHICATION. (⇨ DETERGENTS, SUGAR-BASED)

**Detergents, sugar-based.** Tate & Lyle have recently been testing a new detergent that is made by reacting sugar directly with tallow

– a product made from animal fat. The result is a detergent which is completely biodegradable, contains no phosphates, and is claimed to be just as effective as conventional products.

Toxicity tests indicate that the product is quite harmless. It is hoped that the product will be the forerunner of a whole new chemical industry based on sugar (a renewable source) rather than oil or coal.

**Development-induced displacement.** The eviction of people because of national development is often hidden away. Yet between 90 and 100 million people have been involuntarily resettled over the past decade. In India alone some 23 million have been displaced since 1950. Hydroelectric dam projects each year lead to the involuntary relocation of between 1.2 and 2.1 million people.

In Africa, dams on the Volta (Ghana), Aswan (Egypt), Zambesi (Zambia and Zimbabwe) and Bandama (Cote d'Ivoire) rivers have caused the forcible relocation of thousands of people, mainly farmers and herders, but also townspeople.

In China, water conservancy projects since the 1950s have created over 10 million displacees.

*Source:* C. McDonnell, *The Couriers*, No. 150, April 1995.

**Dew-point.** The temperature at which dew first appears on a solid surface whose temperature is steadily reduced below that of the surrounding moist air. (⇨ ACID DEWPOINT)

**Dichlorvos.** ⇨ DDVP.

**Dieldrin.** ⇨ CHLORINATED HYDROCARBONS; ALDRIN.

**Diesel.** Common name for gas/oil which is a petroleum distillate used as boiler and compression ignition engine FUEL. Also called DERV (diesel engined road vehicle).

With the drive to control and reduce environmental emissions from diesel engined vehicles wide variations in key diesel fuel parameters are of concern to engine manufacturers and legislators. A diesel engined vehicle travelling through Europe could find the following variations:

| | |
|---|---|
| Density | 0.82 to 0.86 (5 per cent) |
| Cetane number | 47 to 57 (20 per cent) |
| Viscosities | 1.8 to 3.2 (78 per cent) |
| Sulphur contents | 0.07 to 0.90 (1100 per cent) |

which will potentially result in wide variations in HC, CO, $NO_x$, $SO_x$ and particulate emissions, see Figure 39. (⇨ BIODIESEL, DIMETHYLETHER)

*Source:* Ethyl European Diesel Fuel Survey, Winter 1988–89.

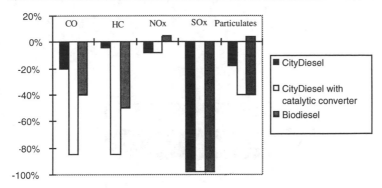

**Figure 39** The effect of citydiesel and biodiesel on emissions.
*Source: IEA Renewable Energy Newsletter*, Issue 3/94.

**Diesel vehicle emissions.** The diesel emissions standards for heavy duty vehicles, extant and proposed for the European Union, are shown in the table below. A bus for example, with a

Current legislation (from 1 July 1992) according to Directive 91/542/ EEC

| Application dates | Mass of CO (g/kWh) | Mass of HC (g/kWh) | Mass of $NO_x$ (g/kWh) | Mass of TPM (g/kWh) |
|---|---|---|---|---|
| 1 July 1992 new vehicles 1 Oct. 1993 all vehicles | 4.5 (4.9) | 1.1 (1.23) | 8.0 (9.0) | 0.36* (0.4*) |
| 1 Oct. 1995 new vehicles 1 Oct. 1996 all vehicles | 4.0 (4.0) | 1.1 (1.1) | 7.0 (7.0) | 0.15* (0.15*) |
| Test Procedures: 1 July 1992 – ECE R49 | | | | |
| 1 October 1995 | ECE R49 | | | |

Figures in parentheses refer to conformity of production.
*In the case of engines 85 kW or less, a coefficient of 1.7 is applied to the limit for particulate emissions.
CO is carbon monoxide, HC is polyaromatic hydrocarbons, $NO_x$ is nitrogen oxides and TPM is total particulate matter.
*Source: IEA Renewable Energy Newsletter*, Issue 3/94.

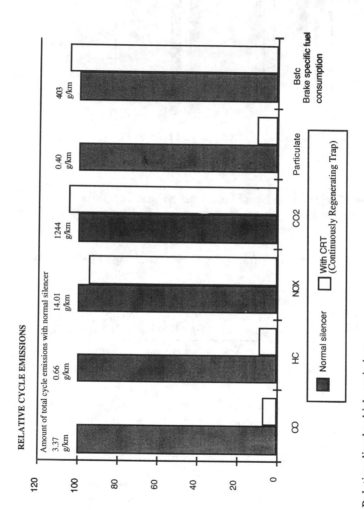

**Figure 40**   Relative diesel vehicle emissions.
*Source:* P. N. Hawker, 'Diesel emission control technology', reprinted from *Platinum Metals Review*, Vol. 39, No. 1.

200 kW engine, which meets the 1993 standards for particulates will, if it conforms to the average emission levels measured in the steady-state test by which its engine is certified for production, emit around 80 grams of particulates per hour, Therefore, a large fleet of vehicles, which may frequently be older, or may simply not conform to current type-approval standards, will emit a significant mass of particulate into urban environments. See Figure 40.

**Digester[1].** A sealed container or vessel in which ANAEROBIC digestion of ORGANIC MATERIALS can take place under temperature controlled conditions.

**Digester[2].** A heated pressure vessel in which wood chips are reacted with chemicals in the first stage of PULP production.

**Dilute and disperse.** A philosophy of liquid industrial WASTE disposal which relies on its absorption onto other waste materials (usually DOMESTIC REFUSE) and in the underlying strata especially any UNSATURATED ZONE. The liquid waste undergoes biological, chemical and physical attenuation (e.g. filtering and BOD reduction) during its downward percolation. When the attenuated liquid encounters an underlying AQUIFER, it is assumed that there has been both sufficient treatment before and subsequent dilution in the aquifer for any resulting pollution to be environmentally inconsequential.

The dilute-and-disperse philosophy has been strongly criticized by the EC on the grounds that there is no ultimate guarantee that environmental damage will not result. An alternative method is to seal the LANDFILL SITE with clay or a synthetic liner and contain the liquid waste within the receiving refuse mass and allow the attenuation processes to take place in the refuse mass alone. These processes may be aided by extraction of the liquids from the mass and spraying them on the landfill surface.

It is possible that the co-disposal of industrial and household waste will continue in the UK on the grounds that 'The UK's landfill methods are well researched and have not resulted in the disasters that have occurred elsewhere.' (*Toxic Waste – The Government's reply to the Second Report from the Environment Committee*, HMSO, April 1989). The term may also apply to air pollution dispersal.

**Dimethylether (DME).** A claimed replacement for diesel fuel in urban locations because of low PM10 and nitrogen oxide emission levels.

The production process developed by Danish company

Haloror, Topsoe, uses steam reforming of natural gas, followed by conversion from the reformed gas into DME using a proprietary CATALYST.

**Dioxins.** A family of chlorinated organic compounds, a major member of which is 2,4,7,8-tetrachlorodibenzo-*p*-dioxin (TCDD), which is a manufacturing impurity in certain classes of HERBICIDES, disinfectants and bleaches. Both TCDD and PCDD (polychlorinated dibenzo-*p*-dioxin) are formed (along with variants including FURANS) when compounds containing chlorine are burnt at low temperature in improperly operated/ designed domestic refuse and industrial waste incinerators where the PCDDs and TCDDs can be found in both the flue gases and the FLY ASH. (⟹ POLYVINYL CHLORIDE)

There are 210 dioxins and furans, 17 of which are toxicologically significant. 2,3,7,8-TCDD is taken as the reference standard and dioxin/furan emissions are toxicologically weighted to a TCDD or TE equivalent (TEQ). The sum of the individual TEQs is used in reporting dioxin/furan emissions.

Considerable public alarm on dioxin emissions (and near relatives) has been raised in many Western nations; usually these have been minute except in one or two cases of very poorly designed industrial incinerators. One expert has stated that spontaneous (or deliberate) fires in rubbish tips emit much more dioxins than controlled incineration. 'A 20 kg piece of chipboard impregnated with CHLOROPHENOL (a "glue") creates as much dioxin when it burns as an entire incineration plant in a whole month'. (*Warmer Bulletin*, No. 9, March 1986). 1988 figures for dioxin emissions from Swedish incineration plants (measured as TCDD equivalents) showed 0.5–2 ng/m$^3$ of dry flue gas. The emission requirement for new Swedish plants is being set at 0.1 ng/m$^3$, i.e. 1 g dioxins in at least 10 000 000 000 m$^3$ of flue gases emitted (STP conditions). This is the equivalent of a quarter of a standard 3 g sugar lump dissolved in Loch Ness (volume 7 000 000 000 m$^3$). Sweden incinerates 70 per cent of its domestic refuse with the energy used for DISTRICT HEATING and power generation.

Many old, poorly designed municipal solid waste (MSW) incinerators discharge dioxin at levels greater than 45 ng TEQ/Nm$^3$ (Normal m$^3$) and in 1989 all MSW incinerators then operating were estimated to contribute about one-fifth of the total known man-made releases to the environment.

But the dioxin release levels from modern state-of-the-art

energy-from-waste (EfW) plants are in the range 0.1–1.0 ng TEQ/Nm$^3$ and under new HMIP regulations which include dioxin limits for the first time, older plants which cannot meet this new standard must be closed by 1 December 1996, or modified so that they do meet them.

These standards imply a 98 per cent reduction in dioxin emissions from MSW combustion.

The RCEP (Royal Commission on Environmental Pollution) states that even with very large increases in the amount of MSW going to energy from waste, the total pollution load from modern plants under these new standards would not be a cause for concern. This reduction is made possible because the control of dioxins is now well understood.

*Dioxin control*

Dioxins (PCDD/PCDF) are principally formed in the 250–400°C temperature range in various combustion processes. They can however be formed in improperly controlled MSW incinerators as well.

However, compliance with HMIP requirements for combustion conditions of > 850°C for at least 2 seconds in at least 6 per cent O$_2$ with fabric filters, to clean the gases, ensures that discharge concentrations of 0.2 ng TEQ/Nm$^3$ may be obtained. Still lower levels can be achieved by the injection of activated carbon in the flue gas stream.

The typical state of the art in this area is reviewed by Kunzli & Haltinger, Pelling, and Fahlenhamp.

Some dioxin reformation may take place in the 200–350°C temperature range after combustion when the gases are cooled. This is the so-called 'de novo formation'. For this to take place, there must be unburnt carbon present and/or metal chlorides. As modern well-run incinerators produce approximately 3 per cent unburnt carbon and most of the heavy metals and particulates are trapped in the fabric filter, measured release rates of dioxins are approximately 0.2 ng TEQ/Nm$^3$.

Modern plants operate well within the new standards; independent monitoring of SELCHP in south-east London, the first of Britain's new generation of energy-from-waste plants, has shown dioxin emission levels well below the HMIP limit of 1 ng TEQ/Nm$^3$.

A modern 400 000 tonnes per year MSW energy-from-waste plant emits less than half a gram of dioxin per year. The calculation is shown below.

Summary of UK dioxin releases to air (g TEQ/year)

| Process category | Air emission (g TEQ/y) | Estimated quality | Comments |
|---|---|---|---|
| *Incineration* | | | |
| MSW and industrial waste | 400–700 | H/M | Largely old data, set to drop significantly (Author's note: set to drop to 8-14 I-TEQ/y by 1997, due to obsolete plant closure) |
| Hazardous (chemical) waste | 1.2–12 | M/M | |
| Clinical | 22.1–106 | H/M | Dropping significantly |
| Landfill gas | 1.64–5.46 | M/M | |
| Sewage sludge | 0.7–6 | H/H | |
| Crematoria | 1.05–35 | H/H | Set to drop |
| Cable and electro-motor | NC | | |
| Kraft black liquor boilers | NC | | |
| *Combustion* | | | |
| Oil | 0.5–6 | H/L | Little data |
| Waste oil | NC | M/M | Insufficient data |
| Coal | 16–113 | H/M | More data becoming available |
| Wood | 7.3–15.7 | M/M | |
| Straw | 5 | L/M | Little known on industry and variable emissions |
| Briquettes | NC | H/H | Probably N/A in UK |
| Tyres | 1.8 | H/H | |
| Cement kilns | 0.6–40.6 | H/H | |
| Lime manufacture | 0.03–0.12 | H/M | Steady state data suggests emissions at lower end of range |

| | | | |
|---|---|---|---|
| *Metal industry* | | | |
| Aluminium (secondary) | 9–50 | | Little data available |
| Copper | NC | | Little data available |
| Magnesium | NC | | Little data available |
| Nickel | NC | | Little data available |
| Tin | NC | | Little data available |
| Lead | 0.02–1.4 | | Little data available |
| Zinc | NC | | Little data available |
| Iron ore sintering processes | 90–2220 | M/L | Only overseas data, very variable |
| Steel manufacturing (using scrap) | 3–41 | H/L | |
| *Traffic* | | | |
| Leaded petrol combustion | 0.2–40.5 | H/L | Highly variable emissions data, no UK tests; but dropping with phase out |
| Unleaded petrol combustion | 0.06–0.86 | H/L | Highly variable emissions data, no UK tests |
| Diesel combustion | 23–273 | H/L | |
| *Different high temperature processes* | | | |
| Asphalt mixing | | | |
| Fires | | | |
| Forest/moor fires | 2–15 | L/L | Difficult to quantify |
| Accidental fires | NC | | Could be large |
| Domestic combust on | NC | | Very small |
| Cigarettes | NC | | |
| Use of wood preservatives | NC | | Could be large depending on fate of PCP |
| Secondary releases from chlorinated aromatic compounds (PCP) | | | |
| Re-entrainment of deposits | NC | | Might be large |

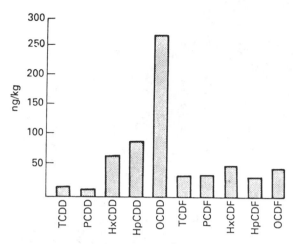

**Figure 41**   Mean dioxin soil concentrations, England and Wales.

*Dioxin emissions*
*Plant refuse*              Throughput 416 000 t/year
*Availability*              7446 h/year
*Dioxin emission*          0.1 ng/Nm$^3$ = 2 g/10$^{10}$ Nm$^3$
*Flue gas flow*            11% O$_2$ dry gas = 289 418 Nm$^3$/h

Plant dioxin emissions are therefore:

$$\frac{2 \times 289\,418 \times 7446}{10\,000\,000} = 0.431 \text{ g/year}$$

or $\frac{1}{16}$ of a 3 g sugar lump per year.

Modern MSW incineration plants destroy 80% of dioxins in the waste, based on input–output measurements, ie state of the art incinerators are dioxin sinks!

Sampling work based on a 20 km square grid in lowland Scotland has shown that PCDDs are ubiquitous in the environment at low levels and that they may be found in trace quantities in snails, human and animal fat, milk and other biological tissues. A summary of the soil concentrations of PCDDs and related PCDFs (FURANS) found in the soil is given in Figure 41. The limit of detection is 0.5 parts per trillion (0.5 ng/kg).

Although MSW incineration has been perceived to be a major source of human exposure to dioxin, measurements in the

US have shown that emissions from MSW incinerators at the national level account for less than 1 per cent of total current input into the environment.

Authors Travis and Blaylock (*Health Risks Associated with Air Emissions from MS Combustors*, American Chemical Society Conference, Washington DC, 21–26 August 1994.) report that MSW combustion, with motor vehicles, hospital waste incinerators, residential wood burning, pulp and paper mill effluent together account for a maximum of 11 per cent of total US TCDD output. Concentrations of dioxin cannot therefore be linked to any one combustion source. Other potential sources include discharges from metal processing and treatment plants, pentachlorophenol production, and coal, forest and landfill fires.

TCDD is known as the Seveso poison, because of an incident in which 2 kg of TCDD equivalent dioxins were released over a 20 minute period from a runaway chemical reaction. This severely contaminated approximately 4 km$^2$ of soil. (It was also present as a contaminant in the US Vietnam War herbicide Agent Orange.) There were no human fatalities from Seveso but several cases of CHLORACNE.

For the determination of the toxic equivalent (TEQ) value,

| Dioxin/furan | Equivalence factor |
|---|---|
| 2,3,7,8-tetrachloridibenzodioxin (TCDD) | 1 |
| 1,2,3,7,8-pentachloridibenzodioxin (PeCDD) | 0.5 |
| 1,2,3,4,7,8-hexachloridibenzodioxin (HxCDD) | 0.1 |
| 1,2,3,7,8,9-hexachloridibenzodioxin (HxCDD) | 0.1 |
| 1,2,3,4,6,7,8-hexachloridibenzodioxin (HxCDD) | 0.1 |
| Octachloridibenzodioxin (OCDD) | 0.1 |
| 2,3,7,8-tetrachloridibenzofuran (TCDF) | 0.1 |
| 2,3,4,7,8-pentachloridibenzofuran (PeCDF) | 0.5 |
| 1,2,3,7,8-pentachloridibenzofuran (PeCDF) | 0.05 |
| 1,2,3,4,7,8-hexachloridibenzofuran (HxCDF) | 0.1 |
| 1,2,3,7,8,9-hexachloridibenzofuran (HxCDF) | 0.1 |
| 1,2,3,6,7,8-hexachloridibenzofuran (HxCDF) | 0.1 |
| 2,3,4,6,7,8-hexachloridibenzofuran (HxCDF) | 0.1 |
| 1,2,3,4,6,7,8-heptachloridibenzofuran (HpCDF) | 0.01 |
| 1,2,3,4,7,8,9-heptachloridibenzofuran (HpCDF) | 0.01 |
| Octachloridibenzofuran (OCDF) | 0.001 |

PCDD is shorthand for polychlorinated dibenzo-*p*-dioxin. PCDF is shorthand for polychlorinated dibenzofuran.

stated as a release limit, the mass concentrations of the following dioxins and furans have to be multiplied with their equivalence factors before summing.

It is certainly true to say that not enough is known about dioxins and world-wide scientific research effort is continuing.

Both vinyl chloride monomer (VCM) manufacture and PVC waste incineration can give rise to the formation of dioxins. Modern VCM plants produce only very small quantities and emit much less to the environment. As an example the Norsk Hydro plant in Norway makes 450 000 tonnes of VCM each year in the course of which 6 g of dioxins are made and 0.42 g escape to atmosphere or the fjord. Measurements in the USA by GEON detected no dioxins in PVC products. In the UK, the MAFF detected traces towards the limit of detection which the Ministry says presents no hazard to the public.

*Source: United Kingdom Comments on the United States Environmental Protection Agency's External Review Draft Reassessment of Dioxins*, Department of the Environment, Toxic Substances Division, January 1995

**Direct energy conversion.** The conversion of chemical, solar or NUCLEAR ENERGY directly into electricity without producing mechanical work in the process (as in the convention boiler-steam turbine-generator system). Direct energy conversion is a very desirable goal since it means that electricity could be made without intermediate equipment and perhaps with greater efficiency.

The classes of direct energy conversion devices are: FUEL CELLS, magnetohydrodynamic generators, THERMIONIC CONVERTERS, and semiconductor THERMOELECTRIC CONVERTERS.

The First Law of Thermodynamics applies to all the above devices as does the Second Law in its general form, but the CARNOT EFFICIENCY restrictions do not apply to all direct energy converters. Thus, in theory, direct energy conversion could offer very attractive conversion efficiencies when realized.

**Directive, EC.** A form of binding European Community legislation either addressed to all Member States or sometimes only to specified ones. A directive states an objective and requires the addressee(s) to take such measures as are necessary to achieve the aim desired by the Community within a stated period. The method by which the aim is achieved depends on the addressee; it may mean new domestic law or the recasting or repealing of existing domestic legal and administrative rules.

**Discharge consent.** An authorization to discharge a liquid effluent to a receiving water course or to a sewer, provided by the UK National Rivers Authority. The consent will normally specify volumetric flow rate, BIOCHEMICAL OXYGEN DEMAND and/or CHEMICAL OXYGEN DEMAND, SUSPENDED SOLIDS content. Other parameters such as pH or TEMPERATURE may also be specified. ($\Rightarrow$ DISCHARGE STANDARDS)

**Discharge Standards (Effluents).** The Eighth Report of the Royal Commission on Sewage Disposal (1912) laid down standards for sewage effluents that are still the UK norm today. Basically these are 30 mg/1 SUSPENDED SOLIDS (SS) and 20 mg/1 BIO-CHEMICAL OXYGEN DEMAND (BOD), the so-called ROYAL COMMISSION STANDARDS. This assumes that the receiving stream can provide a dilution of at least 8 times the volume of effluent. This standard is often adhered to today, except for discharges to non-tidal streams and/or where dilution is much greater than 8:1.

Where potable water is abstracted downstream of the effluent discharge stricter standards may apply and a limit of not more than 10 mg/1 ammoniacal nitrogen may also be imposed. The trend is for a 10:10:10 standard (SS:BOD:ammoniacal nitrogen) in areas where rivers are used as a source of potable water.

**Discounted cash flow (DCF).** A common technique used to evaluate the relative costs of proposed schemes.

Knowing the rate of interest available, we can calculate what value any given sum will have at some future date. Conversely, if we know that, at some time in the future, we are going to have to spend a given sum, then we can easily work backwards to determine the present sum that must be put aside for such a future project. This process is called 'discounting'.

Interest rates gives us the amount by which an investment will grow annually. In discounting, the term 'discount rate' is used to signify the annual rate of interest at which the present value of the required future sum would have to grow over the intervening period in order to become that future sum.

In appraising projects, the term 'rate of return' is also used to indicate the return on investment that the project will generate.

The construction of any project involves the payment of cash at intervals, and its subsequent operation invariably involves further periodic payments (salaries, material, maintenance, etc.). Expressed on an annual basis, such a stream of payments represents a cash flow (outwards in this case, but inwards if you

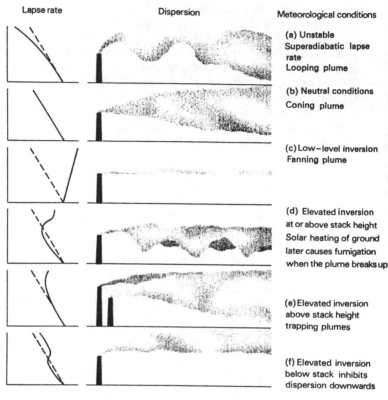

Lapse rate                    Dispersion                    Meteorological conditions

(a) Unstable
Superadiabatic lapse
rate
Looping plume

(b) Neutral conditions
Coning plume

(c) Low–level inversion
Fanning plume

(d) Elevated inversion
at or above stack height
Solar heating of ground
later causes fumigation
when the plume breaks up

(e) Elevated inversion
above stack height
trapping plumes

(f) Elevated inversion
below stack inhibits
dispersion downwards

**Figure 42** Some of the principal types of plume dispersion.

are calculating benefits, e.g. payments for goods or services
resulting from the project).

Now, just as we can calculate the present value of a single
future sum, so also can we obtain the present value of a series of
future sums (costs of benefits), and such a procedure is called
*discounted cash flow*. It is used to evaluate, on purely monetary
grounds, the relative merits of various alternative projects or
alternative ways of implementing a single project. A common
application is in *least-cost analysis*. For example, in laying on a
needed water supply, do we use a large pipe and let the water
flow by gravity (high initial capital investment, low running
costs) or a smaller pipe in association with pumps (lower capital
investment, higher running costs)? Using discounted cash-flow
methods, we can calculate which of the two schemes will cost

least in terms of the present value of the sums to be committed. *Note:* It is very difficult to incorporate environment matters into DCF calculations. (It is customary, in such calculations, to ignore costs that will be common to both schemes, such as pipeline maintenance, repairs, metering, etc.)

**Dispersion.** The dilution and reduction of concentration of pollutants in either air or water. Air pollution dispersion mechanisms are a function of the prevailing meteorological conditions. Figure 42 shows several modes of plume dispersion.

Note that the ADIABATIC temperature profile is shown in solid for the saturated (i.e. wet air) adiabatic LAPSE RATE and dotted for dry adiabatic lapse rate. The dry adiabatic lapse rate is used as a basis for comparison only. The saturated adiabatic lapse rate usually determines the pattern of dispersion. (⇨ INVERSION LAYER; ADIABATIC; LAPSE RATE)

**Disposal Levy.** Levy charged on first purchase of an artefact to cover its eventual disposal. The levy is refundable only at accredited points of disposal where recycling or environmentally sound disposal can be effected.

**Dissociation.** The breakdown of a substance into simpler substances owing to the addition of energy (e.g. heat) or the effect of a solvent.

**Dissolved oxygen (DO).** The amount of oxygen dissolved in a stream, river or lake is an indication of the degree of health of the stream and its ability to support a balanced aquatic ecosystem. The oxygen comes from the ATMOSPHERE by solution and from PHOTOSYNTHESIS of water plants. The maximum amount of oxygen that can be held in solution in a stream is termed the *saturation concentration* and, as it is a function of temperature, the greater the temperature, the less the saturation amount. It is customary to express oxygen concentrations as percentages of saturation concentration at any given temperature. The saturation concentration is also a function of atmospheric pressure and concentration of dissolved salts. As a result of the latter, sea water contains less oxygen than fresh water at the same temperature and saturation conditions.

The discharge of an organic waste to a stream, e.g. sewage, imposes an oxygen demand on the stream. If there is an excessive amount of organic matter, the oxidation of the waste by micro-organisms will consume oxygen more rapidly than it can be replenished. When this happens, the dissolved oxygen is depleted and results in the death of the higher forms of life. In extreme cases the river becomes anaerobic, the oxidation processes cease and noxious odours are given off. (⇨ ANAEROBIC PROCESS)

The oxygen balance of a river is of great importance in controlling the effects of pollution. As soon as organic matter enters the water it exerts a BOD, the dissolved oxygen level falls and as a result an oxygen deficit is created, which means that oxygen transfer from the air to the water take place to try to restore saturation. The rate of oxygen transfer, i.e. the re-aeration rate, depends on the nature of the flow, the stream-bed characteristics, and the water temperature. A mountain stream will have a re-aeration rate many times greater than a stagnant quarry pond, and would thus be able to assimilate a much greater BOD load per unit volume without a serious oxygen deficit.

As the amount of oxygen consumed by organisms in the water increases, the oxygen deficit also increases and in turn the re-aeration rate increases until (if the pollution does not swamp the watercourse oxygen completely) the rates of deoxygenation and re-aeration eventually balance. From then on the re-aeration supplies more oxygen than is consumed by the BOD and the stream eventually recovers to saturation.

This process can be represented by an OXYGEN SAG CURVE, which is a graph of dissolved oxygen against distance (Figure 43). With moderate pollution the dissolved oxygen levels are

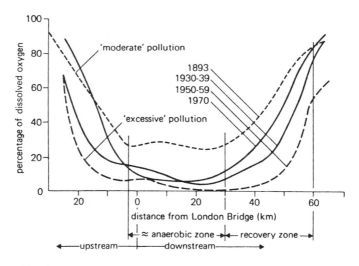

**Figure 43**   Oxygen sag curves for the river Thames.

high enough to support fish life, while with high pollution the stream becomes anaerobic and the fish life disappears, and perhaps the only signs of life are tubificid worms (an indicator of low or zero dissolved oxygen). (⇨ OXYGEN DEMAND)

**Distillation.** A process by which mixtures of liquids are heated and evaporated. Each vapour then condenses back to a liquid at a characteristic temperature and can be separated. Dissolved solids are left behind.

Distillation is a principal DESALINATION process for the production of pure water from saline water. Note that distillation is a two-stage process: first, evaporation and then condensation of the vapour. In the first stage the latent heat of evaporation has to be supplied to the liquid, and in the second stage the latent heat of condensation (virtually identical to the latent heat of evaporation) has to be removed from the vapour for it to condense.

**District heating.** The use of a large, efficient, centralized boiler plant or 'waste' steam from a power station to heat a district and/or supply steam to small industries. The heat is distributed by means of low-pressure steam or high-temperature water mains to the consumers. (⇨ ENERGY)

As the boiler plant is centralized, it combines greater utilization of fuel compared with a myriad of small domestic or industrial

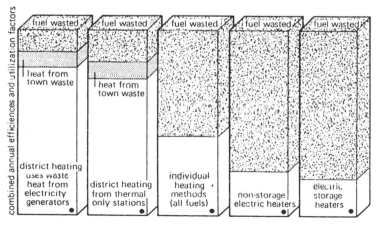

• includes mains losses

**Figure 44** Primary fuel utilization factors for the alternative heating methods.

appliances, together with a considerable reduction in low-level pollution compared with, say, open coal fires. If steam from a power station is used, it gives a much higher thermal efficiency for the system.

District heating has much to commend it in that it leads to much more efficient fuel use (see Figure 44). However, the cost per unit of energy supplied must be competitive with that of alternative fuels, unless subsidies are available to reflect the increased energy efficiency of district heating schemes. This requirement means that the customers must live very close to the boiler plant or power station, otherwise the cost of distribution makes the energy selling price prohibitively expensive. (⇨ COMBINED HEAT & POWER)

**DNA (deoxyribonucleic acid).** An essential compound found in the nucleus of living cells. DNA is a long-chain molecule, which contains the genetic codes necessary for the development and functioning of the human organism. It controls the formation of PROTEIN and ENZYMES. Carbon atoms make up 37 per cent of DNA, and in exposures to IONIZING RADIATION it is entirely possible that a RADIOACTIVE ISOTOPE known as carbon-14 may become incorporated in the DNA structure, thus causing a random change which may have genetic consequences.

Small changes in the complex DNA structure can lead to genetic disorders such as haemophilia, or mental and physical deformity. (⇨ PHTHALATES – DOA)

**Dobson unit (DU).** $1\,DU = 2.7 \times 10^{16}$ molecules per square centimetre.

This unit is used to measure OZONE levels and is the average value of the total ozone column above $1\,cm^2$ of the Earth's surface. It contains no information about the vertical distribution of ozone, only the total number of molecules in the column.

**Dolomite.** A natural calcium magnesium carbonate. It has been proposed as a source of alkali for various methods of removing sulphur oxides from stack gases.

**Domestic refuse (UK); Garbage (USA).** The generic name for waste emanating from households. In the UK, is has an average percentage composition as follows:

| Paper and cardboard | 35% | Dust and ashes | 10% |
|---|---|---|---|
| Vegetable and foodstuffs | 25% | Glass | 8% |
| Plastics | 10% | Metals | 8% |
| Rags | 2% | Unclassified | 2% |

In other words, it has approximately 65 per cent ORGANIC MATTER and 35 per cent INORGANIC MATTER. The organic portion is suitable for PYROLYSIS, HYDROLYSIS, COMPOSTING, or INCINERATION. The inorganics are sources of metals and glass CULLET, while the ash and cinders may also be combustible. The UK (1988) production was 18 million tonnes or approximately 1 tonne per day per 1000 population.

95 per cent of UK domestic refuse is disposed of by LANDFILL where steps must be taken to preserve public health by the prevention of pollution of groundwater, the breeding of flies and rats, and unsightly heaps or offensive odours. Wind-blown litter must also be controlled. Tip space is running out and INCINER-ATION is expected to gradually replace tipping. The adoption of RECYCLING via separate waste collection or at processing centres for materials recovery is expected to grow. The UK has a stated goal of 50 per cent recovery of recyclable materials by the year 2000.

**Domestic waste.** ⇨ DOMESTIC REFUSE; WASTES.

**Dose.** The quantity of a substance or the amount of energy either in a single application or experienced over an interval of time (i.e. the product of the dose rate × time) which produces some specific effect. If the effect is death, this gives rise to the expression $LD_{50}$ (lethal dose), which is the dose large enough to kill 50 per cent of a sample of animals under test. $LD_{50}$ values are normally quoted as mg/kg, but it should be realized that the kg refers to body weight of the animal concerned. (⇨ HALF-LIFE; MERCURY; LETHAL CONCENTRATION; TIME WEIGHTED AVER-AGE). The toxicity of chemicals is usually classified as very toxic, toxic, harmful, and not classified. An $LD_{50}$ greater than 7 g/kg weight is not classified as harmful.

'The right dose differentiates a poison and a remedy' (Paracelsus (1493–1541).

**Downdraught.** The low-pressure region created on the lee side when wind blows over a building. Emissions from stacks that terminate within the wind generated turbulent region are drawn down to ground level in the low-pressure region, giving rise to intense pollution. A general rule for preventing this type of downdraught is that a stack should be at least 2.5 times the height, *h*, of any building situated within a circle of radius 2*h* around the stack.

**Droplet.** A particle of liquid substance of very small mass (up to 20 µm diameter), capable of remaining in suspension in a gas.

**Dry solids.** This is the common basis for measuring the dry matter content of sludges such as sewage sludge and involves drying the sludge, driving off all water and weighing the dry matter left behind. ($\Rightarrow$ DRY WEIGHT).

**Dry weather flow.** The baseline used for sewer design is the average daily sewage flow (dwf) to the effluent treatment works during seven consecutive days without rain. The winter and summer dry weather flows should preferably be obtained to arrive at an average dry weather flow. Storm overflows on sewers are constructed so that they come into operation when the flow exceeds 6 × DWF (DWF denotes daily *peak* flow). Where trade effluents are discharged to sewers, then design DWF is sometimes set at twice the flow between 8 a.m. and 8 p.m. When the overflows come into operation during severe storms, in theory at least, the untreated sewage effluent is diluted six times and this can (presumably) be discharged untreated to a receiving watercourse.

**Dry weight.** The basis used for expressing the concentration of compound in living organisms. For example, fish have a moisture content of 80 per cent, so that a concentration of 5 parts per million dry weight would only be 1 part per million wet weight in fish.

**Duales System Deutschland (DSD).** To reduce German recycling costs, a number of corporations from the retail, packing and filling industries, the producers of PACKAGING material, and the raw material suppliers for the packaging industry founded the Duales System Deutschland (DSD) GmbH in September 1990. The DSD organizes a private waste management network which assures that all primary packaging waste is collected and sorted out by local waste management firms and subsequently reprocessed by domestic or foreign RECYCLING firms.

Objectives of the German Packaging Decree are:

● A radical reduction in volume of packaging by avoiding use and stressing the value of reclamation.
● Manufacturers and retailers will be obliged to take responsibility for the packaging used as they are the originators.
● The local communities will be relieved from this part of waste disposal responsibility.
● Returnable packaging will be widely encouraged.
● Recycling will be given absolute priority over thermal recovery.

The DSD uses a 'green dot' to identify the product as participating in the system. In order to meet the requirements stated by the Verpackungsveer-ordnung (VVO) the DSD grants the 'green

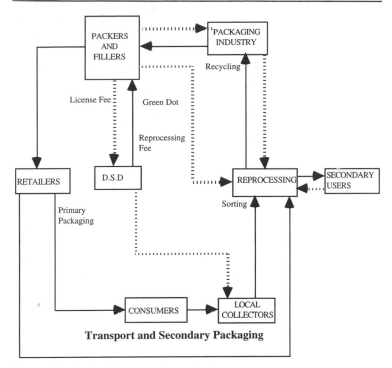

**Figure 45** Packaging waste flows in the DSD.

dot' only to those packers and fillers (or respectively packaging producers) who can present a reprocessing guarantee for the amount and type of packaging which they want to use. As the reprocessing capacities were insufficient, the producers of packaging and packaging materials set up their own one for each type of packaging material. These reprocessing companies issue global reprocessing guarantees for the respective materials. Figure 45 illustrates the packaging waste flows in the DSD.

The reprocessing as well as the collection and sorting activities are costly. Hence, the DSD finances collection and sorting costs through licence fees which have to be paid by each packer or filler who wants to use the 'green dot'. Initially, the licence fees varied according to the volume of the container, not according to the material in question, but in October 1993 the DSD introduced new charges based on the actual cost of collecting and sorting the different materials.

The reprocessing costs are financed differently. For some materials, reprocessing is 'profitable' as the materials sales revenues cover the costs of reprocessing. For the 'non-profitable' materials, the reprocessing costs (net of resale revenues) are

| | |
|---|---|
| Glass | 0.16 DM/kg |
| Tinplate | 0.56 DM/kg |
| Aluminium | 1.00 DM/kg |
| Paper/cardboard | 0.33 DM/kg |
| Plastic | 3.00 DM/kg |
| Compound package | 1.66 DM/kg |

*Source:* Duales System Deutschland GmbH.

allocated among the respective packaging producers. An exception is the case of plastics where the licence fee charged by the DSD includes not only the cost of collection and sorting but also the cost of reprocessing. Hence, the costs of plastics reprocessing are carried by the packers and fillers.

The current version of the EU Directive has a 50–65 per cent recovery of packaging waste within 5 years of implementation (this may be $1\frac{1}{2}$ years after adoption which means $6\frac{1}{2}$ years time-scale and allows a global recycling target of 25–45 per cent within 5 years of implementation with a minimum recycling level of 15 per cent for plastics). All is still to play for, as this is still very much in the 'melting pot', and if Germany complies with it, the reduced quotas shown in the table below will probably apply, on the assumption that German domestic reprocessing capacities will be built up by July 1995. (The EC

Recycling ratios required by the planned amendment to the VVO

| | *Recycling ratios (%) starting on* | | |
|---|---|---|---|
| | *1 Jan. 1993* | *1 Jan. 1996* | *1 Jan. 1998* |
| Glass | 40 | 70 | 70 |
| Tinplate | 30 | 70 | 70 |
| Aluminium | 20 | 70 | 70 |
| Cardboard | 20 | 50 | 60 |
| Paper | 10 | 50 | 60 |
| Plastic* | 10 | 50 | 60 |
| Compound | 10 | 50 | 60 |

*Up to half of the required ratio can be accomplished by energy recovery.
*Source:* Entwurf zur Novellierung der Verpackungsverordnung, December 1993.

target can be exceeded provided domestic reprocessing capacities are available, i.e. the waste is not exported.)

There still remains one 'bone of contention': the role of energy recovery, as it is, is a major gulf between the EU Directive and the proposed amendment of the VVO. The draft EU Directive implies that not less than 10 per cent and not more than 40 per cent of packaging waste can be recycled by energy recovery. The proposed VVO amendment only allows energy recovery in the case of plastics of which 50 per cent are to be recycled by feedstock recycling technology, i.e. converting the plastic back to petrochemical feedstocks, melting or pelletizing. The other 50 per cent may be recycled by energy recovery.

Michaelis has commented that, from an economic point of view such a modification would be desirable since it would reduce the huge amount of excess costs which are presently caused by the strict requirements of materials recycling. He gives the following comments on costs.

> At present, the difference in net-of-revenue-cost between material recycling and energy recovery amounts to about 2.00–6.00 DM per kg of plastic waste. Of course, this cost comparison does not allow for a complete cost-benefit assessment since it neglects the different environmental effects of materials recycling and energy recovery. But in view of the tight German emission standards for waste incineration, it seems to be hardly imaginable that there may exist net environmental benefits from materials recycling that are high enough to outweigh the above mentioned difference in net-of-revenue-cost.

In other words, waste to energy can have a lower environmental impact than materials recycling. Yet, the Germans are set against it.

It is clear that there is still a substantial gulf between Germany and the EU and the worry is that national packaging legislation could still negate the benefits of the single market. The Germans vigorously defend their Duales System Deutschland. Mr Clemens Stroetmann, State Secretary of the German Environment Ministry stated that 'valuable raw materials are collected, assisted, financed by ordinary citizens either by refuse charge by the Municipal Authorities or in the price of the products they buy'. This, of course, also reflects the cost of the green dot (DSD). This financing, however, is for actual services and the setting and operation of collection systems. Stroetmann denied that the export of valuable raw waste materials was subsidized either by the Municipal Authorities or by the DSD.

*Sources:* P. Michaelis, 'Innovation in the management of packaging waste – The Dual System in Germany', *Packaging Waste Flows in the Dual System*, Conference Proceedings, Institute of Wastes Management Conference, Torbay, June 1994.
A. Maitland, 'Germany rebuts criticism of its waste recycling laws', *Financial Times*, 19 October 1993.

**Dumping at sea.** The disposal of authorized wastes (licensed in the UK by MAFF) to sea such as sewage SLUDGE, fly ash from coal-fired power stations, and chemical industry wastes has a long history in the UK. Currently, 30 per cent of UK sewage sludge, amounting to 6 million tonnes (95 per cent moisture content) plus 550 000 tonnes of fly ash annually is disposed of in this fashion. However, sludge dumping is being phased out by 1998.

This still puts the UK out of step, as it is the only signatory to the 1987 North Sea Conference which still supports this method of waste disposal. The official reasoning is that scientific proof of damage to the environment is required before action is deemed to be necessary. This is at odds with the precautionary policy of the EC, which states that by the time zthe damage is shown to exist, it could be too late to take effective remedial measures. (⇨ WASTES)

**Dust.** Often called airborne particulate matter; solids suspended in air as a result of the disintegration of matter. This embraces a wide range of particles but dust is normally taken as between 1 and 76 $\mu$m in diameter (above 76 $\mu$m is termed GRIT). Particles above 10 $\mu$m in diameter are not carried far from their source except by strong winds. The presence of dust particles in the ATMOSPHERE could cause either a net cooling or a net warming, depending on the properties of the particles and the underlying surface. It is commonly supposed that the presence of atmospheric dust will cause the earth to cool, but as well as reflecting radiation from the sun, it will also reflect heat released from earth. Thus particles with a 'white' or 'grey' upper surface and 'black' lower surface would cause cooling of the earth; if this were reversed, warming of the earth would result.

Calculations show that the aerosol extinction coefficient – a measure of how much heat the particles absorb – has a critical value at which there is heating or cooling of the earth's surface. There is a balance in temperature on surfaces that have ALBEDOS in the range 0.35 to 0.60. For albedos greater than 0.6 there is heating; less than 0.35 there is cooling.

Atmospheric science has a long way to go, but the use of the atmosphere as a sink for pollutants should clearly not be encouraged.

### Dust concentration in gases

This is given in terms of milligrams per cubic metre of gas or grains per cubic foot ($2300 \, mg/m^3$ equals $1 \, gr/ft^3$). The concentrations of dust in the gases from the many dust-producing processes range from less than $2000 \, mg/m^3$ to more than $400 \, g/m^3$.

The dust concentration in the cleaned gas from the dust extraction equipment is, however, the really important figure and this can be predicted for any specific duty by relating the grade efficiency curve of the proposed equipment to the particle size distribution of the dust and the inlet concentration.

The table below gives the size range of three typical industrial dusts classified as coarse, fine and ultra-fine respectively.

The table on page 158 gives collection efficiencies for the three typical dusts referred to in the table below for various types of gas-cleaning equipment.

As an example of dust control, British Coal monitors the total dust deposit (including any potential coal content) at every operational location. Monitoring is undertaken using the British

| British Standard sieve no. | Particle size ($\mu m$) | % by weight less than size* | | |
|---|---|---|---|---|
| | | Coarse (CD) | Fine (FD) | Ultra-fine (UFD) |
| 100 (0.006 in.) | 152 | — | 100 | — |
| 150 (0.004 in.) | 104 | — | 97 | — |
| 200 (0.003 in.) | 76 | 46 | 90 | 100 |
| | 60 | 40 | 80 | 99 |
| | 40 | 32 | 65 | 97 |
| | 30 | 27 | 55 | 96 |
| | 20 | 21 | 45 | 95 |
| | 15 | 16 | 38 | 94 |
| | 10 | 12 | 30 | 90 |
| | 7.5 | 9 | 26 | 85 |
| | 5.0 | 6 | 20 | 75 |
| | 2.5 | 3 | 12 | 56 |

*CD = Coarse dust (21% below $20 \, \mu m$); FD = fine dust (45% below $20 \, \mu m$); UFD = ultra-fine dust (95% below $20 \, \mu m$).
From R. Ashman, 'Air pollution control', in A. Porteous (ed.), *Developments in Environmental Control and Public Health*, Applied Science Publishers, 1979.

| Type of equipment | Collection efficiency (%) for three typical industrial dusts | | | Approximate power requirements (kW) for a flow rate of $10^5$ $m^3/h$ |
|---|---|---|---|---|
| | Coarse | Fine | Ultra-fine | |
| Medium efficiency cyclones | 84.6 | 65.3 | 22.4 | 40 |
| High efficiency cyclones | 93.9 | 84.2 | 52.3 | 60 |
| Self-induced spray scrubber | 97.6 | 92.3 | 70.3 | 70 |
| Pressure spray scrubber (medium to high energy) | 99.9 | 99.7 | 99.5 | 150–300 |
| Electrostatic precipitator | 99.5 | 98.5 | 94.8 | 40 |
| High efficiency electrostatic precipitator | 99.96 | 99.85 | 99.35 | 50 |
| Venturi-scrubber (medium to high energy) | 99.97 | 99.9 | 99.6 | 200–370 |
| Shaker-type filter | 99.97 | 99.92 | 99.6 | 50 |
| Reverse jet filter (blow ring) | 99.98 | 99.95 | 99.8 | 130 |
| Reverse jet filter (nozzle) | 99.98 | 99.95 | 99.8 | 150 |
| Reverse pressure filter | 99.98 | 99.95 | 99.8 | 110 |

*Source:* British Coal Opencast Northern, Environmental Review 1993.

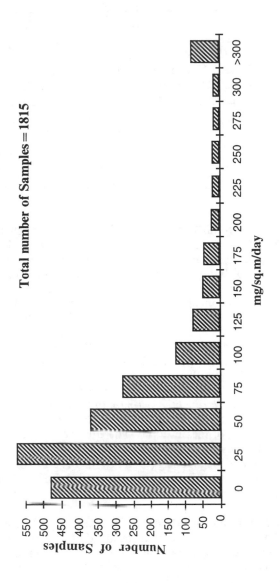

**Figure 46** Monitoring of total dust deposits.

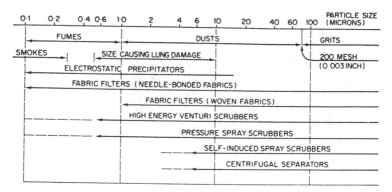

**Figure 47**   Characteristics of gas dispersoids and methods of control.

Standard directional dust gauge, 43 of which are currently installed; 94.5 per cent of their results were below the level of 200 mg/m$^2$/day. The samples above 200 mg/m$^2$/day were short-term incidents at Bates Staithes, Chester House and Keekle Ext. which have been addressed and remedial action taken.

**Duty of care[1].** Legal concept enshrined in pollution and Health & Safety legislation, e.g. Mines & Quarries Act 1954, which places a duty on, for example, a mine manager to take all possible steps to solve foreseeable dangers and apply all the skill and care a manager should use to obviate the danger. (*Mettam* v. *National Coal Board*, 12 July 1988).

Breach of duty of care can constitute negligence (e.g. *Gertsen* v. *Municipality of Metropolitan Toronto*, 1973) in which LANDFILL GAS escaped from a LANDFILL SITE onto adjoining land causing an explosion in garages in 1969. The authorities were found liable for damage and injury caused by the explosion by virtue of NUISANCE and negligence. (⇨ *RYLANDS* V. *FLETCHER*)

Claims based on negligence may have substance if (Hawkins 'A duty not to be dumped', *Surveyor*, 15 September 1988):

1. the defendant has a duty of care towards the plaintiff;
2. there has been a breach of that standard of care imposed, e.g. a failure to act reasonably in a situation where reasonable care is required (this will depend, like a conviction for careless driving, on the facts of the case);
3. damage has been caused which was a reasonably foreseeable consequence of the breach of that duty of care.

In the *Gertsen* case, the defendants were held to have been negligent in: (i) burying the garbage as they did when they knew, or ought to have known, that its decomposition resulted in the production of gas; (ii) in failing to take steps to prevent the escape of gas; and (iii) in failing to warn adjoining owners of the risk, i.e. they had a duty not to cause injury to persons or property.

Hawkins R.G.P., Surveyor, 1988, states that in his opinion, a landfill operator in the UK would have a duty of care to adjoining owners. He also drew attention to *Miller* v. *South of Scotland Electricity Board* (1958):

> A person doing work on land owes a duty of care to all persons who are so closely and directly affected by his work that he ought reasonably to have them in contemplation when he is directing his mind to the task.

A recent UK law case has modified the above ruling. It appears that now the use of the land in question has to be 'non-natural' for strict LIABILITY to apply.

**Duty of care[2].** This is applicable under Section 34 Part II of the UK Environmental Protection Act to 'those who import, produce, carry, keep, treat or dispose of controlled waste or, as brokers, have control of such waste'.

Those subject to the duty of care must take all reasonable measures to achieve the goals summarized below:

(a) to prevent

 - the deposit of controlled waste without a waste management licence or in breach of a licence;
 - the treatment, keeping, or disposal of controlled waste except as licensed in or on any land or by means of mobile plant,
 - the treatment, keeping or disposal of controlled waste in a manner likely to cause pollution of the environment or harm to human health;

(b) to prevent the escape of the waste from their control or that of any other person;

(c) on the transfer of the waste to secure that the transfer is only to an authorized person or to a person for authorized transit; there are six categories of person authorized to receive controlled waste, set out in Section 34(3) of the Act;

(d) on the transfer of the waste to secure that there is transferred such a written description of the waste as will enable each person

- to avoid a contravention of Section 33 of the Act; and
- to comply with the duty at (b) to prevent the escape of waste.

In addition to these aspects, those subject to the Duty must comply with regulations under Section 34(1) as respects the production and retention of documents, that is, they must make and keep records of waste as required by the regulations. They may also be required to furnish documents or copies of documents. Any person who fails to comply with (a)–(d) above is liable to be prosecuted.

Implementation of the Code of Practice will ensure a much needed improvement in the UK's previously lax waste management procedures. Already, responsible waste disposers are implementing them.

An indication of the high level of care exercised by US producers of hazardous wastes is given in the following suggested checklists to be employed before disposing of their waste.

---

**For a landfill:**
1. Are there indications of improper drainage or ponding of water?
2. Is run-on diverted away from all portions of the landfill?
3. Is run-off from active portions of the landfill collected and treated?
4. Is waste that is subject to wind dispersal controlled?
5. Does the disposal operation occur within 200 ft of the property boundary?
6. Are untreated, ignitable or reactive wastes placed in the landfill?
7. Are incompatible wastes placed in the same landfill cell?
8. Are bulk and non-containerized liquid wastes landfilled?
9. Does the landfill have a liner that is chemically and physically resistant to liquids, and a functioning leachate-collection and -removal system with sufficient capacity to remove all leachate produced?
10. Are containers holding liquid waste, or waste containing free liquids, placed in this landfill?
11. Does the operator maintain a map with the exact location and dimensions, including depth, of each cell with respect to surveyed benchmarks?
12. Does the map show the contents of each cell and the approximate location of each hazardous waste type within the cell?

---

**For an incinerator:**
1. What type of incinerator is this (waterwall, boiler, fluidized bed, etc.)?
2. Is the residue from the incinerator a hazardous waste?
3. Does the operating record contain analyses listing the types of pollutants that might be emitted, including the heating value of the waste, halogen and sulphur content, and concentrations of lead and mercury?
4. Does the incinerator have instruments for measuring: waste feedrate, auxiliary-fuel feedrate, air flowrate, operating temperature, scrubber flowrate, and scrubber pH?
5. Are instruments relating to combustion and emission controls monitored every 15 minutes?
6. Is the stack plume observed visually at least hourly for opacity and colour?
7. Are there any signs of leaks, spills or fugitive emissions from pumps, valves, conveyors, pipes, etc.?
8. Are all emergency-shutdown controls and system alarms checked to ensure proper operation?
9. Is the incinerator inspected daily?
10. Is there open burning of hazardous waste?

**For a chemical, physical or biological treatment facility:**
1. Does the treatment system show any signs of ruptures, leaks or corrosion?
2. Is there a means to stop the inflow of continuously fed wastes?
3. Is the discharge-control safety equipment (e.g. waste-feed cutoff, bypass drainage, and pressure-relief systems) inspected daily?
4. Are data gathered for monitoring equipment at least once each day?
5. Are construction materials of the treatment process inspected at least weekly to detect corrosion or leaking of fixtures and seam?
6. Are the discharge confinement structures (e.g. dikes) immediately surrounding the treatment unit inspected at least weekly to detect erosion or obvious signs of leakage (e.g. wet spots or dead vegetation)?
7. Are ignitable or reactive wastes fed into the system treated or protected from any material or conditions that may cause them to ignite or react?
8. Are incompatible wastes placed in the same treatment process?

*Source:* Denise Braker Curtis, 'waste treatment – better safe than sorry', *Chemical Engineering*, May 23 1988.

(⇨ CONTROLLED WASTE; WASTES)

# E

**E. coli (Escherichia coli).** An organism of the coliform group which inhabits the human and animal intestine. If this is absent, water may be passed as safe even if a few other types of COLIFORMS are present: but no coliforms should be present in water which has been chlorinated, i.e. disinfected with CHLORINE.

**Earth Resources Technology Satellites (ERTS).** Orbiting laboratories launched by the National Aeronautics and Space Administration of the USA to monitor natural resources for man. The ERTS are designed to provide systematic repetitive global land coverage. They complete 14 orbits per day photographing three strips 185 kilometres wide in North America, and 11 strips in the rest of the world.

The orbit has been designed in conjunction with the photographic settings to pass over any location on the earth's surface once every 18 days. Thus, the repetition allows changes in the earth's surface features to be monitored over time. The possibilities for monitoring ocean dumping, changes in land use, jungle clearance, waste tips, etc. are endless. The ERTS also have great potential for monitoring crop health and identifying potentially hazardous situations before they develop, and so allow remedial action to be taken.

**Eco Emballages.** The system set up in France for PACKAGING WASTE VALORIZATION. This is a very pragmatic approach in comparison to the highly structured German DSD system.

France has the ambitious target of gradually valorizing (adding value by materials recovery/waste to energy/composting) 75 per cent of household packaging by 2002.

Decree No. 92-377 of 1 April 1992:

> Any producer, any importer . . . or any person responsible for first placing products sold in packaging on the market shall be required to contribute or to provide for the disposal of all his packaging waste . . .

The Decree is a model of verbal economy, the key Articles 2, 3, 4,

5 and 10 are reproduced below. (Regulation No. 92-377/1 April 1992, related to the application of Act No. 75-633/15 July 1975 Republic of France as amended, on the disposal of wastes and recovery of materials from wastes resulting from the discarding of packaging, 1 April 1992.)

**Article 2**  For this Regulation:
- packaging means any container or material used to contain a product, to facilitate its transport or its presentation for sale;
- producer means any person who, for commercial purposes, packages or subcontracts the packaging of its products for further marketing;
- end holder of a packaging means any person that separates the packaging from the product in order to use or to consume such product.

**Article 3**  This regulation applies to the disposal of wastes resulting from the discarding of packaging used to market products consumed or used by households within the meaning of sub-article 2 of Article 2 of the Act of July 15th 1975, above mentioned.

**Article 4**  Any producer, any importer whose products are marketed in packaging to which Article 3 above applies or, if the producer or importer cannot be identified, the person in charge of the first marketing of these products has to contribute to or be engaged in the recovery of all packaging wastes, within the meaning of articles L.373-2 to L.373-5 of the code of collectivities.

To this end, that person identifies the packaging which is taken over by an organisation or an enterprise holding the permit defined under Article 6 under the conditions specified according to Article 5 mentioned below.

**Article 5**  Persons to which said Article 4 applies and who use, for the disposal of their *used* packaging, the services of an authorised organization or company, enter into a contract with that firm which specifies among other things the nature of the identification of said packaging, the estimated volume of annual wastes and the amount which has to be given to that organization or company; these contracts must be consistent with all provisions of the specifications set out in Article 6.

**Article 10** If the persons mentioned in Article 4 above choose to proceed directly to the disposal of wastes resulting from the discarding of *used* packaging, they shall:

    **(a)** set up a system of recovering their packaging which shall carry clear and appropriate markings,

    **(b)** set up for the deposit of said packaging, specific locations for such deposit after having obtained a joint decision of the minister for the environment, the minister for industry and the minister for agriculture, agreeing the methods of controlling the system of disposal which permits the percentages of the marketed material and that disposed of to be determined.

To effect the eventual valorizing of 75 per cent of household packaging waste, Eco Emballages was set up on 12 November 1992. Any company can enter into an agreement with Eco Emballages by signing a contract and paying a contribution for each packaging unit.

The 1993 contributions scale for each Unit of packaging marketed in France is given below:

---

Contribution Scale:

Each company has to contribute for each unit of packaging being marketed in France under its brand names and consumed by households or assimilated if these packagings are of the same nature of those intended for households (option).

Contribution Scale:

PRODUCERS/IMPORTERS' CONTRIBUTION
Scale for 1993
(exclusive of tax)

**Common System**        **Special Dispensation System**

| Volume Contribution per Unit | | | | Weight Contribution per Material | |
|---|---|---|---|---|---|
| >30,001 | cm³ | 10 | centimes | 79 ASS | 1 cts per Kg |
| 30,001 to 30,000 | cm³ | 2.5 | centimes | STEEL | 10 cts per Kg |
| 201 to 3,000 | cm³ | 1 | centimes | PLASTICS | 50 cts per Kg |
| 151 to 200 | cm³ | 0.5 | centimes | ALUMINIUM | 50 cts per Kg |
| 101 to 100 | cm³ | 0.25 | centimes | PAPER, CARDBOARD, WOOD | 30 cts per Kg |
| 50 to 150 | cm³ | 0.10 | centimes | TEXTILES & OTHER MATERIALS | 30 cts per Kg |
| <50 | cm³ | Material weight Maximum 0.10 centimes | | | |

For all packaging consisting of "hollow rigid containers", there is only one possibility: the volume contribution per unit.

For all other packaging, companies have the choice between volume contribution per unit or weight contribution per material as long as the derogation for weight contribution is applied for through the profession and not individually. (This decision has been agreed upon with French representative Associations for Mainland France & Overseas Departments.)

**Ecobalance.** A condition of equilibrium between ecosystems, thus allowing their coexistence.

**Ecohouse.** ⇨ AUTONOMOUS HOUSE

**Eco-label.** An EC scheme (with approved logo awarded by national bodies) to promote products with reduced environmental impacts during their life cycle. The impacts are evaluated on a matrix similar to that below using LIFE CYCLE ANALYSIS.

The devil is the detail, e.g. paper products can be made from natural (managed forestry resources) or recycled paper or both. The ENERGY used can be finite or RENEWABLE or both. Air pollutants may be restricted to SULPHUR OXIDES only. COD and AOX are the main water pollutant parameters to be considered. All these are still bones of contention since the scheme was mooted in 1992. Certainly, European paper mills use more finite energy than their Scandinavian counterparts and it is not unlikely that paper products will founder on this comparison. The Nordic (Scandinavian) weighting scale for AOX and COD is given below. European manufacturers have not yet signed up.

By comparison, a washing machine has its greatest impact during use (in energy and water) and is very easy to analyse. Paper products will remain a major bone of contention in the eco-label systems. (⇨ ECO-MANAGEMENT and AUDIT SCHEME)

Weighting scale for parameters from paper manufacture

| Parameters | Points | | |
|---|---|---|---|
| | 1 | 2 | 3 |
| AOX (kg/t paper) | <0.1 | $0.10 \leq AOX < 0.3$ | $0.30 \leq AOX < 0.50$ |
| COD (kg/t paper) | <20.0 | $20.0 \leq COD < 50.0$ | $50.0 \leq COD \leq 65.0$ |
| S (kg/t paper) | <1.0 | $1.0 \leq S \leq 1.50$ | $1.50 \leq S 2.50$ |

Eco-label environmental impact matrix for complete product life cycle

| Environmental fields | Product life-cycle | | | |
|---|---|---|---|---|
| | Production | Distribution | Utilization | Disposal |
| Waste relevance | | | | |
| Soil pollution and degradation | | | | |
| Water contamination | | | | |
| Air contamination | | | | |
| Noise | | | | |
| Consumption of energy | | | | |
| Consumption of natural resources | | | | |
| Effects on ecosystems | | | | |

**Ecological efficiency.** The ratio between the amount of energy flow at different points along a food chain.

**Ecological indicators.** Organisms whose presence in a particular area indicates the occurrence of a particular set of water, soil and climatic conditions. (⇨ BIOTIC INDEX; GLADIOLI; LICHENS)

**Ecological niche.** Each organism has a special task in an ecosystem – known as a niche. No two species of plant or animal can occupy the same niche for long; competition ensues and one species eventually adapts to occupy a different niche, or dies out.

**Ecological pyramids.** Diagrams which show the overall flow of energy through an ECOSYSTEM. The producer (green plant) level forms the base and the successive trophic or feeding levels (herbivores, carnivores, etc.) occupy the remaining tiers. There are various types of pyramids:

1. *The pyramid of numbers*, which shows the numbers of organisms at each level of a food chain.

2. *The pyramid of biomass*, which is constructed using the total weight of organisms at each level of the pyramid of numbers. It shows the BIOMASS at a particular time, not over a period of time. The pyramid sometimes looks as if it is the wrong way up (see Figure 49), as in the case of phytoplankton and zooplankton in the English Channel. This is because the phytoplankton are eaten almost as soon as they are formed. (The rate of production of the phytoplankton is very much in excess of the growth rate of the zooplankton. The mass of zooplankton can therefore be in excess of the mass of phytoplankton, which are consumed almost immediately.)

3. *The pyramid of energy*, which presents the best overall picture of energy flow. It shows the rate at which food is produced, as well as the total amount. The pyramid is not affected by the size of the organisms, nor by how quickly energy flows through them. (⇨ FOOD CHAIN, FOOD WEB)

**Ecology.** The study of the relationships between living organisms and between organisms and their environment, especially animal and plant communities, their energy flows and their interactions with their surroundings.

**Eco-Management and Audit Scheme (EMAS).** EMAS is a voluntary scheme for individual industrial sites, introduced in April 1995. The scheme is established by European law. In the UK, the Secretary of State for the Environment is responsible for its administration.

It is designed to provide recognition for those companies who

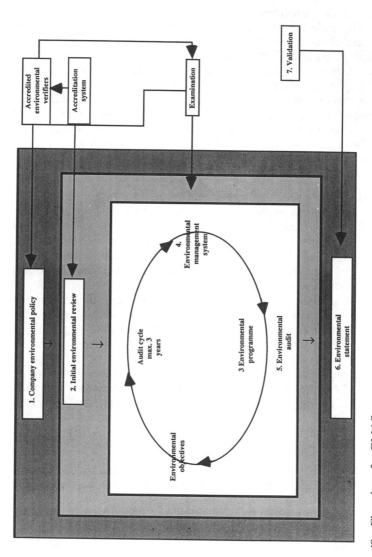

**Figure 48**   Flowchart for EMAS.
*Source: Eco-management & Audit Scheme – An Introductory Guide,* Department of the Environment, 1995.

have established a programme of positive action to protect the environment, and who seek continuously to improve their performance in this respect.

To register a site under EMAS a company should have a clearly defined strategy for environmental management, complete with quantified objectives. Initially, EMAS will apply only to Europe's industrial sites. However, in the UK, the scheme has been extended to include local authorities.

The following must be implemented at the relevant site:

1. An environmental policy, for the whole organization, which commits it to compliance with existing legislation and to reasonable, continuous improvement of environmental performance.
2. An environmental review, which covers all aspects of the industrial site or local authority service registered.
3. An environmental programme, that sets quantified objectives.
4. An environmental management system, to give effect to the policy and programme.
5. An environmental audit cycle, to provide regular information on the progress of the programme.
6. An environmental statement, a concise and comprehensible statement of progress, prepared with the general public in mind.
7. Validation, by an independent verifier.

(⇨ ENVIRONMENTAL REPORTS)

**Economics.** The study of the ways in which people choose what to make of whatever resources they can get, and the ways in which they conduct and organize matters of exchange. The basic propositions of economics relate to matters of human psychology – such as 'values', offers to buy (DEMAND), offers to sell (SUPPLY), tendencies to consume or to delay consumption, and the effects of different inducements is changing these kinds of behaviour. Among those who have criticized the role that economics has come to play in our society are a number of eminent economists, including Galbraith, Mishan, Boulding and Schumacher. One can do no better than quote E. F. Schumacher in this context *Small is Beautiful*, Blond & Briggs, 1973, Ch.3.:

It is hardly an exaggeration to say that, with increasing affluence, economics has moved into the very centre of public concern, and economic performance, economic growth, economic expansion,

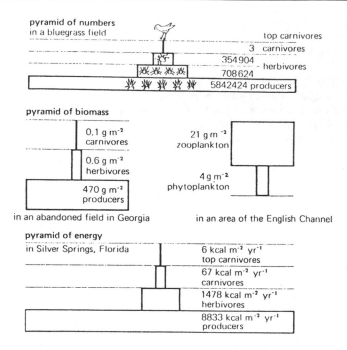

**Figure 49** Ecological pyramids. A pyramid of numbers shows the totals of individual organisms found in an area at a particular time. A pyramid of biomass shows the total weight, in grams per square metre, or organisms in an area at a particular time. A pyramid of energy shows the amounts of energy, in kilocalories per square metre per year, available to other organisms in a year's time.

and so forth have become the abiding interest, if not the obsession, of all modern societies. In the current vocabulary of condemnation there are few words as final and conclusive as the word 'uneconomic'. If an activity has been branded as uneconomic, its right to existence is not merely questioned but energetically denied . . .

In fact, the prevailing creed, held with equal fervour by all political parties, is that the common good will necessarily be maximized if everybody, every industry and trade, whether nationalized or not, strives to earn an acceptable 'return' on the capital employed. Not even Adam Smith had a more implicit faith in the 'hidden hand' to ensure that 'what is good for General Motors is good for the United States'.

However that may be, about the *fragmentary* nature of the judgements of economics there can be no doubt whatever. Even within the narrow compass of the economic calculus, these judgements are necessarily and *methodically* narrow. For one thing, they give vastly more weight to the short than to the long term, because in the long term, as Keynes put it with cheerful brutality, we are all dead. And then, second, they are based on a definition of cost which excludes all 'free goods', that is to say, the entire God-given environment, except for those parts of it that have been privately appropriated. This means, that an activity can be economic although it plays hell with the environment, and that a competing activity, if at some cost it protects and conserves the environment, will be uneconomic.

. . . it is inherent in the methodology of economics *to ignore man's dependence on the natural world.*

*Sources:* T. Congden and D. McWilliams, *Basic Economics: A Dictionary of Terms*, Arrow, 1976.
J. K. Galbraith, *The New Industrial State*, André Deutsch, 1972; Penguin Books, 1968.

**Ecosystem.** The plants, animals and microbes that live in a defined zone (it can range from a desert to an ocean) and the physical environment in which they live comprise *together* an ecosystem. The ecosystem embraces the FOOD CHAIN through which energy flows together with the biological CYCLES necessary for the recycling of essential nutrients. Thus an ecosystem has the means of producing both energy and materials for life going on continuously. Taken on a global basis, all the separate ecosystems are the life-sustaining processes on which our survival depends. Their integrity must be preserved, otherwise biological communities can die out, or essential services not be performed, such as the self-purification of a river or the control of greenfly by ladybirds. Most ecosystem are extremely complex: for example, a deciduous forest can support over 100 species of birds, as well as many wild flowers, grasses and shrubs. Man's intervention by means of planting conifers reduces bird species, animal variety and the types of flowers, ferns and grasses. In general, ecosystems are extremely resilient and where a consumer has more than one source of food, then if one becomes less readily available the consumer can still survive. However, if the consumer depends on or only has one source of food available, then its survival can be in jeopardy. (⇨ FOOD CHAIN)

Figure 50 shows an aquatic ecosystem where the primary

inputs are energy, by PHOTOSYNTHESIS fixed by the primary
producers algae and green plants, *plus* the organic and inorganic
materials carried by the river. The primary and micro-consumers
eat dead organic matter; the intermediate consumers, worms
and insect larvae, feed on the primary consumers and ALGAE.
The herbivores eat the green plants. The carnivores (fish) eat the
intermediate consumers. The scavengers eat the bottom debris
and dead organic matter too, so that virtually nothing is wasted.
Thus, the ecosystem is a complex interlinking arrangement with
its own form of *equilibrium,* i.e. a balance is struck between the
total production of living material and the rate of death and
decay over a period of time. This is ecological equilibrium. The
more complex the ecosystem, the greater the stability. (⇨
DECOMPOSERS; PROTOZOA)

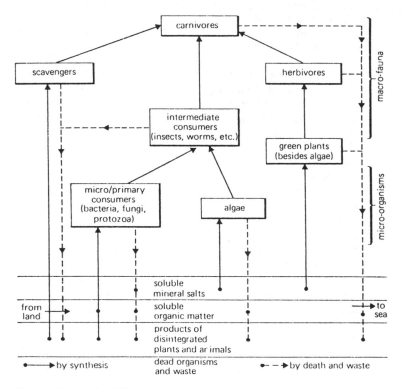

**Figure 50**  A simplified aquatic ecosystem for a river-bed community.

Man's intervention often simplifies ecosystems dramatically, for example, by mono-crop agriculture where some fields grow wheat for up to 30 years with the result that diseases tend to build up. These are often kept at bay by synthetic chemicals whereas the traditional rotation of crops would have ensured (usually) that disease did not build up. Other common examples of the dangers involved in man's modification of ecosystems are the conversion of prairie land to wheat, which destroys stable grassland ecosystems, the logging of deciduous forests to leave barren land which can lead to soil erosion, or, if replanted with conifers, to a modification of forest ecosystems. Large-scale changes in the use of land also result in a change in the ALBEDO, often increasing it, which in turn affects the MEAN ANNUAL TEMPERATURE and thus, cumulatively, the earth's climate. The spraying of crops with INSECTICIDES can be a major violation of an already simplified ecosystem. The animals at the top of the food chain are smaller in number than those below. Predators which eat pests, e.g. carnivores, are more likely to be destroyed in a blanket-spraying programme compared with the pests as there are many more pests than predators. Thus, once spraying is initiated and the equilibrium upset, it can become a self-perpetuating cycle, spraying then becoming a necessity as the predators have been destroyed. Pests seem to develop resistance to pesticides more quickly than their predators. ($\Rightarrow$ DDT)

In accepting the undoubted short-term benefits of modifying ecosystems, we may in the long term lay ourselves open to the risk that the original ecosystems have been destroyed or gene banks of *naturally* resistant crops or animals lost for ever.

**Effective chimney or stack height.** The effective height of a chimney ($H$) is the sum of its actual physical height ($h$) and the rise of the emitted plume caused by its BUOYANCY and EFFLUX VELOCITY as shown in Figure 51. The effective chimney height is used in air pollution dispersion calculations for the calculation of maximum ground level concentrations (Figure 52). It should be noted that a negative plume rise can occur under some conditions.

**Efficiency[1].** The ratio of energy output to energy input of a process. For boiler plants, the ratio of useful output to total energy input. (*Note:* this can be approximated to fuel input but there are often substantial energy inputs from pumps, feed heaters, etc. as well.) ($\Rightarrow$ THERMAL EFFICIENCY, CARNOT EFFICIENCY)

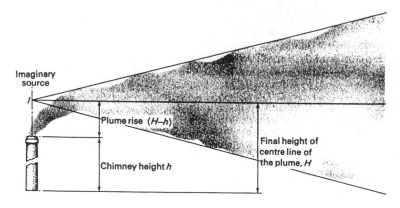

**Figure 51** A diagram to show the imaginary source, *I*, of an effluent plume originating above a chimney exit.

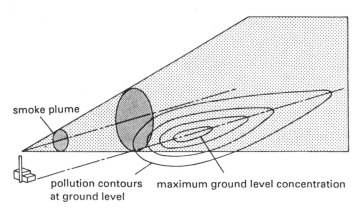

**Figure 52** A pollutant dispersing under conditions of neutral stability.

**Efficiency².** For gas filters, dust separators, and droplet separators, the ratio of the quantity of particles retained by a separator to the quantity entering it.

**Effluent¹.** Any fluid discharged from a source into the environment.

**Effluent².** Any liquid which flows out of a containing space, but more particularly the sewage or trade waste, partially or completely treated, which flows out of a treatment plant. For

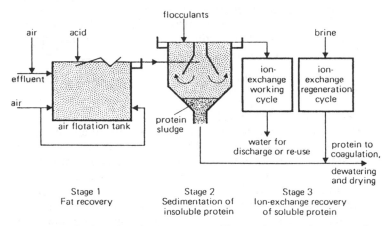

**Figure 53** Physico-chemical treatment of livestock processing wastes for fat and protein recovery.

example, sewage effluent is the liquid finally discharged from a sewage treatment works.

**Effluent, physico-chemical treatment.** The effluents from food-processing industries are often extremely noxious and costly to treat; yet they contain fats and proteins (soluble and insoluble) whose recovery can yield valuable products. The separation of the fats and proteins can be done by flotation and flocculation followed by ION EXCHANGE, i.e. by physico-chemical means. One such plant is shown in Figure 53 for the treatment of poultry-processing wastes.

The incoming effluent is acidified to pH 3 and a coagulating agent added, so that when air is bubbled through the first chamber the fat floats and is skimmed off. The protein-rich liquor is neutralized (pH 7) and flocculants added which make the insoluble protein 'floc' or come together. The protein then settles and is sent for stabilization. The soluble protein can only be removed by ION-EXCHANGE techniques and the problem is to obtain the correct material for the solution to be treated. (⇨ pH)

Cellulose ion-exchange media can selectively remove the high-molecular-weight protein molecules and the final clear water from the plant can be re-used or discharged. The ion-exchange medium must be recharged by brine which removes the protein molecules from the medium in a protein-brine solution which can then be coagulated and dried. The recovered protein has potential for animal feedstuffs, thereby recovering a valuable by-product and reducing effluent biochemical oxygen demand by as much as 90 per cent.

*Source:* Paul Butler, 'Processes in action', *Process Engineering*, May 1975, p. 65.

**Effluent Biological treatment.** Treatment plants are designed and engineered to create an optimized environment in which micro-organisms can grow. The liquid growth medium is the waste water and as this enters the plant it becomes mixed with the existing micro-organisms or recirculated ones as in the activated sludge process. The pollutants represent a food source for the micro-organisms and so these are rapidly broken down through biodegradation and thereby removed from the water. This provides a source of energy and building blocks so that further microbial growth and replication can occur. Thus soluble organic compounds polluting the waste water are biodegraded and converted into microbial cells, i.e. biomass. Biodegradation also produces carbon dioxide, water and mineral nutrients as by-products. Biomass is removed from the treated waste water, as a sludge by natural sedimentation processes. Many inorganic compounds, such as metals, are also removed from the water as a consequence of absorption and adsorption to the biomass. (⇨ SEWAGE TREATMENT)

There are four variations:

- aerated lagoon
- trickling filter
- activated sludge
- rotating biological contactor.

All provide oxygen, extended surfaces for micro-organism proliferation and a means of removing the sludge. The four processes are shown in Figures 54–57. (⇨ REED BEDS)

**Efflux velocity.** The velocity with which gas leaves a stack, which is equal to the volume of gas issuing from the stack mouth per second divided by the cross-sectional area of the mouth. This is a major parameter in air pollution dispersion. If the efflux velocity if low, DOWNDRAUGHT can take place. If it is too high and a wet de-duster used, droplets can be stripped from the stack.

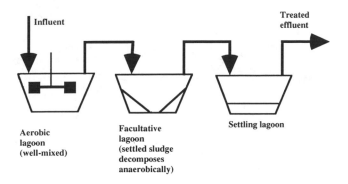

**Figure 54** Aerated lagoon.

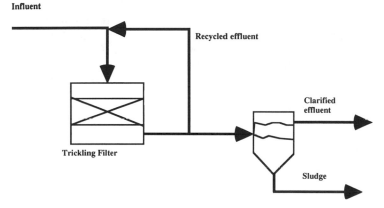

**Figure 55** Trickling filter.

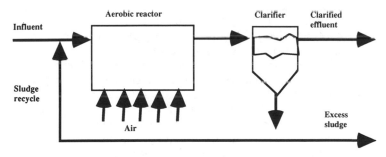

**Figure 56** Activated sludge.

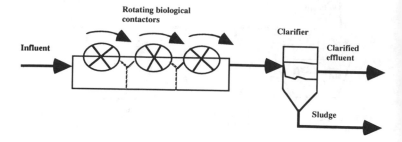

**Figure 57**   Rotating biological contactor.

**Electrodialysis.** The separation of ionic components from solutions by means of ion-selective membranes under the influence of an electric field. Electrodialysis is used in DESALINATION of low-salinity waters for human consumption, kidney machines, and specialized effluent-treatment processes.

**Electrolysis.** The chemical splitting of an electrolyte – a solution that can carry an electric current – by the passage of an electric current through it. The solution is ionized into positively- and negatively-charged IONS which move towards the oppositely charged electrodes immersed in the solution. Once at the electrode, they give up their charge and can be collected. Thus, the electrolysis of water gives HYDROGEN and OXYGEN as separate components. The hydrogen can then be used in a FUEL CELL.

The electrolysis of brine – sodium chloride (NaCl) and water ($H_2O$) – gives free CHLORINE (Cl) plus hydrogen ($H_2$) plus a solution of caustic soda (NaOH). The electrolysis of brine is the main process for CHLORINE manufacture.

**Electron.** A negatively charged elementary particle which is present in the orbital structure of all ATOMS.

**Electrostatic precipitator.** The gases from combustion processes may contain large burdens of particulate matter. Before discharge to atmosphere, they must be cleaned. An electrostatic precipitator is one method. The dirty gases are passed through an intense electric field and become electrically charged as shown in Figure 58. The charged particles are attracted to the collector electrodes which have an opposite polarity where they accumulate. A

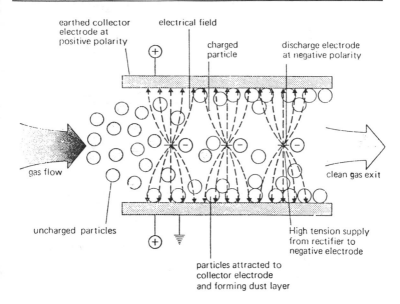

**Figure 58** The principles of electrostatic precipitation.

mechanical handling or rapping system dislodges the accumulated dust which by this time has agglomerated and it falls out of the gas stream. The product from power stations is called FLY ASH and is used in a variety of land reclamation and building materials manufacture.

Electrostatic precipitators can remove 97–99.5 per cent and more of the initial grit burden. They can also be used to remove acid or oil mists and fumes and are thus extremely versatile and efficient gas-cleaning devices. For higher efficiencies, water can be injected to cool the gases and the solid liquid particles separated by the electric field. The choice of construction materials in wet electrostatic precipitators is very important.

The effect of collection efficiency is shown in Figure 59 and shows the importance of fractional increases in efficiency in dust-collection equipment, e.g. if an electrostatic precipitator has an efficiency of 97.5 per cent it will emit twice the amount of dust of an 98.75 per cent efficient installation. Precipitators of 99.7 per cent efficiency are now used on new power station and incinerator installations.

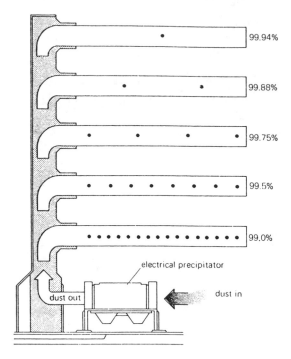

**Figure 59** Dust collection efficiency (each dot represents a unit of mass flow per unit of time).

**Element.** A simple substance which cannot be broken down into simpler substances by chemical methods. An element consists of ATOMS all with the same ATOMIC NUMBER.

A element may be represented by a CHEMICAL SYMBOL which is used in formulae and equations. The symbol is composed of the initial letter of the name, plus another letter where necessary. Usually the initial letter of the English name is used, but sometimes it is that of the Latin name. The symbol is used to represent an atom of the element as well as the element itself. See Appendix II.

The symbols representing the elements may be combined to form chemical formulae which represent compounds: for example, sodium chloride (formula (NaCl) consists of equal numbers of atoms of sodium (symbol Na, from the Latin name *Natrium*) and chlorine (symbol Cl). When a molecule of the compound contains different elements in unequal proportions, their ratios are indicated by subscripts: for example, water (formula $H_2O$) consists of HYDROGEN (symbol H) and OXYGEN (symbol O) in the ratio 2:1, two atoms of hydrogen joined to one atom of oxygen. These are examples of molecular formulae in which the individual and total number of the constituent atoms in the molecule are indicated. (⇨ ISOTOPE, PERIODIC TABLE)

**Elutriation.** A method of separating particles using specific gravity differences between the particles when they are suspended in a fluid. In practice, the particles are usually allowed to settle against an upward-moving flow of fluid (e.g. water or air); heavier material settles to the bottom, while fine material remains suspended and is removed with the fluid, e.g. air classifiers used in REFUSE-DERIVED FUEL production use this principle to separate the light combustibles from the heavy non-combustibles.

**Emission.** The amount of pollutant discharged per unit time. Or the amount of pollutant per unit volume of gas, or liquid emitted.

**Emission factor.** The mass of a given pollutant produced relative to each unit of process. This may be per unit mass of fuel consumed, or per unit of production (e.g. per million litres of petrol produced) or per item of production (e.g. per motor car off the assembly line).

**Emission inventory.** The compilation either by measurement or (more usually) by estimations, of a map of the distribution of emissions over a given area, showing the positions of the more important sources and the amounts they emit. The emission per unit area for each for smaller sources is usually given as well.

**Emission source.** Any factory, furnace, chemical process, etc. that discharges pollutants into the air. A *point source* is plant whose total emission of a pollutant is sufficiently large to warrant separate consideration in an emission inventory. (An emission of the order of 100 tonnes per year or more may lead to plant being considered as a point source.) In the USA any stack is usually considered a point source; stacks with emissions exceeding 100 tonnes per year are considered major point sources.

**Emission standard.** The amount of pollutants not to be exceeded in the discharge from a pollution source. Emission standards are commonly described in one or more of the following ways for liquid, solid or gaseous pollutants:

1. Mass of pollutants over a certain time: e.g. kilograms per hour, tonnes per day, or pounds per hour.
2. Mass of pollutants per unit mass of material processed: e.g. 20 kilograms per tonne (approximately 2 per cent).
3. Mass of pollutant per unit volume of discharged gas at specified conditions of temperature and pressure: e.g. milligrams per cubic metre, or grains per cubic foot.
4. The concentration specified as the volume of pollutants (if gaseous) per unit volume of discharged gas, again at specified conditions of temperature and pressure.

The emission standards for NOISE and IONIZING RADIATION are dealt with under their respective entries. (➪ THREE-MINUTE MEAN CONCENTRATION; CONCENTRATION)

**Endosulfan.** An ORGANOCHLORINE insecticide. It can be toxic to fish in concentrations as low as 0.00002 parts per million, i.e. 1 million litres water would require 0.02 g of Endosulfan, if evenly distributed, to render it toxic to fish. This amount (and more) can easily be left in discarded chemical containers. The potency of many agro-chemicals is such that empty containers should be returned to the supplier for environmentally acceptable disposal. In fact, the disposal of the containers for all such potent chemicals should ideally be registered. (➪ RED LIST)

**Endrin.** ➪ CHLORINATED HYDROCARBONS; ALDRIN; RED LIST.

**Energy.** The ability to do work. Energy has many forms: mechanical, chemical, thermal (heat), nuclear, electrical.

It is common practice to consider energy from thermal sources, such as the heat released from the combustion of fossil fuels or nuclear fission, to be primary energy as it can only be used for heating and must be converted by means of an engine such as a steam turbine before it can do work. The conversion of heat to electrical or mechanical energy is governed by the second law of thermodynamics. (➪ LAWS OF THERMODYNAMICS)

Approximate energy conversion ratios for heat to work for a thermal power station range from 20 per cent (poor) to 40 per cent (excellent). Thus a 33 per cent efficient power station requires an input of 3 kilowatt-hours thermal ($3\,kWh_{th}$) for 1 kilowatt-hour electrical ($1\,kWh_e$) output. Thus the form in which energy is supplied must be specified when ENERGY ANALYSIS is being performed. A process which uses $2\,kWh_{th}$ per unit output is much more energy efficient than a competitive process which requires $1\,KWh_e$ per unit output, as the competitor process in effect requires a thermal input at the power station of about $3.3\,kWh_{th}$ to make $1\,KWh_e$. ($\Rightarrow$ ENERGY RESOURCES; WATT; JOULE)

**Energy analysis.** Every operation carried out on materials, in the mining of fuels or metals, refining or working of metals, transporting, manufacturing and construction of all kinds, involves the use of energy. Hence, any object, from a power station to a milk-bottle, from a barrel of oil to a newspaper or a hole dug in the ground, requires the expenditure of energy to bring it into existence – and also to dispose of it.

It is possible to examine the details of the processes by which something is made and to determine the amount of energy needed to produce it. This involves adding up the amounts of energy needed to make each part and that needed to assemble the product. For each part the energy required to form it and the energy required to provide the necessary materials are included, and allowances must be made for some fraction of the energy needed to make any machines involved in the various processes. This is called 'energy analysis by process analysis'.

Other methods, usually less accurate since they involve the use of non-physical variables such as financial costs, may be used for energy analysis.

Energy analysis constitutes a valuable supplement to economic analysis when attempts are made to estimate the effects of changes in the availability of energy upon a national economy. Energy analysis is based extensively upon physical reasoning; this enables it to evade many of the difficulties that confront economic analysis, and which derive from such matters as absolute and relative variations in price. Nevertheless, energy analysis can only be carried out in accordance with one or another of several possible sets of conventions – concerning, for example, how much of the energy expended should be attributed to each of the products of a process with several outputs.

*Source. S. Nilsson, 'Energy analysis', Ambio, vol. 3, no. 6, 1974.*

Differentiation between whether the energy flows are from finite or renewable resources is also necessary. For example, it is claimed that:

> In energy-balance terms, the energy saved through recycling many materials exceeds the energy which could be produced through energy recovery.

Paper recycling leads to energy savings of 28–70% or 6–15 MJ/kg on the basis of the total energy consumed in the manufacture of paper in the UK in 1989, whereas the calorific value of dry newspaper is 16 MJ/kg which could generate up to 4 MJ of electricity.

However, the basis of the calculation is on primary (FUEL) energy saved, hence the 4 MJ electrical energy is irrelevant. Furthermore, the recycling of PAPER in the UK consumes finite fossil fuels whereas pulp/paper manufacture in Scandinavia uses renewable fuels (forest and plant residues). Hence, paper recycling in the UK cannot really be said to be saving energy at all if the boundary is drawn around the UK.

(a) Differentiation between renewable and finite sources of energy.
(b) All calculations to be done on a primary (fuel) basis. Purists may add in a factor for the procurement of the finite energy (e.g. oil) from the ground and the subsequent processing.

As an example, the US container glass industry purchased $118.5 \times 10^{15}$ J in 1985: 78 per cent was in the form of natural gas, 19 per cent in the form of electricity, and the remaining 39 per cent in the form of distillate oil. When the efficiency of electricity generation is taken into account, these figures imply primary energy consumption of $164 \times 10^{15}$ J. Accounting for transportation and raw-material production energies gives a grand-total primary energy consumption of $192 \times 10^{15}$ J for 1985. (⇨ BOTTLES; PAPER; RECYCLING)

*Sources:* The Local Authority Recycling Advisory Committee's (LARAC) Policy Statement on 'Recycling, Energy Recovery & the National Waste Management Strategy', September 1994.
E. Babcock *et al., The US Glass Industry: An Energy Perspective,* Energetics, Inc., prepared for Pacific Northwest Laboratory, September 1988.

**Energy conservation.** ⇨ ENERGY EFFICIENCY.

**Energy demand.** It is precisely those technologies which are regarded as giving us our productive efficiency, measured in economic terms, that are responsible for the alarming rate of growth of energy consumption in the western world. For example, modern agricultural methods require large fossil fuel subsidies (in the form of OIL, PESTICIDES, artificial FERTILIZERS, etc.). In the USA this has been estimated (Perelman, 1972) as equivalent to an input–output ratio of 5:1 in energy terms. Furthermore, as populations continue to grow, demands for increased food output will require large increases in the use of nitrate fertilizers.

But present methods require about 2 kg of coal equivalent to fix 1 kg of nitrogen. The energy demands of sea water DESALINATION plants, required for increasing water production in the many arid areas of the earth, are also very high.

Future prospects are of increasing energy demands as essential industrial materials become scarce. The table gives the total direct energy inputs for producing certain vital metals, neglecting transport costs for both products and raw materials. Bear in mind that the combustion of coal yields about 29 megajoules per kilogram (MJ/kg):

| Production process | MJ/kg |
|---|---|
| Copper from 1.0% sulphide ore in place (1940s) | 54 |
| Copper from 0.3% sulphide ore in place (1980s) | 98 |
| Aluminium from 50% bauxite in place (1970s) | 204 |
| Magnesium from seawater (anytime) | 360 |
| Titanium from ilmenite in place (1970s) | 593 |

From J.C. Bravard, H.B. Flora II, and C. Portal, *Energy Expenditures Associated with the Production and Recycle of Metals*, ORNL-NSF-EP-24, Oak Ridge National Laboratory, November 1972.

As these ores become progressively more scarce and the market price rises, it will become economically attractive to work leaner sources, with a corresponding requirement for much higher energy inputs. Unfortunately, it appears likely that for most non-structural metals there is not a continuum of progressively poorer ores in growing quantities but rather an abrupt grade gap, i.e. there is good ore and then there is rubbish. Extracting metals from the 'rubbish', even if economically feasible because of scarcity, is almost certainly impossible in terms of total energy availability.

UK Energy Paper 65 gives six projections for low, central and high scenarios for total final energy demand as given in the table below:

A major feature of the projection of the electricity supply industry is a shift towards more efficient and cleaner modes of electricity generation (primarily CCGTs (Combined Cycle Gas Turbine)), resulting in a lower level of primary electricity demand for any given level of final demand. Largely as a result

Total final user demand, Million tonnes coal equivalent (Mtce)

| Scenario | 1990 | 2000 | 2010 | 2020 |
|----------|------|------|------|------|
| LL | 147 | 162 | 176 | 195 |
| LH | 147 | 158 | 168 | 187 |
| CL | 147 | 166 | 185 | 210 |
| HL | 147 | 161 | 178 | 203 |
| HH | 147 | 164 | 185 | 215 |

of this, primary demand is expected to grow more slowly than final demand.

In the long term, the projections suggests that, given favourable assumptions about future fuel price relativities and about generators' concerns to maintain a diverse portfolio of plant, there could be scope for new generating technologies burning coal to establish themselves.

Total final user demand (Mtce)

| Scenario | 1990 | 2000 | 2010 | 2020 |
|----------|------|------|------|------|
| LL | 221 | 231 | 245 | 262 |
| LH | 221 | 226 | 237 | 253 |
| CL | 221 | 237 | 257 | 283 |
| HL | 221 | 240 | 266 | 298 |
| HH | 221 | 235 | 258 | 289 |
| EP59 Range | 221 | 216–245 | 237–311 | 256–381 |

The energy ratio measures the ratio between energy use and output or income. In its most aggregated form, it is based on total primary energy demand and gross domestic product. There has been a long-term trend decline in the energy ratio, reflecting not only improvements in energy efficiency, but also the declining importance of energy-intensive industries. This downward trend is expected to continue throughout the period to 2020 (Figure 61).

Annual demand for energy in China will climb to 1.94bn tonnes of oil equivalent by 2015 from 750m in 1993 if present policies of decentralization and progressive price liberalization continue (according to a study by DRI/McGraw Hill, 1995). The investment required to meet this demand could reach $1000bn, of which slightly more than half would go on electric power generation. Foreign capital would account for some 20 per cent of the total, the report forecast.

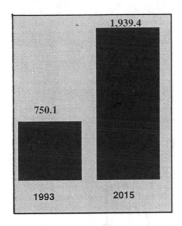

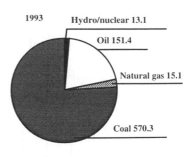

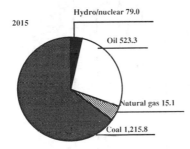

**Figure 60** Primary energy demand by fuel for China (million tonnes oil equivalent) (1993 & 2015).

*Source:* 'China's Energy in Transition', DRI International Energy Consulting, 8–10 rue Villédo, 75001 Paris, France, as reported in *Financial Times* 5/4/95 and Department of Trade & Industry, Energy Paper 65, March 1993.

*Related UK carbon dioxide emissions scenario*
Energy-related $CO_2$ emissions can be obtained using estimates of the carbon content of individual fuels to the projections of energy demand, already referred to. The table below shows the $CO_2$ projections associated with the six energy demand scenarios.

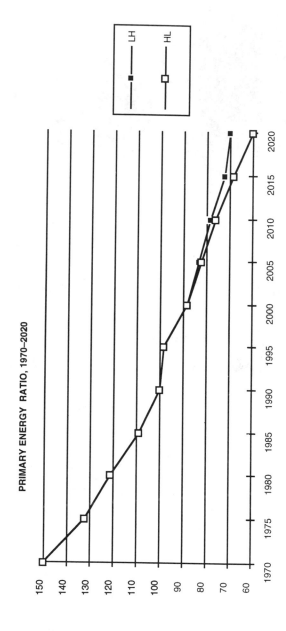

**Figure 61**    Primary energy ratio 1970–2020.

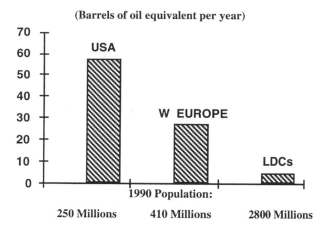

**Figure 62** Per capita consumption of primary energy.
*Source:* J. B. Wyebrew, Shell UK Ltd., Paper presented at IMechE: Power Generation and the Environment, 13–14 November 1990.

UK carbon dioxide emissions, 1920–2020 (Mtc)

| Scenario | 1990 | 2000 | 2005 | 2010 | 2020 |
|----------|------|------|------|------|------|
| LL | 158 | 147 | 157 | 155 | 171 |
| LH | 158 | 144 | 154 | 154 | 173 |
| CL | 158 | 150 | 162 | 162 | 184 |
| CH | 158 | 148 | 159 | 161 | 188 |
| HL | 158 | 152 | 165 | 167 | 193 |
| HH | 158 | 151 | 163 | 165 | 197 |

*Energy efficiency case study*

Up to 20 per cent of the energy used by industry in the UK is expended on water removal or drying. There are two basic drying mechanisms, mechanical dewatering and evaporation by thermal means.

In terms of energy expended per unit mass of water removed mechanical dewatering is always preferable to the use of evaporation. Hence, industrial drying processes should be designed for maximum mechanical moisture removal.

The air knife (or air wipe) is primarily a mechanical method. It is a precisely controlled jet of air delivered by a discharge at high velocity and relatively low pressure. It is used for moisture

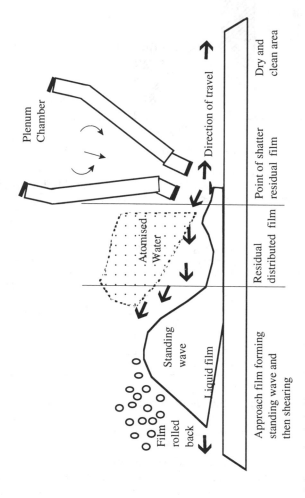

**Figure 63** Schematic air knife action of the high velocity, close proximity jet.
*Source: Introduction to Air Knife Technology*, Midlands Electricity plc, Halesowen.

removal from a wide range of materials and also for the removal of unwanted liquid and solid films and other surface deposits. A good air knife can dry a product using one-tenth of the power, one-tenth of the time and in one-tenth of the space of a conventional drying system.

The 'blade' of the air knife is a thin sheet of air travelling at speeds of approximately 100 km/h. Drying action is primarily kinetic, not thermal, as the jet wipes moisture from the surface. The shape of the blade is dictated by the plenum orifice and whilst most blades are thin flat sheets of air, any shape can be created by suitable orifice design.

The action of the air knife in the simple case of a rigid flat sheet is shown schematically in Figure 63.

It is usual to fix the air knife in position and pass the material to be dried through the blade. The horizontal component of the air stream rolls back the oncoming liquid. This creates a standing wave at the point where the bulk of the liquid is sheared off, leaving a thin residual layer. This residual layer is shattered by the momentum of the air stream to form minute liquid droplets that are then carried away by the air stream to leave a clean, dry surface. A small proportion of the total flow is set up as a high velocity air stream close to the surface to be dried and in the direction of the process material. This effect can be beneficial in providing a high-speed evaporative or warming process if required.

*Source:* A. B. Lovins, *World Energy Strategies*, Earth Resources Research for Friends of the Earth, 1973.
M. J. Perelman, 'Farming with petroleum', *Environment*, vol. 14, no. 8, October 1972, p. 8.

**Energy, effect of conversion on climate.** The influence of man-made heat as a result of our energy conversion activities is already significant on a local scale and will soon become significant globally if present trends continue. Most large industrial areas already add 10 per cent to the sun's heat input, producing significant local variations in climate, and such areas are likely to grow and proliferate.

An extreme example of such 'heat islands' is Manhattan Island in New York City, with a man-made power density of over 700 watts per square metre compared with 93 watts per square metre from the sun. (⇨ TEMPERATURE INVERSION)

On a global scale, significant changes in climate may be

triggered by relatively small variations in the heat balance in critical areas such as the floating Arctic pack ice. At present growth rates, one can foresee the possibility of such man-made changes in the next century, resulting from the combined effects of increases in man-made heat, and increased levels of CHLORO-FLUOROCARBONS and METHANE which are both GREENHOUSE GASES.

It is unfortunate that most of the basic questions about climatic mechanisms are at present unresolved, and it is therefore almost impossible to make assertions on sound scientific bases regarding the specific climate effects of human activity. However, present rapid growth rates are viewed with considerable disquiet by many climatologists, and a prudent policy of energy conversion on a global scale, while perhaps impossible politically, is evidently urgently required.

**Energy efficiency.** The rational use of ENERGY RESOURCES as part of an overall energy policy is now a pressing problem, especially as a means to combat the GREENHOUSE EFFECT. A wide range of options are available, ranging from RECYCLING of non-fuel materials to the avoidance of competition between various sources such as coal, gas and oil. Some areas where improvements can be made or new techniques applied are shown in Figure 64. A study by the UK Department of Energy in 1989 conceded that the scope for energy savings could be as high as 60 per cent, although a more realistic scope for saving is 20 per cent. Of this, typically 10 per cent energy consumption may be reduced by 'good housekeeping' steps. These include minimizing waste by dealing with leaks, checking thermostat and time-control settings, checking combustion efficiency, cleaning lights and heat emitters, and switching off equipment not in use. All are essentially no-cost improvements, but depend on individual motivation. A further 10 per cent saving is often possible by investing in 'retrofit' systems or equipment which will often give a pay-back within two years. Insulation of building fabric, adding time controls and installing pipe and tank insulation are examples.

More energy-efficient factories means reduced energy costs for manufacturing and thus greater long-term product com-petitiveness in the market-place. In the domestic sector, the immediate benefit is reduced fuel costs. For government and public utilities it means a reduction of ENERGY DEMAND.

**Energy recovery.** A form of resource recovery in which the combustible fraction of waste is converted to some form of

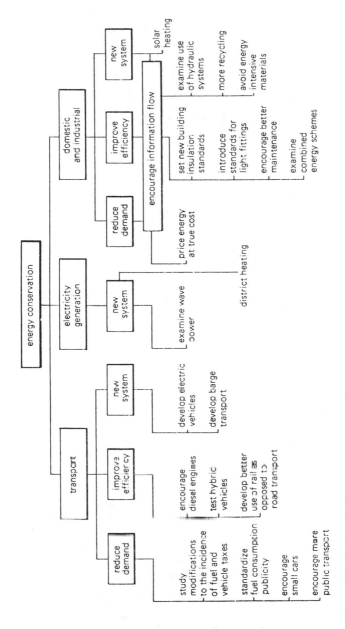

**Figure 64** Proposals for energy conservation.

usable energy, or the recovery of waste heat from industrial or domestic premises/processes for further use. ($\Rightarrow$ INCINERATION)

**Energy resources.** There are two forms of resources, income and capital or renewable and non-renewable. Once a capital resource is spent, it cannot be recovered; oil, coal and uranium are capital resources. SOLAR ENERGY is an income source and comes in continually. The derivatives of solar energy such as grass, trees, etc., are forms of income but are finite in the rate at which they can be exploited. A finite or capital resource is one which has been laid down or formed over geological time. Many such resources are being exploited and quite probably will be depleted over the time-scale of several generations. World production of crude oil, rose from almost zero in 1880 to around 3000 million tonnes per year in 1988 (approx. 23 000 million barrels). This is an example of exponential growth (with a few hiccups when there was an oil crisis). When the production rate is divided into the amount of the proved available reserves in the ground, the RESERVES–PRODUCTION RATIO is obtained. Figure 65 shows the curve of the upper and lower estimates for world crude-oil production (based on the virtual exponential growth since 1880) for values of ultimate total oil production of $1350 \times 10^9$ and $2100 \times 10^9$ barrels respectively. Eventually the rate of production cannot be increased any more and rate of discovery is less than production and so the production will decline to zero – in other words, it is a finite cycle. Figure 65 shows that the middle 80 per cent of oil production on the lower estimate ($1350 \times 19^9$) would occupy a mere 58 years; for the more optimistic estimate ($2100 \times 10^9$) it would occupy 64 years. One can argue about the rate of decline as new resources are discovered or new technology allows further extraction from existing fields, but the ultimate end result will still be a substantial curtailment in oil supplies in the 21st century. ($\Rightarrow$ EXPONENTIAL CURVE)

The long-term availability of usable sources of energy and our ability to use them wisely are the factors which will be predominant in deciding the length and nature of the human race's habitation of our planet. In considering the availability of energy sources one must remember that social, economic, political and geographical factors can be as important as the actual abundance or scarcity of a particular resource. As an example, it seems likely that Britain's North Sea oil reserves, instead of being properly husbanded, will, because of economic

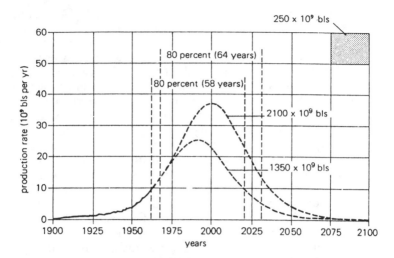

**Figure 65** World crude-oil production cycle for total ultimate production estimates of $1350 \times 10^9$ bls and $2100 \times 10^9$ bls respectively.

demands (a guaranteed minimum return on capital invested) be used up as rapidly as they can be extracted. *The Times*, 20 March 1989, in an article on North Sea oil output cutbacks stated 'Last year, Britain lost 61 million barrels of output from the North Sea and this year an estimated 91 million barrels will be lost. In Government revenues, £600,000 in North Sea taxation will be lost and as much as £2 billion in export earnings.' Such a policy appears quite rational to the conventional wisdom in economic policy, which tends to regard money and goods as equivalent. *The Times* article also stated:

> Since March 1983, Opec has been trying to get Britain to reach some sort of agreement over cutting output.

> Opec has argued that while it has cut its output by half since the days of the second oil crisis to about 18 million barrels per day, Britain at one point was producing 2.7 million barrels per day.

Economic attitudes to energy sources also disguise the very nature of these resources, i.e. that they are a *degradable* resource which, to be used effectively, must take into account the continuous chains of conversion from a high-energy form to low-temperature heat (e.g. car exhausts). It is economic considerations which prevent, for example, the UK Central Electricity Generating Board from designing power stations to make the most thermodynamically efficient use of their waste heat. And, in energy terms, it hardly makes sense to burn gas to generate steam to produce electricity to boil a kettle of water or heat a house, when those operations could be carried out so much more energy-efficiently by using gas directly. (⇨ DISTRICT HEATING)

The way in which we use our energy supply is extremely patchy. The USA, with only 6 per cent of the world's population, uses over one-third of the world's energy. This gross inequality of distribution is one of the most disturbing features of the present situation. Further, it is in those countries with the largest GROSS NATIONAL PRODUCT that the rate of growth of energy consumption is greatest due to increased material standards of living and the use of energy-intensive technologies. It is these countries whose economies are at greatest risk when the realities of energy limitations become apparent. (⇨ GAS; COAL; OIL; HYDROELECTRICITY; NUCLEAR ENERGY; SOLAR ENERGY; OIL SHALES; TAR SANDS)

**Entrainment.** The entrapment and transportation of a material by a flowing fluid.

**Environment.** All of the surroundings of an organism, including other living things, climate and soil, etc. In other words, the conditions for development or growth.

**Environment Act 1995.** An Act to provide for the establishment of a body corporate to be known as the Environment Agency and a body corporate to be known as the Scottish Environment Protection Agency; to provide for the transfer of functions, property rights and liabilities to those bodies and for the conferring of other functions on them; to make provision with respect to contaminated land and abandoned mines; to make further provision in relation to National Parks; to make further provision for the control of pollution, the conservation of natural resources and the conservation or enhancement of the environment; to make provision for imposing obligations on certain persons in respect of certain products or materials; to make provision in relation to fisheries; to make provision for certain enactments to bind the Crown; to make provision with respect to the application of certain enactments in relation to the Isles of Scilly; and for connected purposes. [19th July 1995].

*Source:* Environment Act 1995, Chapter 25.

**Environment Agency.** The Environment Agency in England and Wales (currently a draft bill) will combine the present functions of the NATIONAL RIVERS AUTHORITY (NRA), HER MAJESTY'S INSPECTORATE OF POLLUTION (HMIP) plus the London Waste Regulation Authority and 82 other regulation authorities including Waste Regulation Authorities (WRAs).

Its counterpart in Scotland, the Scottish Environmental Protection Agency (SEPA), will take on the functions of Her Majesty's Industrial Pollution Inspectorate (HMIPI), the River Purification Boards (RPBs), and the WRAs. Local Authority Air Pollution Control (LAAPC) functions under Part 1 of the 1990 ENVIRONMENTAL PROTECTION ACT (EPA) will also be handed over to SEPA, although they will remain with the local authorities in England and Wales.

The draft Bill and Management Structure set out the Agency's aims and objectives. These include the provision of effective environmental protection, management and enhancement covering all aspects of the environment, while imposing the minimum burden upon industry. To this end, the Agency will be required, when deciding how to exercise its powers, to '. . . take into account the costs which are likely to be incurred, and the benefits which are likely to accrue', although this should not affect its obligation to discharge its duties. While there is no absolute requirement for the Agency to undertake only those actions which will show an overall benefit, actions which do not accord with this may leave the Agency open to challenge in the courts. Questions will undoubtedly be raised over ways in which cost/benefit analysis can be undertaken in view of the problems and dangers of giving monetary valuations to the environment.

The Agency will also have a duty to 'have regard to the desirability of conserving and enhancing natural beauty and of conserving flora, fauna and geological or physiographical features of special interest'. This may be seen as a weakening of the NRA's current requirement to further conservation and enhancement.

In order for the Agency to meet needs on a local level, it will be split into regions, each with a separate Environmental Protection Advisory Committee (EPAC), Fisheries Advisory Committee, and one or more Flood Defence Committees. It will be for the Agency to determine the composition of these committees, although the chairman of each will be appointed by the Secretary of State for the Environment or the Secretary of State for Wales. The Agency will determine the boundaries of the regions.

*Sources:* M. Baxter, *Environmental Assessment*, Vol. 2, p. 126, December 1994.
UK Government Draft Environmental Bill, 1995.

**Environmental audit.** This is an account by manufacturers of the products produced and their effects on the environment – energy use policies, materials use policies, waste output and their effects on the environment. Purchasing procedure monitoring to obtain environmentally sound goods and services, etc. may also be adopted as part of an environmental audit.

One offshoot is waste auditing in which waste producers keep records of the composition and volume of wastes produced and the disposal routes used for them. This is a 'cradle to grave' practice which is widely employed in the USA. (⇨ POLLUTION PREVENTION PAYS; WASTES – SOLID AND HAZARDOUS. VII – MINIMIZATION)

**Environmental damage – monetary valuation.** Various methods have been proposed for placing monetary value on environmental damage.

*Preventative expenditure:* the amount paid to prevent or ameliorate unwanted effects; an example is expenditure on insulation and double-glazing to keep out noise.

*Replacement/restoration cost:* the amount public bodies or individuals spend to restore or replace a lost amenity or landscape.

*Property valuation:* differences in the market value of similar properties which reflect differences in the local environment due to traffic or industrial activity.

*Loss of earnings:* loss of productive output through injury or ill-health.

*Changes in productivity:* the monetary value of a reduction in renewable resources (crop or fishery) yield due to environmental damage.

*Contingent valuation:* the amount people say they would be willing to pay to avoid unwanted effects.

*Maintenance expenditure:* the additional cost of maintaining capital structures such as bridges and buildings or the additional costs imposed to control deterioration.

Most studies indicated a cost between 0.4 and 0.7 per cent of Gross Domestic Product (GDP) attributable to nitrogen oxides and volatile organic compounds from transport.

In view of the higher level of exposure to noise in Britain, and its higher level of population density, the cost of noise and vibration is likely to be higher than in many of the countries covered in these studies. A small allowance for the cost of vibration damage and disturbance also has to be added. A range of 0.25 to 1.0 per cent of GDP has therefore been taken as a likely range of costs for noise from transport.

*Source:* Royal Commission on Environmental Pollution, 18th Report, 1994.

**Environmental Impact Assessment (EIA).** Industrial processes involve the conversion of raw materials and the consumption of water/energy/air to provide the output goods/services.

As the conversion can never be total (LAWS OF THER-MODYNAMICS), residues in the form of solid wastes and gaseous/liquid effluents as well as WASTE HEAT, NOISE and vibration will be produced. If the residues are not utilized, they are discharged into the BIOSPHERE and can become POLLUTANTS.

As well as affecting the physical environment, industry and industrialization can have impacts on society. These are generally more complex and often cannot be determined at the initial stages because of interactions which do not follow any fixed guidelines or laws of nature. However, they also need to be considered when the new plant or process is considered especially where there is no existing industrial or established infrastructure.

The principal method of ensuring that environmental considerations are taken into account at the planning stage is to conduct an *environmental impact assessment* or EIA, which is then embodied in an *environmental impact statement* (EIS) which is now often required by law before a new project can proceed. Typically, an EIS will embody the following information:

(a) a description of the development proposed, comprising information about the site and the design and size or scale of the development;
(b) the data necessary to identify and assess the main effects which that development is likely to have on the environment;

(c) description of the likely significant effects, direct and indirect, on the environment of the development, explained by reference to its possible impact on human beings, flora, fauna, soil, water, air, climate, the landscape, the interaction between any of the foregoing material assets, the cultural heritage, employment, transport, education resources, housing, etc.;

(d) where significant adverse effects are identified with respect to any of the foregoing, a description of the measures envisaged in order to avoid, reduce or remedy those effects;

(e) a summary in non-technical language of the information specified above, where public participation is involved.

This is not an insignificant task and three outline examples of EIA applications are given below for a tidal barrage, coal and nuclear power generation, and toxic waste incineration respectively to illustrate what needs to be done.

(i) *Tidal Barrage (for Severn Estuary)*

A proposed BARRAGE for the Severn Estuary may involve a 16 km dam fitted with 200 turbine generators. Clearly this can have a substantial environmental impact.

The first step is to thoroughly understand the behaviour of the existing waves, tides and currents as these influence the nature and amount of sediment deposited, the water quality including OXYGEN content and SALINITY. These factors in turn influence estuarine plant and animal life. The natural and human being centred environments are then studied along with their interlinkages, e.g. some of the problems posed by the barrage if constructed could cause changes in patterns of human activity which could lead to more developments which in turn could lead to higher pollutant levels and/or increased public health risks for those in contact with the water. The relationships between the major subject areas involved in the environmental impact of the proposed barrage are shown in Figure 66.

(ii) *Power generation*

The major environmental impacts (all figures approximate) of coal and nuclear power stations are tabulated opposite:

Impacts of coal and nuclear power stations

| Emissions (limestone/gypsum flue gas desulphurization fitted) | Coal (1800 MW output) | Nuclear (1150 MW output) |
|---|---|---|
| Gaseous | 11 000 000 tpy* carbon dioxide<br>16 000 tpy sulphur dioxide<br>27 000 tpy nitrogen oxides<br>1000 tpy dust | —<br>—<br>—<br>— |
| Solids | 21 000 tpy sludge<br>500 000 tpy gypsum<br>1 000 000 tonnes ash | 530 m³ low level radioactive waste<br>100 m³ intermediate level radioactive waste<br>3.5 m³ high level waste |
| Inputs | 5 000 000 tpy coal<br>250 000 tpy limestone<br>1 200 000 000 m³ cooling water | 186 tpy uranium<br>1 000 000 000 m³<br>cooling water |
| Land | 80 hectares coastal or riverside plus 90 during construction | 20 hectares isolated coastal or river site |

*Note:* Construction and later on fuel traffic movements are not included. For coal, this aspect is substantial and can be a major factor in the environmental impact as well as the visual intrusion of high (100+) stacks.
*tpy = tonnes per year.
*Source:* Council for the Protection of Rural England, 1989.

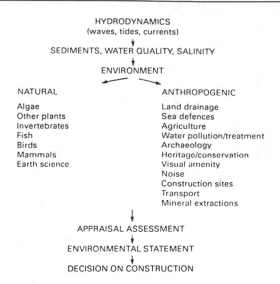

HYDRODYNAMICS
(waves, tides, currents)
↓
SEDIMENTS, WATER QUALITY, SALINITY
↓
ENVIRONMENT

| NATURAL | ANTHROPOGENIC |
|---|---|
| Algae | Land drainage |
| Other plants | Sea defences |
| Invertebrates | Agriculture |
| Fish | Water pollution/treatment |
| Birds | Archaeology |
| Mammals | Heritage/conservation |
| Earth science | Visual amenity |
| | Noise |
| | Construction sites |
| | Transport |
| | Mineral extractions |

↓
APPRAISAL ASSESSMENT
↓
ENVIRONMENTAL STATEMENT
↓
DECISION ON CONSTRUCTION

**Figure 66**   The relationships between the major subject areas involved in the environmental impact of the proposed barrage.
*Source: Quarterly Journal of Renewable Energy,* Issue 6, Jan. 1989, Dept. of Environment.

(iii) *Regional Waste Treatment Centre (RWTC) (Courtesy of Leigh Environmental Ltd,* 1989)
This proposed centre (a public enquiry turned down this proposal on the grounds of potential contamination of water resources; the technology was not queried) is designed to treat and recycle a wide variety of industrial residues by high temperature incineration and a variety of physical and chemical processes.

The plant briefly comprises:

(a) a thermal treatment unit for liquid and solid industrial residues
(b) a solvent recovery plant
(c) an acid neutralization plant.

*Thermal unit:* This is a single stream, thermal unit capable of treating 20 000 tonnes per year of chemical solid, liquid and sludge residues. The plant will initially operate 35 weeks of the year on a three-shift system, 24 hours per day, 7 days per week. It comprises the following elements:

(a) material handling systems for solids, liquids and sludges
(b) rotary kiln thermal unit
(c) secondary combustion chamber
(d) gas cleaning systems
(e) residuals handling system
(f) instrumentation and control system.

*Materials handling system.* The materials handling systems receive incoming residues, which are stored and blended so that specific mixtures will burn under optimum conditions in the rotary kiln. Solid materials are fed into a hopper and enter the rotary kiln via an airlock and feeding chute. Sludge is atomized by steam in a lance and injected into the rotary kiln. Liquids are atomized by air in special burners and lances and injected into the rotary kiln and secondary combustion chamber.

*Rotary kiln.* The rotary kiln is a near-horizontal refractory lined drum which rotates about its axis at between 0.1 and 0.6 revolutions per minute. Residues and fuel are introduced at one end and exhaust gases leave the other end. Solvents and other high calorific residues are atomized in special burners and lances as the main source of combustion. Temperatures in the kiln up to 1250°C (sufficient to destroy POLYCHLORINATED BIPHENYLS, for example) are maintained by the addition of fuel oil through auxiliary burners.

*Secondary combustion chamber (SCC).* The SCC is designed to create a large reaction volume for the secondary combustion of flue gases leaving the rotary kiln and also to make the complete oxidation of material in the SCC possible. The SCC has auxiliary burners operating with fuel oil to ensure that combustion temperatures are maintained. Past experience with this design has shown that there is about five seconds residence time for flue gases. Under these conditions the destruction efficiency of poly'chlorinated biphenyls (PCBs) will meet the IIMIP requirement of not less than 99.9999 per cent.

*Flue gas cleaning.* The exhaust gas cleaning system consists of the following stages:

(a) saturation venturi to reduce the exhaust gas temperature;
(b) variable throat venturi to remove particulate matter;
(c) first scrubber to remove hydrofluoric, hydrobromic and hydrochloric acid;

(d) second scrubber to remove sulphur and phosphorus oxides and reduce the water burden on the outlet gases;

(f) mist eliminator.

After cleaning the system the exhaust gases are diluted with preheated ambient air and discharged from the chimney by an induced draft fan.

*Residuals handling system.* The residual solid material, i.e. inert slag and ash is removed from the end of the rotary kiln through a water quench and conveyed into a skip for disposal off-site in a licensed facility.

*Instrumentation and control.* A number of plant safety and reliability features are built into the design of the control system to ensure that emissions to the atmosphere are kept within the design limits. The control systems include automatic emission monitoring linked to the residue feed auxiliary burners and scrubber dosing systems. In the event of an abnormal situation occurring the plant will be shut down automatically. An auxiliary generator covers loss of mains power and standby scrubber liquor pumps will maintain gas cleaning efficiency in the event of a pump failure.

*Solvent recovery plant.* Deliveries of solvents in drums and in bulk will arrive at the reception area where they will be transferred to reception tanks by vacuum system. Solvent will then be transferred to the feed tanks and then on the vacuum evaporator. After distillation the various fractions will be blended for sale. Residues are passed to the thermal unit for disposal. The solvent recovery plant is generally held under negative pressure by a vacuum pump, the exhaust from these pumps is scrubbed in an oil scrubber before being discharged to ATMOSPHERE. Emissions of non-methane HYDROCARBONS to the ATMOSPHERE will probably be of the order of 5–10 milligrams per second averaged over the day.

*Acid neutralization plant.* The acid neutralization plant will treat acids as well as the exhausted scrubber liquors, adjust their pH and remove the precipitated solids. The latter will be

**Figure 67** (*opposite*)   The scope for public involvement prior to producing the environmental statement.
*Source:* Council for the Protection of Rural England (CPRE) Environmental Statement, November 1990.

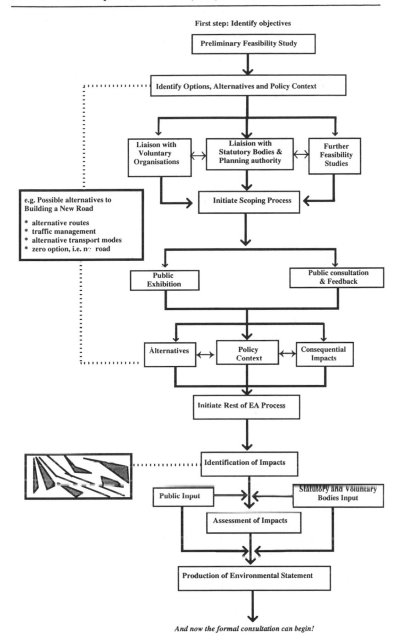

First step: Identify objectives

**Preliminary Feasibility Study**

**Identify Options, Alternatives and Policy Context**

**Liaison with Voluntary Organisations**

**Liaision with Statutory Bodies & Planning authority**

**Further Feasibility Studies**

**Initiate Scoping Process**

e.g. Possible alternatives to Building a New Road

* alternative routes
* traffic management
* alternative transport modes
* zero option, i.e. no road

**Public Exhibition**

**Public consultation & Feedback**

**Alternatives**

**Policy Context**

**Consequential Impacts**

**Initiate Rest of EA Process**

**Identification of Impacts**

**Public Input**

**Statutory and Voluntary Bodies Input**

**Assessment of Impacts**

**Production of Environmental Statement**

*And now the formal consultation can begin!*

prepared for removal and disposal at a landfill site. The cleaned effluent will be discharged to sewer.

One of the major impacts of the plant is its air pollution potential. The table on page 210 gives a comparison between the expected ground level concentrations (glc) and air quality standards and existing air quality determined by monitoring and also illustrates the great care taken in the preparation of this EIS.

Similar assessments for land use, water protection, visual impact and effects on the community, noise, traffic increase and employment were also performed to enable this project to proceed to the formal UK planning application stage. As with the Severn Barrage, the most important impacts are identified and appraised. This is followed by public scrutiny. Clearly, this is a much better way of proceeding with major projects than in the past when waste was buried first, and questions asked afterwards.

The inputs and outputs to the proposed RWTC are shown in Figure 68.

*Environmental Impact Appraisal (EIA) – developing countries*
The techniques developed for EIA generally depend heavily on data which is a major difficulty for developing countries (DCs). Techniques employing such things as value judgements, the experience of local communities and public participation – which may be more appropriate for developing countries, where socio-economic and socio-cultural impacts are often the greatest – are less developed (or non-existent).

Trade-offs have to be established between various sectoral impacts, and between environmental impacts and project modification costs in order to arrive at the environmental cost of a project and determine mitigating measures. So values have to be assigned to environmental resources, and there needs to be further work on valuation methodologies to make this possible.

EIA rarely takes account of broader socio-economic factors. Too often, assessments begin after the core development components have been identified. EIA must be incorporated into policy, programme and project design at the earliest planning stage. Reconciling competing physical, economic, ecological, social and other factors in development decision-making remains the key challenge in designing EIA tools towards development sustainability. This has a major relevance to DCs as often mining or logging is the principal activity.

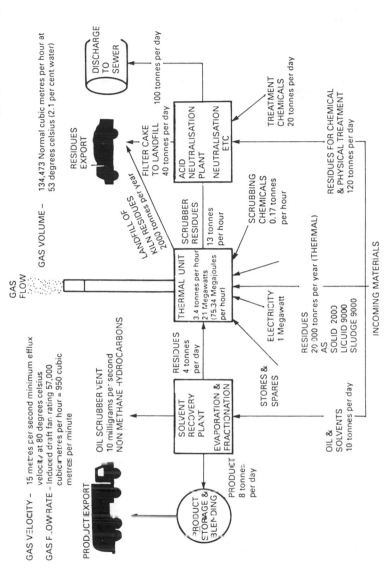

**Figure 68** Inputs and outputs to proposed Doncaster regional waste treatment centre.

Comparison between the ground level concentrations and air quality standards and existing air quality

| Substances | Existing glc ($\mu g/m^3$) | Additional future glc ($\mu g/m^3$) | Air quality standards ($\mu g/m^3$) |
|---|---|---|---|
| Hydrogen chloride | —* | 7.4 | 6 |
| Hydrogen fluoride | | 0.4 | 62.5 |
| Sulphur dioxide | 40 | 7.4 | 100 (mean) |
| Smoke | 17 | 3.6 | 100 (mean) |
| Nitrogen dioxide | 68 | 37 | 200** |
| Lead | — | <0.3 | 0.5 |
| Zinc | — | <0.3 | 125 |
| Mercury | — | <0.3 | 0.3 |
| Cadmium | — | <0.3 | 3 |
| Copper | — | 0.2 | 25 |
| Chromium | — | 0.3 | 1.25 |
| Nickel | — | 0.06 | 25 |
| Tin | — | 1.5 | 50 |
| Cobalt | — | 0.06 | 2.5 |
| Silver | — | 0.0096 | 0.25 |
| Arsenic | — | <0.3 | 5 |
| Hydrogen bromide | — | 0.3 | 250 |
| Antimony | — | <0.3 | 12.5 |
| Phosphorus | — | <0.3 | 50 |
| Silicon | — | <0.3 | 20 |
| Deposited dust (mg/m²/day) | 20 | 0.6 | 200 |

*Data not available. **98 percentile limit.

Just as EIA has tended to be an appendage in the planning process, so its socio-economic component has tended to be appendaged to the rest of it and is frequently not reflected in it. But people are the very centre of the process of sustainable development. The Hague Symposium of 1991 established a guiding principle: 'Every environmental measure must be tested against that yardstick: to what extent it adds to the human welfare of the majority of the world's population'.

There is no single best practice applicable to all countries, and EIA must be tailored to national needs and capabilities. Expertise in its management is lacking in developing countries. Capacity-building for sustainable development should be based on local requirements and socio-economic conditions in developing countries and countries in transition. It should begin with multifaceted and dynamic assessment of needs. Developing country practitioners must be involved in actual activities so as to enhance their skills in conducting, managing and monitoring EIAs and be aware that they do not convey certitude as they may be based on suspect data or value judgements, or both.

*Source: Our Planet*, Vol. 7, No. 1, 1995.

**Environmental Management Systems (EMS).** The British Standards Institution has now established EMS standard BS 7750. The main requirements are:

● Definition and documentation of the organizational structure.
● Drawing up an inventory of releases, wastes, energy. Raw materials usage to be documented.
● Inventory of legislative and regulatory requirements.
● Environmental effects; assessment.
● Setting objectives and targets.
● Environmental management plans.
● Management, documentation and records.
● Environmental audits; audit plan plus reports and follow-ups.
● Verification and testing.
● Personnel factors of awareness, training and qualifications.

**Environmental policy formulation.** The first step for a company wishing to improve its environmental performance is to carry out an assessment of the effects of a company's operations and, possibly, also its products, on the environment.

This should take account of both short and long-term effects and include assessment of the effects of future developments such as new production facilities.

The assessment should be comprehensive and include a review of energy usage and, where applicable, the production and disposal of waste.

The principal purposes of the review are to:

● provide the basis for formulating a statement of environmental policy,
● establish performance targets.

To be effective, such statements should set numerical targets.

**Environmental Protection Act (EPA) 1990.** The law to make provision for the improved control of pollution arising from certain industrial and other processes; to re-enact the provisions of the Control of Pollution Act 1974 (Her Majesty's Inspectorate of Pollution) relating to waste on land with modifications regarding the function of the regulatory and other authorities concerned in the collection and disposal of waste and to make further provision in relation to such waste. A major feature is INTEGRATED POLLUTION CONTROL.

**Environmental reports.** A major feature of environmentally aware industry is the annual publication of their environmental state of health which contains information on waste minimization, recycled or reused energy savings, emissions reduction, water consumption reduction, etc.

A new breed of 'environmental verifier' has emerged as a consequence prompted by the EU's ECO-MANAGEMENT AND AUDIT SCHEME.

The verifier conducts the following operations:

- Confirming whether data are gathered in an appropriate way.
- Ensuring they are reproducible.
- Noting whether data are presented in a fair way; for example, are diagrams and other statistical methods of presentation telling the real story or are they deliberately misleading.
- Assessing what methods company laboratories use and ensuring instruments have traceable calibration records.
- Assessing what calculation methods are used and whether they are applied correctly at the different sites.
- Confirming that the calculation methods and factors are recognized by the company's industrial sector and hence are defensible.

It should be noted that verification is not necessarily a guarantee of soundness at all times, and in all matters, but a prudent measure for acceptability.

*Source:* D. E. Evans and A.-M. Warris, 'Verification of Corporate Environmental Reports', *Environmental Assessment*, Vol. 3, No. 1, March 1995.

**Enzymes.** Proteinaceous substances that catalyse microbiological reactions such as decay or fermentation. They are not used up in the process but speed it up greatly. They can promote a wide range of reactions, but a particular enzyme can usually only promote a reaction on a specific substrate. (⇨ CATALYSIS; ENZYME TECHNOLOGY)

**Enzyme technology.** Enzyme technology covers a wide commercial area which embraces drug manufacture, beer malting, stabilization of foods, and protein production. Thus the enzymes known as cellulases ('-*ase*' indicates an enzyme; '-*ose*' is the substrate or product, e.g. cellulose) are the class of enzymes that promote the decomposition of cellulose, in this case to glucose. There are some 13000 micro-organisms that live on cellulose and they accomplish this by secreting enzymes which break down the substrate for a food source. One organism (*Trichoderma viride*) can produce a cellulase rich in a component capable of breaking down crystalline cellulose.

Using cellulose as an example, a commercial process for cellulose exploitation (CELLULOSE ECONOMY) requires that the micro-organism (in this case *Trichoderma viride*) is grown on a culture medium containing spruce pulp and nutrient salts. The culture is filtered leaving an enzyme solution which is then available to hydrolyse cellulose substrates, that is, convert the cellulose to sugars (HYDROLYSIS). So far tests have been conducted on milled newspapers; the cellulase solution and newspapers are placed in a reaction vessel under strictly controlled conditions (pH 4.8, 50°C). The time taken to break down the cellulose is of the order 20–80 hours and a product of glucose syrup obtained. Any non-utilized cellulose is recycled along with the enzymes. The yield of glucose is roughly 50 per cent of the original cellulose. The glucose is then used for FERMENTATION products and/or the production of SINGLE-CELL PROTEIN. Figure 69 shows a flow diagram for the process based on the US Army Natick Laboratory work.

Enzymes require strictly controlled conditions and are easily destroyed or inactivated by many substances. Feedstock purity and processing conditions must be closely monitored. Although not consumed in the reaction, they eventually degrade and must therefore be continually synthesized. In the cellulase reaction, product inhibition can eventually occur and the reaction cease, with the result that all the cellulase cannot be used in the one go. Multi-charging must then be used if all the cellulose is to be converted to glucose.

This area of enzyme technology has great promise for the production of single-cell protein from low-grade starchy materials, thereby upgrading animal feedstocks.

**Epidemiology.** The study of categories of persons and the patterns of diseases from which they suffer so as to determine the events or circumstances causing these diseases. If a cause is discovered,

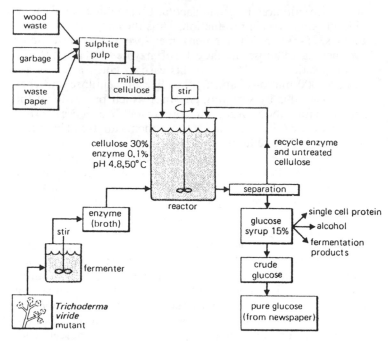

**Figure 69** Natick process for enzymatic hydrolysis of cellulose to glucose.

then those responsible for PUBLIC HEALTH policy can take appropriate steps to prevent the disease in question.

A classic case of epidemiology was Snow's identification of contaminated water from the Broad Street pump in 1854 as being responsible for the spread of CHOLERA in London.

Another example of the role of epidemiology is the evidence that smoke and SULPHUR DIOXIDE act together in causing and aggravating bronchitis. The Clean Air Acts brought a striking reduction in extra deaths due to these pollutants as shown in the data for London in the following table.

The epidemiologist sifts the evidence and identifies patterns and causes. We therefore know with reasonable certainty that in the UK, the CLEAN AIR ACT has done the job it was intended to do.

*Source:* J. J. Snow, *On the Mode of Communication of Cholera*, 2nd edn, Churchill, 1885.

| Year | Maximum daily concentration (micrograms per cubic metre) | | Extra deaths in Greater London |
|------|--------|----------------|----------|
|      | Smoke  | Sulphur dioxide |          |
| 1952 | 76000  | 3500           | 4000     |
| 1962 | 3000   | 3500           | 750      |
| 1972 | 200    | 1200           | —        |

From Royal commission on environmental Pollution, Fourth Report: *Pollution: Progress and Problems*, Cmd 5780, HMSO, December 1974.

**Erosion.** The lowering of the land surface by weathering, corrosion and transportation, under the influence of gravity, wind and running water. It can also apply to the eating away of the coastline by the sea. Raindrops can hit exposed soil with an explosive effect, launching soil particles into the air. In most areas, raindrop splash and sheet erosion are the dominant forms of erosion. It is intensified on sloping land, where more than half of the soil contained in the splashes is carried downhill. (⇨ DESERTIFICATION)

**Error.** The difference between an experimental result and the 'true' value. The uncertainty of an experimental result. If used, the meaning should be made very clear.

**Estuarial storage.** The storage of water for domestic water supplies in estuaries when suitable inland sites for reservoirs have been exhausted. In the UK, Morecambe Bay, the Solway Firth and the Wash have been suggested as potential sites, with a BARRAGE used to control or prevent the entry of sea water.

**Ethanol.** Ethyl alcohol ($C_2H_5OH$). It has a boiling point of 78.4°C and a specific gravity of 0.789. Ethanol is mainly synthesized from ethene in the petrochemical industry. It can also be made by the FERMENTATION of natural sugars, or those produced by HYDROLYSIS of refuse or cellulose waste. Its main use is as a solvent or chemical raw material. It can also be used as a motor fuel. Gross CALORIFIC VALUE 30 MJ/kg.

**EU legislation.** The Commission has the sole right of initiation of proposals. This does not mean that all ideas originate from the Commission, but wherever they come from, they must be turned into proposals by them.

The European Parliament's role is relatively limited. Under the main environment article in the Treaty (130) it is consulted only once. It has no responsibility for initiating proposals and no influence on the composition of Commissioners who are ultimately responsible for proposals that emanate from the Commission.

EC environmental measures have wide implications. The following are included or are imminent:

- new waste controls, including liability for environmental damage caused by waste;
- product safety controls;
- regulation of biotechnology;
- emission limits for small combustion plants;
- limits for discharges of dangerous substances into water;
- public access to environmental information;
- restrictions on the marketing and use of cadmium;
- environmental labelling of consumer goods.

**Eutrophication.** The natural ageing of a lake or land-locked body of water which results in organic material being produced in abundance due to a ready supply of nutrients accumulated over the years. Eutrophication can be greatly increased by man as a result of nitrates and phosphates from fertilizer run-off and sewage treatment processes. (⇨ MODERN FARMING METHODS; SEWAGE TREATMENT, PHOSPHATES)

A eutrophic lake is highly productive in organic material and can result in algal blooms which are short lived and whose decay imposes a heavy oxygen demand on the water. Thus nutrients from sewage and fertilizers can ruin a lake and cause the loss of a body of water which may be of use to man. This has happened in parts of the Great Lakes system where it has been estimated that Lake Erie received 37 500 tonnes of nitrogen from run-off and 45 000 tonnes from sewage in 1968 alone. The result of these man-made nutrients is supposed to have aged the lake 15 000 years quicker than if it were left to its own devices. (⇨ ALGAE; DISSOLVED OXYGEN; OLIGOTROPHIC)

**Excess air.** The additional COMBUSTION air required to ensure complete combustion of a FUEL. This is always stated as a percentage of the theoretical air required.

This ensures virtually complete combustion as measured in the exhaust gases by the low levels of CARBON MONOXIDE (whose presence indicates incomplete combustion).

**Exponential curve.** In many natural phenomena, in which, say, yeast organisms are reproducing, their rate of increase is not constant but rather the rate of increase itself is continually

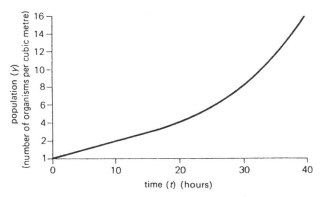

**Figure 70** Exponential growth curve for the number of organisms doubling every ten hours.

increasing because the organisms themselves double in number, in say 10 hours, and double again from the 10-hour number in 20 hours. Thus, at time zero, if the number of organism per cubic metre is 1 million, then in 10 hours there will be 2 million, in 20 hours 4 million, in 30 hours 8 million and in 40 hours 16 million.

Such a rate of growth is called *exponential growth* and a curve such as Figure 70 is known as an exponential growth curve. It is plotted as:

$$y = e^{kt}$$

where $y$ is the number measured at time $t$, and $k$ is a constant of proportionality determined by experiment; e is the base number of what are known as natural LOGARITHMS and has the value 2.7183.

Using the world population as our model, let

$n_o$ = population at time zero
$n_t$ = population at time $t$
$k$ = growth constant
$t$ = time in years

then

$$N_t = N_o e^{kt}$$

i.e. the population at any time $t$ is a function of the population at time $t = 0$ (that is, $N_0$) and the exponential constant $k$.

If we wish to know how long it will take for the world

population to double given that the annual growth rate is 2 per cent (this gives us $k$), we proceed as follows:

$$N_t = N_0 e^{0.02t}$$

and for population doubling

$$\frac{N_t}{N_0} = 2$$

that is,

$$2 = e^{0.02t}$$

Using natural logarithm tables where the natural logarithm of $e = 1$, we find that value which gives us 2, so that:

$$\text{nat. log } 2 = 0.02t$$

From the tables, nat. log $2 = 0.6931$, therefore

$$t = \frac{0.6931}{0.02} = 34.65 \text{ years}$$

Thus the doubling time for a 2 per cent increase is 35 years. This gives rise to a handy rule of thumb that for percentage growth rates less than 10 per cent, i.e. doubling time (years) = 70/(annual growth rate %). For example, with an annual growth rate of 5%, the doubling time would be 14 years.

We have dealt with exponential growth, but we can have exponential decay as well, which is represented by the equation

$$y = e^{-kt}$$

and is just as important in natural systems. For example, the atoms of radioactive elements emit particles (radiation), and in doing so decay to atoms of a different mass. The rate of decay at any time is proportional to the total number of unchanged atoms at that time and is characterized by the HALF-LIFE which is the time taken for half the number of atoms of any radioactive element to decay from an initial condition. The concept of half-life is illustrated in Figure 71.

**Exponential decay.** ⇨ EXPONENTIAL CURVE.

**Exposure–dose effect relationships.** The effect of exposure to a pollutant is a function of the pollutant, the type of target (e.g. animal or vegetable), and the concentration of pollutant and duration of exposure.

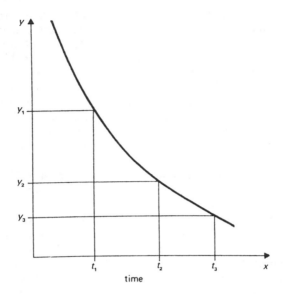

**Figure 71** Exponential decay of a parameter $y$. The figure shows a graph of $y = e^{-kt}$ upon which three points, $(t_1, y_1)$, $(t_2, y_2)$ and $(t_3, y_3)$, are picked out. The values of $y_1$, $y_2$ and $y_3$ are related by $y_1 = 2y_2$ and $y_2 = 2y_3$. The properties of the exponential curve are such that for this case $t_3 - t_2 = t_2 - t_1$. In other words, the time taken for $y$ to be reduced of half its initial value is the same, whatever that initial value $y$ is taken to be.

Short-term exposures to high concentrations are not usually equivalent to one-hundredth of the concentration for 100 times as long. The low-concentration long-period exposure may well have minimal effect, whereas the former may have a serious effect. It is for this reason that MAXIMUM EXPOSURE LIMITS are set. The converse can also be true in some circumstances. (⇨ TIME-WEIGHTED AVERAGE)

**Exposure limits.** Substances hazardous to health may cause adverse effects (e.g. irritation of the skin, eyes or respiratory tract, narcosis or cancer) through accumulation of the substances in the body or through the development of increased risk of disease with each contact. It is important to control both short-term and long-term exposure so as to avoid both types of effect. Two types of exposure limit are recognized. The long-term exposure limit is concerned with the total intake over long periods and is therefore appropriate for protecting against the effects of long-term exposure. The short-term exposure limit is aimed primarily at avoiding acute effects or at least reducing the risk of their occurrence. Specific short-term exposure limits are listed for those substances for which there is evidence of a risk of acute effects occurring as a result of brief exposures. For those substances for which no short-term exposure limit is listed, it is recommended that a figure of three times the long-term exposure limit averaged over a 10-minute period be used as a guideline for controlling exposure to short-term excursions.

Both the long-term and short-term exposure limits are expressed as airborne concentrations averaged over a specified period of time. The period for the long-term limit is normally eight hours; when a different period is used this is stated. The averaging period for the short-term exposure limit is normally 10 minutes; such a limit applying to any 10-minute period throughout the working shift. Ideally, monitoring should be continuous using fast response instruments. Ref. UK Health & Safety Executive 1991. (⇨ OCCUPATIONAL EXPOSURE STANDARD STANDARD)

**External combustion engine.** An engine where the heat source or fuel combustion is outside the engine as opposed to inside the engine as in the INTERNAL COMBUSTION ENGINE. External combustion allows a wide variety of fuels to be used efficiently to supply thermal energy to the engine. However, this class is mainly dominated by steam turbines and steam engines which are hardly suitable for motor cars but very suitable for ships, power stations, etc. An exception is the Stirling engine, where air or gas is trapped in a dual piston cylinder. When the gas is heated (externally), the working piston moves, doing work. As the motion continues a displacer piston moves hot gas to the cool end of the cylinder where, on cooling, it is compressed by the working piston and transferred by the displacer back to the hot end. This method of using two pistons in the one cylinder causes complexities and expense, but the Stirling engine is quiet, virtually non-polluting, can use any fuel and has prospects of increased thermal efficiency compared to the internal combustion engine.

**Extraction of oil from shales.** The gasification or extraction of oil from shales is a technique that may well increase as energy reserves decline. The process is carried out in retorts in four separate stages (from the bottom of Figure 72).

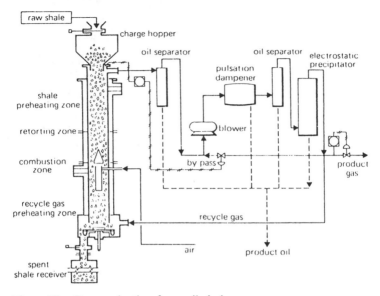

**Figure 72** Gas production from oil shales.

1. The recycled gas stream is preheated by the spent shale.
2. The gas is ignited and the heat of the combustion causes the oils in the shale to be driven off in the retorting zone.
3. The hot gases and oils are used to preheat the incoming shale.
4. The product gases and oils are cleaned in oil separators and an electrostatic precipitator is used to remove oil mist if necessary.

Part of the gas is recycled to run the retort and the remainder plus the oil is sold.

For questions concerning the applicability of the process on a scale sufficient to provide a significant amount of oil, see under 'oil shales'. ($\Rightarrow$ OIL SHALES)

# F

**Fallout (radioactive).** (i) The descent to the earth's surface of the radioactive particles produced in a nuclear explosion or accidental release. (ii) The radioactive particles themselves. (Fallout is sometimes broadened to include wash-out, i.e. the removal of any particulate matter from the ATMOSPHERE by rainfall.)

**Fatal Accident Rates (FAR).** This is a measure of fatalities per 100 million hours spent on one activity.

Highest fatal causes of death

| Cause of death | FAR |
|----------------|-----|
| Disease 40–44 age group | 17 |
| Car (all ages) | 15 |
| Home | 1.5 |
| Coal mining | 8 |
| Oil and gas rigs | 20 |
| Skiing in France (1974–76) | 130 |
| Rock climbing (rock faces) | 4000 |

The UK Health and Safety Executive classifies RISKS as intolerably high, as low as reasonably practicable (ALARP) and broadly tolerable. The FAR values for tolerable risks are:

Workers of all occupations 1/1000 per year (FAR 48)
Public 1/10000 per year (FAR 1.15)
Nuclear industry 1/100000 per year (FAR 0.1)

Such a computation puts the nuclear industry's stringent safety measures in perspective. Similarly, the risk of cancer from DIOXINS from INCINERATION to a maximally exposed individual (i.e. maximum exposure 24 hours per day 365 days per year) is

$$3 \times 10^{-7} \text{ or } 3/10000000 \text{ per year}$$

i.e. a FAR of 0.00345 or negligibly small and well below the Department of the Environment's acceptable annual increment of 1 in 1 million. However, perception is another matter. But, if rationality were to prevail, smoking would be totally banned and cars severely curtailed, yet the number of motor vehicles is predicted to double in the UK by ca. 2025.

*Sources:* L. Roberts, 'The Public Perception of Risk', Lecture at RSA 26 April 1995.
C. C. Travis, and B. P. Blaylock, 'Health Risks Associated with Air Emissions from MS Combustors', *American Chemical Society Conference*, Washington DC, 21–26 August 1994.

**Feed-conversion ratio.** The ratio of weight of feed to gain in weight (in kilograms) in fattening cattle or poultry, etc. Typical values lie between 3 and 10, depending on animal and feed analysis. The use of hormones to promote growth is one method of reducing the feed-conversion ratio. However, growth hormones are suspected CARCINOGENS. Even if this is not the case, it raises the question of just how far man can tamper with natural systems and living creatures for short-term gains but with unknown long-term consequences.

**Fermentation.** The decomposition of organic substances by micro-organisms and/or enzymes. The process is usually accompanied by the evolution of heat and gas, and can be aerobic or anaerobic. Examples:

*Alcoholic fermentation* (*anaerobic*).
$$C_6H_{12}O_6 = 2C_2H_5OH + 2CO_2$$
sugars         ethyl         carbon
               alcohol       dioxide

*Lactic fermentation* (*microaerobic* or *anaerobic*) – a common process for the production, preservation and seasoning of food.
$$C_6H_{12}O_6 \rightarrow 2CH_3CH(OH)COOH$$
sugars         lactic acid

*Acetic fermentation* (*aerobic*).
$$C_2H_5OH + O_2 = CH_3COOH + H_2O$$
ethyl alcohol       acetic acid

Aerobic fermentation of glucose solutions (and other substrates, e.g. HYDROCARBONS) by yeasts also leads to yeast growth which can then be harvested as a source of SINGLE-CELL PROTEIN.

The crucial parameters in industrial aerobic fermentation are the oxygen requirement per tonne of substrate and the removal of the heat evolved during fermentation. If the oxygen requirement per tonne of substrates is taken as 1 for molasses, then *n*-paraffins need 2.5 and methane 5 times as much oxygen. Thus a cheap substrate may have high energy costs due to oxygen transfer requirements. (⇨ AEROBIC PROCESSES; ANAEROBIC PROCESS)

**Fertilizer.** A chemical which promotes plant growth by enhancing the supply of essential NUTRIENTS such as nitrogen, phosphate and potassium. Fertilizers can be inorganic, such as ammonium sulphate ($(NH_4)_2SO_4$) or lime, or organic, such as SEWAGE sludge or manure. They can be added to the soil or, at low concentrations, sprayed on foliage as a foliage-feed.

Fertilizer run-off can be a major threat to watercourses and has already caused enrichment in the Great Lakes (⇨ EU-TROPHICATION)

Where natural fixation of nitrogen is taking place, as in a clover crop, the addition of inorganic nitrogenous fertilizers can inhibit the natural fixation. (⇨ NITROGEN CYCLE)

Fixation of nitrogen by man for fertilizers now equals the natural fixation rate, and the attendant increase in NITROGEN OXIDES may be a threat to the OZONE SHIELD. Such effects are cumulative, and farmers often find that, with time, they are using increasing amounts of fertilizer to achieve the same yield. In addition, nitrate leaching from fertilizers has been blamed for increasing nitrate levels in water supplies, so much so that the EC limit of $50\,mg/m^3$ has been substantially breached in parts of the UK. A novel solution of paying farmers not to use nitrate fertilizers in so-called water protection zones has now been implemented. (⇨ METHAEMOGLOBINAEMIA)

Plant breeders are now attempting to develop plants which will fix nitrogen directly from the atmosphere, and so eliminate or decrease the need for nitrogenous fertilizers.

Although some agronomists believe that further increases in farm production are possible by increased use of fertilizers, some leading agricultural chemists are convinced that we may be close to the economic limit of fertilizer use and that, with the exception of grassland, the economic limit may already have been exceeded. Furthermore, there is reason to suspect that artificial fertilizers alter the soil ecology in a detrimental manner.

**Fertilizer supplies.** The basic raw material of nitrogen fertilizers (e.g. ammonium sulphate, ammonium nitrate) is ammonia

($NH_3$), manufactured from atmospheric NITROGEN and HYDROGEN obtained from HYDROCARBONS such as methane or naphtha (a petroleum derivative). The price of nitrogen fertilizers has recently risen in line with the increasing price of oil, and developing countries are having great difficulty in maintaining their essential supplies without external aid.

**Fibrosis.** A scarring of the lung tissues, caused by dust inhalation. Almost all dust diseases, such as pneumoconiosis (coal-dust disease) and silicosis (stone-dust disease caused by mining, quarrying, shot blasting and stone-dressing operations) are characterized by a scarring of the lungs.

Other dusts which have been implicated in lung disorders include talc, fireclay, mica, china clay, graphite and ASBESTOS. Flax and hemp dust give rise to a disabling lung disease called byssinosis.

**Field moisture capacity.** The equilibrium amount of water retained in the SOIL after excess water has drained away. In the UK the soil generally reaches the field moisture capacity in winter and early spring. Thereafter, evaporation and plant growth removes moisture and causes a SOIL MOISTURE DEFICIT.

**Film badge.** A method of detecting individual exposure to IONIZING RADIATION. Consists of a masked photographic film which becomes progressively more 'exposed' on increasing or prolonged exposure to radiation.

**Filter medium.** This is the material used for example in water filtering such as sand or anthracite.

**Fines.** In the coal industry, coal having a maximum particle size usually less than 1.5 mm and rarely above 3 mm. In general, that fraction of a material that has been broken down into particles too small for a given use. If not properly contained, fines can cause dust pollution or the possibility of dust explosions.

**First law of thermodynamics.** ⇨ LAWS OF THERMODYNAMICS.

**Fission.** The spontaneous or induced splitting of heavy atoms (uranium, plutonium) into two roughly equal parts, thereby releasing large quantities of energy. (⇨ NUCLEAR ENERGY; NUCLEAR REACTOR DESIGNS)

**Fixation.** ⇨ NITROGEN CYCLE.

**Fixed carbon.** A measure of the primary productivity of an ecosystem based on the amount of carbon fixed by PHOTOSYNTHESIS per unit area. (⇨ CARBON CYCLE, PROXIMATE ANALYSIS)

**Flare.** The flame produced by the burning of surplus and residual gases at the top of a flame pipe at an oil refinery or other chemical industry factory or LANDFILL SITE. The gases cannot be released into the atmosphere owing to their unpleasant odour and to the explosion risk that would result. Flares can produce a great deal of smoke; this can be reduced by injecting steam or a water spray close to the point of ignition (this procedure, however, renders the flame noisy).

**Flash point.** The temperature at which a flammable liquid gives off sufficient vapour to catch fire when ignited.

**Flocculation.** The bringing together of divided material in sewage effluents so that a floc is formed (a gelatinous mass) which allows the solids to separate from the sewage effluent by settling. (⇨ SEWAGE TREATMENT; POLYELECTROLYTE)

**Flue gas desulphurization (FGD).** Now that ACID RAIN is on the political agenda, FGD is to be implemented on a large scale. The UK emitted 4.67 million tonnes of SULPHUR DIOXIDE ($SO_2$) in 1980 of which 2.87 million tonnes (61.5 per cent) was emitted by coal-fired power stations. The EEC has set an $SO_2$ target reduction of 60 per cent of 1980 baseline by 2003. Drax and Ratcliffe-on-Soar power stations have now been retrofitted (each 4000 MW output). These FGD costs are estimated at £1 billion (1988).

The wet limestone/gypsum FGD process is shown in Figure 73 in which the sulphur dioxide is absorbed by a limestone slurry spray to form calcium sulphite which is then oxidized to calcium sulphate where, together with water it forms gypsum ($CaSO_4 \cdot 2H_2O$). This FGD option is much to the fore for power stations as it is less costly than alternative methods where there is a ready supply of limestone available. It also has

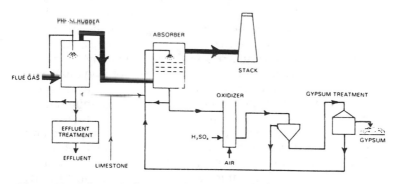

**Figure 73**  Schematic limestone/gypsum FGd plant.

gypsum as a possible saleable end product for wallboard manufacture.

*Source:* W. D. Halstead, *Flue gas desulphurisation – problems and potential*, CEGB Research, No. 22, Sept. 1988.

To get some idea of the dimensions of the problem, and FGD plant or plants that might be needed, consider Drax coal-fired power station with its output of almost 4000 MW. It consists of six boilers, each of 660 MW. Each boiler burns around 72 kg of coal per second, that is 432 kg of coal per second for the whole station. The coal contains, on average, 0.3 per cent chlorine and 2 per cent sulphur. In other words, the station produces 8.6 kg of sulphur per second, which is around 17 kg of sulphur dioxide per second, and about 1.3 kg of hydrochloric acid per second. Each boiler produces 1000 actual cubic metres (acm) of flue gas per second (the volume of a large four-bedroomed detached house and garage).

If we envisage each boiler having two FGD units, each FGD unit with its prescrubber has to handle 500 acm of flue gas per second carrying about 1.5 kg of sulphur dioxide and just over 100 g of hydrochloric acid, together with whatever fly ash makes it past the ELECTROSTATIC PRECIPITATORS at the exit to the boiler.

In terms of input for the whole plant: 17 kg of sulphur dioxide per second will require 27 kg of limestone per second on a stoichiometric basis, that is, 850 000 tonnes of limestone per year, which produces, in theory, around 1.5 million tonnes of gypsum cake per year. In practice, ca. 800 600 tonnes gypsum per year will be produced as the station will not be on 100 per cent load continually and not all the $SO_2$ will be removed. The actual limestone requirements will approximate 0.5 million tonnes per year, which is to be compared with the ca. 200 million tonnes of limestone quarried in the UK annually.

The possible environmental disbenefits of the gypsum FGD process are the need for high grade limestone (up to 1.5 million tonnes per year for the currently planned programme) and the need for tipping space for the calcium sulphate if there is not enough wallboard manufacturing capacity or if the manufacturers' specifications cannot be met. The alternative (regenerative FGD process) which produces $H_2SO_4$ as a by-product, suffers from possibly high running costs rendering the sulphuric acid produced perhaps too expensive in an already established market.

The continued growth of gas-fired power stations and flue gas desulphurization use should ensure a progressive reduction in sulphur dioxide emissions to one-fifth of their 1980 level by 2010. Coal burning in the UK will probably stabilize at 30 million tonnes per year, roughly half of it consumed in the Drax and Ratcliffe-on-Soar power stations, both of which have flue gas desulphurization fitted. The UK is well in advance in this area and when the gas runs out, clean coal gasification can take over.

*Source:* J. Redman, 'FGD. The Wet Limestone/Gypsum Process', *The Chemical Engineer*, Oct. 1988, 29–36 (and subsequent correspondence).

**Flue gas recirculation.** A means of effecting NITROGEN OXIDE reduction in exhaust gases by recirculating a proportion (usually 20 per cent) of the exhaust gas with the COMBUSTION air.

**Flue gas scrubbing.** The removal of acidic pollutants from the flue gases from installations such as incinerators. A single, two- or multi-stage Venturi scrubber using large quantities of cleaning fluid can be utilized. The gases are cooled and saturated with slaked lime or sodium hydroxide depending on the treatment required.

There are other methods available using ammonia as well as so-called regenerative methods.

*Advantages:*
—high degree of removal of acid gases.

*Disadvantages:*
—extensive equipment required
—flue gas reheating is necessary
—waste water may require treatment
—highly visible wet plume
—greater running costs than dry or semi-dry systems (which can have lower removal efficiencies).

The flue gases from the furnace are conveyed through the boilers to the flue gas cleaning system by means of induced-draught fans. Before entering the Venturi scrubber, the hot flue gases are usually circulated through a regenerative glass tube exchanger and thereby transfer some of their heat to the cold flue gases.

Inside the Venturi scrubber the hydrochloric acid (HCl), sulphur dioxide ($SO_2$) and hydrogen fluoride (HF) contents are

reduced by their interaction with the dosing chemicals and in
the process they are cooled down to saturation temperature and
cleaned. Following this, the flue gases are reheated in the
regenerative gas preheaters before being discharged through
the chimney. Then necessary limestone powder is mixed in the
mixing tank with filtrate from the gypsum drain and pumped to
the absorbers. Typical reactions in the scrubber are given below.

Acid gas removal with slaked lime ($Ca(OH)_2$):

$$Ca(OH)_2 + 2HCl \rightarrow CaCl_2 + 2H_2O$$
$$Ca(OH)_2 + 2HF \rightarrow CaF_2 + 2H_2O$$
$$Ca(OH)_2 + SO_2 + \tfrac{1}{2}O_2 \rightarrow CaSO_4 + H_2O$$

This may be more widely used in the future as emissions
legislation is progressively tightened.

Alternatives to the wet method include the semi-dry process
which uses a limestone water SLURRY. A typical semi-dry layout
is given in Figure 74.

Or the dry method may be used, in which the reaction agent
(limestone) is either mixed with the fuel or refuse and introduced
separately into the furnace.

*Advantages:*
—little apparatus needed
—no reheating necessary
—no waste water

*Disadvantages:*
—degree of acid gas removal relatively low compared with
  semi-dry or wet methods
—high limestone demand
—the reaction product is mixed with ash

**Fluff.** The shredded rubber, plastics and carpets residues from car
recycling. Currently, it is mainly used as a FUEL in INCINERATION
plants (or landfilled in the UK).

The increasing use of PLASTICS in car manufacture means
that RECYCLING efforts could be increased. However, it is
important to realize that there are limited markets for low grade
plastic materials and the use of waste to energy can be both a
cost and environmentally effective solution and much better
than LANDFILL. Figure 75 gives a typical breakdown of the
components used in car manufacture, about 16 per cent of
which can result in fluff production.

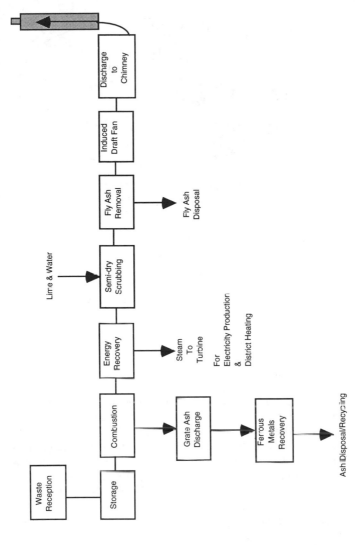

**Figure 74**  Layout for semi-dry process of flue gas scrubbing.

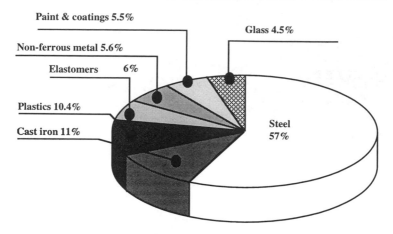

**Figure 75**   Breakdown of components used in car manufacture.
*Source: Financial Times*, 26 April 1995.

**Fluidized bed combustion (FBC).** A form of solid fuel or low
calorific value waste COMBUSTION system in which the fire bed,
composed of inert particles to which the fuel is added, is
fluidized by the combustion air blown upwards through it. This
produces highly efficient combustion and allows the use of
low-grade fuels and combustible wastes that are not suitable for
conventional combustion plant designs.
    Fluidized beds have evolved as:

1. *Bubbling fluidized beds* (BFB) where the combustion is in a
   conventional bubbling bed.
2. *Circulating fluidized beds* where the bed medium is entrained
   and circulated with the combustion gases. A CYCLONE
   separates the bed material and returns it back to the main
   chamber. This enables excellent combustion efficiency at low
   excess air levels, as well as under-staged combustion and low
   furnace temperature conditions, qualities which ensure
   complete combustion and extremely low $NO_x$ emissions.
   The vigorous fluidization and intensive mixing achieved in
   the furnace ensure highly uniform distribution of the fuel/air
   mixture across the combustion zone, minimizing the risk of a
   reducing atmosphere and the attendant risk of furnace wall
   corrosion. The large quantity of circulating particulates

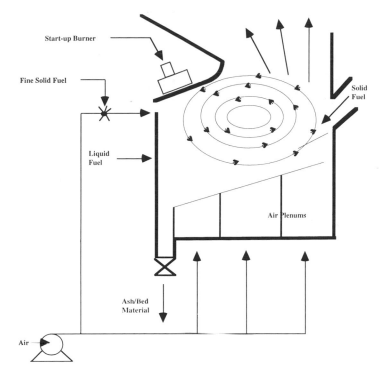

**Figure 76** Revolving fluidized bed.

effectively equalizes the vertical temperature gradients, creating a uniform temperature throughout the furnace. In addition, the bed acts as a thermal 'flywheel' to level out the effect of variations in the fuel quality on the combustion process.

3. *Revolving fluidized bed* where the use of differential air pressures and special bed geometry enables the bed to 'revolve'. This ensures high efficiency combustion of difficult materials such as municipal solid waste. Figure 76 illustrates a revolving fluidized bed.

FBC is receiving considerable interest as part of solid waste management solutions. The Solid Waste Authority of Palm Beach, Florida, is installing this option as it allows optimal RECYCLING followed by the FBC of the residual WASTE

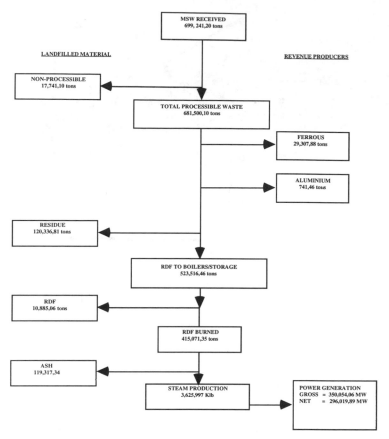

**Figure 77** Palm Beach, Solid Waste Authority mass balance (US tons)/ performance data (1991).
*Source:* The Solid Waste Authority of Palm Beach, Florida, 1991.

DERIVED FUEL. The (1991) mass balance and performance data are given in Figure 77.

**Fluoridation.** The addition of a fluoride salt in trace quantities to public drinking water for improving the resistance to dental caries. The aim is to adjust the fluorine content to the optimum amount of 1 part per million. Many studies have shown that when a population is supplied with water containing such a concentration the incidence of dental decay is at a minimum in the population.

**Fluorides.** Salts of hydrofluoric acid. They are released into the environment by ceramic works, aluminium smelting and phosphate works. The fluoride particles can be deposited on the herbage in the vicinity of process plants, and there have been many recorded cases of cattle suffering fluorosis which leads to loss of teeth and bone growths at joints, giving rise to lameness. Fluorides can also affect plants by entering the stomata and then moving to leaf margins where they accumulate. Extremely small concentrations in air, as low as 0.005 part per million, will blight maize, and 0.001 part per million lowers citrus productivity (US data). (⟹ GLADIOLI)

The release of HYDROGEN FLUORIDE in concentrations in excess of 1 microgram per cubic metre ($1\,\mu g/m^3$) can lead to herbage fluoride levels in excess of 30 to 35 parts per million, at which level dairy cattle are at risk.

**Fluorine (F).** A greenish-yellow gas, and the most reactive element known. It is highly poisonous. Fluorine is used in the manufacture of HALOGENATED FLUOROCARBONS or CHLOROFLUOROCAR-BONS used in AEROSOL PROPELLANTS.

**Fluorosis.** ⟹ FLUORIDES.

**Flux.** The rate at which a particular quantity is transferred per unit area, e.g. $m^3/m^2/day$.

**Fly ash.** The finely divided particles of ash readily entrained in the flue gases arising from the combustion of fossil fuels (mainly coal). The particles of ash may contain unburnt fuel. (⟹ ELECTROSTATIC PRECIPITATOR)

**Fog.** Microscopic water droplets, varying from 2 to 20 micrometres in diameter, suspended in air and reducing visibility. One form, ice fog, occurs in regions where the ambient temperature is less than $-20^\circ$C. It is formed by the spontaneous nucleation and freezing of water vapour in combustion gases from power stations or vehicular exhausts in Arctic regions.

Both forms of fog substantially reduce visibility. Ice fog has been known to cut visibility down to less than 2 metres, as can an extremely dense 'pea-souper'.

In urban areas the density of fog is closely related to the amount of particulate material present in the air and upon which moisture can condense easily as droplets producing SMOG.

**Food chain.** A series of organisms through which energy is transferred. Each link feeds on the one before it (except for the first one which is herbage – see Figure 78) and is eaten by the one following it. Herbage is said to belong to the class known as

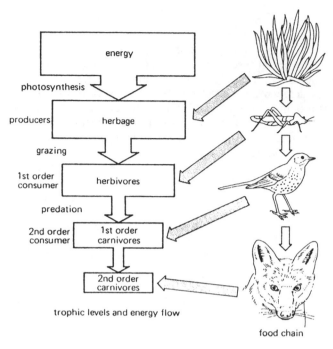

energy

photosynthesis

producers | herbage

grazing

1st order consumer | herbivores

predation

2nd order consumer | 1st order carnivores

2nd order carnivores

trophic levels and energy flow

food chain

**Figure 78** Trophic levels and energy flow, and an example of an associated food chain.

producers, which transform SOLAR ENERGY and CARBON DIOXIDE into sugars via PHOTOSYNTHESIS. The remaining links are consumers. Consumers are ordered first, second, and so on. Thus a three-stage food chain would be grass consumed by cattle consumed by man, i.e. producer, first-order consumer, second-order consumer. A food chain is essentially an energy conversion scheme and as in any energy conversion device there are conversion efficiencies and transfer losses. Hence, the energy fixed by producers will always be greater than that fixed by first-order consumers, which in turn will always be greater than that fixed by second-order consumers, and so on (Figure 78).

Those organisms whose food is obtained from green plants and have the same number of links in their chain are said to occupy the same trophic or energy level. Thus a cow and a rabbit are both at the same trophic level. The shorter the chain, the more food there is available, e.g. 5 kilograms (5 kg) grain may at best produce 1 kg live-weight gain on cattle, and 10 kg meat may produce only 1 kg gain in man. The feed-conversion efficiency is usually very much less than this. If a link can be cut out of the chain, e.g. by man consuming grain directly (he may still need a protein supplement), then 5 kg grain may produce 1 kg in man instead of 0.1 kg as formerly. (⇨ FOOD WEB, ECOLOGICAL PYRAMIDS, ECOSYSTEM)

**Food web.** A system of interlocking FOOD CHAINS. A food web comprises all the separate food chains in a community, including DECOMPOSERS whose role is vital in recycling essential materials and nutrients. (CARBON CYCLE; NITROGEN CYCLE)

**Forest management.** The Forest Stewardship Council, a Scandinavian consortium of forest owners and some environmental groups has drawn up forest stewardship council guidelines.

Forest management operations should:

1. respect the laws of the country it is operating in and international treaties and agreements to which the country adheres;
2. define and legally establish long-term tenure and rights to use land and forest resources;
3. respect the legal and customary rights of indigenous peoples to own, use and manage their lands and resources;
4. support the social and economic well-being of forest workers and local communities;
5. encourage the optimal use of forest products and services to ensure economic viability;
6. maintain the critical ecological functions of the forest and minimize adverse impacts on biological diversity, water resources soils, non-timber resources and ecosystems and landscapes;
7. write, implement and keep up-to-date an appropriate management plan;
8. conduct regular monitoring that assesses the conditions of the forest, the yields of forest producers, the chain of custody and management operations;
9. not replace natural forests by tree plantations. Plantations should complement natural forests and reduce pressures on them.

*Source: Tomorrow, Spring, 1995.*

**Fossil fuels.** Term for coal, oil or natural gas, i.e. fuels derived from

ORGANIC MATTER deposited over geological time scales. (Uranium and vegetable materials are not fossil fuels.)

**Free radical.** A molecular fragment or an ION that has one or more unpaired electrons, rending it highly reactive. Free radicals are very short-lived in gaseous systems, but their high reactivity enables them to take part in chemical reactions that would not otherwise occur, e.g. certain free radicals play a significant role in the production of the constituents of PHOTOCHEMICAL SMOG.

**Freezing.** ⇨ DESALINATION.

**Freon.** A proprietary brand of CHLOROFLUOROCARBONS; not a generic name.

**Frequency.** A measure of rate of vibration or oscillation given by the inverse of the period of simple harmonic motion. It represents the number of complete oscillations or cycles per second. The frequency of sound is measured in hertz (Hz).

**Froth flotation.** Process of using a froth of water and oil to effect the separation of finely divided minerals. Also proposed for separating ground glass for CULLET purposes from ash and other fines in DOMESTIC REFUSE.

**Fuel.** A source of thermal ENERGY. Fuel can be bacterial, fossil, vegetable or nuclear in origin. (⇨ CELLULOSE ECONOMY; COAL; ENZYMES; ETHANOL; GAS, NATURAL; HYDROGEN; METHANE; METHANOL; NUCLEAR ENERGY; OIL; OIL SHALES; TAR SANDS) The list below shows the calorific value of various fuels.

| | |
|---|---|
| Gas oil | 45.5 GJ/t |
| Light fuel oil | 42.5 GJ/t |
| Heavy fuel oil | 41.8 GJ/t |
| Anthracite (rank 100) | 30 GJ/t |
| Industrial boiler (rank 80) | 25 GJ/t |
| Butane | 49 GJ/t |
| Natural gas | 38.60 MJ/$m^3$ |

**Fuel, authorized.** An authorized fuel, which under regulations made under the UK Clean Air Act 1956, can be burned in smoke control areas. The combustion of such fuels is not necessarily completely smokeless.

**Fuel cells.** One of a class of devices known as DIRECT ENERGY CONVERTERS. An individual fuel cell consists of an electrolyte sandwiched between an anode and a cathode. At the anode, the hydrogen splits into ions and electrons. The electrons released pass to the cathode via the external load. The ions are transferred through the electrolyte, completing the electric circuit, reacting with the oxygen supplied to the cathode to produce water. The fuel cell directly converts chemical energy to electrical energy without the intermediate step of random molecular energy, e.g. the raising of steam which needs a turbine and electrical generator before electricity is available. Thus, the fuel cell is *not* subject to the CARNOT EFFICIENCY restriction of the Second Law of Thermodynamics which means that conversion efficiencies of 90 per cent and greater can in theory be obtained. The fuel cell requires hydrogen or a hydrocarbon fuel and oxygen from the air. In a simple hydrogen/oxygen fuel cell, there is an electrolyte and two non-consumable electrodes which catalyse the ionizing reactions. Thus at the anode, or positive electrode, hydrogen $(H_2)$ decomposes to $2H^+ + 2e^-$ (2 hydrogen ions + 2 electrons) and at the cathode, or negative electrode, 2 electrons and 2 hydrogen ions combine with a half-molecule of oxygen (written as $\frac{1}{2}O_2$) to form water, i.e.

$$2e^- + 2H^+ + \tfrac{1}{2}O_2 - H_2O$$

The voltage from the reaction is 1 to 1.5 volts per cell and any voltage can be obtained by connecting fuel cells in series as with ordinary batteries. ($\Rightarrow$ LAWS OF THERMODYNAMICS)

Fuel cells using coal or oil have been mooted but so far the hydrogen/oxygen cell is currently the only practicable one. The costs of power from fuel cells is not cheap and the hydrogen is usually produced by ELECTROLYSIS which demands primary energy consumption to produce the electricity, which means in turn that a system comprising a nuclear reactor and boiler plus turbine plus fuel cell has an *overall efficiency* of 30 per cent or less when the allowance is made for the component efficiencies.

**Fuel cycle.** The sequence of steps involved in supplying and using fuel for nuclear power generation. The main steps are mining and milling, extraction, purification, enrichment (where required), fuel fabrication, irradiation (burning) in the reactor, cooling, reprocessing, recycling, waste management and disposal. ($\Rightarrow$ NUCLEAR POWER)

**Fume.** There is no generally accepted definition of the word, but it is usually taken to mean minute particles less than 1 micrometre in diameter suspended in air or flue gases or, in the UK Clean Air Act 1956 definition (Section 34) 'fume is any airborne solid matter smaller than dust'. Fumes are usually released as a result of certain metal working and chemical processes.

The term is often used to describe the vapours given off by a liquid, especially if it is offensive or toxic. (⇨ FUMES, EFFECTS OF; VOLATILIZATION)

**Fumes, effects of.** When certain metals are heated, they tend to volatize and fumes (i.e. particles less than 1 micrometre in size) can spread easily. Those of manganese and zinc can produce an effect called metal fever.

Fumes released in plastics manufacture can also cause numbness, cramps and impotency. Teflon (PTFE) has been so documented, and recently vinyl chloride disease has been identified as being linked to the fumes from vinyl chloride monomer (VCM), as has ANGIOSARCOMA, a rare form of liver cancer. (⇨ POLYVINYL CHLORIDE)

**Fuming (liquids).** Liquids which emit vapours which dissolve in water to form a mixture with a lower VAPOUR PRESSURE than pure water.

**Fungi.** Simple plants either unicellular or made up of cellular filaments; they contain no CHLOROPHYLL. They are agents of decay in all natural organic materials, food, timber, plant debris, etc.

**Furans.** In an environmental context this term refers to a group of 135 compounds known as chlorinated dibenzofurans. They resemble the chlorinated dibenzo-*p*-dioxins (DIOXIN) in their chemical structure, toxicity and behaviour in the environment. Chlorinated dibenzofurans are formed during the incomplete combustion of PCBs and are thought to be responsible for some of the adverse health effects of these compounds. Below is illustrated the structure of dibenzofuran: chlorine can be added to the carbon atoms marked X.

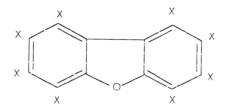

**Furfural ($C_4H_3OCHO$).** Organic solvent or intermediary which can be obtained by acid HYDROLYSIS of PENTOSANS contained in waste agricultural products to pentoses and thence to furfural. Almost any plant material can be used and the production of furfural from these sources is receiving considerable attention as a means of utilizing 'income' resources. In the USA, over 2 million tonnes could be produced from maize cobs alone. Almost any petrochemical can be made from it including nylon intermediates. (The potential yield from sugar cane alone is 50 million tonnes per year.) (⇨ SOLAR ENERGY)

**Fusion.** ⇨ NUCLEAR ENERGY; NUCLEAR FISSION; NUCLEAR REACTOR DESIGNS.

# G

**Garbage.** ⇨ DOMESTIC REFUSE.

**Gas, inert.** A gas that does not react with materials with which it is in contact, e.g. nitrogen, which is widely used in industry because of its inert properties.

**Gas, liquefied petroleum (LPG).** A gaseous mixture of light hydrocarbons, whose maintained principal components are propane, propene, butanes and butenes, liquefied by increased pressure or lowered temperature.

**Gas, natural.** Gas that occurs naturally, usually in association with oil reserves (though decreasingly so as gas exploration techniques improve). It is an exceptionally clean and convenient fuel whose use for low-grade purposes (e.g. raising steam) will probably be banned in most countries within a few years for the simple reason that it is too valuable to be burned, except in economies where unrestrained market forces prevail. The energy content of ultimately recoverable world resources is of the same order for gas as for oil. Cheap reserves of gas may be slightly closer to depletion than those of oil, however, because growth in demand has been more rapid. Gas now provides one-third of energy in the USA and is the sixth largest industry. The USSR apparently holds about 39 per cent of ultimately recoverable world gas resources, and North America and the Middle East about 37 per cent (1988 proven reserves).

Gas is far more costly to transport overseas than oil. Great efforts are being made to augment pipelines with liquefied natural gas (LNG) marine carriers, so as to export more gas from surplus to deficit areas and to help eliminate the flaring that still wastes a substantial fraction of the gas. (⇨ LANDFILL GAS)

**Gaseous pollutants, control.** ⇨ THRESHOLD LIMITING VALUE; THREE-MINUTE MEAN CONCENTRATION; EMISSION STANDARD.

**Gases, properties of.** Gases are compressible and their density is proportional to pressure (all other things remaining constant). They expand in proportion to temperature and thus the greater the temperature, the lighter or less dense is the gas. A minimum gas exit temperature may be specified in air pollution discharge consents, to increase plume buoyancy (plus avoiding ACID DEWPOINT effects).

**Gasification.** The production of gaseous fuels by reacting hot carbonaceous materials with air, steam or oxygen. ($\Rightarrow$ PYROLYSIS)

The process takes place at high temperature. The gasification product is a mixture of combustible gases and tar compounds, together with particles and water vapour. Depending on the gasification method, the proportion of components varies, but common to all the processes is that the gas has to be purified before it can be used directly in a gas engine or a gas turbine.

Gasification of wastes makes it possible to achieve high electrical efficiency. In smaller plant sizes 30 per cent electrical efficiency can be reached by means of gas engines. As far as larger plants are concerned, combined cycles (gas turbines in combination with heat recovery boilers and steam turbines) could achieve efficiencies of 45 per cent or more, provided that the gas purification is of a very high order.

There are three main parts in a gasification system: the gasifier, gas clean-up and the power producer. Thus, end use dictates the amount of clean-up.

GASIFIERS can be divided into two main types, namely fixed beds and fluidized beds. In the fixed bed large amounts of fuel are constantly present in the gasifier. In the fluidized bed the residence time for the fuel in the gasifier is short. The fixed bed technology is more suitable for small/medium size (1–20 MWt) and non-homogenous fuel, the fluidized bed for medium/large size (20–200 MWt) and well-defined and finely partitioned fuel. Typically, a 200 tonne per day plant operating on screened municipal solid waste will produce 8300 $Nm^3$ (Normal $m^3$) with a calorific value of 5 $MJ/Nm^3$.

The untreated gas has a high tar and particulate content and has to be purified before being used in a gas engine. If the gas is burned in a boiler or chamber, a simple CYCLONE could be sufficient, whilst for use in a gas turbine it is necessary to have a high quality gas without tar, particles and alkaline metals. ($\Rightarrow$ PYROLYSIS, FLUIDIZED BED COMBUSTION)

*Source:* A. Donati, 'Different technologies for waste treatment', *Waste to Energy*, March 1995.

**Genetic engineering.** A popular term for techniques which interchange DNA sections between individuals of the same or different species. (⇨ RECOMBINANT DNA TECHNOLOGY)

**Genetic load.** ⇨ IONIZING RADIATION.

**Genetic pollution.** The process by which genes from RECOMBINANT DNA organisms may escape and become incorporated into wild species. For instance, genes conferring resistance to herbicides may escape, in pollen, from a crop plant and be expressed in a related weed species, rendering the herbicide ineffective for controlling that weed.

**Geographic Information Systems (GIS).** Method used to assess sites for WIND TURBINES and other large landscape features. Ordnance Survey digital contour data are superimposed by zones of planning restraint, e.g. green belt, Areas of Outstanding Natural Beauty, thereby filtering out the no-go areas and allowing the position of individual wind turbines and their height in relation to the surrounding landforms to be assessed. These can be changed to assess how moving the columns or varying their height may reduce their visual impact from cherished viewpoints and main routes.

**Geothermal power.** The use of energy from the earth's interior conducted to the surface in a few areas of the globe where igneous rocks are in a molten or partly molten state usually within 10 km of the earth's surface. To be useful the energy must be available in superheated water or steam form and it may then be used in a conventional power plant. Italy, USA, Iceland, USSR and New Zealand all have suitable geothermal energy fields. This method of power generation is a strictly local and usually small-scale affair although very useful to those areas where it occurs. As a fraction of global energy requirements it is very small.

The use of so-called 'hot rocks' up to 10 km under the earth's crust could also be considered as a free energy source – but at what price to extract the thermal energy? Just as with other renewable sources of energy, economic feasibility is all important, and careful evaluation is vital as the money spent on prestigious projects may be more effectively applied in say domestic energy saving through improved house insulation.

**Gladioli.** Flowers which are useful indicators of the presence and concentration of airborne FLUORIDES. The leaves mottle and turn yellow-brown in the presence of concentrations as low as 0.5 micrograms per cubic metre. Skilled interpretation is of course required. The variety 'Snow Princess' is most commonly used as the most sensitive, while other varieties are progressively more resistant to fluoride damage. ($\Rightarrow$ BIOLOGICAL INDICATOR)

**Glass.** Common (window) glass is a mixture of 70 per cent silica ($SiO_2$), 14 per cent lime and magnesia ($CaO + MgO$), 12 per cent soda ($Na_2O$) and 1–2 per cent alumina, ferric oxide and trioxide. For glass bottle manufacture, CULLET may be added both as a means of reducing energy consumption per unit of product and to encourage the development of a market for recycled empty bottles collected via BOTTLE BANKS. ($\Rightarrow$ RECYCLING)

European glass recycling 1991

| Country | Tonnes collected | Share of national consumption (%) |
|---|---|---|
| Austria | 155 700 | 60 |
| Belgium | 223 150 | 55 |
| Denmark | 60 000 | 35 |
| Finland | 15 400 | 31 |
| France | 986 940 | 41 |
| Germany | 2 295 430 | 63 |
| Greece | 26 000 | 22 |
| Ireland | 16 250 | 23 |
| Italy | 763 000 | 53 |
| Netherlands | 360 000 | 70 |
| Norway | 9 600 | 22 |
| Portugal | 50 370 | 30 |
| Spain | 310 000 | 27 |
| Sweden | 57 000 | 44 |
| Switzerland | 198 960 | 71 |
| Turkey | 54 110 | 28 |
| United Kingdom | 385 390 | 21 |

*Source:* British Glass.

**Glucose.** Often called the key sugar as it is important to both plants and animals as an energy-producer. It is a monosaccharide hexose ($C_6H_{12}O_6$) and is present naturally in fruits. It also results from the HYDROLYSIS of other sugars and starches such as sucrose, lactose, maltose, cellulose and glycogen. An example of this in commercial use is the enzymic conversion of maltose and sucrose to glucose by yeast in the brewing process. ($\Rightarrow$ SUGARS; ENZYME TECHNOLOGY)

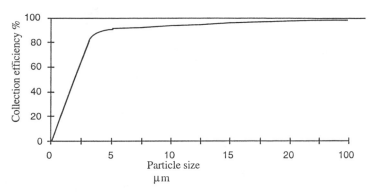

**Figure 79**   Grade efficiency curve versus particle size.

**Grade.** Measure of the quality of an ore. The term is also used in expressing the size of particulate matter as a 'grade distribution'.

**Grade-efficiency curve.** Figure 79 shows a typical curve for a SPRAY TOWER used to remove particulates from a gas stream.

**Granulomas.** Multiple accumulation of cells distributed in nodules in the lung. It is not particularly common. BERYLLIUM, tungsten carbide and zinc dusts have been responsible.

**Gravel pack.** A gravel screen around a perforated well casing. This increases the effective well diameter and filters out sediments. Gravel packs are also used in LANDFILL GAS extraction wells.

**Gray (Gy).** ⇨ IONIZING RADIATION DOSE MEASUREMENT.

**Greenhouse effect.** The mechanism whereby incoming solar radiation is trapped by a glass sheet or the presence of CARBON DIOXIDE and other GREENHOUSE GASES in the atmosphere. As these gases are transparent to solar radiation, the short-wave incoming radiation is transmitted. However, they are opaque to long-wave re-radiation from the earth's surface or from any other object underneath, thus heat is trapped and the underlying surface is thereby warmed.

This effect is put to good use in a greenhouse, but fears have been expressed that the earth's surface temperature could rise due to the build up of carbon dioxide and other greenhouse gases in the atmosphere. The attendant solar heat gain could disrupt CLIMATE patterns. The best estimates by the UN Panel on Climate Change (1990) for temperature changes by 2030 are:

Central North America: increased temperature between 2°C to 4°C in winter and 2°C to 3°C in summer. South-east Asia: warming between 1°C and 2°C throughout the year. Southern Europe: up 2°C in winter and between 2 and 3°C in summer. Australia: up 1°C to 2°C in summer and 2°C up in winter. Sahel (area south of Sahara): up between 1°C and 3°C generally.

There are sufficient reserves of fossil FUELS which when burnt will increase the atmospheric carbon dioxide concentration by up to 8 times. (⇨ ATMOSPHERE; SOLAR HEATING)

**Greenhouse gases.** Collective term for those 'gases' that have influence in the GREENHOUSE EFFECT, i.e. CHLOROFLUOROCARBONS, CARBON DIOXIDE, METHANE, NITROUS OXIDE, OZONE and WATER VAPOUR. This term links all gases together, although some components are much more damaging than others, e.g. a molecule of methane is ca. 30 times more powerful in 'greenhouse' terms than a molecule of carbon dioxide, and CFCs are estimated to be up to 10000 times more powerful than carbon dioxide. Yet public debate tends to focus on the latter.

The best estimate of the relative contributions to the greenhouse effect of these gases is that carbon dioxide is the greatest single contributor (50 per cent), CFCs 15 per cent, methane 20 per cent, ozone 10 per cent, water vapour and nitrous oxide 5 per cent.

Much attention has been focused on carbon dioxide and the claim that a greatly expanded nuclear energy programme will safeguard the environment by containing the greenhouse effect. The reality is that coal-burning emissions contribute around 15 per cent of the global greenhouse effect (power generation 7 per cent). In the UK context, annual consumption of coal is ca. 35 million tonnes out of a total world-wide production of 3 billion tonnes (1989 figures).

Carbon dioxide and nitrogen dioxide emissions from aircraft engines are currently thought to be responsible for about 5 per cent of the greenhouse effect, and air transport world-wide is expected to double in 15 years. However, the Intergovernmental Panel on Climate Change (IPCC) recommended that international air traffic be excluded from national targets for greenhouse gas emissions as it is too difficult to allocate emissions from international flights to individual countries.

The IPCC has issued an update on methane and nitrous oxides as shown in the table. Note that time-scales are very important and figures can be fudged by using 100 year time-scales as some UK bodies are prone to do.

Greenhouse warming-up potential (GWP) relatived to 1 kg carbon dioxide

|  | Lifetime (years) | GWP (20 years) | GWP (100 years) | GWP (500 years) |
|---|---|---|---|---|
| Methane | $14.5 \pm 2.5$ | 62 | 24.5 | 7.5 units |
| Nitrous oxide | 120 | 290 | 320 | 180 |

*Source: Greenhouse Issues*, IEA Publication No. 17, March 1995.

The UK coal-fired power industry contributes less than 0.5 per cent to the total global greenhouse effect. The time-scale for the construction and full commissioning of a nuclear power station is ca. 10 years, so we had better not hold our breath for nuclear power to come to the rescue of the greenhouse effect!

Measures which can be taken fall into the short term and long term. Action on CFCs (and related HALONS) production and use would have a very substantial impact on both the greenhouse effect and ozone layer depletion. A plethora of conferences have pledged the Western world to cuts of 50–100 per cent in CFC usage by 1999. Production of CFCs has now ceased (1995) and only supplies from CFC recycling will be available. The end is in sight!

The UK is well on the way to stabilizing or even reducing its $CO_2$ emission due to gas fired generation (40 per cent reduction CO coal fired/kWh) and nuclear power generation.

As far as the UK is concerned, if a 'business as usual' scenario is envisaged, an overall increase of 15 to 20 per cent in $CO_2$ emissions by 2005 is predicted. This represents an extra 84–108 million tonnes of $CO_2$ over the 1989 level of 542 million tonnes. If, however, existing energy conservation solutions are employed, a potential reduction of 127 million tonnes is claimed to be possible by 2005.

The relevant figures (in million tonnes) by sector, produced by the UK Association for the Conservation of Energy (ACE, 1989) are as shown in the following table. These figures cover ca. 92 per cent of the total energy use in the UK. They do not cover agriculture and some industrial processes such as coking or refining.

Some $CO_2$ facts:

● The generation of 1 kilowatt hour of electricity from a coal-fired power station produces 1 kg of $CO_2$ (1000 kilos =

| | Growth in $CO_2$ by 2005, 'Business as usual' (mt) | Reduction in $CO_2$ (2005) by Conservation (mt) | Current levels (mt) |
|---|---|---|---|
| Electricity | +36 | −21 − 41 | 205 |
| Domestic | −11 | −37 | 90 |
| Industry | +22 | −13 | 94 |
| Commercial/public | +13 | −18 | 33 |
| Transport | +24 − 48 | −18 | 120 |
| Total | 84 − 108 | 107 − 127 | 542 |

1 tonne) (a kilowatt hour = 1000 watts of electricity supplied for 1 hour or a 100 W bulb lit for 10 hours).

- The burning of 1 THERM of natural gas produces 6 kilos of $CO_2$ (1 therm will boil a pint of water on a gas burner 160 times, or keep a burner on full for 9 hours).
- The combustion of 1 litre of petrol produces around 2.5 kg of $CO_2$.
- For the same amount of useful energy, oil emits 38 to 43 per cent more $CO_2$ than natural gas, and coal emits 72 to 95 per cent more.

Data of this nature can form the basis of a carbon tax. On this basis coal would be taxed more than oil, which would be taxed more than gas. However, resource conservation considerations may dictate that some coal is used in order to conserve oil and natural gas. The Paris-based International Energy Association has suggested the following carbon tax levels: $50 on a tonne of coal, $8 per barrel of oil and $1 per million British Thermal Units of natural gas. As well as increasing consumer energy prices by 20 per cent, these moves would slow the rate of growth of carbon dioxide emissions in the OECD by half to the year 2005. The agency is reported as stating that it is impracticable to implement a massive switch to nuclear power over the next 20 years: 'Even if such a switch were made, it would barely come close to keeping energy related carbon dioxide emissions in OECD constant by 2005.'

*Source:* 'Steep tax on carbon fuels urged by agency', *Financial Times*, 1 Feb. 1990.

The 1989 ACE report also states that the market on its own is unlikely to deliver more than a small fraction of the full potential for energy efficiency because of barriers such as:

- The lack of information on how to improve energy efficiency;
- The fact that frequently the people who pay the fuel bills (tenants) are not those responsible for capital improvements (landlords);
- The significant difference between the 2 to 3 year payback periods required before businesses or individuals will invest in energy efficiency, and the 20 year payback period on which the energy supply industries often plan. Because of this difference, a disproportionate amount of investment is going into expanding supply (to meet the inefficient demand) rather than into improving the efficiency of energy use.

It would appear that climate changes are now inevitable and action to contain the growth of the greenhouse effect (and a parallel ozone layer depletion) may 'mean disruption to the world's economic system'. Certainly, fossil fuel combustion and other industrial $CO_2$ releases are up by a factor of 3 since 1950. Globally, these total ca. 5 billion tonnes and any substantial reduction may mean disruption. ($\Rightarrow$ CARBON DIOXIDE; CHLORO-FLUOROCARBONS, HALONS)

*Sources:* D. Everest, *The Greenhouse Effect Issues for Policy Makers*, Royal Institute of International Affairs, London 1988.
T. M. Wigley, *Relative Contributions of Different Trace Gases to the Greenhouse Effect*, Climate Monitor, vol. 16 (1), 1987.
*Solving the Greenhouse Dilemma – A Strategy for the UK*, UK Association for the Conservation of Energy, London, 1989.

**Green Revolution, The.** The most widely recommended means of increasing agricultural yields is through the increased use of FERTILIZERS and the introduction of new 'high yield' varieties of grain.

Fertilizers are easily produced (although the ENERGY required in their production is considerable) and have been used intensively for many years now. However, the environmental consequences of the intensive use of fertilizers and the effects of a really large-scale increase in their use are incalculable. In addition, the difficulties of implementing the proposed increase in fertilizer use on the scale required are immense. Ehlich has calculated that if India were to apply fertilizer at the *per capita level* employed by the Netherlands, India's fertilizer needs alone would amount to nearly half the present world output.

The second proposal for increasing yields is to develop new high-yield or high-protein strains of food crops. Such new strains have had considerable success over the past few years, particularly in Asia. They mature early and are relatively insensitive to the length of day, making the production of two or three crops a year possible. But these new strains usually require high fertilizer inputs in order to realize their full potential, and we have already pointed out the problems involved there. Vast amounts of capital are required for fertilizer production and distribution. Abundant water is also necessary, as are PESTICIDES and mechanical planting and harvesting machinery. It is just those countries that need these crops most who are desperately short of capital. Furthermore, we are unsure how resistant these new strains are to the attacks of insects and plant diseases.

Therefore, although high-yield agriculture is promising, it is unlikely that it will ever fulfil its promise. Lack of capital, expertise, and most of all, lack of time and the will to act quickly will probably mean that at best the Green Revolution will only allow us to keep pace with population growth for a couple of decades.

**Green sand.** A naturally occurring mineral which can be used as filter bed material for iron removal from drinking water.

**Grey list.** Actually List 11 in EC 'Framework' Directive (76/464/EEC) on pollution caused by certain dangerous substances discharged into the aquatic environment. This covers those substances considered less harmful when discharged to water than those in the BLACK LIST/RED LIST.

**Grit.** A general term for coarse particulate matter. BS 3405 defines grit as solid particles retained on a ♯ 200 mesh BS sieve, nominal aperture 76 $\mu$m.

**Gross national product (GNP).** A measure of the total flow of goods and services produced by the economy over a particular period – usually a year. It is obtained by adding up, at market prices, the total national output of goods and services. 'Intermediate' products are not included since it is assumed that their value is implicitly included in the prices of the final goods. To this total figure (frequently termed the 'gross domestic product') is added any income accruing to residents arising from investment abroad, and from it is deducted any income earned *in* the domestic market by foreigners abroad. The final figure is called the 'gross national product' and it is generally supposed to be a measure of economic success.

However, as a realistic guide to a nation's economic well-being there is a lot wrong with the GNP. First, it includes the very considerable expenditure on arms and the military, which is totally non-productive. Second, it includes such things as pollution. When somebody pollutes the environment and somebody else cleans it up, the cleaning-up process is included in the GNP. Similarly, if you are an urban dweller whose health is affected by the pollution, then your hospital bills also contribute to the GNP. But the main defect with the GNP as an indicator of economic well-being is its preoccupation with indiscriminate production.

**Ground-level concentration.** The concentration of a pollutant in air to which a human being is normally exposed, i.e. between ground and a height of some 2 metres above it. It does not mean the concentration in a layer of air in direct contact with the ground, where the concentration may be low if absorption of the pollution by the ground is occurring. The upper limit of height is flexible if concentrations do not change very much with height from 2 m upwards. When the major source of air pollution is a single stack or a small group of stacks the measured ground level concentration at any point will vary very rapidly with time on account of small but rapid changes in wind speed and direction, and 'ground level concentration' is meaningless without specification of the time over which the concentration is averaged.

**Groundwater.** Water occurring within the SATURATION ZONE of an AQUIFER is the only part of all subsurface water which is properly referred to as groundwater (or phreatic water). Groundwater present within this zone can be considered to be occupying a large natural reservoir or series of reservoirs whose capacity is the total volume of the PORES or VOIDS in the rocks that it fills.

Groundwater may be of variable chemical quality ranging from wholesome potable waters to highly mineralized BRINES.

It is the duty of the NATIONAL RIVERS AUTHORITY (NRA) in England and Wales to monitor and protect groundwater and conserve it for water resource usage. It is also the NRA's duty to maintain and conserve surface waters, which in many cases depends upon the proper management of groundwater. (The NRA's powers and duties are set out in the Water Resources Act 1991.)

The NRA can influence planning decisions which may damage groundwater. As the way the land is used and developed is one of the greatest and most consistent threats to the quality of groundwater, land-use planning policies can play a significant role in protecting groundwater. Therefore, the NRA must keep in close contact with the local planning authorities.

As some activities (e.g. LANDFILL) present a particular risk of pollution, the closer an activity is to a well or borehole, the greater the risk of the pumped water being polluted.

Around each groundwater source, the NRA has defined three source Protection Zones. These vary in their size, shape and relationships according to the particular situation at any one place. The type of soil, the geology, the rainfall and the amount of water pumped out of the ground are all taken into consideration. Policy statements dealing with the new developments within each of these three zones, in addition to other areas where groundwater is generally at risk, have been published; for example:

*Policy B (Physical Disturbance)*. Surface mineral exploitation is the main form of physical disturbance. The extraction of minerals, including limestones, sandstones, clay, sand and gravel can mean dewatering activities pose a threat to both the quality and quantity of groundwater.

*Policy C (Waste Disposal)*. Virtually all landfill sites are located at disused mineral extraction sites. The majority of these historical landfill sites have been located and designed on the 'dilute and disperse' principle, without the same regard for groundwater that is at present required.

*Policy E (Disposal of Sludges)*. The application of sewage sludges to agricultural land requires careful control to avoid polluting groundwater by bacteria. (⇨ NATIONAL RIVERS AUTHORITY)

*Source: Policy and Practice for the Protection of Groundwater*, summary document, National Rivers Authority.

**Groundwater vulnerability maps.** A series of maps showing the UK's aquifers, classified according to the properties of the rocks and the overlying soils. (⇨ NATIONAL RIVERS AUTHORITY)

**Guidance Notes.** These are issued for relevant processes and are at two basic levels.

(a) Those covered by Her Majesty's Inspectorate of Pollution (HMIP);
(b) Those covered by Local Authorities (Secretary of State).

Process Guidance Notes issued by the chief inspector of HMIP include for example:

Merchant and In House Chemical Waste Incineration, IPR5/1
Municipal Waste Incineration, IPR5/3
Clinical Waste Incineration, IPR5/4
Furnaces, IPR1/1 (large boilers = 50 MW thermal or more).

Requirements to be met are specified, such as:

(i) Releases into air ⎫ quantities, concentrations, monitor-
(ii) Releases into water ⎪ ing, record-keeping and reporting
(iii) Releases into land ⎬ are all prescribed for specified re-
⎪ leases. In addition, specific plant
⎭ operating conditions must be met.

The notes are issued with the objective that BATNEEC will be used and met, e.g. for MSW incineration.

By 1 December 1996, existing municipal incineration plant with a capacity of at least 6 tonnes per hour must comply with the following combustion conditions: the gases resulting from the combustion of the waste must be raised, after the last injection of combustion air and even under the most unfavourable conditions, to a temperature of at least 850°C for at least two seconds in the presence of at least 6% oxygen. However, in the event of major technical difficulties, the provisions concerning the two-second period must be implemented at the latest when the furnaces are replaced.

By 1 December 1995, other existing municipal waste incineration plant must comply with the following conditions: the gases resulting from the combustion of the waste must be raised, after the last injection of combustion air and even under the most unfavourable conditions, to a temperature of at least 850°C in the presence of at least 6% oxygen for a sufficient period of time to be determined by the Inspector.

The guidance notes may be used by applicants and other interested parties as guidance on the criteria against which judgement will be made on:

(a) the acceptability of an application or variation
(b) the conditions to be included in an authorization or variation notice.

*Source:* Waste Disposal & Recycling, Chief Inspector's Guidance Notes, IPR 5/3.

**Gypsum.** Hydrate calcium sulphate $CaSO_4 \cdot 2H_2O$ used in the manufacture of plasterboard. Also a by-product of the wet limestone FLUE GAS DESULPHURIZATION process. For example the proposed FGD plant for the 4 GW UK coal-fired Drax power station will consume 500 000 tonnes of limestone and produce around 860 000 tonnes of gypsum each year for the building industry.

*Source: The Chemical Engineer*, April 1989.

# H

**Haber process.** ⇨ NITROGEN CYCLE.

**Haemolysis tests.** Haemolysis is the breaking of blood corpuscles by the action of a poisonous substance. It may be used as a means of testing for the biological activity of a suspected substance by incubating the substance with red blood cells, and then measuring the amount of haemoglobin released by the breaking of cells.

Haemolysis tests on PVC powder suggest that PVC dust is as biologically active as blue asbestos or crocidolite. *If* this is the case, it has serious ramifications concerning the handling of PVC powders which are used in many plastics plants. (⇨ ASBESTOS)

**Half-life.**
1. Time needed for half a quantity of ingested material to be eliminated from the body naturally. Also radiobiological half-life: time needed for ingested radioactive material to deliver half its radiation dose. This allows for both decay in activity and the time in the body.
2. The time taken for half the quantity of a substance to disappear from the environment, e.g. by biodegradation or by discharge from a biological system. For example, inorganic MERCURY has a half-life of six days in humans, organic mercury compounds have a half-life of 70 days. Thus, the ingestion of mercury in food will result in totally different body burdens depending on the form (see Figure 80).
3. Radioactive half-life is the time taken for half the number of atoms of a radioactive substance to decay to atoms of a different mass. Different RADIONUCLIDES have different half-lives: plutonium-239 has a half-life of 24 400 years; strontium-90 has a half-life of 28 years; xenon-138 has a half-life of 17 minutes. (⇨ EXPONENTIAL CURVE)

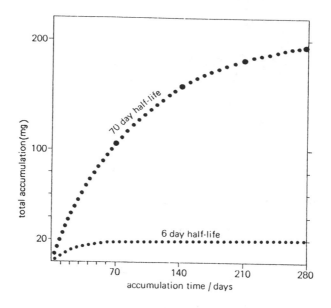

**Figure 80**   The effect of different periods of half-life on accumulation in the body. Organic mercury reaches ten times the level of inorganic mercury in nine months. At the 2 milligram per day ingestion level shown here symptoms of severe poisoning from organic mercury would in fact appear before the third month.

*Radiological half-life:* In estimating the dose of radiation from radioactive matter that has been ingested, both the radioactivity of the matter and the residence time in the body are important and give rise to the concept of radiological half-life.

Various organs have different half-lives for the presence of the same compound or element. For example, an ingestion of organic mercury may have a half-life of 50 days in the liver, but 150 days in the brain. Thus, even though the brain will get a much smaller proportion of the input, the bulk going to the kidneys, at the end of approximately a year the brain concentration is higher.

This accounts for the effect of heavy metal poisoning on the central nervous system as at Minamata. (⇨ MINAMATA DISEASE)

*Source:* A. Tucker, *The Toxic Metals*, Pan/Ballantine; Earth Island, 1972.

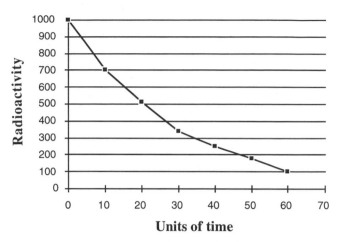

**Figure 81** Radioactive decay.

*Radioactivity decay:* The example in Figure 81 shows a 'half-life' – that is, half the radiactivity fades away – of 20 'units of time'. Some radioactive matter has a very long half-life and the units would be commensurately large, e.g. for plutonium a time 'unit' would scale 1200 years on the curve.

**Halogenated fluorocarbons.** Group name for ethane- or methane-based compounds, in which some or all of the hydrogen in their structures is replaced by chlorine, bromine and/or fluorine. CHLOROFLUOROCARBONS (CFCs) are a major part of this family. They are used as refrigerants, aerosol propellants in fire-fighting and as foam expanders because of their low boiling point. They break down when subjected to high levels of ultraviolet radiation, to chlorine monoxide which reacts with ozone and turns it into oxygen, leading to depletion of the ozone layer. Examples include the refrigerant Freon-12 ($CCl_2F_2$) which has a boiling point of 28°C and TRICHLOROFLUOROMETHANE ($CCl_3F$) used as an aerosol propellant and designated Freon 11.

**Halogenation.** Reaction with a member of the halogen group: namely, fluorine, chlorine, bromine, iodine.

**Halons.** A HALOGENATED FLUOROCARBON which contains bromine used especially in fire extinction. Ozone damage potential relative to CFCs is 3–10 times greater. (⇨ HALOGENATED FLUOROCARBONS)

**Hardness.** A property of water usually manifested as 'needing more soap to get a lather' that is classed as either temporary or permanent. Temporary hardness can be removed by boiling and deposits a carbonate scale. Non-carbonate hardness cannot be removed by boiling and is classified as permanent.

Hardness values are expressed as an equivalent amount of calcium carbonate ($CaCO_3$). Water with a hardness of less than 50 ppm is soft. Above 200 ppm, domestic supplies are usually blended to reduce the hardness value.

**Hazard.** A circumstance that poses a threat. This will need RISK assessment.

**Hazardous pollutants.** This is a classification used by the United States Environmental Protection Agency. A hazardous pollutant is one to which even slight exposures may cause serious illness or death. MERCURY, ASBESTOS and BERYLLIUM have all been so classified.

**Hazardous waste.** ⇨ WASTES.

**Hazardous waste incineration.** This is the INCINERATION under very strictly controlled conditions of WASTES deemed to be hazardous (in UK called 'special' wastes). They are highly dangerous to life, when present in the environment e.g. bioaccumulates. They include spent solvents, PCBs, and pesticide residues.

The hazardous waste incineration sector performs a valuable service as these wastes are destroyed to values of 99.999 per cent destruction and removal efficiency. Typically a ROTARY KILN, and multiple stage SCRUBBING of the flue gases is employed, as shown in Figure 82.

There is now a new and stricter EC Directive (94/67/EC) for emissions from hazardous waste incinerators. (This excludes domestic/municipal waste incinerators whose requirements are currently covered under 89/369/EEC but some tightening up is predicted.)

**Haze.** Dust, salt or smoke particles often suspended in water droplets in suspension in the atmosphere. Haze can also be caused by substances called terpenes given off naturally by vegetation and forests.

**Hazen number.** A unit of measurement for colour in water. (Based on the colour produced by 1 mg platinum per litre in the presence of a cobalt-based compound.)

**Head.** The total height of water against which a pump has to work.

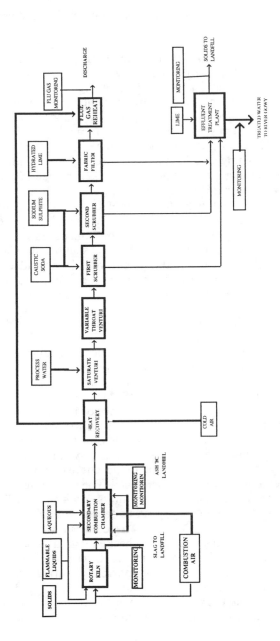

**Figure 82** Hazardous waste incineration flow diagram for Cleanaway Ellesmere Port Plant (Courtesy of Cleanaway Limited).

Comprises the sum of the static head plus friction and velocity heads.

**Hearing.** The ear responds in varying degrees to sound of different frequencies. The range of frequencies it can encompass is roughly from 20 to 20 000 hertz (Hz).

For a given sound pressure level a high frequency sound appears to be stronger than a low frequency one. The loudness of sounds as judged by the ear depends on sound pressure level and frequency. Thus the measurement of sound is weighted in a way similar to the frequency response of the ear and the standard weighing used is called the A-scale. Meter readings on such A-weighting are called sound levels measured in decibels on the A-scale or dB(A). (⇨ DECIBEL)

There is only one important source for which dB(A) are not normally used, and that is aircraft. The sound generated by aircraft engines has predominant components in particular frequency bands, and these can have a significant *subjective* effect for which A-weighted sound levels do not adequately account. Measurements are therefore made (in decibels) in each of a number of restricted frequency bands: from these a total level is calculated, giving due emphasis to the predominant components. The calculated total is called the 'perceived noise level', the units of which are PNdB. The precise numerical difference between PNdB and dB(A) varies depending on which frequency bands contain the predominant sound. The bands differ with different types of aircraft engine. However, values of PNdB are higher than values of dB(A) would be for the same sound, and as a rough guide a difference of 13 can be assumed. Certain types of new aircraft are now subject to noise certification before they come into service. For this purpose the perceived noise level is adjusted to take account of any pure tones (single frequency notes) in the noise, and of the length of time for which the higher noise levels are experienced. The result is termed the 'effective perceived noise level', the units of which are EPNdB. (⇨ NOISE; SOUND; ROAD TRAFFIC NOISE; AIRCRAFT NOISE; INDUSTRIAL NOISE MEASUREMENT; NOISE INDICES)

**Hearing loss.** A person's hearing is checked with an audiometer. The audiometer, in fact, compares the threshold of hearing of the subject under test with that of a normally hearing individual. Sounds of different frequency (usually 0.5, 1, 2, 3, 4 and 6 kHz) are produced and the subject of the test is asked to pinpoint the intensities at which the sounds are just audible. So if the

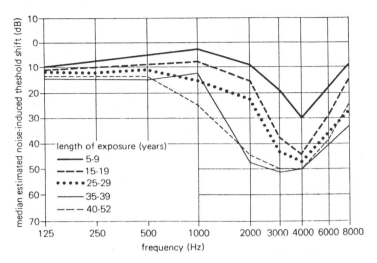

**Figure 83**   Noise-induced hearing loss as a function of years of
exposure for weavers. Note that the initial loss is at 4 kHz.

audiometer records a hearing level of 30 dB at 2000 Hz, the
threshold of the individual being tested differs by 30 dB from a
statistical 0, which is held to represent normal hearing at that
frequency.

   Results of an audiometric test are recorded on an audiogram
which indicates the hearing loss. Figure 83 shows audiograms
indicating hearing losses typical of workers in the weaving trade
after various durations of exposure to the noise of the looms.
(Note the typical initial damage at about 4000 Hz and the way
the damage extends over a wider frequency range as exposure
continues.)

**Heat island.** ⇨ TEMPERATURE INVERSION.

**Heat pump.** A device for pumping heat from a low temperature
source, e.g. the surrounding ground or a stream, and delivering
it at a higher temperature; in other words, a refrigerator in
reverse. Heat pumps require a mechanical or electrical energy
input and therefore the performance coefficient or ratio of heat
produced to electricity consumed is very important. As explained,
under the Second Law of Thermodynamics, the conversion of
heat to electricity in most power stations is about 3 heat units
for one of electricity, so to obtain wide applicability heat pumps

must have COEFFICIENTS OF PERFORMANCE (c.o.p.) greater than 3, otherwise they would consume a greater heat equivalent in electricity than they were producing. Theoretically, heat pumps can have c.o.p. of 8–10, but practically these are much less. (⇨ LAWS OF THERMODYNAMICS)

**Heavy metals.** MERCURY, LEAD, CADMIUM, ZINC, CHROMIUM and PLUTONIUM are among the so-called heavy metals – those with a high atomic mass. (The term is rather loose and is taken by some to include arsenic, BERYLLIUM and SELENIUM, which are not really metals, in addition to those listed above.) Some of the above, if ingested in minute quantities, can be harmful. In the case of plutonium a few micrograms can be lethal. With the exception of plutonium, the others are in common use in industrial processes and therefore they may be discharged to the environment to be mobilized by the action of air and water and concentrated by the action of living organisms and as in the classic case of MINAMATA DISEASE, can cause deformity and death.

As some of these elements are neither heavy nor true metals, the term toxic metals is sometimes preferred.

Because of the tendency to heavy metals to accumulate in selective organs such as the brain or liver, average intake estimates are of little use. The dose to each individual organ must be considered separately as is the case with IONIZING RADIATION, because some parts of the body are more vulnerable than others. (⇨ HALF-LIFE)

Ecosystems can be affected by their discharge, e.g. marine and freshwater diatoms can lose 50 per cent of their growth rate at concentrations as low as 1 part per thousand million ($10^9$) of mercuric compounds. Because of their environmental importance, each is discussed individually.

The damage that can be caused by heavy metals entering the food chain is illustrated by the November 1989 UK cattle feed incident in which ca. 1000 farms had their milk cattle quarantined because of cattle feed pellets which had become contaminated with lead, after contact with lead sulphate during transport, resulting in contaminated milk production.

**Hepatitis.** A disease that can be caught drinking sewage-contaminated water, and which causes liver inflammation. (⇨ PUBLIC HEALTH)

**Her Majesty's Inspectorate of Pollution (HMIP).** (The Role of Her Majesty's Inspectorate of Pollution (HMIP) abridged from Slater, with permission).

HMIP protects the environment by enforcing regulations to prevent pollution by:

- authorizing, enforcing, inspecting and monitoring industrial processes with a greater potential to pollute, under the relevant legislation;
- consulting openly and widely and reporting on its performance;
- providing expert advice to Government;
- initiating R & D and disseminating the results, and
- working cost effectively to the highest professional standards.

The main regulatory functions of HMIP are effected by integrated pollution control (IPC) under Part I of the Acts of Parliament Environmental Protection Act 1990 (the Act); and regulations prepared under it.

*Integrated pollution control (IPC)*
The prescribed processes to be controlled under IPC, the timetable for their introduction and the prescribed substances are detailed in the Environmental Protection (Prescribed Processes and Substances) Regulations 1991.

HMIP is required to grant an authorization, subject to any conditions which the Act requires or empowers it to impose, or to refuse it. HMIP must refuse to issue an authorization unless it is considered that the operator will be able to carry on the process in compliance with the conditions in the authorization.

In setting conditions, the Act imposes a duty on HMIP to meet certain objectives. Conditions should ensure that:

- the best available techniques (technology operation and management practices) not entailing excessive cost (BATNEEC) are used to prevent or, if that is not practicable, to minimise the release of prescribed substances which are released and any other substances which cause harm,
- releases do not cause, or contribute to, the breach of any direction given by the Secretary of State to implement European Community or international obligations relating to environmental protection, or any statutory environmental quality standards or objectives, or other statutory limits or requirements,
- when a process is likely to involve releases into more than one environmental medium (the case for most prescribed processes), the best practicable environmental option (BPEO) is achieved, i.e. the releases from the process are controlled through the use of BATNEEC to give the least overall affect on the environment as a whole.

An important element is the requirement to maintain a register of applications, authorizations issued and information received from operators as a result of conditions in authorizations. The register is available for public scrutiny at all reasonable times.

Waste disposal and recycling (which embraces incineration) is one of the main categories of process prescribed by the regulations. Specifically:

(a) waste chemicals and plastics (arising from their manufacture),
(b) specific chemicals and their compounds, and
(c) any other waste including animal remains, at a rate of 1 tonne or more per hour.

Comprehensive guidance has been prepared, the most useful being Chief Inspector's Guidance Notes (CIGNs) which provide inspectors and (industry) with the main emission standards for prescribed substances arising from each process. They specify the minimum standards to be attained by existing plant and what constitutes BATNEEC for new plant. They take into consideration the results of BAT research reviews commissioned by the Inspectorate.

*The Process Guidance Notes*

- Merchant and In House Chemical Waste Incineration (IPR/51)
- Clinical Waste Incineration (IPR/52)
- Municipal Waste Incineration (IPR 5/3)
- Animal Carcass Incineration (IPR 5/4)
- The Burning Out of Metal Containers (IPR 5/5)
- Making Solid Fuel from Waste (IPR 5/6)
- Sewage Sludge Incineration (IPR 5/11)

They provide guidance on:

- the general provision of the Act, the Regulations, applications and authorizations,
- the process descriptions to which they apply,
- a suggested timetable for upgrading plant to current standards,
- definitions of substantial change,
- BATNEEC/BPEO process/abatement techniques, and
- release levels corresponding to those techniques.

The importance of the Guidance Notes cannot be underestimated. A developer proposing to build an incineration plant should ensure that the proposal meets, at least, current best practice.

Consideration should also be given to known or likely future standards. HMIP will continue to review international technical developments and regularly up-date its guidance. (⇨ GUIDANCE NOTES)

*Sources:* D. Slater, HMIP, 'Incineration the best practicable option', HMIP's Viewpoint, Paper presented at IBC Conference, Manchester, January 1994.

**Herbicides.** Also known as defoliants. Chemicals used for the elimination of unwanted plant or the total elimination of all plant growth. They are extensively used in agriculture, ranging from selective control of couch grass or wild oats, to ground clearance of overgrown areas. Other uses are in roadside maintenance, railway track clearance, etc. There are many chemical classes of herbicide. Of the phenoxyaliphatic acids, 2,4-D and 2,4,5-T are common, and cause changes in plant metabolism. A mixture of these two herbicides was used as Agent Orange, of which certain samples were found to be contaminated with DIOXIN. Other important herbicides include the substituted ureas (e.g. diuron) and the heterocyclic nitrogens (e.g. the triazines such as simazine), all of which interfere with photosynthesis. Other classes include carbamates, thiocarbamates, etc.

While the direct effect on animals may be minimal, herbicides have enormous ecological implications by destroying food sources, habitats, etc. Their use is yet another example of man changing ECOSYSTEMS to suit himself. The use of defoliants, as in Vietnam or in jungle clearance for agriculture, can permanently destroy tropical forests. Once the tree cover is removed, the soil is subjected to EROSION and precious nutrients are rapidly leached away (⇨ LEACHING)

**Herring catch.** ⇨ MAXIMUM SUSTAINABLE YIELD.

**Hertz (Hz).** SI unit of frequency, i.e. number of occurrences or cycles per second, as used in the frequency of radio waves, sounds, vibrations, etc. (⇨ NOISE INDUCED HEARING LOSS)

**Heterotrophic organism.** An organism that requires organic material (food) from the environment, i.e. all animals, fungi, yeasts, and most bacteria are heterotrophic. For their food supply these organisms eventually rely on the activities of the AUTOTROPHIC ORGANISMS.

**Hexa-chrome.** ⇨ CHROME WASTE.

**High rate filter.** A BIOLOGICAL FILTER which operates at a loading in excess of 3 cubic metres of effluent per cubic metre of bed per day.

**High volume sampling.** A means of measurement of airborne particulate matter in which large samples of air are sucked through appropriate filters and volumetric or flow metering equipment. This provides samples for subsequent analysis.

**Hub height.** The height of a wind turbine tower from the ground to the centre-line of the turbine rotor.

**Humic acid.** Omnibus term for a wide range of organic compounds resulting from the decay of vegetable matter such as peat. They make water acidic, frothy, and impart discoloration.

**Humus.** Biologically stable dead organic material resulting from aerobic decomposition of sewage and other organic matter. (⇨ SLUDGE, SEWAGE)

**Hydrated lime.** Alternative description to SLAKED LIME, for the chemical CALCIUM HYDROXIDE $Ca(OH)_2$.

Hydrated lime is used extensively in the chemical and water treatment industries for process chemistry, water pH control, effluent treatment, acid and gas scrubbing. (⇨ SCRUBBER)

**Hydraulic conductivity (permeability).** The ease with which water moves through an aquifer, i.e. the rate of flow of GROUND WATER through unit cross-sectional area of an aquifer under unit HYDRAULIC GRADIENT. It can be measured in feet per second, centimetres per second or inches per day. Commonly, hydraulic conductivity is measured in metres per day. Hydraulic conductivity is usually denoted as $k$ and possesses a wide range in values from clays ($k = 10^{-5}$ to $10^{-7}$ metre per day) to gravels ($k = 1$ to $10^{-4}$ metres per day or more). Thus, a LANDFILL SITE lined with clay will mainly contain LEACHATE whereas one on gravel will allow virtually instant dispersal of any pollutants. EC draft landfill regulations posit a maximum hydraulic conductivity (or permeability coefficient) of $1 \times 10^{-9}$ metres per second for new landfill sites. This could be the death knell for the common UK practice of DILUTE AND DISPERSE landfill sites for hazardous waste disposal.

**Hydraulic gradient.** Hydraulic gradient is the difference in hydraulic head divided by the distance along the fluid flow path. Ground water moves through an AQUIFER in the direction of the hydraulic gradient.

**Hydrocarbon.** An organic compound containing the elements carbon and hydrogen. The carbon atoms may be arranged either in open-ended chains which may or may not be branched, or in closed rings. Some may also have minor or trace quantities of oxygen, nitrogen, sulphur and other elements. There are many groups of different hydrocarbon compounds but the three commonest in the FOSSIL FUELS are the paraffins, naphthenes and benzene-type compounds. Some are solid, others liquids or gases.

**Hydrochlorofluorocarbons (HCFCs).** Used as a replacement for CFCs in refrigeration, foam blowing and aerosols. OZONE DEPLETION POTENTIAL relative to CFCs is 0.02–0.1. ICI is concentrating on HFA-134a ($CF_3CH_2F$) as a replacement for CFC-12 ($CF_2Cl_2$), and on HCFC-123 ($CF_3CHCl_2$), as a replacement for CFC-11 ($CFCl_3$).

**Hydroelectricity.** ⇨ WATER POWER.

**Hydrogen ($H_2$).** Colourless, odourless gas, normally existing as the molecule $H_2$. It is the lightest substance in existence, atomic mass 1. It is extremely flammable and when burned combines with oxygen to form water with heat liberated in the process. It can be readily converted to liquid fuels, e.g. METHANOL, ammonia, hydrazine, and can of course be pumped readily by pipeline. These attributes have made many people propose the 'hydrogen fuel economy' as an alternative source of energy when conventional ENERGY resources are no longer available. The hydrogen would be manufactured electrolytically or by hydrogenation of remaining coal supplies. The hydrogen fuel economy is based on the premise that large quantities of NUCLEAR ENERGY will be available for the generation of the necessary power for the electrolysis. (⇨ ELECTROLYSIS)

**Hydrogenation.** The chemical combination of HYDROGEN with another substance, usually by the action of heat and pressure in the presence of a catalyst, e.g. the hydrogenation of coal combines its CARBON with the hydrogen to produce HYDRO-CARBON.

**Hydrogen chloride (HCl).** A colourless gas with a choking odour, highly soluble in water to form hydrochloric acid. Both the gas and its solution are poisonous, corrosive, and phytotoxic (although much less injurious to plants than hydrogen fluoride). (⇨ ACID)

**Hydrogen fluoride (HF).** A colourless FUMING gas or liquid (boiling point 19.5°C). The aqueous solution is known as hydrofluoric acid. Both are highly corrosive and toxic. Many salts of hydrofluoric acid (fluorides) are also highly poisonous. (⇨ FLUORINE; FLUORIDES)

**Hydrogen ion concentration.** ⇨ pH.

**Hydrogen peroxide ($H_2O_2$).** Powerful disinfectant and bleaching agent which rapidly gives off OXYGEN. Measured in volume strength, i.e. 20 volume $H_2O_2$ will give off 20 times its volume of oxygen.

**Hydrogen sulphide ($H_2S$).** Dense colourless gas with a smell of rotten eggs. It is extremely toxic. It is produced under anaerobic decay conditions and can accumulate in sewers. This has accounted for several fatalities. It is also produced in industrial processes such as oil refining, chemical manufacture and wood pulp processing. Gas absorption is often employed to effect removal. (⇨ TOXIC WASTES)

**Hydrographs.** Graphical representations of either surface stream discharges or water level fluctuations in wells. For example, a stream hydrograph portrays the characteristics of the flow of a stream. Discharge is plotted vertically and time horizontally.

Hydrographs of water levels in wells can be computed from measurements of water depth below ground surface. Long-term records may allow estimates of the ultimate yield of the aquifer and the rate of its replenishment.

**Hydrological cycle.** The means by which water is circulated in the biosphere. Evapo-transpiration from the land mass plus evaporation from the ocean is counterbalanced by cooling in the atmosphere and precipitation over both land and oceans (see Figure 84). The hydrological cycle requires that on a world-wide basis the evaporation and precipitation are equal. However, oceanic evaporation is greater than oceanic precipitation, thus an excess precipitation is given to the land. Eventually this land precipitation ends up in lakes and rivers and thus eventually returns to the sea, so completing the CYCLE. (⇨ TRANSPIRATION)

Man intercepts the land precipitation by means of RESERVOIRS or river and ground-water abstraction, and so obtains his water supplies, but after use by man this abstracted water still ends up in the sea – its arrival there is merely delayed. Water is also required for PHOTOSYNTHESIS but the fraction of water so used compared with that which is transpired by green plants is less than 1 per cent.

**Hydrology.** The science concerned with the occurrence and circulation of water in all its phases and modes and the relationship of these to man.

**Hydrolysis.** The breaking down of a substance by interaction with water. The hydrolysis of carbohydrates, e.g. conversion of

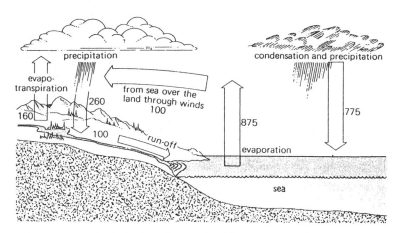

**Figure 84**  Hydrological cycle in cubic kilometres per day.

starches and sugars, is of major importance in the food and brewing industries. Carbohydrates may be hydrolysed biologically with the aid of ENZYME TECHNOLOGY or chemically at high temperatures with an acid or alkali present as a catalyst. The hydrolysis of CELLULOSE produces fermentable sugars which can be processed by fermentation to yield ethyl alcohol, butanol, acetic acid or PROTEIN. Hydrolysis has been proposed as a refuse disposal process and is a means of using renewable resources to obtain organic molecules such as ethanol, one of the building blocks of the chemical industry. It could also be used as a source of ethene, e.g. for PVC manufacture. Protein production is economically feasible. Laboratory work is proceeding in both the UK and the USA on the processing of all classes of cellulosic materials by both biological and chemical means.

**Hydropulper.** A wet pulping device for converting dry pulps and waste papers into a fibrous slurry by the addition of water. It is used in the recycling of PAPER and is the basis of a recycling process for DOMESTIC REFUSE. The pulp recovered from domestic refuse is of a very low quality and so far has little commercial value except for low-grade uses such as roofing felt where the pulp is subsequently tarred or coated to make it waterproof.

**Hydrosphere[1].** The portion of the earth's crust covered by the oceans, seas and lakes. When the complete earth's crust is meant, it is often referred to as the hydro-lithosphere. (⇨ LITHOSPHERE)

**Hydrosphere[2].** The body of (liquid) water on or near the surface of the earth. The oceans are the largest single component in the hydrosphere, cover over 70 per cent of the earth's surface and have an average depth of nearly 2.5 miles.

**Hyperactivity** (associated with behavioural disturbances). Over-activeness, especially in young children and youths. Many experts believe LEAD poisoning to be a major cause of this syndrome in many cases and recommend 'de-leading' as a cure. (⇨ CHELATING AGENTS)

# *I*

**Ideal gas law.** One mole of an ideal gas has a volume of 22.4143 litres at a temperature of 0°C and a pressure of 1 atmosphere (STP). For example, the molecular mass of air is 29, so 29 g of air has a volume of 22.4143 litres and its density is $29/22.4143 = 1.29$ g/l at 0°C and 1 atmosphere.

For many gases this approach works well. The density can be corrected for temperature and pressure using the ideal gas law: $pV = nRT$.

Physical properties of gases and symbols

| Property | Symbol | Unit |
|---|---|---|
| Density (gas) | $\rho$ | kg/m$^3$ |
| Density (liquid) | $\rho$ | kg/m$^3$ |
| Heat capacity (gas) | $C_p$ | J/kg K |
| Thermal conductivity (gas) | $\lambda$ | W/m K |
| Thermal conductivity (liquid) | $\lambda$ | W/m K |
| Dynamic viscosity (gas) | $\mu$ | Pa s |
| Dynamic viscosity (liquid) | $\mu$ | Pa s |
| Kinematic viscosity (gas) | $\nu$ | m$^2$/s |
| Kinematic viscosity (liquid) | $\nu$ | m$^2$/s |
| Vapour pressure | $P_{vp}$ | bar |
| Heat of vaporization | $H_{vp}$ | J/mol |
| Normal boiling point | $T_b$ | K |

**IGCC.** Integrated Gasification Combined Cycle.

IGCC is expected to be the most promising technology for generating electricity from coal giving an increase of 10 per cent in thermal efficiency over conventional coal burning stations. Coal consumption will be some 30 per cent less than normal and $CO_2$ emissions will be reduced by nearly 25 per cent.

In an IGCC plant, coal is pulverized, dried and fed to the gasifier where steam and oxygen are added to oxidize it partially to hydrogen and carbon monoxide. Heat from gasification produces steam used in a steam turbine driving generator. The gas is cooled to solidify the ash and generate more steam. Sulphur in the process is converted to sulphuric anhydride from which it is recovered as sulphur for sale. The gas and steam turbines constitute the combined cycle of electricity production. The turbine exhaust gases pass to a recovery boiler to produce more steam for the turbine. (⇨ COMBINED CYCLE POWER STATION, GASIFICATION)

**Imhoff cone.** A graduated glass cone (1 litre capacity) used in the laboratory for the measurement of settleable solids in a sewage effluent.

**Incineration.** A waste volumetric reduction process that relies on combustion under suitable controlled conditions to reduce the volume and/or mass of material for disposal. Incineration falls into two main classes, MUNICIPAL SOLID WASTE and HAZARDOUS WASTES. Each has differing design requirements.

The two most important aspects of municipal solid waste (MSW) as a FUEL are that it has a low CALORIFIC VALUE typically 30–40 per cent of that of an industrial bituminous coal and a density, as fired, of about 200 kg/m$^3$ or 20 per cent of that of coal.

In order to illustrate the factors involved in MSW incineration the material analysis of 'as received' MSW, by weight, is given in the table below.

Material analysis of sample of MSW

| *Material* | *Percentage by weight* |
|---|---|
| Dust and cinder | 9.0 |
| Vegetable matter | 24.0 |
| Paper and cardboard | 30.5 |
| Metals | 7.8 |
| Textiles | 4.9 |
| Glass | 11.2 |
| Plastics | 8.3 |
| Unclassified | 4.3 |

For fuel purposes, the ULTIMATE ANALYSIS is given in the table on page 275, with the gross calorific value plus the breakdown of moisture, combustibles and inerts.

It should be noted that the composition of MSW varies seasonally with regard to the amount of vegetable matter and

Ultimate analysis and gross calorific value of MSW

| Material | Percentage by weight |
|---|---|
| Carbon | 22.6 |
| Hydrogen | 2.8 |
| Oxygen | 15.9 |
| Nitrogen | 0.5 |
| Sulphur | 0.2 |
| Water | 31.2 |
| Ash and inerts | 26.8 |
| Gross calorific value as fired, 10 120 MJ/kg (4350 Btu/lb) | |
| Moisture | 31.2 w/w |
| Combustibles | 42.0 w/w |
| Inerts | 26.8 w/w |

the moisture content (also related to the amount of vegetable matter). Hence, calorific values can fluctuate by as much as $\pm 6$ per cent of the mean and due allowance should be made for this in well regulated energy recovery incinerators.

*Combustion air requirements:* Assuming an excess air supply of 80 per cent to ensure complete combustion and the prevention of reducing conditions (which promote corrosion), an air supply of 5.14 kg air/kg MSW is needed. If it is anticipated that industrial waste will also be burnt in the incinerator, it is necessary to also know the composition and throughput and the effect of this industrial waste on the waste composition.

The peculiarities of MSW combustion are such that a secondary or overfire air supply is necessary to burn the volatiles released in the primary combustion stage. The underfire air both dries the refuse and supplies the air necessary for the primary or bed-based combustion. For calculation purposes, the working assumption can be made that the primary : secondary air ratio is 3:1. The gas, moisture and solid flows to be calculated per tonne of MSW are shown in Figure 85 (numbers rounded off for ease). ($\Rightarrow$ CALORIFIC VALUE; ULTIMATE ANALYSIS, MSW, INCINERATION PERFORMANCE)

The disposal costs associated with incineration are typically greater than short haul landfill costs, but where there is a market for power and heat, as is widely practised in Europe, costs approaching or bettering those of long haul landfill are

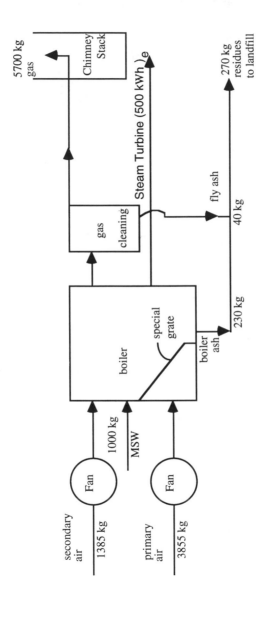

**Figure 85** Schematic mass burn waste to energy plant

(a) MSW incineration costs – privately financed 220 000 tonnes per year installation (design capacity 35 tonnes/hour)

|  | Cost (£ million) | Disposal cost (£/tonne) |
|---|---|---|
| *Capital cost* (to full EC standards + site work) | 45 | |
| Operational costs | 3.1 | |
| Capital charges (20%, over 20 years) | 9.3 | |
| Total cost without power generation | 12.4 | 56.3 |
| *Income* Electricity sales (550 kWh/t at 3.5 p/kWh) | 4.2 | |
| Nett cost | 8.2 | 37.00 |
| Income on NFFO tariff of 6.5 p/kWh | 7.7 | |
| NFFO net cost | 4.7 | 21.3 |

Comments: (i) Up to 600 kWh/t may be available depending on MSW calorific value. (ii) Electricity tariff and return on investment crucial to viability. (iii) NFFO – Non fossil fuel option. Tariff can vary: (iv) Larger plants have significantly lower costs.

possible. Table (a) gives outline cost data. Indeed, incineration may be the only MSW disposal option readily available due to the nature of a locality and the difficulty of having new landfill sites accepted by the public.

*Hazardous waste incineration* requires features such as air locks through which the hazardous substances (e.g. PCBs) are loaded, supplementary fuel and perhaps a ROTARY KILN for thorough destruction of wastes. In addition stringent air pollution control measures are necessary, including gas scrubbing stages and ACTIVATED CARBON filtration

Incineration is now practised to the highest standards. Table (b) gives the UK regulatory details, and Table (c) gives percentage of total emissions to air. These are now negligible.

Table (d) gives the composition, quantities and energy content of UK wastes, all of which can be used for energy recovery purposes prior to ultimate disposal of the ASH left over from combustion.

Also, incineration of waste to energy can cope with any level of recycling as set out in Figure 86. Waste to energy is totally compatible with any level of recycling.

(b) *HMIP emission standards*

| | Requirements under IPR 5/3 | Typical design emission levels for new plants | Proposed EU requirements |
|---|---|---|---|
| Particulate (mg/Nm³) | 30 | 15 | 10 |
| HCl (mg/Nm³) | 30 | 30 | 10 |
| HF (mg/Nm³) | 2 | 1 | 1 |
| $SO_x$ as $SO_2$ (mg/Nm³) | 300 | 70 | 50 |
| $NO_x$ as $NO_2$ (mg/Nm³) | 350 | 300 | 200 |
| Metals | 0.1 Cd | 0.1 Cd and Ti | 0.5 Cd and Ti |
| | 0.1 Hg | 0.1 Hg | 0.05 |
| | 1.0 (As, Cr, Cu, Pb, Mn, Ni, Sn) | 0.5 (As Ni, Pb, Cr, Mn, Cu) | 0.5 |
| THC (VOC) (mg/Nm³) | 20 | 20 | 10 |
| Dioxins (ng/Nm³) | | | |
| limit | 1.0 | 0.1 | 0.1 |
| target | (0.1) | — | — |
| CO (mg/Nm³) | 100 | 50 | 50 |

Nm³ = gas volumes referred to 'normal' conditions.

(c) *Emissions to air: estimated 1991 emissions from UK incineration plants as a proportion of total UK emissions of each pollutant, and author's estimate for the year 1997*

| | Emissions to air from UK incineration plants in 1991 (tonnes) | Percentage of total emissions to air | |
|---|---|---|---|
| | | 1991 | Estimated year 1997 comparison (by author) |
| Total particulate matter | 7400 | a | b |
| Carbon monoxide | 5000 | 0.1 | Negligible |
| Volatile organic compounds (excluding particulates; expressed as total carbon) | 100 | Negligible | Negligible |
| *Acidic gases* | | | |
| Sulphur dioxide | 5140 | 0.1 | Negligible |
| Hydrogen chloride | 10 500 | 3.0 | Negligible |
| Hydrogen fluoride | 20 | a | Negligible |
| Nitrogen oxides (as $NO_2$) | 875 | Negligible | Negligible[c] |
| *Heavy metals* | | | |
| Cadmium | 10 | 32 | 5[d] |
| Mercury | 4 | 11 | 7[d] |
| Lead | 161 | 2 | 0.5 |
| Copper | 22 | 2 | 0.4 |
| Chromium | 16 | 2 | 0.4 |
| Nickel | 10 | 2 | 0.4 |
| Manganese | 6 | 2 | 0.4 |
| Arsenic | 3 | 2 | 0.4 |
| Dioxins | 668 (gTEQ) | b | 14 (gTEQ) |
| Carbon dioxide[e] | 2 750 000 | 1 | Negligible |

[a] Percentage not known because no estimate of total UK emissions was available.
[b] Approx. 90% reduction from 1991 levels (and less than 1/1000 of road transport particulate emissions (1992) from RCEP 18th Report).
[c] Approx. 60% reduction from 1991 levels.
[d] Includes estimate for battery recycling and low Cd and Hg products.
[e] Carbon dioxide from WTE plants is virtually environmentally neutral as most of the components would decay to $CO_2$ naturally or methane and $CO_2$ if landfilled.
*Source:* Royal Commission on Environmental Pollution 17th Report, *Incineration of Waste*, HMSO, May 1993.

*(d) Breakdown by waste type, quantities and energy content (gross)*

| Waste type | UK arisings (Mtpa)* | Energy content (Mtcepa)** |
|---|---|---|
| Domestic | 30 | 9.6 |
| Commercial/industrial | 25–30 | 15 |
| Hospital | 0.4 | 0.25 |
| Scrap tyres | 0.4 | 0.5 |
| Chicken litter | 1.8 | 0.9 |
| Surplus straw | 10 | 5 |
| Woodwaste | 1.5 | 0.85 |
| Fragmentizer waste | 0.5 | 0.4 |
| Sewage sludge | 20 | — |

*Million tonnes per annum.
**Million tonnes coal equivalent per annum.
*Sources:* Energy Technology Support Unit, Harwell (Communication S. Dagnall and Reports).
Department of Trade & Industry, *Renewable Energy Planning for the Future*, 1993.

**Incineration grates.** The properties of crude unsorted (MSW) are such that thorough controlled bed agitation is required. In addition, both undergrate and overfire air are necessary, as well as a reliable means of ash removal. One well tried arrangement is given in the VKW roller grate which is installed worldwide.

Figure 87 shows the layout of a typical roller grate incinerator (along with allied boilers and gas cleaning plants in cross section). It comprises a series of six hollow rollers which transport the MSW downwards through the furnace, whilst an angle of about 30° ensures sufficient agitation for complete combustion. Primary air is supplied to assist the combustion through holes in the grate. Each roller is about 1.5 m diameter and wide enough to carry the design throughput. The design of the grate segments permits expansion and contraction, and allows small particles to drop through whilst allowing efficient distribution of the primary under-bed combustion air supply. The speed of rotation of each roller is separately controlled through a variable speed gearbox to ensure efficient combustion. The speed range varies from 5 revs/hour near the entry point to 0.5 rev/hour at the final roller, in order to match the changing waste densities. This system allows a high degree of combustion control as feed and combustion rates can be evenly balanced.

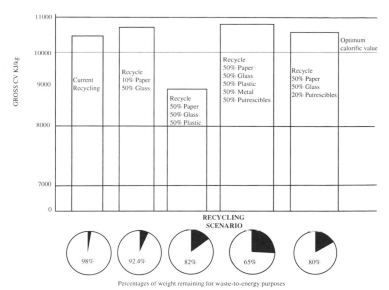

Percentages of weight remaining for waste-to-energy purposes

Study sponsored by Hampshire County Council, UK, in 1991 and carried out by Warren Springs Laboratory, UK.

Source: Hampshire County Council, UK.

**Figure 86**   The effect of recycling on the gross CV of household waste. (*Note:* percentages are percentages of potential recyclables, as not all paper is recyclable (usually one-third is)).

*Ash removal:* Ash removal (see Figure 87) is usually effected by quenching the hot ash. This causes disintegration of the clinker and gives a granulated ash suitable for further processing and ferrous metals extraction. Incinerator ash is extremely abrasive and requires extremely robust handling systems.

*Fluidized beds:* Several types of fluidized bed grates have also been developed for burning low calorific value fuel including lignite, wood waste, peat, pulverized MSW (with metals extracted) and others. The main operating principles are that particles of fuel which are usually required to be smaller than 25 mm, are fed into a sand bed which is agitated either by means of primary air entering from beneath or by a combination of a mixing motion and primary air, such that the fuel is virtually floating in a bed of fluidized hot sand during combustion. This reduces corrosion and erosion problems and increases the efficiency of combustion.

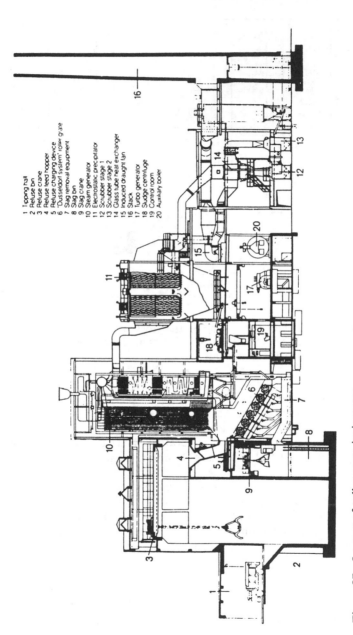

1  Tipping hall
2  Refuse bin
3  Refuse crane
4  Refuse feed hopper
5  Refuse charging device
6  "Dusseldorf system" roller grate
7  Slag removal equipment
8  Slag bin
9  Slag crane
10 Steam generator
11 Electrostatic precipitator
12 Scrubber stage 1
13 Scrubber stage 2
14 Glass tube heat exchanger
15 Induced draught fan
16 Stack
17 Turbo generator
18 Sludge centrifuge
19 Control room
20 Auxiliary boiler

**Figure 87**  Layout of roller grate incinerator.
*Source: Motherwell Bridge Engineering, PO Box 4, Logans Road, Motherwell, ML1 3NP, Scotland.*

The sand bed is recycled for reuse and it is claimed that the sand can easily be cleaned after mixing with the fine particles of grit and dust resulting from the combustion of waste. Crushed limestone can also be added to the bed to reduce acid emissions. Rotary kilns (Figure 88) can also be used for difficult wastes, e.g. clinical or hazardous wastes.

## Incineration performance.

*Waste reduction:* Based on the MSW weight composition given in the table in the INCINERATION entry, incineration can achieve a 75 per cent weight reduction, and given that complete combustion has been achieved, an 85 per cent volume reduction (see Figure 89). However, the degree of BURN OUT must also be assessed and this involves measuring both the amount of fixed carbon and putrescible matter in the residue by chemical analysis and fermentation respectively. The normal performance guarantees are no more than 3 per cent carbon and 0.03 per cent putrescible matter by weight in the ash. This means that the ash is virtually sterile and, in fact, is scavenged in some European plants for both ferrous metals recovery and clinker for road construction or as a sand and gravel replacement. This considerably reduces the amount of residues for ultimate disposal.

*The South East London Combined Heat and Power Plant* (SELCHP), opened in 1994, is a typical modern installation, details of which are given overleaf. The table gives the SELCHP emissions performance.

Waste incineration plants in England and Wales are controlled by public authorities from four main aspects:

(a) by the local planning authority (for this purpose, the county or metropolitan district council in England and the district council in Wales), as a development of land
(b) by the waste regulation authority (the county or metropolitan district council in England and the district council in Wales), as a form of waste management
(c) by Her Majesty's Inspectorate of Pollution (HMIP) (Part A processes) or for smaller plants (Part B processes) by the district council and other authorities, as a source of pollution
(d) by the Health and Safety Executive (HSE) where there is a potential hazard to workers on the site and persons outside it.

| | |
|---|---|
| *SELCHP* | 420 000 tonnes per annum in 2 × 29 tonnes per hour refuse burning streams. |
| *Storage capacity* | 4 days of full plant capacity – 5000 tonnes. |
| *Number of tipping bays* | 11. |
| *Steam output* | 144 tonnes of steam per hour at 395°C and 46 bar with refuse nett calorific value of 8500 kJ/kg. |
| *Flue gas treatment* | Each stream fitted with CNIM semi-dry lime scrubbers followed by high performance Bag House type filters, ejecting into a double flue 100 metre chimney. |
| *Availability* | Guaranteed 85% each refuse burning stream. |
| *Operating staff* | 55 persons. |
| *Site area* | 5.5 acres. |
| *Design and construction cost* | £85 million (1992 basis). |
| *Combustion temperature* | In excess of 850°C with 2 seconds residue time for dioxin destruction. |

Comparison of current operational mean plant emissions with authorization limits for SELCHP – 1994

| *Parameter* | *Concentration limit* $(mg/mg^3)$ | *Unit 1* $(mg/m^3)$ | *Unit 2* $(mg/m^3)$ |
|---|---|---|---|
| Total particulate matter | 20 | 0.2 | 0.2 |
| Hydrogen chloride | 30 | 3 | 1 |
| Hydrogen fluoride | 2 | <0.1 | <0.1 |
| Sulphur dioxide | 80 | 73 | 44 |
| Nitrogen oxides | 450 | 371 | 415 |
| Volatile organic compounds | 20 | <3 | <3 |
| Cadmium | 0.1 | 0.02 | 0.03 |
| Mercury | 0.1 | <0.01 | 0.01 |
| Other metals total* | 2.0 | 0.15 | 0.15 |
| Carbon monoxide | 80 | 14 | 23 |
| Dioxins (ng m$^{-3}$)(TEQ) | 1.0 | 0.12 | 0.40 |

All concentrations are expressed at reference conditions of 273 K, 101.3 kPa, 11% $O_2$ dry gas.
*Other metals are arsenic, chromium, copper, lead, manganese, nickel and tin taken together.
*Source:* G. Atkins, 'Future of incineration for waste disposal', presented at *CEA Conference*, Hinckley, 12–13 October 1994. Taken from UK AEA Report AEA/CS/18350062/REMA-003, May 1994.

The Secretaries of State for the Environment and for Wales decide appeals under (a), (b) and (c) above, and their Departments

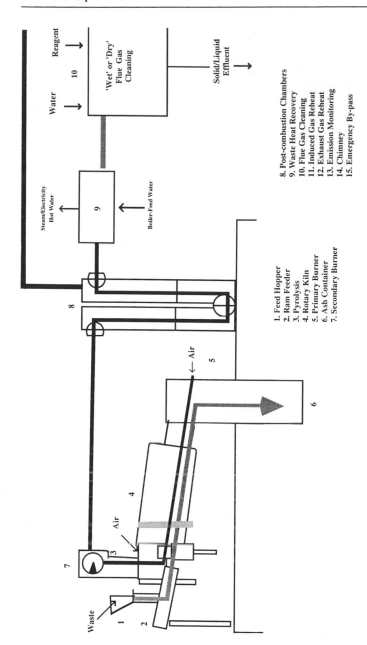

**Figure 88** Typical diagram of rotary kiln system (Courtesy: Motherwell Bridge Envirotec).

1. Feed Hopper
2. Ram Feeder
3. Pyrolysis
4. Rotary Kiln
5. Primary Burner
6. Ash Container
7. Secondary Burner
8. Post-combustion Chambers
9. Waste Heat Recovery
10. Flue Gas Cleaning
11. Induced Gas Reheat
12. Exhaust Gas Reheat
13. Emission Monitoring
14. Chimney
15. Emergency By-pass

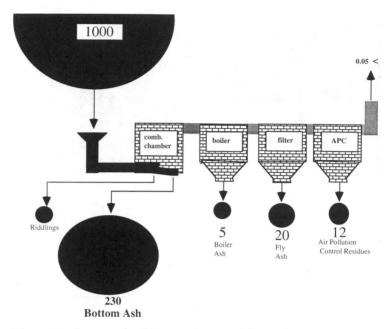

**Figure 89** Streams of solid masses in a municipal solid waste incinerator with wet flue gas scrubbing (values in kg/Mg waste).

provide advice and guidance to local authorities. In important respects, the legal provisions applying to incineration plants are now, or will probably in future be, contained in EC legislation, rather than purely national legislation. If a plant is designed to produce electricity for public supply and has a net capacity of more than 50 MWe, its construction and operation require the consent of the President of the Board of Trade.

In Scotland, the legislation is broadly similar. The planning authority may be the regional, district or islands council; the waste regulation authority is the district or islands council. There is no direct Scottish equivalent to HMIP but Her Majesty's Industrial Pollution Inspectorate (HMIPI) controls emissions to air from large incineration plants; the district or islands council is responsible for smaller plants. Discharges of liquid effluents to the environment are controlled by the river purification authorities. References in this entry to particular functions of regulatory authorities or Ministers should not normally be taken to include the arrangement in Scotland.

HMIP regulates as Part A processes for waste disposal the incineration of:

(a) wastes arising from the manufacture of chemicals or plastics
(b) specified waste chemicals (bromine, cadmium, chlorine, fluorine, iodine, lead, mercury, nitrogen, phosphorus, sulphur, zinc)
(c) any substances if the incinerator is rated at 1 tonne an hour or more
(d) chemicals contaminating reusable metal containers.

Other processes involving wastes are currently regarded by HMIP as having the primary purpose of providing energy. These include combustion processes with a net rated thermal input of greater than 3 MW such as waste-derived fuel, tyres, wood waste, straw or poultry litter.

The 1990 Act also made improvements in the legislation on local authority control of air pollution. District and metropolitan district councils regulate emissions to air from smaller incineration and combustion plants and from crematoria (Part B processes). However, incinerators not regulated by HMIP and designed to dispose of 50 kg an hour or less of substances other than clinical waste, sewage sludge, sewage screenings and municipal waste, do not require an authorization from the local authority.

Modern MSW incineration plants are now DIOXIN sinks as they destroy more dioxins than they create.

Fate of dioxins in energy-from-waste plant

| | Old plant (%) | Modern plant (%) | |
|---|---|---|---|
| Waste in 50 $\mu$g/mg | 100 | 100 | |
| Bottom ash | < 10 | < 10 | 70% dioxin |
| Boiler residue | 32 | < 3 | reduction can |
| Gas cleaning | 800 | < 20 | be achieved |
| Flue gas | 100 | < 1 | |

*Source:* IAWG, December 1994.

Vogg has stated:

> Thermal waste treatment as a pollutant sink is no longer an illusion but already reality in a number of cases. If one has the basic statements expressed in this paper interact in a reasonable manner

and combines them intelligently under the aspect of reasonableness, a substantial amount of pollutants are undoubtedly eliminated from the ecosystem by thermal treatment. This applies both to inorganic and to organic noxious chemicals.

The future of a new generation of incineration facilities has already come.

*Source:* H. Vogg, 'Thermal waste treatment reasons and objectives', *Waste Management* 95, Copenhagen, 16–17 March 1995.

By 1 December 1996, existing municipal incineration plant with a capacity of at least 6 tonnes per hour must comply with the following combustion conditions: the gases resulting from the combustion of the waste must be raised, after the last injection of combustion air and even under the most unfavourable conditions, to a temperature of at least 850°C for at least 2 seconds in the presence of at least 6 per cent oxygen. However, in the event of major technical difficulties, the provisions concerning the 2 second period must be implemented at the latest when the furnaces are replaced.

By 1 December 1995, other existing municipal waste inciner-ation plant must comply with the following conditions: the gases resulting from the combustion of the waste must be raised, after the last injection of combustion air and even under the most unfavourable conditions, to a temperature of at least 850°C in the presence of at least 6 per cent oxygen for a sufficient period of time to be determined by the Inspector.

*Source:* G. Atkins, 'Future of incineration for waste disposal', presented at *CEA Conference*, Hinckley, 12–13 October 1994. Taken from UK AEA Report AEA/CS/18350062/REMA-003, May 1994.

**Industrial noise measurement.** The measurement used in the UK for industrial noise is the corrected noise level (CNL). This is based on a measure in dB(A) (⇨ DECIBELS A-SCALE) to which a specified correction is added if a definite continuous tone (whine, hum, etc.) is present. Another correction is added if there are any impulsive irregularities (bangs, clanks, etc.); and if the noise is not continuous, a further correction is applied which depends on the proportion of time for which the bursts of noise last.

Some values of CNL and the situations in which they are found are listed below.

| CNL (dB(A)) | Situation of measuring point |
|-------------|------------------------------|
| 85 | At 10 metres from a building housing an air/steam drop forge hammer |
| 75 | At 10 metres from an air compressor housed in a building with louvred doors |
| 60 | At 15 metres from a can-making factory. Continuous noise from stamping machinery and from handling of thin sheet metal |

**Industrial waste.** ⇨ WASTES.

**Infections, water-borne.** These include bacterial infections such as CHOLERA, the enteric diseases, TYPHOID and para-typhoid, and Weil's disease; the viral infections poliomyelitis and hepatitis, protozoan infections such as amoebic dysentery, and SCHISTOSOMIASIS (bilharzia) caused by a trematode worm. All the agents responsible are water-borne; bilharzia is parasitic in some snails in irrigation ditches in developing countries and Weil's disease may be transmitted by rats in sewers. (⇨ DAM PROJECTS)

PUBLIC HEALTH practice is intended to prevent, among other things, the spread of disease. Thus, drinking water and sewage are kept separate. However, in less sanitary countries, or even in a crisis in more developed countries, if life-support systems break down, water withdrawal from some sources must cease immediately.

**Infiltration[1].** The downward flow of water through the soil surface into subsurface strata.

**Infiltration[2].** The term is also used to denote the contamination of wells by salt water or the spoiling of an aquifer by pollutants, e.g. by LEACHATE from a LANDFILL SITE.

**Infrared radiation.** Radiation whose wavelength is in the infrared radiation spectrum. It can be used to conduct energy surveys using a special camera and so enable sources of waste heat to be pinpointed.

**Infrasound.** Low-frequency sound which usually emanates from low revving ship or heavy transport engines in frequencies below 100 Hz. A typical living-room will resonate around 12 Hz thus amplifying heavy traffic noise. One noise measurement in London recorded a value of 85 DECIBELS at 5 Hz coming from ships in the docks. Infrasound is not catered for on the dB(A) scale which is weighted to represent the response of the human ear and therefore takes little account of noise below 100 Hz.

The only way of cutting down infrasound is to stop it at source as insulation is rather ineffective in this frequency range.

**Inhibitor.** A substance, naturally occurring or added, whose presence in small amounts retards or prevents the occurrence of certain phenomena considered undesirable, e.g. gum formation in stored gasolines, colour change in lubricating oils, corrosion in turbines, scale formation in boilers, etc.

**Inorganic matter.** Matter which is mineral in origin and does not contain carbon compounds, except as carbonates, carbides, etc. (⇨ ORGANIC MATTER)

**Insecticides, synthetic.** ⇨ CHLORINATED HYDROCARBONS; ORGANOPHOSPHATES.

**Insolation.** The amount of direct solar radiation incident per unit horizontal area at a given level, measured in $mW/m^2$.

**Integrated pollution control (IPC).** Introduced in the UK Environmental Protection Act 1990 so that all major emissions are considered simultaneously and not in isolation, i.e. the reduction of pollution in one environmental medium can have effects on another BEST AVAILABLE TECHNIQUE NOT ENTAILING EXCESSIVE COSTS – all are considered together. The BATNEEC is required to minimize pollution of the environment as a whole.

**Integrated waste management.** A strategy for the management of waste utilizing a range of environmentally sound systems and processes. Typically it would include the promotion of waste minimization, materials recycling, resource recovery with landfill as a last resort, i.e. the first three options would be optimized and landfill minimized.

**Interceptor.** A trap which intercepts and separates oil, grease, grit from sewage or surface water.

**Interflow.** The lateral movement of water at a soil–rock interface. This contributes to surface runoff in permeable catchments.

**Intermediate processor.** A company which purchases source-separated materials, for further processing. (⇨ RECYCLING)

**Internal combustion engine.** The internal combustion engine is characterized by combustion of the fuel inside an enclosure, usually a cylinder, which causes rapid gas expansion which in turn forces a piston to move and do work. The characteristics of an internal combustion engine are that there is a need for a homogeneous easily ignited fuel (petrol or diesel oil) which, on ignition, does work due to the expansion of the gases which are then released to the atmosphere. The cyclic nature of the operation plus poor fuel distribution in the cylinder means that the THERMAL EFFICIENCY is low, combustion is often incomplete and thus unburnt HYDROCARBONS, CARBON MONOXIDE and CARBON DIOXIDE are released to the atmosphere. NITROGEN OXIDES are also released at concentrations which depend on the temperature and duration of the combustion flame. Despite this, the internal combustion engine is in mass production and has market dominance. Attempts to replace it will meet with strong opposition due to the capital invested in its manufacture and aftercare. (⇨ EXTERNAL COMBUSTION ENGINE; AUTOMOBILE EMISSIONS)

**International Committee on Radiation Protection (ICRP).** Committee charged with setting permissible DOSE levels of IONIZING RADIATION to both the population at large and the workforce (who are 'permitted' greater doses). (⇨ IONIZING RADIATION)

**International Maritime Organisation (IMO).** IMO standards require installations in less than 75 m of water (and weighing less than 4000 tonnes) to be completely removed.

All post 1 January 1998 installations will also have to be removed entirely if they are in less than 100 m of water and be designed to be removable if placed in deeper water.

All other installations will be partially removed or left in place so long as there is at least 55 m of clear water above any submerged remains and regular maintenance to any remains above sea level in order to prevent structural failure.

The cut-off portions are likely to be dumped on the fringes of the North Sea, in depths of 2000 m where they can disintegrate slowly.

The dispersive powers of the ocean should ensure that minimal environmental damage occurs despite high decibel claims to the contrary from well-intentioned pressure groups. (⇨ OIL RIGS)

**International Nuclear Event Scale (INES).** Launched on 1 November 1990, this classifies nuclear events at seven levels. Basically, the lowest three levels will be for incidents of increasing seriousness and the upper four apply to accidents of increasing seriousness. Events which have no safety significance are classified as below scale or level 0 and may not be given much (if any) publicity.

**Intrinsic energy value.** Energy either directly consumed in the manufacture of a product or the energy needed to produce feedstock. See table below for comparison of the intrinsic energy and calorific values for various commodity plastics.

| Material | PVC | PET | PP | PS | LDPE |
|---|---|---|---|---|---|
| Intrinsic energy value (MJ/kg) | 53 | 84 | 73 | 80 | 69 |
| Calorific value (MJ/kg) | 18 | 22 | 41 | 38 | 43 |

*Source:* Association of Plastics Manufacturers.

**Inversion layer.** ⇨ TEMPERATURE INVERSION.

**Ion.** An electrically charged atom or molecule produced by a loss or gain of electrons. Gases can be ionized by electrical discharge or ionizing radiation. Ions are also formed in solution.

**Ion exchange.** The removal of IONS from solution. It can be carried out by the use of a suitable ion-exchange medium or bed, which has the power to remove or capture the desired ions from solution. The choice of medium determines the type of ion removed (positive or negative). Once the medium is saturated it must be taken off-stream and recharged by passing an alkaline or acidic solution through the bed which replaces the captured ions from the recharging solution. Ion exchange has many uses. For example, in water softening, calcium ions are replaced by sodium ions, and the water is thereby softened. The bed is then recharged with a brine solution which replaces the captured calcium ions and the bed is then ready for re-use. Normally one bed is 'on-stream' while the other is 'off-stream' for re-charging.

Ion exchange is used in SEWAGE TREATMENT, water softening, water purification, SOLUTION MINING, metallic effluent treatment, etc.

**Ion-selective electrode.** An electrode which has a high degree of selectivity for one ION over other ions which may be present in a sample. A potential difference is obtained that is proportional to the LOGARITHM of the activity of a particular ion. Ion selective electrodes are available for $H^+$, $Na^+$, $K^+$, $NH^+_4$, etc.

**Ionizing radiation.** Certain radiations in their passage through matter, are capable of causing ionization, i.e. they can 'knock' electrons out of atoms or molecules or create ions either directly or indirectly. The radiations that do this directly are either fast-moving particles (for example, electrons, protons and $\alpha$-particles) or electromagnetic rays such as X-rays or $\gamma$-rays. ($\gamma$-rays are similar to X-rays but have much shorter wavelengths and higher energies. The difference is more one of name than of physics.) There are numerous sources of naturally occurring ionizing radiation including the cosmic rays that arrive continuously from outer space, but the sources of environmental significance are certain radioactive substances liberated into the environment by man.

**Ionizing radiation, dose measurement.** Ionizing radiations may be divided into two main groups:

1. *Electromagnetic radiations* (X-rays and gamma rays), which belong to the same family of electromagnetic radiations as visible light and radio waves.
2. *Corpuscular radiations*, some of which – alpha particles, beta particles (electrons) and protons – are electrically charged, whereas others – neutrons – have no electric charge. This distinction between the two groups becomes blurred, however, when their mode of absorption in materials is considered.

Whilst the exact nature of the biological effects of these radiations is not fully understood, they are related to the ionization that the radiations are capable of producing in living tissue. Thus, the biological effects of all ionizing radiations are essentially similar. However, the distribution of damage throughout the body may be very different according to the type, energy and penetrating power of the radiation involved. α-particles from radioisotopes have ranges of only about 0.001–0.007 centimetres in soft tissue and less in bone. β-particles have ranges in soft tissues of the order of several millimetres, i.e. much greater than those of α-particles in such tissues. X-rays and γ-rays are not stopped by tissues.

The dose of radiation is the amount of energy absorbed per unit mass of material. The unit is the RAD (radiation absorbed dose (rad)), which is 0.01 joules per kilogram ($10^{-2}$ J/kg). The SI unit of absorbed dose is the GRAY (Gy) and corresponds to an energy absorption of 1 J/kg of matter. So,

$$1 \text{ Gy} = 100 \text{ rad}$$

However, the rad or gray is a physical measure of energy absorbed and does not indicate the biological effects of irradiation. These vary according to the type and energy of the radiation involved.

The most significant factor in estimating the biological effectiveness of radiation is the LINEAR ENERGY TRANSFER, LET. Significant penetration into soft tissues (typically up to

2 m) is achieved by X- and $\gamma$-rays, which have lower LETs compared to $\beta$-particles, which can typically penetrate a few millimetres, or to $\alpha$-particles which usually have very high LETs and can only penetrate a few hundredths of a millimetre. Thus $\alpha$-particles are likely to produced a much greater amount of damage than $\beta$-particles or $\gamma$-rays but this damage is usually summarized by a quality factor, $Q$ (before 1962 this was known as the relative biological effectiveness or RBE). This is defined as the ratio of the effectiveness of the radiation to the effectiveness of 200 kV X-rays. For $\beta$-particles, X- and $\gamma$-rays, $Q = 1$, for $\alpha$-particles $Q = 20$, and for neutrons, $Q = 10$. Thus for an absorbed dose $R$ (in rad or Gy), the biologically effective dose $D$ is given by

$$D = RQ$$

In addition to the absorbed dose and the quality of the radiation absorbed, biological changes can be influenced by a large number of other factors; some organs are more susceptible to damage than others, internally ingested radioactive materials may concentrate in certain organs and some organs may be regarded as more vital than others.

To allow for these different circumstances, the use of a dose equivalent is recommended. The dose equivalent is defined as the absorbed dose $R$ (in rad or Gy) multiplied by the quality factor $Q$ and any other modifying factor $N$; hence

$$\text{Dose equivalent} = RNQ$$

Dose equivalent is measured in REM (Rœntgen Equivalent Man) if absorbed dose is measured in rad. It is measured in SIEVERT (Sv) if the absorbed dose is measured in gray.

By way of illustration, for neutrons impinging on the eye $Q = 10$ and $N = 3$. Hence an absorbed dose of 0.01 Gy (1 rad) corresponds to a dose equivalent in this instance of 0.335 Sv (30 rem). The same absorbed dose falling on the skin has $N = 1$ and so has a dose equivalent of 0.1 Sv (10 rem) indicating that a dose to the eye is three times more damaging than the same dose to the skin. In both cases, the $Q$ factor indicates that neutrons produce 10 times the effect of 200 kV X-rays. Neither the quality factor, $Q$, nor the modifying factor, $N$, in the above equation can be calculated; both are empirical constants. ($\Rightarrow$ GRAY; LINEAR ENERGY TRANSFER; RAD; REM; SIEVERT)

The data below show the relationships between the SI units

associated with radiation and those preceding 1978, but which are still in common use.

| Quantity | New named unit and symbol | In other SI units |
|---|---|---|
| Exposure | — | $C\,kg^{-1}$ |
| Absorbed dose | Gray (Gy) | $J\,kg^{-1}$ |
| Dose equivalent | Sievert (Sv) | $J\,kg^{-1}$ |
| Activity | Becquerel (Bq) | $s^{-1}$ |

| Quantity | Old special unit and symbol | Conversion factor |
|---|---|---|
| Exposure | roentgen (R) | $1\,C\,kg^{-1} \approx 3876\,R$ |
| Absorbed dose | rad (rad) | $1\,Gy = 100\,rad$ |
| Dose equivalent | rem (rem) | $1\,Sv = 100\,rem$ |
| Activity | curie (Ci) | $1\,Bq \approx 2.7 \times 10^{-11}\,Ci$ |

**Ionizing radiation, effects.** Exposure to ionizing radiation can be harmful as the radiation can cause cancers in the living population and genetic changes that may produce heritable defects in future generations. Ionizing radiation causes mutations, i.e. random changes in the structure of DNA, the long molecule that contains the coded genetic information necessary for the development and functioning of the human being. As the mutations are random events, they are almost certainly harmful to some degree.

The outcome of radiation exposure may be that the cell may suffer so much damage that it dies, or that the cell may continue to function but in a modified way. The second outcome can lead to uncontrolled growth of a colony of cells derived from the affected one. This is a cancer, and the upset is called somatic (bodily) mutation.

For reasons not fully understood, cancer resulting from irradiation appears only after a long delay. Only recently was the incidence of cancer beginning to show up as abnormal among the survivors of the nuclear explosions at Hiroshima and Nagasaki. An abnormal incidence of leukaemia (the uncontrolled growth in numbers of white blood cells), which has a much shorter latency period, was in evidence long ago in the studies of the Atomic Bomb Casualty Commission.

Now if the mutation is in a germ cell (sperm or egg), the entire organism arising from the germ cell together with its progeny will be affected. Many mutations are sufficiently severe to prevent their carriers from living, or at least from reproducing and handing on the defect to later generations. Many others, though, have their effects masked if they are inherited from only one parent. Such mutations are recessive. Their effects appear only in about a quarter of the off-spring of parents both of whom carry the mutant gene.

There are many undesirable recessive genes and most of us carry a few. Consequently from time to time a child of normal healthy parents shows a terrible defect such as gross malformation. The hidden undesirable genes in a population are often called its genetic load. The tendency of ionizing radiation is to increase the genetic load.

Not all mutations have such dramatic effects. They may result merely in a loss of vigour, a susceptibility to disease, or a reduction in life-span. Such genes may spread widely in human populations that are sustained by medical attention.

Ionizing radiation also tends to produce a reduction in the life-span of the animals actually irradiated. The effect is often regarded as premature ageing. The following figures, published in 1957, are *possible* examples of this premature ageing.

| Group | Average age at death |
|-------|----------------------|
| USA population | 65.6 years |
| Physicians with no known contact with radiation | 65.7 years |
| Radiologists | 60.5 years |

The radiologists referred to in the table were practising their profession over a period in which the dangers of X-rays were far less clearly appreciated than they are today. The age distribution of the control was also different from that of the radiologists which casts some doubt on drawing *absolute* conclusions from the data. Nowadays much more sensitive X-ray film is used which requires less intense radiation, and the relevant equipment is carefully shielded to protect its users from undesirable radiation.

It is usual to compare levels of man-made radiation to that of the natural background radiation from cosmic rays and other natural sources. The inference is then drawn that as man is currently contributing radiation amounting to around 0.1 per cent of the natural background, we therefore have a lot of scope for increasing nuclear installations before we need become concerned about the level of man-made radiation. However, because the natural background is irreducible, it does not necessarily follow that this sets an allowable scale for man-made radiation emissions as radionuclides can be concentrated in the food chain, for example.

For the UK, natural background radiation accounts for 87 per cent of the average annual dose to the UK population (2.5 million sieverts in all), artificial (medical X-rays 12 per cent, nuclear discharges 0.1 per cent, fallout 0.4 per cent, work and miscellaneous 0.6 per cent). There are large variations from the average; the actual dose received by any individual will vary depending upon altitude (because of an increased contribution from cosmic rays) and the type of rocks occurring in the locality: granite for example emits some five times more radiation than clays.

The effects of an increase of radiation are proportionately very small indeed, but applied to a total population can give rise to large numbers of cancer incidences and genetically defective births. For example, the genetic handicap risk (Maryland Academy of Sciences) for a radiation dose of 1.7 mSv (170 millirem) per year is estimated as $4.5 \times 10^{-5}$. The number of cases per annum is then (for the US population of 200 million) $4.5 \times 10^{-5} \times 200 \times 10^6$, and for a birth-rate of 4.6 million per year, the absolute numbers of genetically unsound births per 1000 births is therefore

$$\frac{4.5 \times 10^{-5} \times 200 \times 10^6}{4.6 \times 10^6} \times 1000$$

or 2 per 1000 births.

It is worth emphasizing that knowledge of the effects of ionizing radiation on mutation rates is slight and growing only slowly. Estimates for human populations are based on readily observed cases, but mutations that give rise to slightly impaired individuals whose characteristics are nevertheless within the ordinary range of variation are an even more unpleasant prospect. The attrition of human capacities and the deterioration of health and vigour could proceed unnoticed for indefinitely long periods.

It is now generally agreed that, at least for the purposes of estimating hazards, any dose of ionizing radiation, no matter how small or how slowly delivered, must be regarded as prospectively harmful, able to induce genetic mutations and cancer and generally to erode the life-span of its recipient.

The costs of increased exposure to radiation (as in any form of pollution) must be balanced against the social benefits to be derived from the activities leading to the increased exposure. The debate is whether these risks are justifiable and whether we have the right to saddle future generations with our long-life radioactive wastes. (⇨ COST-BENEFIT ANALYSIS; NUCLEAR ENERGY; NUCLEAR REACTOR DESIGN; NUCLEAR REACTOR WASTES; IONIZING RADIATION, DOSE MEASUREMENT)

*Source:* National Radiological Protection Board, *Radiation Doses – Maps and Magnitudes*, Harwell, Oxon, undated.

**Ionizing radiation, maximum permissible dose and dose limit.** The maximum permissible dose of ionizing radiation was originally regarded as one that, in the light of the knowledge available at the time, was not expected to cause appreciable bodily injury to any occupationally exposed person at any time during his life. The phrase 'appreciable bodily injury or effect that a person would regard as being objectionable and/or competent medical authorities would regard as being deleterious to the health and well-being of the individuals' is used in the International Commission on Radiological Protection, 1966.

The ICRP retains the term 'maximum permissible dose'* for the exposure of radiation workers. By monitoring the doses received by individual workers and by controlling their environment it is possible ensure that the maximum permissible doses are not exceeded, except in the case of accidents. It is not possible, however, to determine the doses received by members of the public. It is therefore necessary to control the sources giving rise to exposure and to assess the average dose received by a specific group or an entire population, both by means of sampling procedures in the environment and, in appropriate

---

* Observe that this term is likely to be misleading and is often used in a misleading way, because it begs the question, 'What is a permissible dose?' As knowledge of the effects of ionizing radiation has increased over the years, the dose described as 'permissible' has been repeatedly revised downwards. The current value is only a small fraction of earlier estimates of what was tolerable. One may reasonably continue to ask whether the 'maximum permissible dose' is permissible in any commonly understood sense of 'permissible'.

cases, checks on the doses received by a few individuals in the group or population involved. The ICRP believes that the term 'maximum permissible dose' is seldom meaningful in relation to the individual members of the public and has recommended that, in their case, the term 'dose limit' should be used. The current values of the maximum permissible doses and dose limits for specified organs and tissues, as recommended by the ICRP, are given in the table overleaf.

The dose rates are the suggested maximum exposures and represent those dose rates that should never be exceeded by any person occupationally exposed to ionizing radiations. The recommended maximum average dose for the same workers is one-third of the tabulated value. Any worker receiving the maximum should be removed from exposure until his annual average rate has been reduced to the currently recommended level.

*Note:* These limits are under review in the UK and may well be substantially reduced from the current maximum radiation exposure dose permitted of 50 millisieverts per year for a radiation worker and 5 millisieverts per year for a member of the public. Very few radiation workers in the UK actually receive doses exceeding 15 millisieverts per year. Following worries about a possible cause of childhood leukaemia being caused by irradiation of fathers employed in the UK nuclear fuel reprocessing industry, demands are being made for a new statutory limit of 10 mSv per worker per year with no more than 5 mSv in any six months. ($\Rightarrow$ STOCHASTIC EFFECTS)

*Source:* M. Morris. Union press for lowest radiation risk, *Guardian*, 10 March 1990.

**Ionization.** $\Rightarrow$ IONIZING RADIATION, EFFECTS.

**Ionosphere.** The region of the upper atmosphere (starting from 90 km above the Earth) which includes the highly ionized Appleton and Kennelly–Heavyside layers, the existence of which enables intercontinental radio transmission to be achieved.

**Irradiation, uses of.** The major potential uses of irradiation in the environment are:

1. *Insect sterilization.* Irradiated males can be released and, as the mating in some species only takes place once, it is possible to reduce population of pests without recourse to pesticides.

2. *Digestibility aids.* Many cellulosic wastes can be used by ruminants as an energy source especially if the cellulose structure is broken down. Irradiation is one means of doing this.

Dose equivalent limits (DEL) in mSv in a year given or implied by CIRP Publication 26 (1977)

| Tissue/organ | Worker | | Member of the public[a] | |
|---|---|---|---|---|
| | Stochastic[b] | Non-stochastic[b] | Stochastic[b] | Non-stochastic[b] |
| Whole body | 50[c] | – | 5[d] | – |
| Tissues/organs irradiated singly: | | | | |
| gonads | 200 | (500) | 20 | (50) |
| breast | 330 | (500) | 33 | (50) |
| red bone marrow | 417 | (500) | 42 | (50) |
| lung | 417 | (500) | 42 | (50) |
| thyroid | (1670) | (500) | (167) | (50) |
| bone surfaces | (1670) | 500 | (167) | 50 |
| lens | – | 300 | – | 50 |
| other single organ | (833) | 500 | (83) | 50 |
| bone | – | – | – | – |
| skin | – | 500 | – | 50 |
| Planned special exposure single | 2 × DEL | | | |
| lifetime | 5 × DEL | | | |

Pregnant woman after diagnosis of pregnancy
0.3 × DEl pro rata[e]

*Notes:* [a] No DEL for populations is recommended but application of the DELs for individuals, as well as observance of the Commission's general principles, is likely to ensure that the averager dose equivalent to a population will not exceed 0.5 mSv per year.

[b] The DEL is the stochastic or the non-stochastic limit, whichever is the lower. (The higher value is shown in brackets.)

[c] This limit applies both to uniform irradiation of the whole body and to the weighted mean of the doses to indiviudal tissues. For external irradiation, when the dose distribution is unknown, the limit applies to the deep dose equivalent index.

[d] As exposures at the DEL are not likely to be repeated over many years, this limit is likely to ensure that the lifetime dose to a member of the public will not exceed 1 mSv per year.

[e] e.g. 0.175 × DEl in 7 months.

3. *Food preservation.* The storage life of iced fish on board ship can be doubled by a dose of 100 krad without any apparent adverse effects. This technique may enable trawlers to use distant fishing grounds, extend the duration of voyages and deliver a more hygienic product to the customer. Food preservation is an area in which major advances are expected as irradiation has not been shown to produce dangerous radioactivity in foods when a cobalt-60 radiation source is used. (This produces $\gamma$-rays of 1.17 meV (electron-volt is the general unit of energy of moving particles) which are suitable for food processing.) However, there are serious doubts about its use in treating food in which decay has already commenced and the subsequent sale of the produce as sound. Should this method proliferate, the subsequent disposal of the spent cobalt sources, whose half-life is ca. 5 years, will produce yet more radioactive waste for secure long-term disposal.

4. *Radioactive tracers.* The pathways of pollutants can be traced by releasing radioactive monitors. This has been done in studies of air pollution and effluent dispersal. They are also of great use in following fluid flow and leak detection in industry, the ingestion/digestion of food in animals, monitoring engine wear and many other applications.

5. *Disinfestation of seeds.* The removal of pests from stored grains.

6. *Sterilization.* Large doses of radiation will produce a completely sterile product, e.g. pharmaceutical products and material for transplants.

**Irrigation.** The application of water to arable soils, so that plant growth may be initiated and maintained. Irrigation is used not only in many arid and semi-arid lands, but in the UK to rectify SOIL MOISTURE DEFICITS (SMDs). In the past, eastern England could have benefited from irrigation every three years in ten. The practice may be expected to grow in the UK due to the high SMDs now being experienced.

Irrigation with brackish water over long periods can eventually render soils unfit for crop growth, as plant evapo-transpiration acts as a distillation process and leaves the dissolved solids in the soil where they accumulate. The spread of Middle Eastern deserts has been partially attributed to early irrigation attempts which resulted in salinization of once fertile soils.

**Isomer.** A substance which has identical molecular counterparts but differs in constitution or structure from other members of the group by the different arrangement of atoms within the molecule.

**Isotope.** Atoms of the same ELEMENT which have the same number of protons in their nuclei but differing numbers of neutrons. Hence, they have the same atomic number but different MASS NUMBERS. Isotopes are written as mass number (i.e. number of protons and neutrons) followed by the symbol, e.g. $^{235}$U, which is the fissile isotope of uranium and is capable of sustaining chain reactions in a nuclear reactor.

Isotopes of the same element have identical chemical behaviour but differing physical behaviour, e.g. cobalt-60, used as a $\gamma$-ray source for food irradiation, is manufactured in a nuclear reactor by irradiating ordinary cobalt-59 with neutrons. ($\Rightarrow$ IRRADIATION, USES OF; NUCLEUS; RADIOACTIVE ISOTOPE)

# J

**Jet, overfire.** A jet of air or steam designed to promote turbulence above a fuel bed in a furnace and so reduce the formation of smoke.

**Joule (J).** International System of Units (SI) unit of work and energy. It is defined as the work done when a force of 1 NEWTON acts through 1 metre. Commonly, the megajoule (MJ) is used because it is a more convenient size for the measure of energy supplied. It is 1 million joules.

The measure of power is the WATT, which is 1 joule per second, or $1 \text{ J/s}^1$.

There are several forms of ENERGY, e.g. thermal, chemical, electrical, and joule is applicable to all of them. The form of the energy is important, however, because thermal energy and electrical energy are very different in their ability to do work. For this reason, in ENERGY ANALYSIS, it is very common to denote the form the energy is in, i.e. whether it is primary (fuel energy) or electrical energy. The conversion factor for fuel energy to electrical energy is commonly taken as 30 per cent but may be less where internal combustion engines are in use. ($\Rightarrow$ LAWS OF THERMODYNAMICS)

# K

**Kelvin (K).** The SI unit of temperature interval. The kelvin or absolute scale of temperature starts from absolute zero temperature. 1 kelvin = 1 Celsius degree = 1.8 Fahrenheit degrees. $273.15\,K = 0°C = 32°F$.

**Kerbside collection.** The regular pick up of recyclable materials and DOMESTIC REFUSE from residential kerbsides.

**Kilogram mole.** The mass in kg occupied by 1 mole of a substance, e.g. 1 kg mole $CO_2$ is 44 kg.

**Kilowatt-hour (kWh).** Unit of energy equal to 1000 WATT-HOURS or 3.6 megajoules commonly taken to mean electrical energy. However, the kilowatt hour can be used as a measure of ENERGY input or output in energy analysis, in which case, $kWh_{th}$ is used for energy in the form of fuel (or thermal energy). If in the electrical form, the designation $kWh_e$ is used which, depending on the efficiency of conversion, is equal to $3-4\,kWh_{th}$. ($\Rightarrow$ WATT; EFFICIENCY)

This concept can best be illustrated by considering the energy value of a tonne of high grade coal. It can be stated as $8000\,kWh_{th}$ or, if burnt in a high efficiency power station, it provides around $2600\,kWh_e$. This is the basis of ENERGY ANALYSIS which looks at the energy required for say, the production of a tonne aluminium on primary (fuel) energy basis and thus enables the identification of those areas where energy saving is possible. In the case of ALUMINIUM, RECYCLING of used beverage containers can save 95 per cent of the energy required to manufacture the metal from the bauxite ore. This provides a national resource conservation argument for either recycling aluminium or substituting other less energy intensive materials (if available) if recycling is not possible.

# L

**LC$_{50}$.** ⇨ LETHAL CONCENTRATION.

**LD$_{50}$.** ⇨ DOSE.

**Leq.** ⇨ NOISE INDICES.

**L$_{10}$.** ⇨ NOISE INDICES.

**L$_{90}$.** ⇨ NOISE INDICES.

**Laminar flow.** A fluid flow pattern in which each layer of liquid or gas flows smoothly and is not broken up. When the flow becomes mixed up it is said to be turbulent. (⇨ TURBULENCE)

**Landfill gas (LFG).** LFG is generated in landfill sites by the anaerobic decomposition of domestic refuse (municipal solid waste) (⇨ METHANE). It consists of a mixture of gases and is colourless with an offensive odour due to the traces of organosulphur compounds. A typical landfill gas composition for mature refuse (% by volume) is given below:

| Component | Typical value (%) |
|---|---|
| Methane | 63.8 |
| Carbon dioxide | 33.6 |
| Oxygen | 0.16 |
| Nitrogen | 2.4 |
| Hydrogen | <0.05 |
| Carbon monoxide | <0.001 |
| Saturated hydrocarbons | 0.005 |
| Unsaturated hydrocarbons | 0.0009 |
| Halogenated compounds | 0.00002 |
| Organosulphur compounds | <0.00001 |
| Alcohols | <0.00001 |
| Others | 0.00005 |

Aside from its unpleasantness, it is highly dangerous as methane is explosive in concentrations in air between 5 per cent, the Lower Explosive Limit (LEL), and the Upper Explosive Limit (UEL) of 15 per cent.

*Waste Management Paper* 27 'Control of Gas', Revised Draft
3rd Peer Review Group Meeting, 01/08/88, gives the following
recommendations for action should LFG migrate to dwellings
or buildings.

| Gas level | | Procedures |
|---|---|---|
| Per cent gas in air | Per cent lower explosive limit | |
| <0.5 | <10 | No action |
| 0.5 | 10 | Audible alarm |
| | | Verification of concentration with a portable instrument |
| | | Ventilation of buildings using windows and doors |
| | | Turn off/extinguish potential ignition sources |
| 1.0 | 20 | Evacuate building |
| | | Notify emergency services |
| 5.0 | 100 | Lower explosive limit (LEL) |

The volume of LFG generated in time is given in Figure 88.
Both the timescale and the amounts are variable, but 1 tonne of
domestic refuse can generate up to 450 m³ (in addition, putrescible
wastes that are co-disposed in the landfill site can substantially
increase the volume of LFG generated). LFG will appear within
3 months to 1 year after waste deposition, peaking at 5–10 years
and tapering off over 20–40 years. However, many factors
interplay and it is for this reason that closed landfill sites must
be monitored for up to perhaps 30 years after site closure to
ensure that there is no repetition of the infamous Loscoe UK
LFG explosion which took place on 24 March 1986 destroying
a bungalow and severely injuring the occupants. This occurred
$3\frac{1}{2}$ years after the Loscoe landfill site was closed.

Landfill gas must be controlled at all operational landfill
sites, wether actively or passively vented or both especially in
the case of deep sites.

Figure 89 shows schematic venting systems for shallow and
deep sites respectively. It should be noted that barrier systems to
prevent lateral gas migration are not enough and that for deep
sites extraction will also be required.

LFG utilization in the UK centres on using it for cement or

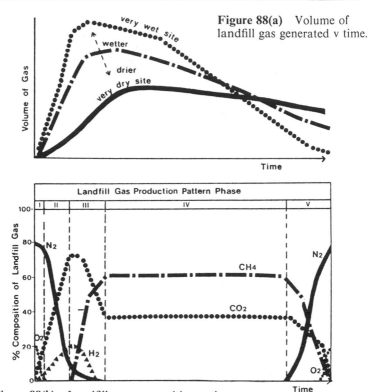

**Figure 88(a)** Volume of landfill gas generated v time.

**Figure 88(b)** Landfill gas composition v time.

brick kilns or on-site electricity generation in modified reciprocating engines or gas turbines. Good results have been achieved with all these uses. However, it must be said that only around one-third of the available refuse energy can be utilized by LFG extraction because of losses. INCINERATION with energy recovery is much more energy efficient than LFG collection as it uses all of the refuse for fuel purposes. Also, waste incineration prevents the formation of the LFG in the first place, thereby stopping the powerful greenhouse gas, methane, being formed. This has a greenhouse effect of up to 30 times that of carbon dioxide.

As many old landfill sites are now considered to be prime candidates for development, the options are either waste removal and redeposition elsewhere (£20–£30 per m³ in 1990) or

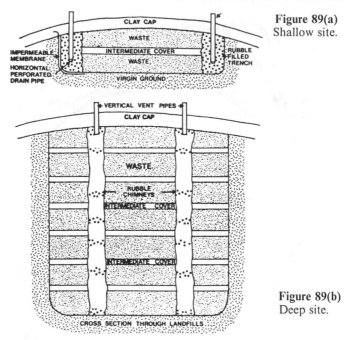

Figure 89(a)
Shallow site.

Figure 89(b)
Deep site.

piling of the site plus provision of a ventilation space between the floor slab (sealed against gas) and the landfill cover.

The extent of the UK's LFG problems are only now coming to light. The first Annual Report (1989) of the UK HM Inspectorate of Pollution reveals that 1390 active or recently closed landfill sites pose a potential gas risk, as set out below.

| | Total | Distance to housing or industry (metres) | | Gas control installed |
|---|---|---|---|---|
| | | 0–250 | 250–500 | |
| Active landfills: | | | | |
| England | 581 | 319 | 70 | 135 |
| Wales | 21 | 5 | 3 | 2 |
| Total | 602 | 324 | 73 | 137 |
| Landfills closed since 1977: | | | | |
| England | 761 | 426 | 67 | 81 |
| Wales | 27 | 6 | 5 | 0 |
| Total | 788 | 432 | 72 | 81 |

It should be noted that over 50 per cent of the sites lie within 250 metres of housing and that less than 30 per cent had even rudimentary gas control measures at the time of the report's compilation.

The seriousness (both to lives and financial consequences) with which methane gas migration must be viewed is exemplified by the UK Abbeystead, Lancashire, pumping station explosion in 1984. Sixteen people were killed and 28 injured. Compensation claims totalling up to £4 million (by those involved and North-West Water) have resulted against the firm of consulting engineers who designed and supervised the construction of the pumping station into which naturally occurring methane gas had migrated. An explosion was triggered when the pumps were switched on.

Landfill gas is also a powerful GREENHOUSE GAS by virtue of its 60 per cent methane content.

**Landfill gas surface emissions monitoring.** As discussed in relation to LANDFILL GAS, the US EPA now requires landfill gas surface emissions to be monitored. This has been done in California since April 1985.

Californian landfills must be monitored in several categories: ambient air and weather station monitoring; instantaneous and integrated surface sampling; landfill gas migration perimeter probe monitoring; and landfill gas collection system sampling.

The air samplers are placed at pre-determined locations. Ambient air samples are taken monthly over a 24-hour period; wind data are continuously recorded.

To prohibit photochemical reactions, ambient air bag samples are enclosed in light-sealed containers, usually cardboard boxes. Within 72 hours of collection, the Tedlar bag samples must be analysed for total organic compounds (TOC) and toxic air contaminants.

Composite surface sampling, also called integrated surface sampling, is required each month. The landfill surface is divided into monitoring/sampling grids, each approximately 50 000 square feet (4600 m$^2$).

Sampling a portion of the landfill area is required each month; the sample size depends on the total landfill area. California's limit for TOC concentrations is 50 ppm, measured as methane.

In addition, the Californian Air Quality Management District (AQMD) system includes probes installed outside the refuse

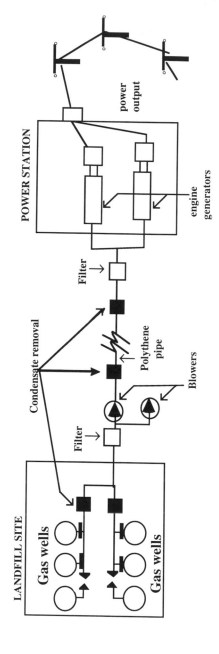

**Figure 90**   Schematic of a landfill gas plant.
*Source:* Waste Management International plc, News Release, January 1995.

deposit area and parallel to the landfill's perimeter; probe locations are more than 1000 feet (320 m) apart. At least one probe is installed at a pre-approved depth at each location.

Probe sampling is required monthly. Methane concentration limits of 5 per cent methane are set and enforced by the California Integrated Waste Management Board. The depth of each probe is determined by the refuse depth (within 160 m of the probe), as follows:

First depth: (3 m) below surface.
Second depth: 25 per cent of refuse depth or (8 m) below surface, or whichever is deeper.
Third depth: 50 per cent of refuse depth or (16 m) below surface, or whichever dimension is deeper.
Fourth depth: 75 per cent of refuse depth or (24 m) below surface, or whichever is deeper.

**Landfill site.** A disposal site in the UK (licensed under the CONTROL OF POLLUTION ACT 1974) for the disposal of controlled wastes. Figure 92 shows an 'ideal' landfill operation with both LANDFILL GAS and LEACHATE collection systems installed. Once the site has been selected, planned and engineered it should comply with guidelines similar to those listed below so that the environmental impact is minimized:

1. All deposits of waste made in individual layers which are compacted on deposition.
2. Layers no more than 2.5 metres in depth.
3. Each layer to be covered with earth, or similar, at least 225 mm thick.
4. Waste to be covered within 24 hours.
5. No waste to be tipped in water.
6. Screens erected to collect windblown rubbish.
7. Precautions taken to prevent fire and vermin.
8. Organic waste covered with 600 mm of earth.
9. Each deposit to be kept tidy.
10. Adequate competent labour to be available.
11. Each layer allowed to settle before the next layer is started.

(⇨ WASTES CLASSIFICATION)

The US EPA now requires landfill gas surface emissions to be monitored. This has been done in California since April 1985.
(⇨ LANDFILL GAS MONITORING)

UK landfill practices are undergoing a steep change with

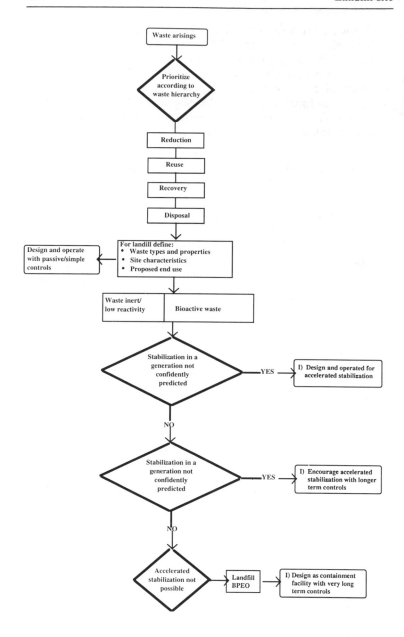

highly engineered landfills now the norm for new landfills. Unfortunately there is a continuing legacy of 'leaky sieves' which will continue to pollute groundwaters and emit landfill gas for many decades. A LANDFILL TAX (or levy in which the revenues are returned for environmental improvement) can do much to minimize waste arising by encouraging RECYCLING and INCINERATION (with energy recovery). Times are changing in the UK waste industry and a systematic approach (Figure 91) is now being adopted.

**Landfill tax.** A levy per tonne or cubic metre of WASTE sent to LANDFILL. This is used for example in Sweden to encourage the use of RECYCLING and WASTE MINIMIZATION.

A Landfill Tax has been proposed by the UK Chancellor (1994) who said:

Landfill tax kills three birds with one stone:

- by deterring landfilling of waste, we can help our environment,
- by encouraging recycling, we can save resources,
- by raising revenue from waste disposal, we will be able to make further cuts in employers' national insurance contributions and so create more jobs.

As an illustration, if it were to raise £3500 million, that could be used to finance a cut in the main rate of employer National Insurance Charges of 0.2 per cent or a reduction of more than 1 per cent in the lower rates of employer NICs.

The tax, to be collected by Customs and Excise, is designed to use market forces to protect the environment by increasing the cost of disposing of waste in landfill sites. Waste disposal companies should be in a position to pass the additional costs on to waste producers who, in turn, will be made aware of the true costs of their activities and so have an incentive to reduce waste and make better use of the waste they produce.

The tax will provide an additional incentive for recycling, which reinforces the Government's general policy of sustainable waste management. It reflects Ministers' views that taxation should play an important role in protecting the environment.

It has now been decided that the tax would be charged per

**Figure 91**
*Source:* Waste Management Paper 26B, Draft for consultation, *Landfill Design, Construction and Operational Practice – What difference will it make?*, Department of the Environment, February 1995.

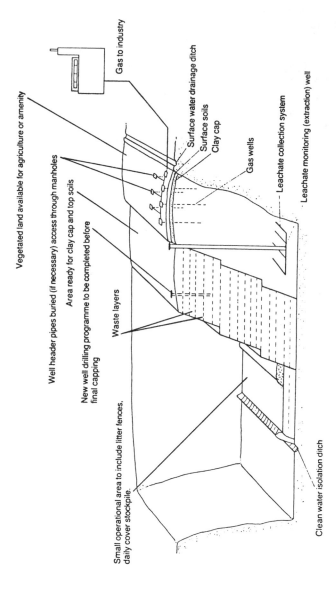

**Figure 92**   Cut away section of 'ideal' landfill operation with gas utilization.

tonne of 'active' waste and that 'inactive' waste will be subject to a reduced rate of tax (HM Customs and Excise, August 1995).

*Source:* Communiqué from HM Treasury, 21 March 1995.

**Lapse rate, temperature.** The rate of change of air temperature with increasing height: usually an average rate over a distance (commonly 100 metres). It is often stated as the ADIABATIC lapse rate of dry air, which is very close to that of moist, unsaturated air. Its value is about 1°C/100 m. (There is also the saturation adiabatic lapse rate, the adiabatic lapse rate of air that is saturated with water vapour. This is less than that of the dry adiabatic lapse rate.)

Super-adiabatic lapse rate is a temperature lapse rate which is greater than the dry adiabatic lapse rate. This occurs over a land strongly heated by the sun, and is usually very favourable to the dispersion of air pollutants. However, it causes looping of elevated PLUMES and can lead to pollutants being brought down to ground level in high concentrations for short periods.

**Laws of thermodynamics.** The laws of thermodynamics apply to all known phenomena.

*The First Law of Thermodynamics.*
This law is a statement of the conservation of ENERGY, i.e. energy can neither be created nor destroyed. Energy has many forms – electrical, chemical (the combustion of coal is release of chemical energy), thermal, nuclear, etc. The first law tells us that it is the total energy in all its various forms which is a constant. Physical processes only change the distribution of energy, never the sum. However, the first law only tells us energy is conserved, it does not tell us the direction in which processes proceed. This is the province of the Second Law of Thermodynamics.

*The Second Law of Thermodynamics.*
This law specifies the direction in which physical processes proceed. One form of this is that heat transfer takes place spontaneously from a hot body to a cooler body. Note that the opposite process, spontaneous heat flow from a cold body to a hot body (the hot body becoming warmer; the cold body becoming colder) *does not violate the first law.* It does not occur because it violates the second law.

The second law is also frequently formulated in terms of order and disorder. It tells us, for instance, that concentrations will decrease, e.g. a sugar lump will dissolve in water, or in general that order becomes disorder. The general statement of the second law is that everything proceeds to a state of maximum disorder, and by implication is less useful in the event of the disorder occurring, e.g. a sugar lump is more useful when it is not dissolved in the water. A kettle of boiling water is more useful than the same amount of water mixed in a bath of cool water. All these examples are manifestations of the second law. It applies to all biological and technological processes and cannot be circumvented. It explains that the final outcome of energy consumption is the eventual production of heat at very low and near useless temperatures, such as the waste heat discharge from power stations or heat losses from the body.

The laws of thermodynamics are often used to determine the ideal efficiency for a given process and explain why the conversion of heat to work cannot be done on a one-for-one basis, i.e. why 1 kilowatt-hour of electrical energy output from a generator requires 3 or 4 kilowatt-hours of thermal energy input. (⇨ CARNOT EFFICIENCY)

**Leachate.** The seepage of liquid through a waste disposal site or spoil heap. Leachates from municipal waste landfill sites (early years of decomposition) can be characterized by high BODs (up to 20000 mg/l), high BOD:COD ratios, several hundred mg/l ammonia and several hundred mg/l organic nitrogen, high concentrations of volatile fatty acids, acidic pH and an unpleasant smell.

The final 'methanogenic' or 'stabilized' phase is characterized by:

- BOD < 200 mg/l; COD several hundred mg/l;
- low BOD:COD ratio; $NH_3$ several hundred mg/l;
- very low concentration of fatty acids and neutral to alkaline pH.

If household wastes have been landfilled, the leachates will have a high oxygen demand which requires treatment either on site or at a sewage treatment works before being discharged into any receiving body of water.

Landfill site leachates have been recorded in strengths ranging from 100 to 54000 mg/l total organic carbon. These values show that regular leachate monitoring is required both during operation and after closure of landfill sites in order to guard against water pollution incidents. This is particularly important after a period of heavy rainfall.

A typical analysis of the UK Pitsea waste disposal site leachate (1984) is given below (all in mg/l except pH).

| | |
|---|---|
| pH | 8.0–8.5 |
| Total organic carbon (TOC) | 200–650 |
| Chemical oxygen demand (COD) | 850–1350 |
| Biochemical oxygen demand (BOD) | 80–250 |
| Ammonia nitrogen, $NH_3$-N | 200–600 |
| Organic nitrogen (as N) | 5–20 |
| Oxidized nitrogen (as N) | 0.1–10 |
| Alkalinity (as $CaCO_3$) | 2000–2500 |
| Phosphate (as P) | 0.2 |
| Total suspended solids (105YC) | 100–200 |
| Volatile suspended solids (550°C) | 50–100 |
| Fatty acids, $C_1$–$C_6$ (as C) | 20 |

*Source:* K. Knox, Leachate production control treatment, Ch. 4 in *Hazardous Waste Management Handbook*, A. Porteous (Ed.). Butterworths, 1985.

Leachate values from the German (Dusseldorf) landfill before and after treatment, are tabulated below.

| Parameter | Raw leachate (mg/l) | Plant effluent (mg/l) |
|---|---|---|
| pH | 8.7 | 4.6 |
| Conductivity | 17 000 mS/cm | 33 mS/cm |
| COD | 3900 | < 15 |
| BOD | 1400 | 2.4 |
| AOX | 2150 | 47 |
| $NH_3$-N | 1066 | 1.3 |
| Chloride | 2200 | 1.8 |
| Magnesium | 210 | < 0.1 |
| Calcium | 190 | < 0.5 |
| Lead | 0.1 | 0.02 |
| Chromium | 0.6 | 0.017 |
| Copper | 0.08 | 0.01 |
| Zinc | 0.4 | 0.4 |
| Cadmium | 0.007 | < 0.0002 |
| Mercury | 0.0003 | < 0.0002 |
| Nickel | 0.27 | 0.001 |

These extremely high discharge standards are achieved by the use of REVERSE OSMOSIS. Currently UK landfill sites do not come near this treatment performance. Croft and Campbell have stated:

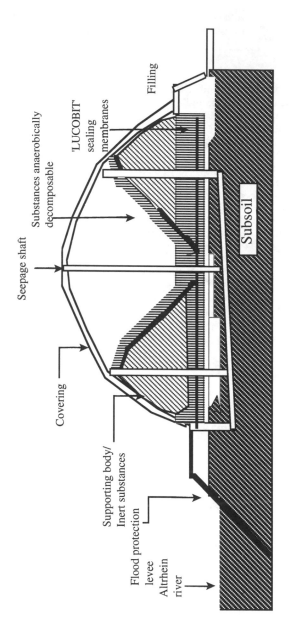

**Figure 93**  Double sealing quilt for landfill site.

Most sites (62%) took no measures to prevent ground water ingress, and 54% performed no monitoring of ground water.

Most [UK] sites (62%) had no means of intercepting or collecting leachate.

The standards now adopted in Germany are typified by that at Flotzgrün, where BASF operates an industrial waste landfill. In 1986 Bilfinter + Berger was awarded the contract for the construction of the sixth section of the landfill site. The contract was for the construction of the base of the site using a system developed by Bilfinger + Berger and BASF; it included the seepage shafts and the piped drainage system as shown in Figure 93.

The site base is formed by a double sealing 'quilt' comprising 2.2 mm thick plastic membranes of LUCOBIT (civil engineering grade of BASF's ECB plastic raw material) spaced at a distance of 40 cm, with an intermediate layer of drainage gravel. This package is provided with pipes for leak detection, ventilation and injection if necessary (see Figure 94). The 'quilt' is divided into individual fields of approximately 50 × 50 m, which are linked to seepage shafts in groups of four. The shafts are made of individual sections which can slide into one another telescopically and have been designed for a final height of 55 m. The pipes within the 'quilt' and the overlying leachate collection pipes drain into these shafts. The leachate is fed through closed pipes into the collectors which, laid underneath the site base in protective tubes, interconnect the individual shafts. The design of the protective tubes and the collectors is such that they can follow ground settlements without suffering damage. During operation of the site the leachate is pumped from a storage tank and transported to the BASF wastewater treatment in Frankenthal.

In order to protect the waste brought to the facility against floods, the new section, like the existing ones, is protected by a levee. When the waste has reached its final height it will be covered with another lucobit plastic membrane and landscaped soil.

Eventually, UK landfill standards will get there. It is to the credit of major UK companies that this is now recognized, as the Waste Management International statement below shows, and leachate management is now firmly on the agenda as well as the highest operating standards.

Waste Management International plc (WMI) and its operating

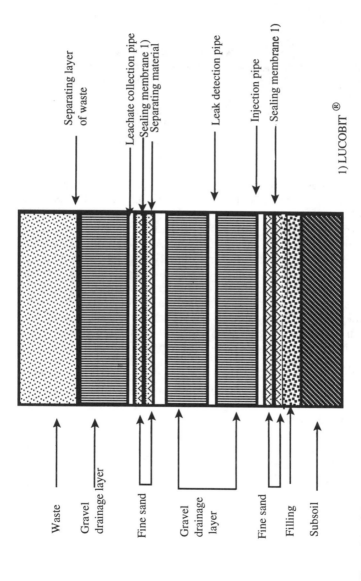

**Figure 94** CONTREP base sealing system.

subsidiary, UK Waste Management Ltd (UK Waste) have made public evidence submitted in November 1994 to the Royal Commission on Environmental Pollution. The evidence included the results of a survey of UK landfill sites undertaken between 1991 and 1994.

The survey covered 158 UK landfill sites and suggested that many UK landfills:

- do not operate to current best practices
- have inadequate containment systems
- have little or no landfill gas or leachate management systems
- have inadequate environmental monitoring.

The findings demonstrate that:

- 96 per cent of the closed sites were not designed to minimize risks of groundwater pollution.
- 71 per cent of operating sites were similarly 'non-engineered'.
- 54 per cent of the sites surveyed had no means of controlling the leachate (liquid percolating through waste). Of the sites with no leachate controls, 71 per cent were non-engineered.
- 50 per cent of the sites had no gas control systems in place.

UK Waste Management stated:

> In our view, this is a double standard and we feel that a programme to upgrade existing sites to meet current best practice should urgently be implemented. Those sites that cannot meet these standards within two years should be closed.
>
> This action will raise the costs of landfill to pay for higher standards of environmental protection and thereby provide the incentive for waste generators to recycle and reduce waste. We believe this to be a better option than an indiscriminate tax on landfill which has little environmental benefit.

*Sources: The Chemical Engineer*, Number 551, Oct. 1993.
B. Croft, D. Campbell, 'Characterisation of 100 Landfill Sites', paper presented at *Harwell Waste Management Symposium*, 1990.
Bilfinger + Berger promotional leaflet for CONTREP Base sealing system for BASF site at 'Flotzgrün'.
Waste Management International plc, News Release, January 1995.
J. G. Dobrowlski, 'Controlling LFG Surface Emissions', *World Wastes*, March 1995.
Waste Management International, 17 January 1995.

**Leaching.** Commonly, the removal of water of any soluble constituent from the soil or from waste tipped on land. Generally, the gradual dissolution of a material from a solid containing it, e.g. metal from ore. Leaching of domestic refuse tips can give rise to a LEACHATE with a high BIOCHEMICAL OXYGEN DEMAND which is offensive and can spoil the amenity of neighbouring water-courses if untreated. Leaching often occurs with soil constituents such as nitrate fertilizers with the result that NITRATES end up in potable waters. Changes in farming practice such as not ploughing up grass land and programmed fertilizer application can lead to stabilized nitrate levels. However, in 1990, over 1.6 million Britons drank water that exceeds EC limits for nitrates.

Leaching of calcium ions, both from the soil and from the leaves of plants which have already absorbed the calcium ions through their roots, is accelerated in areas where there is acidic rainfall due to industrial pollution. When this occurs the plant suffers as calcium is an essential material for cell structure and an enzymatic activator. (⇨ ACID RAIN; ENZYMES)

**Lead (Pb).** A HEAVY METAL. A soft, blue-grey, easily worked metal, which has a multitude of uses, from lead acid accumulators to early water pipes, printer's metal and glazes. It is manufactured by roasting the ore (galena) in a furnace. World production is over 3.5 million tonnes annually.

As a pollutant, lead is a systemic agent affecting the brain. It has been suggested that it is associated with mental retardation and hyperactivity in infants living in soft-water areas where old lead piping is still in use, as the lead readily enters into solution. Anti-knock additives in petrol may also be implicated in infant retardation, and the implementation of lead-free motoring in Europe and the USA over the next decade will ensure removal of this pollutant source. In addition, lead which is stored in the bones replacing calcium can be remobilized during periods of illness, cortisone therapy, and in old age. Research indicates that lead may blunt the body's defence mechanisms, i.e. the immune system. If this is so, then both the rate of infections by bacteria and viruses and the incidence of cancer can be enhanced by severe exposure. (⇨ SYNERGISM)

An approved code of practice for the control of lead at work requires a worker to be suspended from work if a blood level of $80 \mu g/dl$ was found, or $40 \mu g/dl$ in the case of a woman. Even more stringent precautions apply in the US because of the recognized CHRONIC effects. Under EEC surveys of blood lead concentration carried out in the early 1980s, reference levels for triggering action to identify and reduce exposure were set at:

- no more than 2 per cent of any group to have blood lead $> 34\,\mu g/dl$
- no more than 10 per cent of any group to have blood lead $> 30\,\mu g/dl$
- no more than 50 per cent of any group to have blood lead $> 20\,\mu g/dl$

Lead intake can come from drinking water (EEC Maximum Admissible Concentration in Drinking Water is $50\,\mu g/l$), from food, and from inhaled particles from motor exhausts. The particles are less than $1\,\mu m$ in size, and therefore can enter the lungs easily. A person breathing traffic-congested air may absorb close to the toxic level, excluding any intake from food and drink. Other sources are lead-based paints, and for some children, PICA results in them swallowing paint flakes or contaminated dust. ($\Rightarrow$ CHELATING AGENTS)

**Lean combustion (or lean burn).** Technique for the reduction of AUTOMOBILE EMISSIONS which makes use of gasified petroleum as the engine fuel compared with the conventional method of fine droplets. This enables the fuel to be burnt in air to petrol ratios of around 20 : 1 as opposed to the conventional 15 : 1, thus greatly reducing emissions of $NO_x$ and CO. $\Rightarrow$ UNBURNT HYDROCARBONS without resort to a CATALYTIC REACTOR.

**Lethal concentration.** The concentration of a substance in air or water that can cause death. This is usually expressed as the concentration that is sufficient to kill 50 per cent of a sample within a certain time, and is abbreviated to $LC_{50}$.

The toxicity of materials is commonly investigated by the use of survival curves which plot the numbers of population of, say, fish surviving at various times with the concentration of the suspected toxic agent. For a short time almost all fish survive even at high poison concentrations. Very few survive for long periods, hence the $LC_{50}$ concept is very useful as it spans the two extremes. Figure 95 shows survival curves for various concentrations of a toxic material. These curves are plotted as percentage survival against time. The time taken to kill half the fish is the intersection of the broken line and the concentration curves, e.g. a 75-week $LC_{50}$ would be 20 grams per cubic metre.

**Lethal dose.** $\Rightarrow$ DOSE.

**Leukaemia.** A CANCER associated with changes in the lymphatic system. Acute lymphoblastic leukaemia is common in children

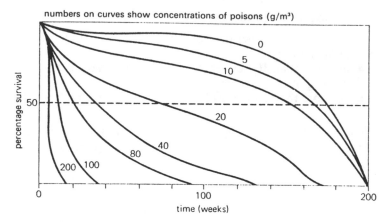

numbers on curves show concentrations of poisons (g/m³)

**Figure 95** Percentage survival against time for toxicity tests on various concentrations. The points of intersection of the broken line with the curves are lethal concentrations for 50 per cent of the population ($LC_{50}$).

and myeloid leukaemia is associated with workplace exposure to BENZENE.

Childhood leukaemia has been blamed variously on IONIZING RADIATION discharges and 'new town' viral infections when young families are gathered from diverse areas in close proximity to a new environment.

Statistically significant clusters have been found in new towns as well as near SELLAFIELD where nuclear fuel reprocessing takes place in the UK, but which itself has undergone a major population shift with large numbers of newcomers living in the area. It would be difficult to draw firm conclusions on the causes.

Recent work by Dr L. Kinlen of the Cancer Research Campaign's Epidemiology Unit has, on the basis of detailed examination of large rural construction sites, reinforced the hypothesis that leukaemia and non-Hodgkins' lymphoma are infectious diseases unconnected with radiation.

Dr Kinlen stated in the abstract of his joint paper:

Objective – to determine whether population mixing produced by large, non-nuclear construction projects in rural areas is associated with an increase in childhood leukaemia and non-Hodgkin's lymphoma.

Design – A study of the incidence of leukaemia and non-Hodgkin's lymphoma among children living near large construction projects in Britain since 1945, situated more than 20 km from a population centre, involving a workforce of more than 1000, and built over three or more calendar years. For period before 1962 mortality was studied.

Setting – Areas within 10 km of relevant sites, and the highland counties of Scotland with many hydro-electric schemes.

Subjects – Children aged under 16.

Results – A 37% excess of leukaemia and non-Hodgkin's lymphoma at 0–14 years of age was recorded during construction and the following calendar year. The excesses were greater at times when construction workers and operating staff over-lapped (72%), particularly in areas of relatively high social class. For several sites the excesses were similar to or greater than that near the nuclear site of Sellafield (67%), which is distinctive in its large workforce with many construction workers. Seascale, near Sellafield, with a nine-fold increase had an unusually high proportion of residents in social class I. The only study parish of comparable social class also showed a significant excess, with a confidence interval that included the Seascale excess.

Conclusion – The findings support the infection hypothesis and reinforce the view that the excess of childhood leukaemia and non-Hodgkin's lymphoma near Sellafield has a similar explanation.

*Source:* L. J. Kinlen, M. Dickson, C. A. Stiller, *Childhood leukaemia and non-Hodgkin's lymphoma near large rural construction sites, with a comparison with Sellafield nuclear site*, internal report for Cancer Research Campaign Epidemiology Unit, University of Oxford, Radcliffe Infirmary, Oxford.

**Levies.** Charges made against products to provide resources to support the recycling of the component materials. ($\Rightarrow$ RECYCLING, FINANCIAL INCENTIVES FOR)

**Liability.** Most (UK) liability is fault based, i.e. requiring proven negligence of a liable party. As it is often very difficult to prove that any one party is liable, especially in CONTAMINATED LAND cases where sites have been subject to multiple contaminative uses or multiple ownership, the European Commission would like to complement any fault-based system with a 'strict liability system' where no fault need be proved.

This is akin to the US Superfund approach where clean-up can be mandated.

See below for an example of strict liability and the consequences:

*R* v. *Yorkshire Water Services Limited* (July 1994)
Court of Appeal

Yorkshire Water Services Limited ('YWS') pleaded guilty to offences under Section 107 (1) (c) of the Water Act 1989, namely

causing sewage effluent to enter controlled waters. YWS did not
play any direct part in the discharges to the controlled waters as
the day-to-day management had been delegated to its agent,
Scarborough Borough Council. YWS accepted that it was
strictly liable pursuant to Section 107 (1) (c) of the Act, but
appealed against the amount of the fines imposed by the
Magistrates Court, which were £50000 for the more serious
incident and £25000 for the second incident.

The Court of Appeal accepted YWS's case and stated that the
fines should have been substantially lower, particularly, it
seems, as this was a case of strict liability on the part of a
principal in respect of actions of his agent. The fines were
reduced to £10000 and £5000 respectively.

*Source: Time Law Reports, 19 July 1994.*

**Lichens.** A compound plant formed by the symbiotic association of
two organisms: a fungus and an alga (⇨ SYMBIOSIS). They occur
on a variety of surfaces, e.g. tree trunks, rocks, walls and the
ground. Nutrition is derived from the dissolved solids in rain
which are absorbed through the whole body surface. For this
reason lichens are intolerant of poisonous substances and are
therefore sensitive to air pollution, especially to sulphur oxides.
Classification of surviving lichen flora has been used to monitor
maximum air pollution levels reached over the country (⇨
BIOLOGICAL INDICATOR) and also enabled levels of radioactivity
from CHERNOBYL to be monitored in Poland as they absorbed
radioactive CAESIUM-137; an increase of 165 times previous
levels was found.

**Lidar.** A method for the detection of cloud patterns using the
particle scatter radiation in a tuned laser beam. The method is
also suitable for the detection of atmospheric pollutants (including
hydrocarbons) and can deliver instantaneous measurements of
traffic pollution.

**Life cycle analysis (LCA).** LCA is a method for evaluating 'the
whole life of a product', that is all the stages involved, such as
raw materials acquisition, manufacturing, distribution and
retail, use and re-use and maintenance, recycling and waste
management, in order to create less environmentally harmful
products. LCA consists of three parts (Figure 96): inventory
analysis (selecting items for evaluation and quantitative analysis)
(Figure 97), impact analysis (evaluation impacts on ecosystem),
improvement analysis (evaluation of measures to reduce
environmental loads).

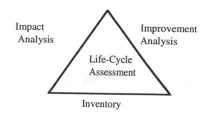

**Figure 96** Sub-systems of LCA.

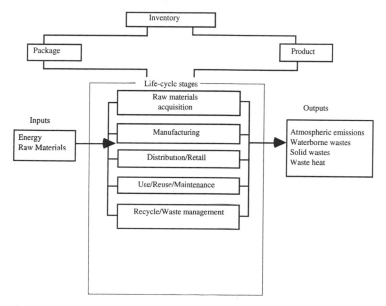

Figure 97   Life-cycle inventory procedures.

Before carrying out an analysis, the boundaries of the operations that together produce the process or product have to be defined. This is important because if any part of the system contained within the boundaries is changed, all the other inputs and outputs will also change. This is shown schematically in Figure 98.

Life cycle analysis is only one of the names for this sort of analysis; it is also known as eco-balance, cradle-to-grave analysis, resource analysis, environmental impact analysis.

The main purpose of an LCA is to identify where improvements

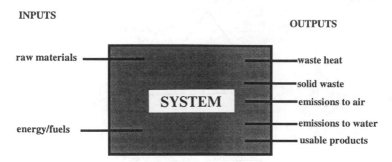

Figure 98

can be made to reduce the environmental impact of a product or process in terms of energy and raw materials used and wastes produced. It can also be used to guide the development of new products.

It is important to distinguish between life cycle analysis and life cycle assessment. Analysis is the collection of the data – it produces an inventory; assessment goes one stage further and adds on an evaluation of the inventory.

An LCA does not define or explain actual environmental effect. For example, an LCA will tell us how many grams of limestone are used to make a bottle for mineral water and how much energy was used to extract it. But it does not tell us the environmental impact of this action, such as whether limestone is a scarce resource or whether its extraction causes pollution.

An LCA is especially applicable in the packaging field because, unlike washing machines, cars or many other consumer goods, packaging is a comparatively simple item made of few materials and follows a fairly simple life cycle.

**Life cycle inventory (LCI).** The end result of a LIFE CYCLE ANALYSIS with aggregate emission such as $NO_x$, $SO_2$ or other impacts determined for a particular product and presented as an inventory of impacts. Great care is needed in appraising LCIs, e.g. in the cumulated LCI of a product, the figure of $NO_x$ emissions is very high. This is a typical short-term, regional problem and the highest single contribution to the overall figure is required (e.g. emissions of a special waste incinerator). See Figure 97.

For the development of possible improvement options the decision maker wants to know:

1. Is the emission at the site we are talking about interpreted as an ecologically relevant problem?

   This would be visible in a low regional threshold, e.g. $NO_x$. Therefore, the location of the site has to be known in order to determine the valid threshold and the share of the whole critical air volume of this step attributable to the $NO_x$ emissions, weighted with the emission threshold and the NOEC (No Observable Effect Concentration).

2. How good is the data quality of the overall figure? For example, is the scale of the $NO_x$ emissions of the special waste incineration only high because the $NO_x$ emissions of most other steps are not known and are set to 'zero' by default, rather than because of reality? This information helps to identify data gaps and is very useful to avoid sub-optimization due to bad data quality or an expensive remediation of a minor ecological problem.

LCAs and the associated LCI compilations are fraught with value judgements. Care must be taken to avoid a 'flavour of the month' approach and certainly the 'hype' which accompanies said analyses in consumer products is to be ignored.

*Source:* H. Brunn, *LCA News*, Vol. 5, No. 2, March 1995.

**Light water reactor.** ⇨ NUCLEAR REACTOR DESIGNS.

**Lignin.** The organic glue that holds the cell walls and cellulose fibres of plants together. It is a phenolic polymer which is resistant to biological attack. It forms 25–40 per cent of the wood of trees and must be dissolved chemically in PULP manufacture to obtain high grade pulp. (⇨ DIGESTOR)

   Its presence in agricultural wastes makes many of them unsuitable for feedstocks unless chemically treated. (⇨ STRAW)

**Lindane.** The active ISOMER of hexachlorocyclohexane is the $\gamma$-isomer, also known as lindane, gammexane, $\gamma$-BHC and $\gamma$-HCH. It is used as an insecticide. It is made by chlorinating benzene, which produces several isomers, but the $\gamma$-form is the only active ingredient of technical grades which include a mixture of isomers. It is these impurities which give lindane its undesirable, characteristic musty odour, which may taint food products. Lindane is a neurotoxicant with mammalian toxicity greater than that of DDT. The UK Advisory Committee on Pesticides, which regulates their use, is now considering new evidence on Lindane, which it cleared from accusations of a link to the human blood disorder, aplastic anaemia, in 1981. ($\Rightarrow$ ENDOSULFAN; CHLORINATED HYDROCARBONS)

**Linear Energy Transfer (LET).** The linear rate of energy transferred (i.e. dissipated) by the passage of particulate or electromagnetic radiation through an absorbing material. It is measured in units of energy deposited per unit length, e.g. eV (electron volt) per micron or joules per micron. ($\Rightarrow$ IONIZING RADIATION)

**Lithosphere.** The earth's crust, that is, the layers of soil and rock which comprise the earth's crust. When the complete earth's crust is meant, it is often referred to as the hydro-lithosphere. ($\Rightarrow$ HYDROSPHERE)

**Load factor.** The average load on an electrical generating system throughout the year expressed as a percentage of the highest load that occurs at any time during the year. The estimation of total pollutant emission by power stations is dependent on the load factor.

$$\text{Load factor} = \frac{\text{Average load over a period}}{\text{Peak load for that period}}$$

**Loading rate, organic.** The ratio of food-to-microorganisms (FM) in a biological treatment plant process.

$$FM = \frac{\text{Mass of BOD added to the system each day (food)}}{\text{Mass of microorganisms present (biomass)}}$$

Factor FM is also called the organic loading rate and its value determines the required capacity of the biological treatment plant.

**Logarithms.** When two quantities, $x$ and $y$, are related by an EXPONENTIAL CURVE, that is, by the equation

$$y = e^x$$

$x$ may be seen to be the power to which e (e = 2.718) must be raised to give $y$. Alternatively, $x$ is called the *natural logarithm* of $y$, and is represented by the symbol 'In $y$' or '$\log_e y$'.

There are many types of logarithms to bases other than e. The logarithms that are most frequently used are those to base 10, in other words, those which result from an equation, $y = 10^x$. These are called *common logarithms* and are represented by the symbol '$\log y$'. Thus:

$$10 = 10^1, \text{ i.e. } \log 10 = 1$$
$$100 = 10^2, \text{ i.e. } \log 100 = 2$$
$$1000 = 10^3, \text{ i.e. } \log 1000 = 3$$
$$10\,000 = 10^4, \text{ i.e. } \log 10\,000 = 4$$

Logarithms to base e and base 10 are chosen to suit the system under consideration. For example, the DECIBEL scale for SOUND measurement uses logarithms to base 10, as does pH (⇨ pH). Logarithmic graphs are in common use for values that span ranges of 1000 or 10 000 units of measurement. 'e' is used for exponential growth considerations, e.g. the time $t$ required for a quantity to double from its original value at a constant growth rate $r$ is given by $t = \dfrac{\ln 2}{r} = \dfrac{0.6931}{r}$. (⇨ EXPONENTIAL CURVE)

**Long-term exposure limit LTEL).** ⇨ TIME-WEIGHTED AVERAGE.

**LPG.** ⇨ GAS, LIQUEFIED PETROLEUM.

# M

**Magnetherm.** This is one of the main methods for producing magnesium and is based on the reduction of magnesium oxide when it is heated in a partial vacuum with oxides of calcium, aluminium and silicon. The oxides form a molten slag which reacts with a metallic reducing agent, such as ferrosilicon, to generate magnesium, which vaporizes in the vacuum. The magnesium is captured by condensing the vapour in a crucible.

The yield of magnesium depends on the temperature and pressure of the furnace, and the composition of the slag.

An unforeseen benefit is that a modified version of the process can make use of certain kinds of hazardous waste as part of the feedstock, such as ASBESTOS and cement waste containing asbestos which, when blended correctly, contains the correct proportions of the various oxides.

*Source:* 'Magnesium smelter puts toxic waste to work', *New Scientist*, March 1995.

**Magnetohydrodynamic generator (MHD).** A device for DIRECT ENERGY CONVERSION which generates electricity by the passage of a gas at high temperature and pressure through a magnetic field. The gas is ionized and this is enhanced by seeding with caesium or potassium. The gas is then passed through a nozzle at right angles to a magnetic filed. The magnetohydrodynamic generation is essentially a 'topping' cycle – that is, a means of extracting greater electrical energy from a given amount of fuel by burning the fuel at as high a temperature as possible (2500°C or greater) followed by a conventional steam-turbine generation at a lower temperature.

Magnetohydrodynamic generation received great attention in the 1950s and 1960s but has now dropped from favour on grounds of capital cost and scale, as it can only be usefully employed in the range from 10 megawatts upwards.

**Magnox.**
1. A magnesium alloy which contains small amounts of aluminium and beryllium used for cladding natural uranium metal fuel used in Magnox reactors.
2. The name given to the type of gas-cooled graphite moderated reactor using Magnox-clad fuel, on which Britain's first nuclear power programme was based.

**Malathion.** ⇨ ORGANOPHOSPHORUS COMPOUNDS.

**Marginal cost.** This is a measure of the cost to society of producing one more unit of output. If the marginal cost is less than the marginal utility (a measure of the increase in social benefit from consuming one more unit of output), it is worthwhile having the extra unit of output, because the additional benefit exceeds the extra cost.

    Marginal analysis is the technique for investigating marginal cost and marginal utility. In a nutshell it may be paraphrased as if we change our economic behaviour in such and such a fashion, would we benefit or not – this type of analysis, of course, assumes that no cataclysmic changes would be made and that business as usual is the *modus operandi.*

**Marl.** A soft rock mixture of clay or silt and calcite (calcium carbonate, $CaCo_3$) mud. Marls are plastic when wet and friable when dry.

**Marsh gas.** ⇨ METHANE.

**Mass balance.** The balancing of the inputs and outputs of a process (e.g. boiler plant) which is operating in the steady state mode.

    For example, a boiler plant burning coal of 0.7 per cent carbon and 30 per cent ash at 50 per cent EXCESS AIR has the inputs and outputs shown in Figure 99.

Another example is the mass balance for an MSW processing plant: the inputs total 13.2 kg; the outputs (including ash) total 13.2 kg, i.e. they balance.

**Mass burning.** ⇨ INCINERATION.

**Mass number.** The number of protons and neutrons in an atom. (⇨ ISOTOPE)

**Materials, energy consumption.** The energy required in the manufacture of some common materials and manufactured products is shown in the table on page 338. The high energy input and its effects on costs are very apparent and it is obvious

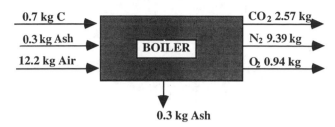

Figure 99

MSW (100t)

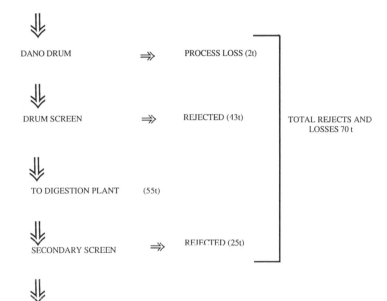

i.e. Digester Accepts c. 30% of incoming material
(of which approx. 25% will be converted to Biogas).

⇒COMBUSTION.

Figure 100  Outline mass balance for 100 tonne MSW input to a Dano
drum processor (courtesy of Motherwell Bridge Envirotec).

Energy consumption in basic materials processing*

| Material | Energy for unit production ($kWh_{TH} tonne^{-1}$) | Machinery depreciation ($kWh_{TH} tonne^{-1}$) | Transportation ($kWh_{TH} tonne^{-1}$) | Total ($kWh_{TH} tonne^{-1}$) |
|---|---|---|---|---|
| Steel (rolled) | 11700 | 700 | 200 | 12600 |
| Aluminium (rolled) | 66000 | 1000 | 200 | 67200 |
| Copper (rolled or hard drawn) | 20000 | 800 | 200 | 21000 |
| Silicon, metal and high-grade steel alloys | 58000 | 1000 | 200 | 59200 |
| Zinc | 13800 | 700 | 200 | 14700 |
| Lead | 12000 | 700 | 200 | 12900 |
| Miscellaneous electrical produced metals | 50000 | 1000 | 200 | 51200 |
| Titanium (rolled) | 140000 | 1000 | 200 | 141200 |
| Cement | 2000 | 50 | 50 | 2300 |
| Sand and gravel | 18 | 1 | 2 (short distance hauling) | 21 |
| Inorganic chemicals | 2400 | 100 | 200 | 2700 |
| Glass (plate finished) | 6700 | 300 | 200 | 7200 |
| Plastics | 2400 | 300 | 200 | 2900 |
| Paper | 5900 | 300 | 200 | 6400 |
| Lumber | 1.47 per board foot | 0.02 per board foot | 0.02 per board foot | 1.51 per board foot |
| Coal | 40 | 2 | | 42 |

*Accuracy ±20 per cent.
From 'Economical use of energy and materials' *Environment*, vol. **14**, no. 5, June 1972, p. 14.

why 'alternative technology' is based on both low cost and low energy input materials.

**Materials recovery facility (MRF).** Site where the plastics, glass, metal and paper components of solid waste are, either mechanically or manually, separated, baled and stored prior to reprocessing.

**Maximum allowable concentration (MAC).** In the USA, the upper limit for the concentration of noxious or toxic emissions in workplaces. Exposure of the MAC of a pollutant should not cause any distress to anyone except the occasional sensitive individual. In the UK, the EC directive (1982) concerning the quality of water intended for human consumption, which quotes MAC values, is used.

**Maximum exposure limits (MEL).** Maximum level of exposure to a harmful substance allowed for a worker (or member of the population at large).

The criteria for setting limits are:

1. A substance not able to satisfy the criteria for an OCCUPA-TIONAL EXPOSURE STANDARD (OES) and which has or is liable to have a serious risk to man including acute toxicity and/or potential to cause serious long-term health effects; *or*
2. Socio-economic factors, which indicate that, although the substance meets the criteria for an OES, a numerically higher value is necessary if certain uses are to be reasonably practicable. (⇨ OCCUPATIONAL EXPOSURE STANDARD; TIME-WEIGHTED AVERAGE)

**Maximum sustainable yield.** The maximum yield that can be obtained from a given crop or species if it is to maintain equilibrium. The management of a forest or fishery or farm should avoid over-grazing, over-cropping or over-harvesting. Thus, fishery catches should be strictly controlled so that the fish population can have a sufficient breeding mass and thus give a sustained yield for future generations. This philosophy is particularly applicable to whaling, herring and salmon fishing, but is relevant to many areas, such as forestry, and is basically the application of good stewardship of natural resources that cannot regenerate if exploited too far. The loss of WHALE species is an example of exploitation of a renewable source.

Counterparts abound everywhere: a river should not be so depleted of DISSOLVED OXYGEN by domestic or industrial pollution that it cannot recover. The land should not be 'mined' for short-term agricultural profit. THE COMMONS must be husbanded and good husbandry enforced.

**Mean, arithmetic, geometric.** Generally similar to the idea of an average, a central value about which individual values, greater or less, are distributed. The arithmetic mean, or the average, of a set of values is their sum divided by their number. The geometric mean of $n$ values is the $n$th root of the product of all of the values.

**Mean annual temperature (global).** This is the mean temperature of the earth (averaged over a year) for individual latitudes leading to a composite value which is a measure of the earth's radiation balance. If more radiation is absorbed by the land mass and oceans this year than was absorbed last year without a compensating change in the outgoing radiation, the mean annual temperature will rise a fraction of a degree. If the opposite process occurs, it will drop. The mean annual temperature has risen $0.5°C$ in this century. An unstoppable rise of $1°C$ is forecast with a possible $4.5°C$ increase by 2030 if GREENHOUSE GAS emissions are not reduced, as the longer measures are delayed, the worse the effect. ($\Rightarrow$ GREENHOUSE EFFECT)

**Means, best practicable (BPM).** According to the Alkali, etc. Works Regulation Act 1906, and subsequently the Health & Safety at Work, etc. Act 1974, a scheduled process must be provided with the *best practicable means* for preventing the escape of noxious or offensive gases and smoke, grit and dust to the atmosphere and for rendering such gases, where necessarily discharged, harmless and inoffensive. The words 'best practicable means' are often used to describe the whole approach of British anti-pollution legislation towards industrial emissions: the cost of pollution abatement and its effect on the viability of industry are taken into account. Thus, while better pollution control can often be achieved, the 'best practicable means' philosophy can allow lower standards of control to keep a sector of industry viable, having regard to established practice, the area in which it is sited, etc. In effect this may mean that no firm is legally compelled to modify its activities or install pollution-abatement equipment until it is economically convenient to the firm to do so. In the meantime, the social costs of the consequent pollution continue to be borne by the community. The firm may well be urged to investigate means of suppressing pollution, but may not be forced to do so. The BPM concept is to be replaced by that of the BEST PRACTICABLE ENVIRONMENTAL OPTION (BPEO) which may well be BPM in practice for many established industries. (⇨ BATNEEC; EXPOSURE–DOSE EFFECT RELATIONSHIP; THRESHOLD LIMITING VALUE)

**Melt-down.** A (so far theoretical) nuclear reactor accident where the coolant escapes and the fuel becomes sufficiently hot that it melts and falls onto the reactor base. The nearest occurrence was in the USA (Three Mile Island incident).

**Membrane.** A thin sheet of ion-exchange material used in the ELECTRODIALYSIS to remove impurities from brackish water. In REVERSE OSMOSIS the membrane is semi-permeable, i.e. it has minute pores through which water can diffuse but through which the passage of salts is prevented, again effecting purification.

**Mercaptans.** A family of foul-smelling sulphur compounds produced by decaying organic matter, emitted at sewage works, food-processing plants, brick-making works and oil refineries. The most offensive, ethyl mercaptan ($C_2H_5SH$; boiling point $30°C$) can be smelt at concentrations as low as 1 part in 50 000 million in air.

**Mercury (Hg).** A HEAVY METAL that exists as a liquid at normal temperatures; atomic mass 200.59. It is extracted from an ore (cinnabar). Mercury is used as a catalyst in many industrial processes as well as in barometers, thermometers, etc. Over 10 000 tonnes are produced annually.

Provisional tolerable weekly intake for man:

| Substance | Intake per week expressed as: | |
| --- | --- | --- |
| | *mg per person* | *mg/kg body-weight* |
| Mercury | | |
|   total mercury | 0.3 | 0.005 |
|   methyl mercury | | |
|     (expressed as mercury) | 0.2 | 0.0033 |
| Lead[1] | 3 | 0.05 |
| Cadmium | 0.4–0.5 | 0.0067–0.0083 |

[1]These intake levels do not apply to infants.

From Food and Agriculture Organization, *Evaluation of Mercury, Lead, Cadmium and the Food Additives Amaranth, Diethylprocarbonate and Octy Gallate*, Rome, 1973.

As a pollutant it is a SYSTEMIC AGENT, affecting the brain, kidneys and bowels (⇨ MINAMATA DISEASE). The organic forms, e.g. methyl mercury, are particularly toxic. World Health Organization food regulations state 0.05 parts per million as the highest allowable concentration of mercury in foodstuffs.

Inshore fish in the Mersey Estuary and Morecambe Bay have been found to contain up to 1.1 mg/kg. Fish from distant waters contain an average 0.06 mg/kg (Ministry of Agriculture, Fisheries and Food, *Survey of Mercury in Food,* HMSO, 1971). It is possible that people with very restricted diets of local fish could ingest more than the provisional *tolerable* weekly intake for man as shown in the data above for mercury, lead and cadmium. (The term 'tolerable' signifies permissibility rather than acceptability since the intake of a contaminant is unavoidably associated with the consumption of otherwise wholesome and nutritious foods: the term 'provisional' expresses the tentative nature of the evaluation (FAO, 1973). (⇨ CRITICAL GROUP)

The provisional tolerable weekly intake that is used for metallic contaminants in food as an ACCEPTABLE DAILY INTAKE cannot be readily determined. (⇨ TOLERABLE DAILY INTAKE)

There is a paucity of information on the effects of sub-lethal doses of heavy metals but, in view of the adverse properties of LEAD, there is obviously a strong case for strict control of all heavy-metal discharges and emissions.

Major emission sources are coal burning (where mercury is always present), the paper industry and chemical plants.

**Metabolite.** Either a simple product formed by the enzymatic decomposition processes of a living organism or a complex product also created enzymatically. (⇨ ENZYME)

**Metals.** Solid materials generally possessing high melting points, which are good thermal and electrical conductors. Most pure metals are usually alloyed with others to improve their mechanical properties (e.g. steel: iron alloyed with carbon). The cost of metals reflects their relative abundance in the earth's crust (as ores) and the cost of extraction of the metal from the ore. Aluminium, though very abundant in the form of alumina ($Al_2O_3$), is expensive owing to the large amounts of electricity needed to extract the metal from its ore (⇨ MATERIALS, ENERGY CONSUMPTION). Iron is very abundant as $Fe_2O_3$, etc. and much less thermal energy is needed in manufacture of the raw metal using the blast furnace. Metals are widely recycled, particularly precious metals, gold, silver and platinum but also iron and steel alloys, owing to their abundant use (car bodies, tin cans, etc.). As with all materials RECYCLING problems are created with alloys.

Some alloying elements are easily removed in the SLAG, particularly highly reactive metals like aluminium, but other, more noble elements like copper and tin remain in the steel. They are known as *tramp elements*. The behaviour of alloy elements and impurities in steel making is shown below:

| | |
|---|---|
| Elements almost completely taken up by slag | Silicon |
| | Aluminium |
| | Titanium |
| | Zirconium |
| | Boron |
| | Vanadium |
| Elements distributed between slag and metal | Manganese |
| | Phosphorus |
| | Sulphur |
| | Chromium |
| Elements remaining almost completely in solution in the steel | Copper |
| | Nickel |
| | Tin |
| | Molybdenum |
| | Cobalt |
| | Tungsten |
| | Arsenic |
| Elements eliminated from slag and metal | Zinc |
| | Cadmium |

The degree to which tramp element impurities are harmful depends on the intended use of the steel. Reinforcing bars, for example, may contain impurities which would make the steel entirely unsuitable for car body sheet.

The economics of recycling metals depends on their purity, cost and degree of dissemination. If the main use of a given metal is only for one product, then recycling is usually straightforward since there is then an assured, continuous supply for the recycling industry. Examples include photographic film (for silver) and car batteries (for lead).

**Methaemoglobinaemia.** A disorder, known as the 'blue baby disease', which affects the oxygen-carrying capacity of the blood. It is associated with drinking water containing nitrogen in the form of nitrates. The amount of nitrate per litre of water at intakes on the River Thames and the River Lee has, since 1972, approached (and for the Lee exceeded) the 50 mg nitrate per litre (or 11.3 mg/1 as nitrogen) limit considered to be safe. Above this limit, bottle-fed babies may develop methaemoglobinaemia; while above a concentration of 20 mg per litre, as nitrogen, the adult population may be at risk. So far, London's water supply has been kept well below the safe limit by diluting the Thames and Lee water with water from reservoirs.

Modern farming techniques involving heavy applications of nitrogen FERTILIZER, extensive cultivation of leguminous plants such as peas and beans, and sewage works effluent are all contributing to the high nitrate levels. The residues from agriculture are present in the ground in large quantities, and modern land drainage techniques mean that they will eventually find their way into the water supply. (⇨ LEACHATE)

**Methane (marsh gas; CH₄).** A gas which can easily be liquefied. It is chiefly used as a fuel. Methane occurs naturally in oil wells and as a result of bacteriological decomposition, e.g. the ANAEROBIC DIGESTION of sewage SLUDGE. It can also be synthesized. (⇨ LANDFILL GAS)

**Methanol (methyl alcohol or wood alcohol; CH₃OH).** A chemical that can be produced by destructive distillation of coal or wood. Methanol is usually synthesized from carbon monoxide and hydrogen or from METHANE. It is used as an intermediary by the chemical industry, and also as a solvent and denaturant.

One important and growing use is a substrate for the production of protein; a class of bacteria has been isolated that can utilize it. It is also being actively promulgated as a motor fuel to conserve oil reserves, either on its own or mixed with petrol in proportions between 15–85 per cent. It is claimed that higher mileages result, and that 50 per cent less carbon monoxide is produced from such fuel mixtures. However, its CALORIFIC VALUE (gross) is 21 MJ/kg compared to PETROLEUM'S 45 MJ/kg.

**Methyl tertiary butyl ether (MTBE).** An unleaded fuel additive considered non-toxic but it can taint drinking water at concentrations of around 3–6 mg/1.

MTBE has already been banned in Alaska because gas station staff have been affected by headaches and nausea.

**Metrolink.** A light rail system operated by Greater Manchester Metro Ltd, using light rail vehicles. There are 26 vehicles in the fleet, each costing £1 million and carrying 86 passengers seated and 120 standing – 206 in total. The vehicles travel up to 50 mph out of the city and 30 mph through the city.

The total cost of construction was £135 million.
Patronage – 1994 figures:

Daily    30 000
Annual   12.2 million passenger journeys.

Patronage under British Rail prior to change of ownership was 7 500 000.
This has significantly cut inner city car use.

**Microbes.** Microscopic organisms. The term usually refers to bacteria, of which some are pathogenic, e.g. *Salmonella*, which is associated with food poisoning in man.

They are essential scavengers of organic material breaking down dead plant and animal remains, sewage and even 'toxic' wastes that are organic in origin (known as BIOREMEDIATION).

Microbes have been shown to be capable of breaking down oil spills and the detritus of the Gulf War with up to 50 per cent of the oil destroyed within 20 weeks in appropriate conditions.
(⇨ BIOREMEDIATION)

**Micron.** One-millionth of a metre, hence the more correct term 'micrometre'. It is commonly used for particle sizing. Symbol $\mu$m in SI units.

**Minamata disease.** Minamata is a town on the west coast of Kyushu Island (Japan) where an extreme case of HEAVY METAL poisoning from methyl MERCURY ingested in the staple fish diet of the inhabitants caused severe disablement and death: 43 deaths and 68 major disablements were recorded between 1953 and 1956. (Environmentalists have claimed that almost 800 people have been killed in all by the discharges.)

The symptoms include numbness in fingers and lips and difficulty in speech and hearing. There is a marked inability to control limbs, followed by seizures. Children and old people are particularly vulnerable, as is the case with all heavy-metals ingestion.

The source of mercury in the bay was eventually traced to a PVC plant which used mercuric sulphate (an inorganic chemical) as a CATALYST and which discharged effluent containing both inorganic and organic mercury. The inorganic form was subsequently converted by marine life to the methyl form.

Japan's supreme court ruled in March 1989 that two former executives of a chemical firm were responsible for the mercury pollution in the 1950s. Each received two year's imprisonment.

The Minamata Disease Centre says there are 18 128 victims

still waiting to be formally acknowledged as Minamata patients. Minamata is now a classic case of industrial pollution and subsequent evasion of responsibility.

**Mineral.** A naturally occurring material with a local constant chemical composition. Usually a solid crystalline substance but can embrace SILT, CLAY and SEDIMENTS.

**Mist.** Microscopic liquid droplets suspended in air. Their diameter is less than 2 micrometres. ($\Rightarrow$ FOG)

**Mixing layer.** The lower part of the atmosphere within which air movement is much affected by the proximity of the earth's surface and where pollutants are dispersed and mixed. For convective conditions the mixing layer may extend up to several kilometres above the surface but for stable conditions its height may be only some 200 m.

**Moisture content.** The quantity of water present in any substance, e.g. waste or SLUDGE. Expressed on a wet basis (as received) or dry basis (which is based on dry weight of sample).

**Mole[1].** (US). The number of atoms or molecules equal to the 'Avogadro constant' (per mole, designated $N_A$).

**Mole[2].** (UK). The basic SI unit of substance which contains as many elementary units as there are atoms in 0.012 kg of carbon-12, e.g. 1 mole of a compound has a mass equal to its molecular weight in grams.

- 1 mole HCl has a mass of 36.46 g
- 1 mole $H_2O$ has a mass of 18 g
- 1 mole of electrons has a mass of $mass_{electron} \times N_A$

Thus elementary or molecular unit used must be specified.

**Mole fraction.** The ratio of the number of MOLES of a component of a mixture to the total number of moles in the mixture.

**Molecule.** The smallest part of an element or compound capable of existing independently which has all the chemical properties of the element or compound.

**Monitoring.** In environmental health, the repetitive and continued observation, measurement, and evaluation of health and/or environmental or technical data to follow changes over a period of time. Measurements of pollutant levels are made in relation to a set standard or to assess the efficiency of regulatory and/or control measures.

**Monomer.** A compound whose simple molecules can be joined together (POLYMERIZATION) to form a giant POLYMER molecule;

e.g. VINYL CHLORIDE is polymerized to form the plastic POLYVINYL CHLORIDE.

**Monthly average.** An average value of the concentration of a component (e.g. sulphur dioxide) on a monthly (30 day) basis.

**Mosses.** Multicellular, CHLOROPHYLL-bearing plants. Mosses are not complex but have a highly-developed reproductive system. They occur in damp conditions, such as in woods and also on walls. The stems contain a central core of elongated cells which act as ion-exchange membranes. The intricate branching system with many leaves provides large surface areas on which particles can be trapped. This structure means that mosses can be used to concentrate airborne pollutants. Thus the presence (or absence) of certain species of mosses (and LICHENS) indicates the absence (or presence) of airborne pollutants such as sulphur dioxide, fluorides, and copper, lead and zinc fumes, etc. Where the mosses are absent, mossbags – i.e. bags of dry dead moss suspended in nylon nets – can be used to assess the presence available. As the accumulation rate of the metals is a function of air concentration of the pollutant(s), the average concentration over the period of exposure can be found by analysis of the moss bags.

**Most probable number (MPN).** This is a method for estimating bacterial numbers in a sample (of water, food, soil, etc.) by multiple tube fermentation. Several tubes containing an appropriate nutrient medium for growth of the particular bacterial group are inoculated with measured quantities of different dilutions of the test sample. The nutrient medium chosen is designed to indicate the presence of the bacteria after an incubation period at a specified temperature. Presence of the bacteria is usually indicated by a colour change in the medium or the production of gas. After the prescribed incubation time, examination of the tubes will reveal 'positive' and 'negative' ones indicating the presence or absence in each tube of the bacterial group in question. The number of 'positive' tubes for each of the sample concentrations originally selected can be converted into an estimate of the most probable number of bacteria in the original sample.

**Mud.** Wet loose mixture of particles less than 60 $\mu$m in diameter. ($\Rightarrow$ CLAY; SILT)

**Multiple flues.** In large power stations or process plants, where there are separate boilers, the best practice for good plume dilution and dispersion is to take all gas discharge streams to one chimney with multiple flues so that emission velocity and temperature is maintained over a wide working range and interference from a multitude of chimneys, each with its own small plume, is avoided. The use of multiple flues means that one flue discharge does not affect the combustion draught conditions of an adjacent flue. (⇨ PLUME RISE)

**Municipal Solid Wastes (MSW).** The common name (US) given to the combined residential and commercial waste material generated in a given municipal area. (⇨ DOMESTIC REFUSE; WASTES; GARBAGE)

**Mutagen.** Something (e.g. a chemical or radiation) that changes the genetic material (chromosomes) that is transferred to the daughter cells when cell division occurs. The result is that the new cells have changed characteristics.

The mutagen can act on a *germ* cell, e.g. sperm of man or that of any other sexually reproducing organism (rats, mice, fruit flies), and some of the offspring will carry the mutant genes in all their cells. A mutagen may also affect *somatic* cells, in which case the effects are to the person concerned and not to offspring. These effects depend on the type of cell affected. One effect could be to make cells grow and multiply faster than they can be removed by the blood; if the white cells are affected, the result is leukaemia. Another effect could be to start up cell divisions in cells that do not normally divide; if the division products displace or invade normal tissues, the result is a cancer. Both types of mutation are discussed further. (⇨ IONIZING RADIATION)

Mutagens may also be carcinogenic or teratogenic, and therefore any substance that is found to be mutagenic must be tested for its carcinogenic and teratogenic properties as well. Although mutagens may not be carcinogens or teratogens, they must be considered suspect until cleared. *This conservative viewpoint is essential if human health is to be protected.* There is often a major time-lag between contact with a mutagenic agent and the onset of cancer in man. For example, VINYL CHLORIDE was only found to be carcinogenic after long use. It is now becoming apparent that many substances in common use are suspected as being carcinogenic, e.g. hair dyes. Because of the latency period before cancer manifests itself, these suspected substance may continue in use for many years. The principal area for research activity and concern is the use of the growing class of chemicals that affect nucleic acids, the basic chromosome component. (⇨ CARCINOGEN; TERATOGEN)

The World Health Organization Scientific Group on the Evaluation and Testing of Drugs for Mutagenicity stated that the following should receive special attention:

1. compounds that are chemically, pharmacologically and biochemically related to known or suspected mutagens;
2. compounds that exhibit certain toxic effects in animals, such as depression of bone marrow; inhibition of spermatogenesis or oogenesis; inhibition of mitosis, teratogenic effects, carcinogenic effects, causation of sterility or semi-sterility in reproduction studies; stimulation or inhibition of growth or synthetic activity of a specific cell or organ; inhibition of immune response; and
3. compounds that are likely to be continuously absorbed into the body and retained by it for long periods.

The testing should be done with mammals, e.g. rodents, so that an indication of the potential effects on man may be determined.

A mutagenic index is derived from the tests which reflect the percentage of dominant *lethal* mutations in the experimental group.

The finding of mutations in one species does not mean that it is mutagenic in all species, but a positive finding should certainly be considered as an indication of potential mutagenic activity in man.

**Mutagenic index.** ⇨ MUTAGEN.

**Mutant genes.** ⇨ IONIZING RADIATION.

**Mutation.** The random alterations of reproductive cells of organisms which result in changes in the other cells of the organism. If a particular mutation protects the organism in a hostile environment, then the mutant has a good chance of survival.

# *N*

**Nanogram.** One thousandth of one millionth of a gram ($10^{-9}$ g).

**National Radiological Protection Board (NRPB).** A UK statutory body created under the Radiological Protection Act 1970 to provide information, advice and monitoring on and of all aspects of IONIZING RADIATION.

**National Rivers Authority (NRA).** The organization set up under the 1989 Water Act to oversee the activities of the regional water companies. The role of the NRA is to enforce standards laid down by Government, to monitor pollution in rivers and to manage water resources. It also has responsibility for flood control and land drainage.

It is the duty of the NRA in England and Wales to monitor and protect groundwater and conserve it for water resource usage. It is also the NRA's duty to maintain and conserve surface waters, which in many cases depends upon the proper management of groundwater.

The NRA's powers and duties are set out in the Water Resources Act 1991.

To protect groundwater quality the NRA has to:

- achieve and maintain specific targets for water quality (Water Quality Objectives);
- control potentially polluting discharges from, for example, industrial and agricultural premises, by issuing special consents;
- prevent pollution wherever possible;
- take remedial action if pollution occurs;
- take action against polluters

To protect the quantity of groundwater the NRA has to:

- make sure water resources are used properly;
- manage groundwater so that acceptable flows are maintained in rivers;
- control the pumping of groundwater by issuing special licences;

- take action against people who remove groundwater illegally;
- take action to re-distribute and increase resources where needed.

The NRA can influence planning decisions which may damage groundwater. As the way the land is used and developed is one of the greatest and most consistent threats to the quality of groundwater, land-use planning policies can play a significant role in protecting groundwater. Therefore the NRA must keep in close contact with the Local Planning Authorities.

Some activities (e.g. LANDFILL) present a particular risk of pollution. The closer an activity is to a well or borehole, the greater the risk of the pumped water being polluted.

Around each groundwater source, the NRA has defined three Source Protection Zones. These vary in their size, shape and relationships according to the particular situation at any one place. The type of soil, the geology, the rainfall and the amount of water pumped out of the ground are taken into consideration. Policy statements dealing with new developments within each of these three zones in addition to other areas where groundwater is generally at risk have been published. The NRA will shortly become part of the ENVIRONMENT AGENCY with the same powers and duties.

**National Survey of Air Pollution.** A UK nationwide survey of SMOKE and SULPHUR DIOXIDE set up in 1960. The National Survey as such ceased in 1982 and is now called the UK Smoke and Sulphur Dioxide Monitoring Network. This smaller network includes a basic urban network monitoring for compliance with EEC air quality limits, as well as some rural sites and *ad hoc* surveys.

**Natural background radiation.** ⇨ IONIZING RADIATION.

**Natural gas.** Natural hydrocarbon gas(es) associated with oil productions, principally METHANE ($CH_4$) and some ethane ($C_2H_6$).

It is a FUEL with a calorific value of approximately 30 MJ/m$^3$N which burns cleanly with minimal SULPHUR DIOXIDE emissions. Typical emission values are 0.7 mg $SO_2$ per MJ compared with up to 910 mg $SO_2$ per MJ for coal-fired boilers.

This clean fuel aspect has given natural gas a major commercial advantage in the UK for power generation purposes using combined cycle power plant with typical CARBON DIOXIDE emissions of 30 grams carbon per MJ (electricity) compared with 65–70 gram carbon per MJ (electricity) for coal-fired plant. Natural gas reserves are finite. The UK's are predicted to last to ca. 2025 and supplies will have to be sought from Norway and Russia. We have, in effect, a short-term fix and the goal of SUSTAINABILITY will have to be pursued vigorously.

**Natural turnover.** The annual throughput of material 'processed' by a natural system in equilibrium. The natural turnover can be used to scale man-made inputs of the same material in order to assess any risks that may occur if the man-made input swamps the natural system and destroys its stability. (⇨ CARBON CYCLE)

**Necrosis.** Death of a cell while attached to a living body, e.g. $SO_2$ damage to leaves.

**Nephritis.** Inflammation of the kidneys which can be caused by drinking water with lead in solution, e.g. rainwater collected from lead-painted roofs.

**Neutral atmosphere.** An atmosphere in which the ADIABATIC lapse rate matches the environmental lapse rate. (⇨ LAPSE RATE)

**Neutralization (of a solution).** The addition of acid to an alkali or alkali to an acid to achieve neutrality, i.e. pH of 7.

**Neutron.** An elementary particle, which has no electric charge. It is a component of atomic nuclei except HYDROGEN. On their own neutrons are radioactive and decay with a half life of 12 minutes by beta-radiation emission to a PROTON.

**Newton (N).** Unit of force, i.e. the force required to accelerate a mass of 1 kilogram at 1 metre per second. Conversion factor $1 N = 0.2248$ pounds force.

**Nickel (Ni).** A silvery-white magnetic metal which resists corrosion; atomic mass 58.91. It is obtained from various ores and used for nickel plating, as alloys and as a catalyst. Metallic nickel and its soluble and insoluble salts are potent skin sensitizers. Also, occupational asthma due to respiratory sensitization has been recorded in the electroplating, metal polishing, catalyst re-processing and stainless steel welding industries. A number of studies have shown clear evidence of lung and sino-nasal cancers in the nickel-refining industry. (⇨ CONTROL LIMITS, OCCUPATIONAL)

**Nitrate (NO₃).** SALTS of nitric ACID which are formed naturally in the soil by micro-organisms from protein and nitrites, and in this form are available as a plant nutrient. Nitrates are also produced industrially and used as fertilizers. However, once spread on the land, not all may be used by crops and so may be leached down into aquifers or run into rivers. (⇨ NITROGEN

CYCLE; EUTROPHICATION, LEACHING).

The EC limit for nitrate in drinking water is 50 mg per litre (or 50 ppm). So-called Blue Baby Syndrome arose from water which contained in excess of 200 mg/litre nitrate. This level is never approached in the UK today.

**Nitric oxide.** ⇨ NITROGEN OXIDES.

**Nitrification.** Conversion of nitrogenous matter into nitrates by bacteria, especially in soil. (⇨ NITROGEN CYCLE)

**Nitrogen cycle.** The atmosphere contains 78 per cent by volume of nitrogen ($N_2$) and 21 per cent of oxygen ($O_2$), yet nitrogen is not a common element on earth. It is an essential component for plant growth and proteins, yet it is chemically very inactive and before it can be incorporated by the vast majority of the biomass, it must be fixed. The fixation of nitrogen is its incorporation in a combined form such as ammonia ($NH_3$) or nitrate ($NO_3^-$), whereby it can be used by plants or animals. This fixation can be carried out industrially or naturally by the action of bacteria.

The industrial fixation of nitrogen now matches the bacterial fixation rate. This is an example of man matching a natural cycle, so both routes must be discussed.

*Industrial fixation (Haber process)*

The fixation of nitrogen industrially is carried out at 500°C and high pressure (200 atmospheres) and involves (a) the re-forming  of methane to provide hydrogen, (b) the introduction of atmospheric nitrogen and oxygen. The oxygen reacts to form carbon monoxide and thereafter a catalytic reaction combines nitrogen and hydrogen to form ammonia, which can then be readily converted to nitric acid and then ammonium nitrate, a fertilizer of wide applicability. Current global industrial nitrogen fixation rates are around 40 million tonnes per year and are increasing.

*Biological fixation*

Nature accomplishes nitrogen fixation by means of nitrogen-fixing bacteria. The components of the natural nitrogen cycle are shown in Figure 101. They form a very intricate chain of interlocking activities and are essential to the maintenance of the atmospheric composition.

We start with plants such as peas, beans and clover (legume family) which have nitrogen-fixing bacteria living symbiotically in nodules on their roots and can fix as much as 0.24 to 0.36

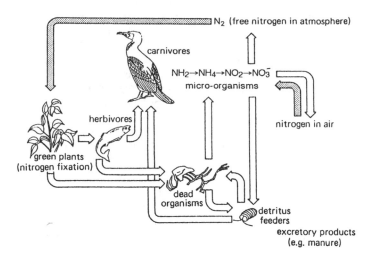

**Figure 101**   The nitrogen cycle.

tonnes nitrogen per hectare. The plant supplies the bacteria with food and energy, and the bacteria fix atmospheric nitrogen in the form of soluble compounds, some of which are excreted into the soil while the rest supply the plant with essential nitrates. The fixed nitrogen is incorporated in plant protein and if eaten becomes re-incorporated as new proteins in the animal. Eventually this protein returns to the soil when the animal of plant dies (there is also a contribution from defecation) and is decomposed by bacterial action into its component amino acids. In aerobic soil conditions many bacteria oxidize these amino acids to carbon dioxide, water or ammonia. Two additional micro-organisms – the nitrifying bacteria – convert the ammonia to nitrite ($NO_2^-$) and thence to nitrate ($NO_3^-$). Once in the nitrate form, it is again available for plant food. Thus, the cycle nitrate to protein, protein to ammonia, ammonia to nitrite, nitrite to nitrate can be repeated. Similarly animal excreta are decomposed to ammonia and converted to nitrate by the nitrifying bacteria.

Superimposed on the cycle are the denitrifying bacteria which decompose (at a smaller rate than fixation) the nitrites or nitrates to molecular nitrogen ($N_2$) to the atmosphere. Without these denitrifying bacteria, most of the nitrogen would be locked up in the soil, as run-off in the oceans or in sediments.

As man fixes 40 million tonnes of nitrogen per year with estimates showing that biological fixation is of the same magnitude, and as the denitrifying bacterial activities are in balance with the biological fixation, it must be assumed that a proportion of the man-made nitrates eventually end up as run-off in lakes, rivers or seas where EUTROPHICATION can result. The benefits of nitrate application are obvious in increased crop yields; however, the ecological implication of interfering with natural cycles has not been explored in sufficient depth. Industrial nitrogen fixation is expected to total 100 million tonnes per year by the year 2000 to cope with ever-increasing demands for food from finite land resources. In fact, man can defeat nature in some cases, as the application of nitrogenous fertilizers stops the action of the nitrogen-fixing bacteria.

Breeding developments are in hand to induce these bacteria to live in symbiotic relation with other food crops such as cereals.

The nitrogen cycle is very closely linked to those of carbon and oxygen, as some of the bacteria involved oxidize organic matter in the soil. There is, thus, a simultaneous cycling of carbon, oxygen and nitrogen.

The nitrogen cycle is very closely linked to those of carbon and oxygen, as some of the bacteria involved oxidize organic matter in the soil. There is, thus, a simultaneous cycling of carbon, oxygen and nitrogen.

In aquatic environments, nitrogen fixation can be accomplished by blue-green algae which make some waters highly productive regions, rich in fish life because of the availability of food. Where nitrogen is the limiting nutrient in a body of water excess nitrates from agriculture or sewage can stimulate the growth of ALGAE resulting in ALGAL BLOOMS in periods of hot weather.

**Nitrogen dioxide.** ⇨ NITROGEN OXIDES.

**Nitrogen-fixing bacteria.** ⇨ NITROGEN CYCLE.

**Nitrogen oxides.** These are collectively referred to as $NO_x$. Three are of importance:

*Nitrous oxide* ($N_2O$) is a colourless gas used as an anaesthetic: background concentration is 0.25 parts per million. It is mainly formed by soil bacteria in decomposing nitrogenous material. It reacts in the upper atmosphere with OZONE as follows:

$$N_2 + O \rightarrow 2NO: NO + O_3 \rightarrow NO_2 + O_2$$

*Nitric oxide* (NO) is colourless, formed during the high

temperature combustion of fuels which allow the nitrogen in the air to combine directly with oxygen. This takes place at temperatures above 1600°C, but if the fuel also contains nitrogen, nitric oxide will be produced at temperatures above 1300°C. NO is oxidized by ozone to $NO_2$:

$$NO + O_3 \rightarrow NO_2 + O_2.$$

*Nitrogen dioxide* ($NO_2$) is a reddish-brown, highly toxic gas with a pungent odour. It is one of the seven known nitrogen oxides which participate in PHOTOCHEMICAL SMOGS and primarily affect the respiratory system. Nitrogen dioxide is extremely poisonous and can pass into the lungs and form nitrous acid ($HNO_2$) and nitric acid ($HNO_3$), both of which attack the mucous lining. Nitrogen dioxide is also thought to act as a plant-growth retardant at normal atmospheric concentrations.

A UN protocol was signed in October 1988 by 24 countries to limit $NO_x$ emissions by 1994 to 1987 levels.

The highest concentrations of nitrogen dioxide (1991) in London were recorded at the urban background monitoring sites at Bridge Place and Earls Court. The WHO guideline of $150 \, \mu g/m^3$ averaged over 24 hours was exceeded, and at Bridge Place the guideline of $400 \, \mu g/m^3$ still averaged over an hour was also exceeded. In this case there was no breach of EC Directive because, although hourly means exceeding $200 \, \mu g/m^3$ were recorded at the urban background sites at which compliance is measured, the requirement in the Directive is for 98 per cent compliance over a year.

The Royal Commission on Environmental Pollution (18th report) gave the 1992 $NO_x$ emissions from transport as:

| | |
|---|---|
| Road | 1 398 000 tonnes |
| Rail | 32 000 tonnes |
| Air | 14 000 tonnes |
| Shipping | 130 000 tonnes |
| Total | 1 574 000 tonnes |

This accounts for 57 per cent of all UK emissions. The remaining 43 per cent emanate from power stations and boiler plants.

In London, ca. 80 per cent of $NO_x$ emissions are transport related and as WHO guidelines can be breached, this means that point $NO_x$ sources have to be strictly controlled. That said, traffic management (is your car really necessary?) will increasingly be called upon. It is unfair to burden industry totally whilst the private motorist and the heavy goods vehicle contribute most of the urban $NO_x$.

There are three methods of controlling power station $NO_x$ emissions: burner control, injection of chemicals into the boiler and selective catalytic reduction (SCR) where the flue gas is passed over a catalyst. The cheapest method is burner control which it is claimed can give 50 per cent reduction; with new plant 70 per cent may be possible. Retrofits will not achieve the same levels. Boiler injection and SCR systems are claimed to give up to 90 per cent reduction.

As coal-fired boilers produce 70–80 per cent of their $NO_x$ from the oxidation of nitrogen in the fuel, burner design is paramount. The aim is to obtain combustion of the volatiles which form most of the $NO_x$ in the 'cool' region of the flame with the char portion of the fuel burned at the 'hot' end of the flame as discussed by Redman (1989). ($\Rightarrow$ AUTOMOBILE EMISSIONS; CATALYTIC CONVERTER)

*Sources:* J. Redman, 'Control of $NO_x$ emission from large combustion plant', *The Chemical Engineer*, March 1989.

**Noise.** SOUND that is socially or medically undesirable, i.e. any sound that intrudes, disturbs or annoys. Very high levels of sound can cause hearing damage. Noise terminology is complex as the physical properties of sound and the mental processes by which it is heard and reacted to must be considered. ($\Rightarrow$ HEARING; ROAD TRAFFIC NOISE; AIRCRAFT NOISE; INDUSTRIAL NOISE MEASUREMENT; NOISE INDICES)

**Noise abatement zone (CoPA 1974).** An area designated by a local authority as one in which the levels of noise produced by an industrial premises must not exceed certain registered levels.

**Noise and number index (NNI).** Index of air traffic noise. For a number of aircraft, N, heard in a day, NNI is given by the equation:

$$NNI = \text{average peak perceived noise level} + 15 \log N - 80$$

where the average noise level is the logarithmic average.

Daytime NNI is the average NNI value during the summertime (mid-June to mid-September) for a daytime period from 06.00 until 18.00 hours. Note that 80 is subtracted from the NNI as this is the supposed zero annoyance level. (AIRCRAFT NOISE; LOGARITHMS)

**Noise at Work Regulations.** UK legislation which came into force on 1 January 1990. They define three action levels. 85 dB(A) or more first action level and 90 dB(A) or more second action level are both daily personal noise exposures; in addition there is a peak action level of 200 pascals (140 dB where 20 $\mu$ Pa is used as the reference sound pressure).

Employers need to take some basic action at 85 dB(A), principally noise assessments, records of assessments, provision of information for employees and availability of personal protection. More will be required at 90 dB(A); there will be a duty to reduce exposure to noise by means other than personal protection so far as reasonably practicable, compulsory use of protectors and marking of ear protection zones. All measures required at 90 dB(A) will also be required where peak pressure exceeds 200 pascals.

Employees will be required to use personal protection above 90 dB(A) and other protective measures and report any defects they discover.

Machinery suppliers will be required to provide adequate information concerning the noise levels likely to be generated if this is 85 dB(A) or more. ($\Rightarrow$ DECIBEL)

**Noise certification.** A procedure by which new aircraft must meet certain noise standards on take-off and landing before they are allowed to operate in the UK. They are based on engine power and unladen weight.

**Noise control.** There are three identifiable areas for noise control: source, path, and receiver. A source generates the noise. The path is the source–receiver separation and is characterized by the nature of the intervening space. The receiver is the listener. Noise control can be attempted at any one or all three of these areas.

1. *Control at source.* This is the most widely used form of noise control, e.g. motor vehicles are fitted with silencers and are required to emit less than a maximum permitted noise level (emission standard). For heavy vehicles in particular, control at source can mean tackling any or all of the components shown below.

2. *Noise control between source and receiver.* In many instances, the source cannot be significantly altered, e.g. traffic noise, and therefore sound insulation such as absorbent reveal, double glazing and fan ventilation instead of open windows

and air bricks has to be used. Noise barriers can also be erected. These should be as close to the source as possible and have a length of at least ten times the shortest distance between source and observer so that the sound waves do not merely bend round the ends of the barrier and so reach the observer. They will bend over the top as well but this can be guarded by use of T-shaped barriers.

| Source | Possible control measures for heavy vehicles | |
|---|---|---|
| Engine casing | Split in two to balance and 'break' transmission vibration | |
| Engine inlet | Engine inlet modification | |
| Engine exhaust | Redesign silencer | |
| Transmission and drive train | Anti-vibration | Total emitted airborne noise |
| | Redesign gear train | |
| Fan | Use variable pit blades | |
| Tyre/roadway | Redesign suspension | |
| | Improve road surface | |
| Aerodynamics | Redesign body shell contours | |

The effectiveness of sound barriers depends on frequency and density. The higher the frequency and the denser the material the greater the reduction of SOUND. The standard for facade exposure to traffic noise in the UK is 68 dB(A) on the $L_{10}$ INDEX. ($\Rightarrow$ NOISE INDICES)

3. *Receiver control.* This means that ear protectors are required and these should be chosen for the particular sound pressure level that is to be reduced to a safe level. Ear protectors are mainly used in an industrial environment and are particularly important if NOISE-INDUCED HEARING LOSS is to be avoided.

**Noise control boundary.** The boundary around premises included in a noise abatement zone on which registered levels are fixed.

**Noise dose.** The total sound energy received at the ears of an exposed individual over the period of interest (time) measured in $Pa^2H$. Sometimes measured as sound exposure level (dBh). Sound exposure level is equal to period Leq multiplied by the time of exposure.

**Noise footprint.** An area on the ground under a flight-path during take-off or landing that is founded by the 90 PNdB or EPNdB noise contour.

**Noise indices.** Measures of the disturbing qualities of noise-loudness, variations in time, whines, bangs, etc. and the average subjective response. Individual responses to noise vary; noise indices are average measures. They are used in planning residential developments, motorways, etc. $L_{10}$ is the level of noise in dB(A) exceeded for just 10 per cent of the time. For the measurement and prediction of noise from traffic the average of $L_{10}$ values for each hour between 6.00 a.m. and 12.00 midnight on a normal weekday has been adopted by the Government as giving satisfactory correlation with dissatisfaction. This is known as $L_{10}$ (18 hour); NNI is the NOISE AND NUMBER INDEX for air traffic; PNdB for individual aircraft; CNL is the CORRECTED NOISE LEVEL for noise from industrial premises; Leq is a measure of the equivalent continuous noise level from a site in energy terms over a specified period; $L_{90}$ level is exceeded for 90 per cent of the period of interest and is specified as a measure of background level in BS 4142, when measured in dB(A) with the industrial source of interest not operating.

**Noise-induced hearing loss.** Damage to the ear not caused by ageing or accident or disease. This is an insidious affliction. Many industries are inherently noisy – foundries, boiler shops, shipyards, press rooms. A worker first entering such an environment spends some time getting acclimatized and then may eventually accept the noise levels to which he or she is subjected. This casual acceptance encourages in both workers and management an attitude that noise is simply part of the job.

The first effects of exposure to excessive noise are ringing in the ears and a dulling of the hearing – temporary threshold shift, i.e. a 'louder' level of noise is required to conduct a conversation. These effects are often temporary, but if there is continued exposure or infrequent gaps between exposures, the ears do not recover and there is permanent damage. The inner ear can be damaged and in extreme cases the eardrum can be ruptured.

Industrial deafness is usually associated with inner-ear damage which may be unnoticed for many years until it is chronic. The usual result is that frequencies above 4000 Hz such as a high-pitched whistle or the full range of music cannot be detected. As speech is usually in the 500–2000 Hz range, it is only when an affected individual cannot follow speech so readily that the damage is detected. To assess noise-induced

(a) Classification of handicap due to hearing loss

| Class | Degree of handicap | Average hearing level (dB) | Ability to understand ordinary speech |
|-------|-------------------|---------------------------|---------------------------------------|
| A | Not significant | Less than 25 | No significant difficulty with faint speech |
| B | Slight | 25 to less than 40 | Difficulty only with faint speech |
| C | Mild | 40 to less than 55 | Frequent difficulty with normal speech |
| D | Marked | 55 to less than 70 | Frequent difficulty with loud speech |
| E | Severe | 70 to less than 90 | Shouted or amplified speech only understood |
| F | Extreme | 90 | Usually even amplified speech not understood |

From W. Burns, *Noise and Man*, John Murray, 1968; revised edn, 1972.

(b) The social effects of hearing loss experienced by weavers with long exposure to noise. A population of weavers who had been working the trade for a number of years is compared with a control population who did not have a history of severe noise exposure

| Social effects | Weavers | Control |
|----------------|---------|---------|
| Difficulty in understanding family/friends | 77% | 15% |
| Difficulty in understanding strangers | 80% | 10% |
| Difficulty in use of telephone | 64% | 5% |
| Difficulty at public meetings, church, etc. | 72% | 6% |
| Own estimate of hearing below normal | 81% | 6% |

From W. Taylor *et al.*, *Proceedings of a Conference on Occupational Noise in Medicine*, National Physical Laboratory, 1970.

hearing loss an audiometer is used which compares the subject's hearing threshold with that of a normal hearing individual. Thresholds are usually measured at 0.25, 0.5, 1, 2, 3, 4, 5, 6 and 8 kHz. Thus at, say, 2 kHz (speech frequency), if the subject is asked to identify the intensity at which he just hears 2 kHz and if this is 30 decibels, then the hearing level is said to be 30 dB at 2 kHz. Now normal hearing intensity at 2 kHz is 0 dB, so that the subject's threshold differs by 30 dB. An audiogram can be obtained which plots noise-induced threshold shift against frequency as shown in Figure 83. (⇨ HERTZ; NOISE-INDUCED HEARING LOSS)

The degree of handicap is given in table (a). Ordinary conversation is not the only criterion: the social dimension of hearing loss is extremely important and this is shown in table (b). Clearly excessive noise is also a damaging pollutant.

**Noise measurement.** ⇨ DECIBEL; SOUND.

**Non-Fossil Fuel Obligation (NFFO).** A requirement of electricity companies in England and Wales (and shortly Scotland) to purchase, from specified producers at a premium price for a fixed period, specified amounts of electricity produced by renewable energy means, including waste to energy.

This has meant that WIND ENERGY has received enhanced revenues of up to 11.5 pence/kWh, and waste to energy 4.5 to 6.5 pence/kWh – compared with more normal price levels of 2–2.5 pence/kWh. (⇨ INCINERATION)

**Non-Fossil Fuel Order (NFFO).** A commitment to produce energy from non-fossil fuel sources as required by the Electricity Act 1989.

This usually results in enhanced revenues for authorized non-fossil fuel power plant operators.

*Source:* 'The Government's Policy for Renewables and the Non-Fossil Fuel Obligation', *Renewable Energy Bulletin* No. 5, p. 5, Department of Trade, October 1993.

**Non-renewable resources.** Resources such as minerals or fossil fuels which are replaced on a geological time scale.

**Normal distribution.** A very common SYMMETRICAL DISTRIBUTION found in statistics which can be used as a working approximation for many prediction purposes. It is easy to use with quite low error when deviations from non-normality are not too severe. It is characterized by two parameters: the MEAN and STANDARD DEVIATION. Both are obtained by sampling.

The graph of the distribution is given by

$$y = \frac{1}{\sigma\sqrt{2\pi}}\exp\left[ -\frac{1}{2}\left(\frac{x - \mu}{\sigma}\right)^2 \right]$$

where $x$ can run from $-\infty$ to $+\infty$: $\mu$ and $\sigma$ are the mean and standard deviation of the distribution. The spread of the curve about $\mu$ depends on $\sigma$. The curve is symmetrical about the point $x = \mu$. (⇨ VARIANCE)

Summary of 1991 and 1994 NFFO contracts for England and Wales

|                          | 1991   | 1994 (*NFFO* 3)         |
|--------------------------|--------|-------------------------|
| Waste                    |        |                         |
| MW DNC                   | 271.48 | 241.87                  |
| No. of schemes           | 10     | 20                      |
| Price (p/kWh)            | 6.55   | 4                       |
| 'Other'                  |        |                         |
| MW DNC                   | 30.15  | 122.06                  |
| No. of schemes           | 4      | 9                       |
| Price (p/kWh)            | 5.9    | 8.75[1]/5.23[2]         |
| Hydro                    |        |                         |
| MW DNC                   | 10.86  | 14.48                   |
| No. of schemes           | 12     | 15                      |
| Price (p/kWh)            | 6.00   | 4.85                    |
| Landfill gas             |        |                         |
| MW DNC                   | 48.45  | 82.07                   |
| No. of schemes           | 28     | 42                      |
| Price (p/kWh)            | 5.7    | 4                       |
| Sewage gas               |        |                         |
| MW DNC                   | 26.86  |                         |
| No. of schemes           | 19     |                         |
| Price (p/kWh)            | 5.9    |                         |
| Wind                     |        |                         |
| MW DNC                   | 84.43  | 145.9[3]/19.7           |
| No. of schemes           | 49     | 31/24                   |
| Price (p/kWh)            | 11.0   | 4.8/5.99                |
| Cumulative capacity[4]   |        |                         |
| MW DNC                   | 472.23 | 626.9                   |
| No. of schemes           | 122    | 141                     |

[1] Energy crops.
[2] Residuals.
[3] Wind has two bands, $> 1.6$ MW and $< 1.6$ MW, respectively.
[4] 340 MW is already operational from NFFO 1 and 2 contracts. It is anticipated that NFFO 3 will double this capacity in practice.
MW DNC = megawatt declared net capacity.

**Normal solution.** A solution which contains the chemical equivalent weight in grams of the dissolved substance (solute) in a litre of solution.

**Normal temperature and pressure (NTP).** A temperature of 273.15 K (0°C) and a pressure of 101 325 Pa; used as a reference standard for gas volumetric measurements.

**North Sea Gas.** (⇨ NATURAL GAS)

**Nuclear energy.** There are potentially two principal processes for obtaining energy from nuclear sources – fission and fusion. (Fusion has not yet been engineered as a continuous process.)

*Fission* uses the energy released in a controlled nuclear reaction whereby an isotope (uranium-235) is split by the capture of neutrons, energy (plus more neutrons) is released and mass consumed in the process. One gram of uranium-235 consumed in a fission reaction will release 81 900 million $(8.19 \times 10^{10})$ JOULES or the equivalent of the heat of combustion of 2.7 tonnes of coal or 13.7 barrels of crude oil – hence a nuclear power station would consume about 3 kilograms of $^{235}$U per day. While seemingly small, this is a significant amount of uranium in relation to the total supplies. The prospects for nuclear power would not be 'infinite' were it not for the various other types of reaction (such as the breeder reactor) and the possibility of an alternative long-term nuclear energy supply – fusion.

*Fusion* relies on the energy release when a heavier element is formed by the fusion of lighter ones. The sun's energy is a fusion process resulting from the formation of helium by the fusion of atoms of one or more of the hydrogen isotopes, i.e. hydrogen (H), deuterium (D) and tritium (T). As in fission, there is a loss of mass and it is this mass difference that appears as energy. Thus, the fusion reaction revolves round the hydrogen isotopes and, indeed, the energy released in an uncontrolled explosive manner by the fusion of D and T is the basis of the H-bomb. The potential of fusion power is illustrated by considering 1 cubic metre of water which ordinarily contains 1 D atom for each 6500 H atoms and has a potential fusion energy (D-D fusion) of 8 160 000 million $(8.16 \times 10^{12})$ joules or the heat of combustion of 269 tonnes of coal or 1360 barrels of crude oil.

1 cubic kilometre = 1000 million cubic metres, so that 1 cubic kilometre of water contains the fusion (D-D) potential of 269 000 million tonnes of coal or 1 330 000 million barrels of crude oil, which approximates to the lower estimate of the world reserves of crude oil. ($\Rightarrow$ ENERGY RESOURCES)

The total volume of the ocean is about 1500 million cubic kilometres, and if 1 per cent of this were used, the potential energy release would be sufficient for 5 000 000 times that of the world's initial coal and oil supplies (see Hubbert, 1969). The reasons for the quest for controllable fusion power are now evident.

For the provision of nuclear energy as currently practised, i.e. fission, ⇨ NUCLEAR REACTOR DESIGNS. (⇨ IONIZING RADIATION; NUCLEAR REACTOR WASTES)

M. K. Hubbert, 'Energy resources', *Resources and Man*, W. H. Freeman, 1969, ch. 8.

**Nuclear fission.** In spite of the intense and very costly development programmes of the last three decades, the much heralded future of power from nuclear fission is still not in sight. Great Britain still generates only one-eighth of electric power from nuclear fission (the extraction of heat either directly from fissile isotopes or indirectly from fertile isotopes). (⇨ NUCLEAR REACTOR DESIGNS (*Fast Breeder*))

The problems of assessing the relative merits and demerits of an extensive nuclear fission technology are of an ethical nature and cannot simply be weighted by cost-benefit analysis. There is a tendency to over-emphasize the short-term benefits and undervalue the long-term problems. In attempting to assess the risks associated with a fissile technology, in which 'no acts of God can be permitted', we cannot call on experience, since the magnitude of the problems confronting us are unique in human history. CHERNOBYL and Three Mile Island pressurized water reactor incidents have both demonstrated this.

**Nuclear power.** Unlike any other human product, nuclear wastes will survive for periods that are more appropriate to geology than history. The HALF-LIFE of plutonium-239 is 24 400 years. For every kilogram of plutonium created today, there will still be 500 g left in 24 400 years, 250 g in 48 800 years, 125 g in 97 600 years, and so on. In view of the extreme toxicity of such a material, we cannot ignore serious doubts about our right to saddle our descendants with it – particularly as we have, as yet, devised no practical means of disposing of it. We have entered into a Faustian bargain whereby we are given an unlimited energy source in return for a pledge of eternal vigilance (Edsall, 1974).

We have no means of knowing whether the safety problems we are setting ourselves are capable of solution. The nuclear industry points with ample justification to the very strict controls and high levels of safety within the industry. But such statements avoid rather than answer the central question of whether the problems of plant safety and of containment and disposal that we are setting ourselves are inherently insoluble.

Unfortunately this dispute is not essentially a technical one and is not therefore resolvable on its technical merits.

> People have to operate nuclear power-plants, no matter how much automation we introduce. People are forgetful; often they are irresponsible; and quite a few of them suffer from deep-seated irrational tendencies to hostility and violence . . . I believe that the confident advocates of the safety of nuclear power-plants base their confidence too narrowly on the safety that is possible to achieve under the most favourable circumstances, over a limited period of time, with a corps of highly trained and dedicated personnel. If we take a larger view of human nature and history, I believe that we can never expect such conditions to persist over centuries, much less over millennia (Edsall, 1974).

> We can only hope that 'the safety of the public . . . will never be made dependent upon almost superhuman engineering and operational qualities (Cottrell, 1974).

Time has, however, moved on, and despite the CHERNOBYL and THREE MILE ISLAND incidents, UK nuclear power stations have an excellent record.

J. Collier (M.D. Nuclear Electric plc) has stated:

> Taken overall, by good fortune as well as by design, nuclear power now seems poised to satisfy, on merit, the broad spread of policy aims set for the British energy sector:
>
> 1. It provides competition and consumer choice – with the full costs already included in the price.
> 2. It expects to face the disciplines of the capital markets – and meet the challenge.
> 3. The highest priority is placed on health and safety and the industry has a record second to none.
> 4. It already makes a major contribution to limiting the environmental impact of the energy sector – and offers the most dependable and cost effective route to energy supply without environmental risk. Energy conservation and energy efficiency are complementary measures not alternatives.

> With its origins in science and nurtured in defence, the author has a vision that historians may one day look back on nuclear power as one of the more tangible examples of a 'peace dividend' from the twentieth century. The case for nuclear power is now very strong and the outcome of the Government's review is eagerly awaited. This really is 'the end of the beginning' and the start of the future clean energy for the twenty-first century.

UK Nuclear Electric has also brought attention to bear on the security of gas supplies and the emissions from conventional power generation methods (Nuclear Electric and The Environment, Jan. 1995).

Gas Demand is forecast to double by 2020. The Government forecasts that by 2020 nearly 60% of electricity will be generated using gas. The consequence will be a period of greatly accelerated increase in demand.

As a result of the rapid increase in gas demand the UK's currently *proven* reserves could be used up early in 2001, though more gas reserves are likely to be discovered. Using the most optimistic Government figures for further discoveries, however, the UK's own gas could last only until 2025 unless supplemented by imports.

To sustain the anticipated increases in demand, gas will have to be imported from Norway and from the European gas transmission systems. But gas demand is increasing rapidly across the whole of Europe. This means that if the UK electricity supply is dominated by CCGT (combined cycle gas turbines) we could ultimately become reliant on gas from further afield – Northern Norway, Russia or the Middle East. This would expose the UK to supply and price risks as a result both of politics (at the well-head and *en route*) and of the economics of long distance pipelines.

Gas is an attractive fuel for power generation in the short term. It is relatively clean, and can be burnt in relatively inexpensive CCGT

Budget apportionment for Sizewell B Power Station

| Project | Value (£ million at April 1987 prices) |
|---|---|
| Nuclear steam supply system | 566 |
| Civil construction | 406 |
| Turbines and other mechanical plant | 254 |
| Control and instrumentation | 129 |
| Electrical plant | 110 |
| First fuel charge | 65 |
| Construction and commissioning | 43 |
| Software including engineering design fees | 449 |
| Setting up and launch costs | 8 |
| Total | 2030 |

*Source: Prospects for Nuclear Power in the UK*, Conclusions of the Government's Nuclear Power Review, Department of Trade & Industry and the Scottish Office, May 1995.

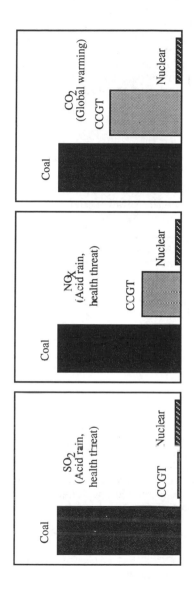

**Figure 102** Relative atmospheric emissions from coal-fired (without FGD) and CCGT electricity generation. Nuclear energy has negligible atmospheric emissions. *Source*: Nuclear Electric.

power stations. Those buying gas need not consider the state of the market beyond the 15 years which their projects are typically financed.

Nuclear Electric and other companies in the nuclear power industry, however, are used to taking a longer view – 40 years or more. Decisions to burn gas liberally, taken on the basis of relatively short-term pay-back, could adversely affect the UK's energy supply and balance of payments in the medium and long-term. Also a valuable primary fuel and chemical feed stock is being rapidly consumed when other sources of electricity with comparable long run economics are available.

Our judgement is that nuclear power will become increasingly attractive with time. The Company, therefore, welcomes the Government's commitment to retaining the nuclear option for the UK.

The UK Department of Trade & Industry Review of Nuclear Power (1995) has given the following data on estimated marginal costs of $CO_2$ emission abatement.

| Date | 2005 | 2010 |
|------|------|------|
| Estimated (marginal costs for 5 MtC reductions) | £15–70/tC | £30–70/tC |
| Estimated cost of investment in new nuclear plant for $CO_2$ abatement** | £100–250/tC* | £100–250/tC* |

*The costs of abating $CO_2$ by investment in new nuclear plant are calculated by comparing the additional $CO_2$ saved with the extra costs incurred by Sizewell C over and above a CCGT.
**Assuming saving from Sizewell C = 2 MtC a year; lifetime costs of Sizewell C = 3.5–5.75 p/kWh.

The above benchmark estimates can be used to judge whether the costs of a new nuclear power station might be justified for $CO_2$ policy purposes. Taking account of the uncertainties, it was judged to be reasonable to use a range of 3.5–5.75 p/kWh for lifetime nuclear generation costs (1990 prices) for comparison purposes. This equates to an additional cost per tonne of carbon abated – over and above the cost of CCGT generation assumed in the EP56 projections – of about £100–250/tC. **DTI estimates for economy-wide $CO_2$ abatement costs in the 2005–2010 period are substantially lower than this range. This suggests, on the basis of currently estimated relative costs of new CCGT and nuclear capacity, that the significant amount of additional capacity provided by new nuclear build is currently too expensive to be justified for $CO_2$ policy purposes alone.** (AUTHOR'S BOLD) Nuclear power in the UK is now on a plateau.

*Sources: Prospects for Nuclear Power in the UK*, Conclusions of the Government's Nuclear Power Review, Department of Trade & Industry and the Scottish Office, May 1995.

A. Cottrell, Letter to the *Financial Times*, 7 January 1974.

J. T. Edsall, *Environmental Conservation*, vol. 1, no. 1, 1974, p. 32.

J. Collier (M.D. Nuclear Electric plc) extract from *Journal of Power & Energy*, vol. 207, 1993: and Nuclear Electric Briefing Note, 6 October 1994.

D. R. Davies, 'Planning and acquiring the UK's first PWR power station', *Proc. Instn. Civ. Engrs.*, Civ. Engng., Sizewell B Power Station, 1995.

**Nuclear reactor designs.** *Burner reactor:* This is essentially a pressure vessel contained in a biological shield to contain radiation release. It has the means for inserting the fuel and removing the waste products; moderators for slowing down the neutrons from their fast velocities to thermal velocities so that fission can take place; control rods for controlling the overall speed of the nuclear reaction; canning, i.e. containers for the fuel; and lastly the main constructional materials. The drawback with burner reactors (and they are the only ones available commercially) is that the fuel is consumed and as uranium is a rare element and $^{235}U$ is less than 1 per cent of natural uranium, the burner reactors are considered a stop-gap measure until breeder reactors are available. Reactor designs vary in the fuels, moderators, coolants, canning, etc., and only the main designs are discussed.

1. Fuel can be natural uranium oxide or $^{235}U$-enriched uranium oxide. The enriched fuel has a much higher energy release per unit mass.
2. Moderators can be light (ordinary) or heavy water or graphite. (Heavy water is water containing a substantial portion of deuterium oxide, $D_2O$.)
3. Coolants come from light or heavy water, carbon dioxide, helium and sodium. (Lithium has been proposed for a possible fusion reactor.) Carbon dioxide is readily manufactured. Helium is obtained from natural gas but could eventually be in short supply. Sodium is used as a coolant in fast breeder reactors and is obtained by electrolysis of molten salt.
4. Cans – fuel containers in metallic and non-metallic varieties: Magnox (magnesium oxide alloy), stainless steel, zirconium, silicon carbide/graphite.

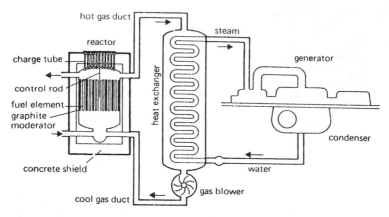

**Figure 103**   Magnox (magnesium oxide) reactor. The first British commercial reactor using metallic natural uranium metal fuel clad in magnesium alloy, Magnox, with graphite moderator and carbon dioxide coolant.

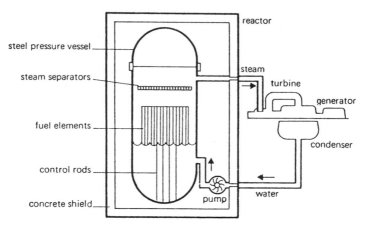

**Figure 104**   Boiling water reactor (BWR) using enriched uranium oxide fuel pellets clad in Zircaloy, with light water moderator and coolant.

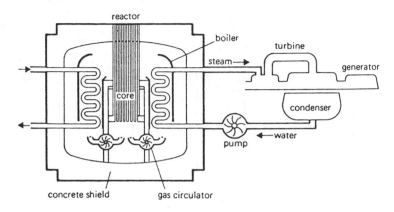

**Figure 105** Advanced gas-cooled reactor (AGR) using enriched uranium oxide fuel pellets clad in stainless steel, with graphite moderator and carbon dioxide coolant.

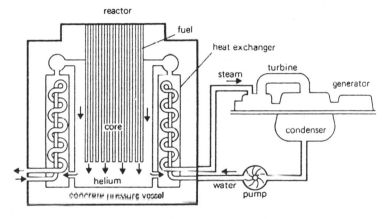

**Figure 106** High temperature reactor (HTR) using enriched uranium oxide or carbide fuel granules clad in carbon-silicon carbide, with graphite moderator and helium coolant.

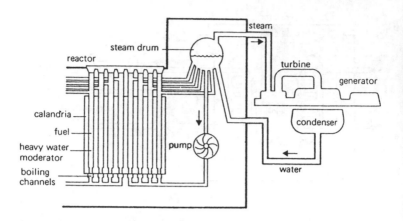

**Figure 107** Steam generating heavy water reactor (SGHWR) using enriched uranium oxide fuel pellets clad in Zircaloy, with heavy water moderator and light water coolant.

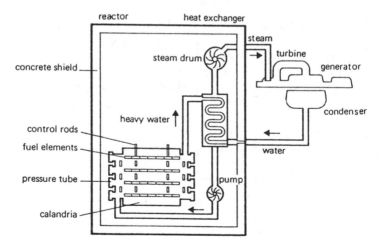

**Figure 108** Candu reactor. Canadian reactor using natural uranium oxide fuel pellets clad in Zircaloy, with heavy water moderator and coolant.

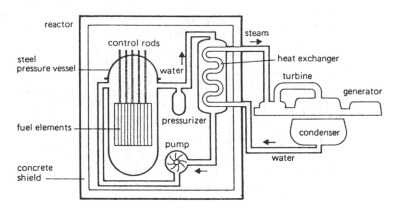

**Figure 109** Pressurized water (PWR) using enriched ranium oxide fuel pellets clad in Zircaloy, with light water moderator and coolant.

5. The materials of construction are steel, cement, copper, etc., and are the normal materials of conventional structures.

A summary of the common fission reactor types is given in Figures 103–109.

*Source:* G. R. Bainbridge and A. A. Farmer, 'Nuclear reactors for the future', *Symposium: Energy Resources or Misuse?*, University of Newcastle upon Tyne, Institute of Chemical Engineering, September 1974.

*Fast breeder reactor:* The fast breeder reactor uses liquid sodium as a coolant, but because of the inherent risks (however small) of a violent sodium–water reaction, special design techniques are employed to contain this reaction, if any, *outside* the reactor. Thus a primary liquid sodium cooling loop is employed which transfers heat to a secondary liquid sodium loop, which transfers heat in turn to a high pressure water–steam circuit to separate steam from the turbines. The arrangement is shown in Figure 110.

A breeder reactor is one in which additional fissionable fuels are produced from the neutrons provided by an initial charge of uranium-235. The raw material for these new fuels is non-fissionable uranium-238 which comprises over 99 per cent of natural uranium, and thorium-232, which is essentially 100 per cent natural thorium. They are called fertile materials. The

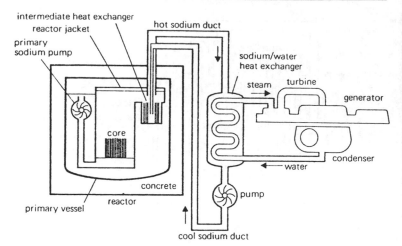

**Figure 110**   Fast breeder reactor with primary and secondary sodium circuits.

neutrons are not slowed down to thermal velocities as in burner reactors, hence the generic name *fast* breeder reactors.

The conversion of fertile materials to fissionable materials is called breeding and takes place for $^{238}$U as follows:

(number of protons and neutrons)

$$\frac{238}{92}U + n \rightarrow \frac{239}{92}U \rightarrow \frac{239}{93}Np \rightarrow \frac{239}{94}Pu$$

(number of protons)

i.e. uranium-238 absorbs a neutron and becomes uranium-239. This changes spontaneously through two short-lived radioactive transformations to neptunium-239 and thence to plutonium-239.

Similarly thorium-232 and uranium-233 are fissionable in an identical fashion to uranium-235. The thermal energy released per gram of both materials is virtually the same as that of uranium-235. In a breeder reactor more fissionable material is produced than is consumed and in theory it is possible to convert all available fertile material, given that there is enough uranium-235 to start off with.

The breeding gain per charge currently achieved is about 1.2, and design improvements can get this up to 1.4. The doubling time is very important and currently about 20 years are required to double the initial amount of fuel. This could be shortened, but safety criteria may dictate otherwise. Prototype fast breeder reactors are now operating in France. The UK programme has been shut as there was little likelihood of the commercialization of the design within a 30–40 year time span.

Nuclear energy was expensive six years ago, at over 5p per unit. UK overall Advanced Gas Cooled Reactor (AGR) costs are now down to 2.5p/kWh, and still improving – including all decommissioning costs. Sizewell B, on the same basis, costs 2.2p/kWh. That is certainly commercially competitive and is one of the reasons for real overseas interest in the design.

*Source:* B. V. George, Executive Director (Engineering) Nuclear Electric, letter published in *Professional Engineering*, 10 May 1995.

**Nuclear reactor economics.** ⇨ SIZEWELL.

**Nuclear reactor wastes.** Radioactive wastes are produced by the controlled release of energy from fissile fuel and to a much lesser extent by neutron bombardment of otherwise neutral materials in the reactor, including reactor metals, coolant fluids, air, carbon dioxide or other gases. The mass of radioactive fission products produced in a reactor is very nearly equal to the mass of fuel consumed. Once a certain fraction of the fuel is spent, the elements are removed from the reactors and reprocessed; that is, the unspent fuel is separated from the fission products and re-used.

Most radioactive waste emanates from the fuel-processing plant in liquid or slurry form and is stored in steel or concrete tanks for a preliminary cooling period prior to ultimate disposal. There are three classifications of radioactive waste slurries or liquids:

1. *High level* – radioactivity greater than 1 curie per gallon (4.5 litres). Current (1988) production $100 \, \text{m}^3$/year.
2. *Intermediate level* – radioactivity in the range $1 \times 10^{-6}$ curie to 1 curie per gallon. Current UK production $5000 \, \text{m}^3$/year.
3. *Low level* – radioactivity less than $1 \times 10^{-6}$ curie per gallon. Current UK production $40000 \, \text{m}^3$/year.

The problems associated with nuclear wastes are highlighted in the extreme case of PLUTONIUM.

Up to the mid-1970s in the UK about 30 tonnes of plutonium had been produced by the nuclear power programme. The planned expansion of fast breeder reactors would mean that this amount would increase to several hundred tonnes by the end of the century. Clearly these quantities represent a very substantial hazard and must be effectively isolated from the biological environment for many thousands of years.

Conventionally, radioactive waste is stored until natural decay processes reduce the activity to acceptable levels. For plutonium, the storage periods are so long that artificial storage is only a means of buying time until more satisfactory solutions, e.g. VITRIFICATION, are found. Other proposals for high-level waste disposal include injecting them down deep boreholes into impermeable strata, or deep underground storage.

Storage, however, cannot be justified as a policy *ad infinitum*. High-level wastes (which have been conveniently ignored in the current debate) would require surveillance over about 250 000 years. At some time, disposal, or at the very least abandonment of surveillance, will become inevitable. Although attractive at first sight, a policy of storage is short-sighted and in the longer term intrinsically unsustainable.

> A boundary must be drawn somewhere. It is not feasible, in the search for perfection, to reject the disposal of radioactive wastes for ever and a day. Society cannot be expected to supervise in perpetuity the wastes it has produced.
>
> We must at some time relinquish control of them in the knowledge that there will be some uncertainty about the adequacy of the technology of disposal. There is no way of escaping the fate of imperfection: the point is to work to minimise uncertainties, not to ignore them.
>
> *Source:* Berkhaut, 'The folly of perfection', *The Independent*, 1 Jan. 1989.

The guiding principles for any long-term programme of nuclear waste disposal were laid down by the US National Academy of Sciences in 1967 as:

1. All radioactive wastes should be isolated from the biological environment during their periods of harmfulness.
2. No waste-disposal practice, even if regarded as safe at an initially low level of waste production, should be initiated unless it would still be safe when the rate of waste production becomes orders of magnitude larger.
3. No compromise of safety in the interests of economy of waste disposal should be tolerated.

This problem of what to do with long-lived radioactive waste poses in an actual form some of the dilemmas we discussed. Apart from radiation, nuclear reactors do not emit serious air pollution as do conventional power stations. They offer mankind a long term prospect of cheap electricity, whereas coal and oil are in limited supply. Many people feel that to hold back on nuclear power because of the unsolved radioactive waste problems would be unduly cautious; technology in twenty years time, they argue will surely have found a way round the difficulty.

The evident danger is that man may have put all his eggs in the nuclear basket before he discovers that a solution cannot be found. There would then be powerful political pressures to ignore the radiation hazards and continue using the reactors which had been built. It would be only prudent to slow down the nuclear power programme until we have solved the waste disposal problem, or until we have developed fusion power or some other type of reactor which does not leave dangerous wastes behind it.

Many responsible people would go further. They feel that no more nuclear reactors should be built until we know how to control their wastes. Since planned demand for electricity cannot be satisfied without nuclear power, they consider mankind must develop societies which are less extravagant in their use of electricity and other forms of energy. Moreover, they see the need for this change of direction as immediate and urgent.

Currently high-level waste disposal is handled strictly in accordance with the three principles listed above. However, both the first and the second principles are infringed by many present (UK and USA) practices involving the disposal of low-level liquid wastes into the sea and the release of gaseous wastes, after removal of most longer lived isotopes, through tall chimneys. Most low-level wastes such as contaminated clothing, concrete, pipes, etc. are containerized and placed in a licensed landfill at the Sellafield UK nuclear reprocessing plant. Long-term plans involve the construction of an undersea repository of 2.5 million $m^3$ volume which will take care of these wastes till 2050.

In the UK, United Kingdom Nirex Ltd is responsible for providing and operating a repository for the disposal, deep underground, of intermediate-level radioactive wastes and of low-level wastes requiring deep disposal. This is in accordance with government policy for wastes arising in the UK. Similar disposal methods are favoured by other countries producing substantial quantities of long-lived radioactive waste.

Following an extensive site selection exercise, which culminated in preliminary geological investigations from 1989 at two sites, an area near the Sellafield Nuclear Reprocessing Plant, West Cumbria was chosen in 1991 as the focus for further in investigations.

An estimated 60 per cent of the waste volume destined for the repository arises at British Nuclear Fuel's Sellafield works.

Before Nirex can build and operate a repository it will have to show that a repository at the proposed site would satisfy the strict safety guidelines laid down by the government.

Results show that the Borrowdale Volcanics Groups of rocks, the top surface of which is 400–600 m beneath the site near Sellafield, continues to hold good promise as an eventual location for the repository.

However, there is still substantial work to be done before a decision could be taken to submit a planning application for the construction of a repository at Sellafield. (⇨ IONIZING RADIATION, EFFECTS; IONIZING RADIATION, MAXIMUM PERMISSIBLE DOSE)

*Source: Pollution: Nuisance or Nemesis*, a report on the Control of Pollution, HMSO, February 1972.

**Nuclear White Paper.** The White Paper on the Prospects for Nuclear Power states:

> It (Goverment) will not provide public money, whether from the taxpayer or the consumer, for the building of more nuclear power stations.
> That means nuclear power should only survive if it is economic in the long run.

**Nucleus[1].** The positively charged core of an ATOM made up of NEUTRONS and PROTONS. Accounts for almost the whole mass of the atom.

**Nucleus[2].** Focus for growth; for example, a dust particle acts as a nucleus of a fog droplet or ice particle formation.

**Nuclide.** Term for the ISOTOPE of an atom. If it is radioactive, it is known as a radionuclide. (⇨ RADIOACTIVE ISOTOPE)

**Nuisance.** In law, that which annoys or hurts, e.g. excessive noise levels, emission of noxious odours, polluting discharges. A large body of case law exists. A recent example (1989): The UK South West Water Authority (1990) is being sued under British Common Law for causing a Public nuisance in the CAMELFORD incident. This carries the prospect of an unlimited fine. ($\Rightarrow$ DUTY OF CARE; RYLANDS V. FLETCHER)

**Nuisance threshold.** The lowest concentration of an air pollutant that can be considered objectionable.

**Nutrients.** The raw materials necessary for life which are consumed during the metabolic process of nutrition. Their type and consumption vary according to the particular plant or animal species. The main categories are proteins, carbohydrates, fats, inorganic salts (e.g. nitrates, phosphates), minerals (e.g. calcium, iron), and water.

**Nymph.** The larval stage of aquatic insects such as may flies, stone flies and dragon flies. They are useful biological indicators of the purity of a water course. ($\Rightarrow$ BIOTIC INDEX; BIOLOGICAL INDICATOR)

# O

**Occupational exposure standard (OES).** The concentration of an airborne substance, averaged over a reference period, for which, according to current knowledge, there is no evidence that it is likely to be injurious to employees if they are exposed by inhalation, day after day, and which is specified in a list approved by the Health and Safety Committee (HSC). For a substance which has been assigned an OES, exposure by inhalation should be reduced to that standard. However, if exposure by inhalation exceeds the OES, then control will still be deemed to be adequate provided that the employer has identified why the OES has been exceeded and is taking appropriate steps to comply with the OES as soon as is reasonably practicable. In such a case the employer's objective must be to reduce exposure to the OES, but the final achievement of this objective may take some time. Factors which need to be considered in determining the urgency of the necessary action include the extent and cost of the required measures in relation to the nature and degree of exposure involved (Ref. UK Health and Safety Executive, 1991).

The criteria for setting limits are:

1. The ability to identify, with reasonable certainty, a CONCENTRATION averaged over a reference period, at which there is no indication that the substance is likely to be injurious to employees if they are exposed by inhalation day after day to that concentration.
2. The OES can reasonably be complied with.
3. Exposures to concentrations greater than the OES, for the period of time it might reasonably be expected to take to identify and remedy the cause of excessive exposure, are unlikely to produce serious short- or long-term effects on health.

**Octane number.** A measure of the knock-resistance of a fuel for spark-ignition engines. Determined in test engines by comparison with reference fuels. (⇨ CETANE NUMBER)

**Odorant.** A distinctive, sometimes unpleasant, odour added to odourless materials to give warning of their presence, e.g. to natural gas.

**Odour threshold.** The concentration of an odour-bearing gas at which only half a panel of 'sniffers' can detect the smell. The odour is usually diluted in a dynamic system and presented to groups of volunteers at various dilutions. (⇨ ODOURS)

**Odours.** The smell(s) produced by, usually, very small concentrations of organic vapours can produce violent aversion reactions in anyone exposed to them. The reactions range from nausea to insomnia. Many of these vapours come from operations such as animal by-products processes, farming, maggot breeding, brick-making and metallurgical processing. What is surprising is that the odour threshold in many cases occurs at very low concentrations of the substance in air and almost invariably these concentrations are well below the THRESHOLD LIMITING VALUE for the substance. For a 'normal' person, there is at least some comfort in knowing that mass poisoning is not taking place. (⇨ MERCAPTANS)

Nevertheless, odour suppressions at source can and should be practised on both public health and amenity grounds. The common methods are:

1. Absorption in a suitable liquid which may oxidize the offending vapour or neutralize it in the process (⇨ ABSORP-TION).
2. Adsorption: ACTIVATED CARBON will adsorb organic molecules in preference to water vapour and can be tailored to provide optimum adsorption for the particular contaminant (⇨ ADSORPTION).
3. INCINERATION, i.e. oxidation at high temperature, will destroy most malodorous, gaseous and organic wastes but is usually very expensive.
4. After-burning of non-inflammable weak mixtures of organic vapours which heats the effluent air stream to temperatures greater than 750°C. (⇨ AFTER-BURNER)
5. Ozonation: the use of OZONE's very powerful oxidizing abilities will deal effectively with some odours such as hydrogen sulphide.

(⇨ ODOUR THRESHOLD)

**Offensive trades.** Under UK legislation offensive trades are subject to regulation by local authorities under the statutory provisions of the Public Health Acts. The processes involve animal products or residues (skin, bones, fat, blood, etc.) and are liable to give rise to odour nuisance.

**Offshore oil extraction, risks of.** The extraction of offshore oil has five danger spots for large or catastrophic oil pollution:

1. Spill from the seabed or wellhead.
2. Pipeline fracture from the seabed to the oil rig.
3. Pipeline fracture or leak from the rig to the shore or a mooring buoy.
4. Spillage at a mooring buoy when tankers couple up and uncouple.
5. Collision with wellheads or rigs.

The results can be a large OIL SLICK which has substantial effects on fisheries, sea birds and coastal amenities.

**Oil.** Our major source of fossil energy resources at the present moment. Roughly two-thirds of the world's known ultimately recoverable reserves are in the Persian Gulf area, and the total effective lifetime of world reserves will probably be 70 to 80 years using currently available technology. With the exception of additional resources in Russia and SE Asia it would appear that there is no geological justification for assuming that there will be further discoveries that will significantly alter the world picture, but the use of technological innovations in extracting more oil from the reservoirs may significantly extend existing oil field lifetimes. Exploration experts predict that total world oil production will peak in the year 2000 and decline slowly thereafter. The proven UK North Sea reserves represent only ca. 3 per cent of ultimately recoverable world reserves (see Figure 111). However, one forecast by the UK Offshore Operators Association predicts that up to 1.4 million barrels per day could be available from the North Sea in 2013 compared with 2 million barrels per day in 1989. The UK would, by then, be a net importer of oil but around 60–70 per cent of demand could still come from the North Sea. Existing North Sea discoveries total 20.8 billion barrels and an expected further 5.8 billion barrels will be found. The Association is quoted by the *Financial Times* as stating that 'actual performance will depend heavily on the final environment, including the degree to which smaller and more complex field development is encouraged by tax arrangements . . .'

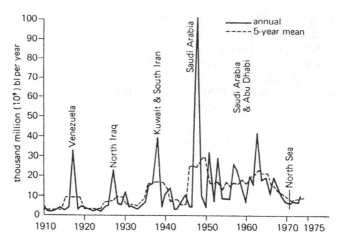

**Figure 111**  World oil discoveries (excuding Communist bloc). Note the importance of North Sea discoveries compared with Saudi Arabia.

Laherrere analysed global reserves and stated:

> We conclude that the ultimate range of oil producible below $25/b is 1,500–1,750–2,300 Gb (Gigabarrels) with the mid number as the mean (expected value). It furthermore looks as if production is set to decline naturally at about 2.7 per cent a year from around 2000. If successful efforts are made to delay the onset of the decline, it simply means that the subsequent slope becomes a cliff. Heavy oil, tar sand and enhanced recovery will become important after 2000, and will mitigate but not reverse the decline, such that by 2050 production will have fallen to what it was in the 1960s.
>
> It is too axiomatic to state that oil production has a huge impact on the political and economic development of the world, including especially transport and agriculture, for which there are no easy substitutes. The weakness of the world's reserve and production data, much held confidential by governments and state companies, is a serious obstacle to the analysis of this important subject which is needed if the world is to have time to adjust and if its leaders are to secure the political consensus for the measures they will be forced to adopt.

*Source:* J. Laherrere, 'World oil reserves – which number to believe?', *OPEC Bulletin*, 9–13, February 1995.

Tahmassebi has analysed future oil demand as tabulated below – it is clear that a crunch is coming in energy supplies.

| | Demand (million barrels per day) | | |
| --- | --- | --- | --- |
| | Low-growth Scenario | Base-case Scenario | High-growth Scenario |
| 1994 estimated | 67.8 | 67.8 | 68.1 |
| 1995 projected | 68.6 | 69.0 | 69.3 |
| 2000 projected | 73.2 | 75.3 | 77.3 |
| 2005 projected | 75.1 | 80.3 | 84.6 |
| 2010 projected | 78.9 | 86.1 | 93.9 |

*Source:* C. H. Tahmassebi, 'The changing structure of world oil markets and OPEC's financial needs', *OPEC Bulletin*, March 1995.

**Oil rigs, removal of.** Over the next 10 years ca. 50 N. sea oil rigs will come to the end of their useful lives and will be taken out of service.

The remainder will be decommissioned by 2080 when the oil runs out. The Brent Spar platform removal (1995) was a *cause célebre* in that environmentalist miscalculation of the platforms residual contents gave rise to unnecessary public concern. – We have been here before!

**Oil shales.** Large oil-bearing shale deposits in the Green River formation of Colorado. Wyoming and Utah have been estimated to contain 1800 thousand million barrels of oil – more than four times the crude oil discovered to date in the USA. However, only 6 per cent of the deposits is accessible, yielding more than 30 gallons of oil per tonne of rock.

Pilot projects have demonstrated the feasibility of recovering shale oil (kerogen) and converting it to crude oil in liquid form. Commercial extraction by surface retorting is now possible. (⇨ EXTRACTION OF OIL FROM SHALES)

Mining of the shale for commercial purposes must be carried out on a huge scale, however – in the region of 500 million tonnes per year. There is the problem of where to tip the spent shale, which is considerably greater in volume than the original rock. There is a danger of LEACHATE from the spent shale polluting watercourses.

One possible limitation on oil-shale production is thought to be the lack of sufficient water in the Colorado watershed as there are already large agricultural demands on the available water. The recovery of oil from oil shales on an industrial scale requires vast quantities of water for cooling and the effects on the immediate locality, whose economic structure depends largely on irrigation, are likely to be considerable.

**Oil slick.** A floating layer of oil on the seas, rivers, canals or lakes, usually as a result of accidental spillage from tanks, pipelines, etc. or deliberate, as in the case of discharge of water ballast. Slicks can cause major local ecological disasters, especially to sea birds, and, if allowed to drift ashore, will foul beaches.
Common methods of eliminating slicks are:

1. To contain the oil slick by floating booms, and then transfer it into tanks by suction.
2. The oil can be absorbed on nylon 'fur' and then squeezed out of the fur for re-use or disposal.
3. Detergents can be used to break up the slick. However, this method has been attacked on the grounds that it does not remove the oil and that the detergents contain toxic ingredients harmful to birds and marine life, particularly shellfish.

The catastrophic nature of oil spills was highlighted when the *Exxon Valdez* oil tanker discharged more than 11 million gallons of crude oil after running aground off the South Alaskan coast in Prince William Sound on 24 March 1989. A 3000 square mile slick which contaminated at least 700 miles of shoreline in an ecologically sensitive area was the result. Unpreparedness for such a disaster and the inability of physical cleaning methods meant that a massive detergent spraying programme was used. The clean-up operation took more than six months with incalculable effects on wildlife from both the oil pollution and the detergents. Rare species such as the humpbacked and killer whales are threatened from ingestion of oily water and eating polluted food.
This incident has highlighted that oil (or other large-scale natural resource extraction) operations can exact a substantial environmental toll. In this instance, the oversight of an oil tanker captain caused the *Exxon Valdez* disaster. The human factor can never be entirely eliminated and the greater the potential environmental insult, the greater the precautions and back-up systems required. (⇨ CHERNOBYL; VALDEZ)

**Oligotrophic.** An aquatic environment which has low concentrations of nutrients present and therefore has low plant and animal life productivity. The opposite of eutrophic. A lake that is oligotrophic, being poor in producing organic matter, is a characteristic of 'young' lakes and reservoirs. (⇨ EUTROPHICATION)

**Open-cast mining.** This is the process of surface mining in which large quantities of mineral-bearing rock are scooped out to produce, in effect, a very large hole, with terraced sides. The largest, in Bingham Canyon, Utah, is $1\frac{1}{2} \times 1\frac{2}{3}$ miles $\times \frac{1}{2}$ mile deep.

The two basic types of open-cast mining are quarrying, in which the rock is cut into large blocks and usually transported away from the site in this form, and metal-mining, where the large quantities of rock are ground to a powder on site and the metal-bearing minerals are removed.

In this way large quantities of poor ores may be processed in order to obtain relatively small amounts of the valuable minerals. This method also leaves a very large amount of useless ground-up rock.

The scale of some of these operations is illustrated by the Anaconda Company's mine at Twin Butte near Tuscon, Arizona, where 236 million tonnes of overburden and rock were removed to get to the low-grade copper ore 600–800 feet below ground. The ore itself has a copper content of 0.5 per cent, i.e. 200 tonnes ore are required to obtain 1 tonne copper, assuming 100 per cent recovery of copper. This leaves 199 tonnes of spoil for disposal. Lower and lower grades of ore cannot be worked as the energy required would escalate out of proportion to the material recovered. For example if (say) 0.1 per cent ore were mined, this would require a minimum of 1000 tonnes of ore for 1 tonne of copper and produce 999 tonnes of spoil. Clearly, there are limits.

**Operator and Pollution Risk Appraisal (OPRA).** A methodology for formal RISK assessment for processes subject to INTEGRATED POLLUTION CONTROL (IPC). OPRA has two components: Operator Performance Appraisal (OPA) which provides an appraisal of the probability of an incident; Pollution Hazard Appraisal (PHA) which appraises the consequences of the incident. This enables the risk to be evaluated ($\Rightarrow$ OPERATOR PERFORMANCE APPRAISAL (OPA), POLLUTION HAZARD APPRAISAL (PHA))

**Operator Performance Appraisal (OPA).** An OPA is carried out by Her Majesty's Inspectorate of Pollution (HMIP) for those processes subject to Integrated Pollution Control against the following indicators:

- Compliance with limits and adequacy of records.
- Knowledge of authorization requirements and implementation.
- Plant maintenance and operation.
- Management and training.

- Procedures and instructions.
- Frequency of incidents and justified complaints.
- Auditable environmental/management systems.

Each is graded on a scale from 1 to 5 (1 worst, 5 best) and then weighted to determine a final score.

It is part of the HMIP OPERATOR AND POLLUTION RISK APPRAISAL process.

HMIP gives the following example of its scale values 1 to 5:

Scale 1 – plant maintenance in operation, no preventive maintenance programme in place, the company relies only on breakdown maintenance, and haphazard procedures implemented by operators.

Scale 5 – a suitable maintenance programme is in place based on industry standards and/or manufacturers' recommendations. This is fully implemented and plant operations are clearly defined and followed.

**Ore.** Any noticeable concentration of a metalliferous mineral (whether or not it is of economic value), e.g. the metal content of the various ores differs widely; currently a workable iron ore contains 20–30 per cent iron, whereas an ore containing only 0.5 per cent copper may be worked profitably.

**Ore dressing.** The processing of raw material won by mining into a marketable form; the crushing, concentration and separation of the ORE mineral(s) from the waste residue. This may involve hand-picking, gravity concentration, magnetic separation, dense-media separation or chemical separation, e.g. separation of gold by the use of CYANIDE (and its attendant pollution risks).

**Organic matter.** Material containing CARBON combined with HYDROGEN often with other elements (OXYGEN, nitrogen), e.g. plastics, vegetable matter.

**Organochlorines.** A major class of chemicals emanating from the organic chemicals industry. It includes INSECTICIDES, AEROSOL PROPELLANTS, PCBs (POLYCHLORINATED BIPHENYLS), PVC, DDT and ENDOSULFAN. Organochlorines are characterized by persistence, mobility and high biological activity. They have very long HALF-LIVES: that for DDT is ten years or more, for DDE many decades.

All organochlorines have a high capacity to injure living systems and allied with their other attributes may possibly constitute the greatest threat to our life-supporting ECOSYSTEMS and associated biological cycles.

In many insecticide applications the organochlorines are being replaced by the ORGANOPHOSPHORUS COMPOUNDS and the new synthetic PYRETHROIDS are likely to be produced at rates which might rise to 20000 tonnes annually world-wide. (⇨ PESTICIDES; CHLORINATED HYDROCARBONS)

**Organophosphorus compounds.** A group of pesticides which embraces such names as azodrin, malathion, parathion, diazinon, trithion and phosdrin. All are chemicals related to nerve gas developed during the Second World War and block the central nervous system by inactivating the enzyme responsible for breaking down a nerve 'transmitter' chemical called acetylcholine. The result is hyperactivity resulting in death.

Organophosphorus compounds are generally very much more toxic to insects than mammals and also have a very much shorter HALF-LIFE than the ORGANOCHLORINES. For these reasons they are labelled as 'safe' insecticides, although they may be CARCINOGENIC if precautions are not taken during use. However, the long-term ecological effects of these chemicals is not known. They have only been in use for one generation which is insignificant on the evolutionary time-scale.

In common with most similar products, these chemicals buy time which must be put to use for the stabilization of populations and resource consumption and planned ECOSYSTEM management. (⇨ PESTICIDES)

**Orimulsion.** A bituminous fuel produced by the Venezuelan state oil company. It comprises 70 per cent bitumen and 30 per cent water with a claimed energy value 9 per cent higher than power station grade coal and can be burned in modified oil-fired boilers. Sulphur dioxide emissions when burning Orimulsion are about 2300 parts per million volume, compared with 1800 parts per million from heavy fuel oil, and typically about 1700 when using coal mined in Britain.

FLUE GAS DESULPHURIZATION will be required and is planned to be fitted out at the 2000 MW Pembroke oil-fired power station which plans to burn between 4 m and 5 m tonnes of the fuel a year from 1998. This will remove 94 per cent of the SULPHUR DIOXIDE emissions.

The suppliers of this fuel are to build a facility in Germany to recover VANADIUM, magnesium and NICKEL from the power station ash which contains 19 per cent vanadium, 13 per cent magnesium and 2.3 per cent nickel. Claimed recovery rates are 99.8 per cent, 92 per cent and 91 per cent, respectively.

*Source: Chemical Engineer,* April 1995, p. 6.

**Overband magnet.** An electromagnet which is positioned crosswise above a conveyor carrying solid waste or incinerated residue in order to recover ferrous metals.

**Oxidation pond.** A basin used for retention of wastewater before final disposal in which biological oxidation of organic material is effected by natural or artificially accelerated transfer of OXYGEN to the water from air.

**Oxide fuel.** Nuclear fuels manufactured from the oxides of the fissile material. They can withstand much higher temperatures and are much less chemically reactive than metals.

**Oxidizing agent.** Chemical which either gives up oxygen in chemical reactions or supplies an equivalent element such as CHLORINE to combine with a REDUCING AGENT. Term also used for the removal of hydrogen from a substance. Atmospheric oxidizing agents include OZONE and NITROGEN DIOXIDE.

**Oxygen (O).** A colourless, odourless gas essential both for all aerobic forms of life and for combustors. It forms 21 per cent by volume of the atmosphere. Chemical symbol O, formula $O_2$ (molecular weight 320). Unstable form OZONE, formula $O_3$.

**Oxygen cycle.** Oxygen is a major component of all living matter and is vital in the free state for the higher animals which require it in their metabolism. Its presence on earth is almost certainly due to the process of PHOTOSYNTHESIS in plants which is the assimilation of carbon dioxide and water for the production of carbohydrates and free oxygen. This free oxygen both supports and comes from life.

The emergence of free oxygen has its origin around 3000 million years ago when simple AUTOTROPHIC ORGANISMS evolved which were able to split water and release oxygen. (There is geological evidence of oxidized sediments around 1500 million years old in the form of ferric compounds, the oxidized form of ferrous rocks.)

With the emergence of free molecular oxygen ($O_2$), the sun's energy split the oxygen molecules into atomic oxygen (O), which is highly reactive and forms ozone ($O_3$) which has built up into the OZONE SHIELD. Thus the earth's atmosphere began to evolve and stabilize. However, oxygen produced by photosynthesis is used up by respiration either by the consumers in the FOOD CHAIN or the decomposers. So, if the oxygen is recycled in this manner, how did atmospheric concentration build up? ($\Rightarrow$ CARBON CYCLE) The answer lies in the carbonate and carbonaceous sediments; calcium carbonate (limestone) is an example of the former and coal an example of the latter. The sediments formed by the deposits of animal and plant bodies removed carbon from the carbon cycle and tied it up for geological time. For every atom of carbon laid down to sediment, two atoms of oxygen are left free. Thus, the bank of free oxygen that we have in the atmosphere was made possible by the formation of carbonaceous sediments. As the sediments were being deposited, the atmosphere evolved, the ozone shield grew, and we now have a stable atmosphere. The biosphere and the atmosphere evolved simultaneously due to the carbon and oxygen cycles operating together.

If photosynthesis were to stop tomorrow the reservoir of oxygen would be sufficient to sustain higher life for millions of years without significant depletion.

The carbonaceous sediments number many thousand million (US billion) tonnes, of which all the coal and oil likely to be used by man is much less than 1 per cent. Therefore, the combustion of coal and oil, while leading to an increase in carbon dioxide content, will not significantly affect the oxygen content, although it may through the GREENHOUSE EFFECT, alter the earth's radiation balance and climate.

The indivisibility of the biosphere processes is clearly illustrated in the oxygen cycle. We should take care not to abuse that which we do not understand and are unable to control.

**Oxygen deficit.** ⇨ DISSOLVED OXYGEN.

**Oxygen demand.** The oxygen demand of liquid waste effluent is of crucial importance and two measures are commonly made.

1. BIOCHEMICAL OXYGEN DEMAND (BOD) is a measurement of the oxygen required by the microbes which reduce the wastes to simple compounds, as in SEWAGE TREATMENT. The BOD is a standard test of the amount of oxygen required by a sample of effluent over five days at $20°C$ (or 7 days for 'difficult' effluents) and is stated as the parts per million of oxygen (milligrams per litre) taken up by the sample of effluent incubated in the dark. This measurement is denoted $BOD_5$ for 5-day tests and $BOD_7$ for 7-day tests.

2. CHEMICAL OXYGEN DEMAND (COD) measures the number of parts per million of oxygen taken up by a sample from a solution of boiling potassium dichromate in two hours. The BOD and COD tests differentiate between materials that can be oxidized biologically and those that cannot, and indicate what types of treatment will be required.

It should be pointed out that the above criteria are not in themselves adequate indicators of pollution. There are extremely toxic solutions of CYANIDE, for example, that have acceptable BOD and COD values. ($\Rightarrow$ DISSOLVED OXYGEN)

**Oxygen sag (curve).** The decline and subsequent rise in percentage saturation of dissolved oxygen in a river downstream from a discharge of effluent containing biodegradable material. A graph of the percentage dissolved oxygen versus distance shows a characteristic dip or sag, and if re-aeration or recovery takes place, the dissolved oxygen content rises again (see Figure 43). The extent of the recovery is a function of the river length, biotic content, BOD loading of the effluent and initial dissolved oxygen of the receiving stream. ($\Rightarrow$ DISSOLVED OXYGEN)

**Ozone ($O_3$).** Ozone contains three atoms of oxygen (O), whereas the atmospheric oxygen molecules contain two atoms ($O_2$). Ozone is formed naturally in the upper atmosphere (15–39 kilometres) by the action of the sun's ultraviolet rays (and also by lightning). This splits $O_2$ into single oxygen atoms which combine with other $O_2$ molecules to form $O_3$. The 'ozone layer' actually consists of a very few molecules of ozone per million molecules of air. The ultraviolet rays both destroy ozone and help create it. This cycle which absorbs much of the harmful ultraviolet radiation has been greatly upset by mass use of CHLOROFLUOROCARBONS.

Ozone is removed by a series of chain reactions involving trace quantities of molecular and RADICAL species. The removal cycles have the general form:

$$O_3 + X \rightarrow O_x + O_2$$
$$XO + O \rightarrow X + O_2$$

where X is OH, NO, Cl or Br. The regeneration of the original reactant, X, accounts for the remarkable efficiency of minute quantities of trace gases in destroying ozone, and the ozone concentration at the parts per million level is moderated by trace gases at the parts per billion level (that is, parts in $10^9$ by volume).

Ozone concentrations in the stratosphere are naturally moderated in this manner such that 75 per cent of ozone is destroyed by the $NO/NO_2$ removal cycle (X = NO). At present about 2 per cent of ozone is destroyed by the Cl/ClO cycle (X = Cl) throughout the atmosphere, caused mostly by chlorine atoms which have been photochemically released from molecules such as chlorofluorocarbons (CFCs) like CFC11 ($CFCl_3$) and $CFCl_2$ ($CF_2Cl_2$).

$$CFCl_3 \rightarrow CFCl_2 + Cl$$
$$CF_2Cl_2 \rightarrow CFC_2Cl + Cl$$

Molecules of chlorine will catalyse the destruction of large numbers of ozone molecules:

$$[Cl] + O_3 = [ClO] + O_2$$
$$[ClO] + [O] \rightarrow Cl_2 + O_2$$

The chlorine atoms are extremely reactive and it has been estimated that each chlorine atom will destroy about $10^5$ molecules of ozone before being removed from the atmosphere by reaction with methane to form hydrochloric acid, which ultimately is precipitated in rain:

$$Cl + CH_4 \rightarrow HCl + CH_3$$

In contrast, inorganic forms of chlorine released at the earth's surface are removed by rain before they can enter the atmosphere.

Ozone is used as an oxidizing agent, in water treatment for example. It is also produced by photochemical reactions involving hydrocarbons from car exhausts and NITROGEN OXIDES where it becomes a dangerous irritant to eyes, throat and lungs. It can be formed in PHOTOCHEMICAL SMOG.

Individuals vary considerably in their response to ozone. Those who are sensitive experience temporary breathing difficulties if they take vigorous outdoor exercise when ozone concentrations are at or above about $160 \, \mu g/m^3$; this level is often exceeded during hot summers, especially in southern England. In terms of lung function, people who suffer from asthma or other respiratory disorders are not more likely to be sensitive to ozone than other members of the population, although laboratory studies have shown that ozone may produce an enhanced inflammatory response in the airways of asthmatics.

During July 1994, periods of hot weather gave rise to prolonged, elevated levels of ozone across Europe. On a number of days the levels recorded at national monitoring sites exceeded the lower bound, and often the upper bound, of the WHO health-based guideline, which is $150–300 \, \mu g/m^3$ averaged over an hour. In many cases, they entered the 'poor' air quality band defined by DOE ($180–358 \, \mu g/m^3$). The peak hourly levels recorded on those days are shown below.

| Date | Place recorded | Levels of ozone ($\mu g/m^3$) |
|------|----------------|-------------------------------|
| Friday, 1 July | Lullington Heath (Sussex coast) | 190 |
| Saturday, 2 July | London (Bridge Place) | 190 |
| Monday, 11 July | Lullington Heath | 204 |
| Tuesday, 12 July | Sibton (Suffolk) | 238 |
| Friday, 22 July | Lullington Heath | 206 |
| Saturday, 23 July | Harwell | 204 |
| Sunday, 24 July | Lullington Heath | 218 |

RCEP 18th Report using data supplied by the National Environmental Technology Centre, Harwell; peak levels were converted from the ppb values supplied using a factor of 1 ppb = $2\,\mu g/m^3$.
*Source:* S. Penkett, 'The changing atmosphere', *Chemical Engineer,* August 1989.

**Ozone depletion potential (ODP).** Measure of the potential for depletion of the ozone layer, e.g. most CHLOROFLUOROCARBONS have an ODP of 1, whereas HALONS can range from 3 to 10. The ODP value can vary as the bromine-based halons react synergistically (⇨ SYNERGISM) with chlorine and therefore any increase in free chlorine levels in the ozone layer will increase the ODP of heavy halons released. (⇨ CHLOROFLUOROCARBONS)

**Ozone shield.** A layer of OZONE surrounding the earth formed by ultraviolet radiation which splits molecular oxygen ($O_2$) to two atoms of oxygen which is highly reactive and forms ozone ($O_3$). The ozone shield acts as a barrier to the radiation and protects the BIOSPHERE. The maximum concentration is found between 15 and 30 kilometres from the earth's surface. If the shield is reduced this may increase the incidence of radiation-induced skin cancer. Recent attention has also focused on nitric oxide (NO) which also attacks the layer. (⇨ OXYGEN CYCLE; NITROGEN OXIDES)

# *P*

**Packaging, definitions.**
*DSD Commercial Recycling Objectives.* Objectives of the German DSD Commercial Packaging Decree:

- A radical reduction in volume of packaging by avoiding use and stressing the value of reclamation.
- Manufacturers and retailers will be obliged to take responsibility for the packaging used as they are the originators.
- The local communities will be relieved from this part of waste disposal responsibility.
- Recycling will be given absolute priority over thermal recovery.

*Group packaging or secondary packaging.* Any packaging conceived so as to constitute at the point of purchase a grouping of a certain number of sales units whether the latter is sold as such to the final user or consumer or whether it serves only as a means to replenish the shelves at the point of sale; it can be removed from the product without affecting its characteristics.

*Economic operators.* In relation to packaging this means suppliers of packaging materials, packaging producers and converters, fillers and users, importers, traders and distributors, authorities and statutory organizations affected by the processing of packaging.

*Non-returnable packaging.* Any packaging for which no specific provisions for its return from the consumer or final user has been established.

*One-way packaging.* Any packaging not being used more than once for the same purpose.

*Packaging.* All products made of any materials of any nature to be used for the containment, protection, handling, delivery and presentation of goods, from raw materials to processed goods, from the producer to the user or the consumer. Non-returnable items used for the same purposes shall be considered to constitute packaging.

*Packaging waste.* Any packaging or packaging material covered by the definition of waste in Council Directive 75/442/EEC.

*Packaging waste management.* The management of waste as defined in Directive 75/442/EC.

*Prevention.* The reduction of the quantity and/or the harmfulness of materials used, packaging and packaging waste at production processes level and at the marketing, distribution, utilization and elimination stages, in particular by developing 'clean' products and technology.

*Recovery.* Any of the applicable operations provided for in Annex II.B to Directive 75/442/EEC.

*Recycling.* The recovery of the waste materials for the original purpose or for other purposes excluding energy recovery; recycling means also composting, regeneration and biomethanization.

*Waste disposal.* The collection, sorting, transport and treatment of packaging waste as well as its storage and tipping above or under ground, the transformation operations necessary for its re-use, recovery or recycling.

*Returnable packaging.* Any packaging whose return from the consumer or final user is assured by specific means (separate collection, deposits, etc.), independently of its final destination, in order to be reused, recovered, or subjected to specific waste management operations.

*Reusable packaging.* Any packaging which has been conceived and designed to accomplish within its life cycle a minimum number of trips or rotations in order to be refilled or reused for the same purpose for which it was conceived; with or without the support of auxiliary products present on the market enabling the packaging to be refilled, such packaging will become packaging waste when no longer subject to reuse.

*Sales packaging or primary packaging.* Any packaging conceived so as to constitute a sales unit to the final user or consumer at the point of purchase.

*Transport packaging or tertiary packaging.* Any packaging conceived so as to facilitate handling and transport of a number of sales units or grouped packagings in order to prevent physical handling and transport damage.

*Used packaging.* The packaging itself left over once it has been emptied or the product has been unpacked.

*Valpak.* The Producer Responsibility Group's name for the industry organization which will implement packaging recycling. Valpak will pay the difference between the market price of the collected materials and the costs incurred in their collection. The aim is to provide reasonably stable prices and market confidence. (⇨ VALPAK)

*Waste.* Any substance or object which the holder disposes of or is required to dispose of pursuant to the provisions of national law in force.

**Packaging-derived fuel (PDF).** PDF is produced from waste which has been 'source separated' in the household. It has a CALORIFIC VALUE of 80 per cent that of coal; while contributing just 20 per cent to the weight of municipal solid waste, packaging contains 40 per cent of the energy.

Such fuel is usually combusted as a sole fuel in dedicated plants. However, it is also possible to co-combust up to 30 per cent of it with coal in conventional facilities.

*Source:* European Energy from Waste Coalition.

**Packaging materials (UK and German statistics).**

Packaging material use in 1991 – UK and Germany

| | UK | | Germany | |
|---|---|---|---|---|
| Material | Million tonnes | % of total | Million tonnes | % of total |
| Paper/board | 3.51 | 46.8 | 5.55 | 43.6 |
| Glass | 1.77 | 23.6 | 4.81 | 37.8 |
| Plastic | 1.41 | 18.8 | 1.51 | 11.9 |
| Tinplate | 0.75 | 10.0 | 0.75 | 5.9 |
| Aluminium | 0.06 | 0.8 | 0.11 | 0.8 |
| Total | 7.50 | | 12.73 | |

Evidence submitted by the British Plastics Federation (based on work by the UK Warren Springs Laboratory) to the House of Commons Environment Committee on RECYCLING of plastics showed that plastics could account for 10.26 per cent by weight of household waste.

Plastics as a proportion of Household Waste

| Plastics source | Percentage of household waste |
| --- | --- |
| *Plastic Film* | |
| Refuse Sacks | 1.1 |
| Other Plastics Film | 4.18 |
| Total | 5.3 |
| *Dense Plastics* | |
| Clean Beverage Bottles | 0.63 ⎤ Recyclable |
| Coloured Beverage Bottles | 0.12 ⎬ portion |
| Other Plastics Bottles | 0.12 ⎦ |
| Food Packaging | 1.9 |
| Other Dense Plastics | 2.14 |
| Total | 4.92 |

This is a point well made in the study recently published by the Packaging and Industrial Films Association, Flexible Packaging Association, Oriented Polypropylene Film Manufacturers Association, *The Management of Waste Plastic Packaging Films* (1993). This demonstrated that for plastic film waste arising at the domestic level the most sound method of disposal was as a high calorific component of the refuse fraction sent to fuels at an incinerator. To merely set a material recovery target whilst ignoring the degree of contamination and, therefore, the major problems (in both cost and quality) that can be incurred in re-processing would be to impose unnecessary burdens on industry and the environment. This is especially so when an environmentally sound and financially viable option, energy recovery, exists to effect VALORIZATION as well. This point was accepted by the House of Commons Environment Committee in its Second Report on recycling when it stated 'We agree that energy recovery is appropriate for parts of the household waste plastics stream'. It is clear then that individual material appraisals need to be made.

*Sources:* Packaging and Industrial Films Association, Oriented Polypropylene Film Manufacturers Association and the Flexible Packaging Association, *Management of Waste and Plastic Packaging Films*, January 1993.

The Consortium of the Packaging Chain (COPAC), *Action Plan to Address UK Integrated Solid Waste Management*, 26 October 1992.
German Federal Bulletin, 'Survey of Sales Packaging Consumption 1991', 27 August 1992.
House of Commons Environment Committee, Recycling Report, July 1994.

*Packaging Recovery Targets (UK)*

Value recovery of UK packaging waste to year 2000

|  | Quantities by year ( × 1000 tonnes) | |
| --- | --- | --- |
|  | 1993 | 2000 |
| Total packaging waste | 7292 | 8051 |
| Domestic | 3600 | 3757 |
| Commercial and Industrial | 3692 | 4294 |
| Total quantity recycled | 2199 | 4003 |
| From domestic sources | 513 | 1306 |
| From commercial and industrial sources | 1686 | 2697 |
| Total recovered by waste to energy | 150 | 650 |
| Total diverted from landfill | 2349 | 4653 |
| Value recovery (%) | 32 | 58 |
| Residual packaging waste to landfill | 4943 | 3398 |

*Source: Report on Public Consultation*, Producer Responsibility Group, Committee Response to the PRG 'Real Value from Packaging Waste', June 1994.

A comparison is provided by the summary in the following table.

**Packaging and Packaging Waste EC Directive.** The Directive on Packaging and Packaging Waste proposed by the European Commission on 15 July 1992 has been adopted (December 1994). The Directive aims to harmonize national measures concerning the management of packaging and packaging waste. This is the first step in a long term process which will increase convergence gradually.

The relevant features are:

*Scope:* the directive covers all packaging placed on the market in the Community and all packaging waste, regardless of the materials used.

*Targets:* specific articles are included on preventative measures and re-use systems, and sets of quantitative targets for recovery

Packaging legislation summary for selected EU countries

| | Germany | France | Netherlands | UK | EC |
|---|---|---|---|---|---|
| Status | Regulation approved June 1991 | Decree April 1992. Effective 1 Jan. 1993 | Signed Covenant | Industry's PRG report published 8 Feb. 1994 for comment. Response May 1994. Funding Report July 1994. | 'Common Position' 15 Dec. 1993. 19 amendments voted. 4 May 94 make no major changes. |
| Scope | Primary (P), Secondary (S), Transit (T) Packaging | Packaging in household waste | All packaging placed on market | All packaging waste | All packaging and packaging waste |
| Collection responsibility | (P) Distributor of '3rd part'. (S) Distributor | Local authorities – first scheme to commence end 1993 | Local authorities/separate collection | Local authorities | Member State decision |
| Material take back responsibility | P & T: Manufacturers and Distributors; (S) Distributors | Producers, importers and distributors | Packaging chain | Industry, where applicable | Member State decision |
| Recovery targets | By 11.1.98: Glass 70%; Tinplate 70%; Aluminium 78%; Paper & board 60%; Plastics 60%; Composites 60% | Objective 75% valorization (materials recovery, electricity, steam, compost) by 2003. Annual progress to be reported 2002 | By 2000: 60% recycle min. 40% incineration max. | 58% (between 50–75%) by year 2000 | 5 year targets: recovery 50–65%; recycling 25–45%; with each material min 15%. Derogations up & down permitted. Landfill last resort |
| Landfill targets | Restrict to inorganics and insolubles only | Main aim to limit landfill disposal | Zero by 2000 for packaging waste | ⅓ less packaging waste to landfill | |
| Permitted uses for used packaging | Reuse. Recycle. 'Chemical Recycling' for remainder. Composting. Limited energy recovery | Reuse. Recycle. Energy Recovery. Compost | Reuse. Recycle. Energy Recovery. Compost | Reuse. Recycle. Compost. Energy Recovery. Landfill | Reuse. Recycle. Energy Recovery. Compost + biomethanization |

| | | | | | |
|---|---|---|---|---|---|
| Funding | From 1.10.93 new fees Dm kg: glass 0.2, alu 1.00, paper 0.33, composites 1.66, tin plate 0.56, natural materials 0.20, plastics 3.0 (inc. recycling costs) | According to pack size e.g. 201–3000cc 1c. alternative option to industry plastics 50c per kg. Gas 5 c per kg. Steel 10 c per kg. Aluminium 50 c per kg | Levy under discussion | Annual funding of £100m believed necessary. Mechanism for collecting this not yet agreed. Govt. asked to provide necessary legislation. To stop 'freeloading' possibly converter levy | Economic instruments permitted |
| Scheme management | Industry (P) DSD (T) RESY and others | (a) Eco Emballages S.A. (b) Adelphe set up for wine & spirit bottles | KAPS- (KRINGLOOP ACTIE PLAN SVM). Glass, compostibles and wastepaper excluded | An industry board for 'Valpak'. Plus sector materials organizations. (MOs) | Member State decision |
| Identification | (P) Green dot. (T) Resy symbol + others | Green dot adopted | | | Materials marking to be specified |
| Timetable | Transit 1/12/9 - Secondary 1/04/92. Primary 1/01/93 | Decree effective 1/01/93. 5m people in 37 schemes now covered | Progressive annual action plans. The 1992 review now available | 1994 | At earliest 1996 |
| Specific items | – DSD in finance problems, cash injections from Waste Management Industry. – Separate regulations in draft for packaging hazardous products and or marking of packaging | – Local authorities to be reimbursed from fund for costs of sorting. – Landfill tax on household and non-toxic waste. (20 FF/tonne). – New draft decree on industrial packaging being finalized – waste holder to be responsible for valorization. | – Weight of packaging in market in 2000 to be below 1986 weight. – Replace materials which cannot be recycled. – New laws may replace Covenant. | Landfill Levy under Discussion. Basis for 'Valpak' now being worked out. 85% of households to have access to a collection system. | 'Small packaging; large packaging materials to be defined |

*Source:* Based on format (data modified) presented in *Environment Matters*, No. 24, published by Courtaulds, May 1994. Updated by A. Porteous in *PIFA International Report*, March 1995.

and recycling of packaging waste. The present targets which are to be reached within five years from the implementation date are: recovery – between 50 per cent and 65 per cent of the packaging waste; recycling – between 24 per cent and 45 per cent of the totality of packaging materials with a minimum of 15 per cent recycling for each individual material.

Not later than 10 years from the implementation date the targets shall be revised with a view to substantially increasing them. Member States are allowed to set programmes going beyond these targets under the conditions that their policies do not create obstacles for the setting up of similar policies in other Member States.

It is up to the Member States to take necessary measures to establish specific return, collection and recovery systems in order to reach the objectives of the Directive. In compliance with the principle of subsidiary, Member States are free to develop their own waste management schemes which have to be in conformity with the Treaty. Harmonized national databases have to be established to ensure a monitoring mechanism for the implementation of the objectives set out in the Directive.

The Directive lays out an important number of areas for standardization, regarding the essential requirements on the composition of re-usable and recoverable, including recyclable packaging. To obtain the objectives the Directive calls for all parties involved – consumers, industry and authorities – to co-operate in the spirit of shared responsibility. To this end the Member States shall ensure that users of packaging obtain the necessary information.

*Source: CBI Environment Newsletter*, Number 16, January 1995.

**PAN.** ⇨ PEROXYACETYLNITRATE.

**Paper.** A matrix of CELLULOSE fibres usually free of non-cellulosic materials. UK (1989) consumption is 9 million tonnes per annum, of which 4.25 million tonnes are produced nationally. The amount of recycled fibres is ca. 50 per cent mainly from specialist suppliers who reclaim the better grades of paper for repulping. The main classes are:

*Newsprint:* made from mechanically ground wood PULP (which contains short cellulose fibres and some LIGNIN) and recycled newspapers. Up to 40 per cent of newsprint is recycled and this factor is expected to increase, provided market stabilization for reclaimed newspapers is established. Currently most recycled newsprint comes from printing plants, newsagents or is imported. Each tonne of recycled newspaper saves around 15 trees. However, the pulp-producing countries claim that trees are planted faster than they are consumed. This neglects the fact that recycling paper saves both water and energy, cf. virgin pulp manufacture. Although regulations differ, most US states and Canadian provinces are committed to phasing in a recycled content in newsprint of 40–60 per cent, availability permitting. (⇨ DE-INKING; RECYCLING)

*Kraft:* a strong brown paper with long cellulose fibres, usually made from a sulphate pulp. It is used for packaging and wrapping. It can be readily recycled.

*Printing and writing paper:* usually bleached, with a fine texture and containing special fillers for ink absorption. Recycling is usually easily accomplished.

*Speciality:* this embraces papers having a high strength when wet, papers treated with resins, paper towels, papers for photographic emulsion, etc. Recycling is often difficult as resin types vary; therefore separation at source is essential.

With justification recycling paper can save water and, it is claimed, energy, compared to pulp manufacture. But as pulp manufacturers use renewable energy (forest and plant residues) and recycling consumes finite energy in the UK, it should be viewed as a cheap feedstock option for the paper industry which can make economic sense for the paper manufacturers.

However, if the wastepaper collection incurs a net change to the public, it is legitimate to ask why it should be subsidized when there are other pressing needs for public monies. (⇨ ENERGY ANALYSIS)

As an example of corporate consciousness in this area, British Telecommunication's (BT) consumption of paper in 1993/94 was approximately 76 000 tonnes. This represents 1.6 per cent of UK newsprint, printing and writing paper consumption.

BT is keen to increase the quantity of recycled paper in its directories to save waste going to LANDFILL and to encourage the growth of recycling. All Phone Books and Yellow Pages now have a minimum recycled fibre content of 20 per cent.

Over 3.6 million Phone Books were produced on paper with 65 per cent recycled fibre content. The minimum recycled fibre

Quantities of paper used in the production of BT Directories

|                              | 1991 (tonnes) | 1992 (tonnes) | 1993–94 (tonnes) |
| ---------------------------- | ------------- | ------------- | ---------------- |
| Yellow Pages                 |               |               |                  |
| 0% recycled fibre content    | 37345         | 30340         | 0                |
| ⩾20% recycled fibre content  | 0             | 6000          | 35656            |
| ⩾65% recycled fibre content  | 0             | 0             | 0                |
| Total                        | 37345         | 36340         | 35656            |
| Phone Books                  |               |               |                  |
| 0% recycled fibre content    | 24196         | 22842         | 0                |
| ⩾20% recycled fibre content  | 0             | 0             | 19726            |
| ⩾65% recycled fibre content  | 0             | 4492          | 2194             |
| Total                        | 24196         | 27334         | 21920            |

content in Phone Books was increased to 23.5 per cent from April 1994 but an increase to 25 per cent was not achieved. Other papers are being investigated, but difficulties in finding a suitable alternative have been encountered.

*Source: BT and the Environment*, Environmental Performance Report 1994.

**Paraffin.** UK name for high grade kerosene (a medium light distillate produced in oil refining) used as a space heating FUEL. Also, the trivial name for 'alkanes' or saturated aliphatic hydrocarbons such as methane, ethane, propane and butane.

**Parasites.** Organisms that live attached to or in living organisms. They gain food and often shelter but the host gains nothing and usually suffers as a result.

**Parathion.** ⇨ ORGANOPHOSPHORUS COMPOUNDS.

**Particle, fundamental or elementary.** This term, used in nuclear physics, refers to any particle of matter that is not composed of simpler units. The electron, proton and neutron, the main components of any atom, were the first to be discovered and researched. However, since then, others have been identified, such as the positron, the neutrino, mesons and hyperons.

Electrons, protons, positrons and neutrinos are stable; the remainder, mesons, hyperons and neutrons, decay spontaneously into fundamental particles of lower mass, accompanied by the liberation of energy. Neutrons will only decay spontaneously when isolated from the atomic nucleus. Mesons and hyperons, of which there are many different types with different properties, have mean half-lives of less than 0.000 003 seconds.

**Particulate pollutants, control.** ⇨ BAG FILTER; ELECTROSTATIC PRECIPITATOR; CYCLONE DUST SEPARATOR; THRESHOLD LIMITING VALUE; THREE-MINUTE MEAN CONCENTRATION.

**Particulates.** Fine solids or liquid droplets suspended in the air. The solids often provide extended surfaces due to their irregularities and therefore other pollutants can be carried along; for example, smoke particles and sulphur dioxide have greater effects on health when in combination than when emitted separately. It is postulated that the smoke particles, especially the PM10s which are less than 10 metres in size, are carried deeper into the respiratory tract with the sulphur dioxide 'attached' (or adsorbed) and thus the medical effects are compounded. (⇨ SYNERGISM)

The term particulates as used in air pollution includes all the separate terms: GRIT, DUST, FUME, AEROSOL, SMOKE, etc.

**Pathogen.** A living organism (usually a micro-organism) that causes disease.

**Payback.** The ratio of the annual income or savings derived from a project to its capital cost. Where a project has a payback of less than 5 years might receive consideration in low inflationary times for many industries less than 3 years is considered desirable.

**PCBs (Polychlorinated biphenyls).** Chlorinated hydrocarbons formerly used as plasticizers and in transformer-cooling oils to enhance flame retardance and insulating properties. They are highly persistent bioaccumulative pollutants found worldwide. Their use was banned in 1979 by law, but many old electrical components such as transformers and capacitators are around which means that PCBs will require specialist disposal by INCINERATION for many years to come. Many Third World countries also have substantial quantities of PCB-filled electrical equipment in use which is coming to the end of its useful life with no facilities for their environmentally effective disposal. The UK has stated at the *Third North Sea Conference* 1990, that it will call in and destroy all the UK's remaining stocks of PCBs by the year 2000.

Great concern has been expressed over PCBs being disposed of in landfill sites. Where this has been done, 'eternal' vigilance would appear to be required unless the PCB contaminated fill is extracted and thermally treated. Current UK guidelines allow the landfill of materials containing less than 20 ppm. However, one major PCB waste disposal site in the UK has been found to be leaking and in order to regularize the position (1989) permission has been granted for reburial of the wastes if they contain no more than 50 ppm. (J. Erlichmann, 'B.I.C.C withhold test results after toxic waste site leak', *Guardian*, 31 August 1989).

PCBs are present in sea-water such as the Clyde Estuary, where in 1969, concentrations of less than 0.01 micrograms per litre were found. Concentration in mussels was between 10 and 200 micrograms per litre, which gives a concentration factor of between 1000 and 20 000. They have been blamed for the death of seabirds in times of stress. PCBs have been found in the tissues of Arctic seals (1989) and maternal milk, and are showing a remarkable longevity in the environment due to their stability. Not only that, but concentrations are rising which perhaps poses serious threats for communities where fish is a major part of the diet. Also seals, walruses, etc. which are at the end of a FOOD CHAIN are at risk. PCBs and other persistent chemicals were thought to have impaired the immune system of the common seals in the North Sea, 70 per cent of whom fell prey to a distemper virus in 1988. This is a classic example of a man-made compound being introduced without assessment of its environmental impact.

**Pelleted fuel.** A form of WASTE DERIVED FUEL in which the shredded waste (often DOMESTIC REFUSE) is compressed into solid fuel pellets by means of a press wheel. Figure 112 shows the flow sheet for a WDF pellet plant, processing 18 t/h input of domestic waste. The fuel properties of UK pellets made from domestic refuse, shredded waste derived fuel and (UK) coal are given in the table below.

Comparison of the properties of a waste derived fuel (WDF) and coal, Warren Spring Laboratory (WSL), 1977

| Fuel | Calorific value as received (MJ/kg) | Moisture | Ash (%) | Volatile matter (%) |
|------|------|------|------|------|
| WSL Pellets ex Doncaster refuse | 15.7 (mean) | 16.8 (mean) | 14.6 (mean) | 64 (mean) |
| Shredded WDF produced in USA | 11.2 (mean) | 26 (mean) | 27 (mean) | |
| Typical UK household coal | 30 | 6 | 6 | 34 |

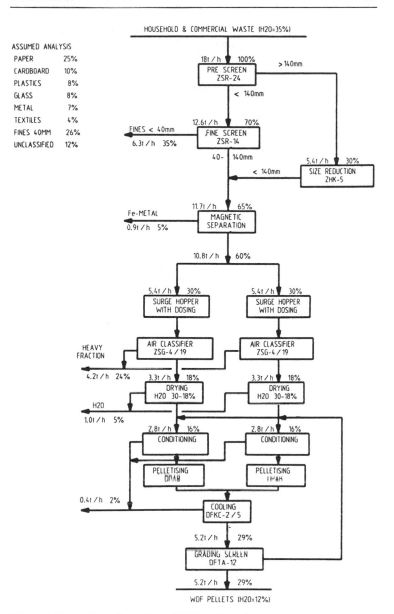

**Figure 112** A flow sheet for a WDF pellet plant.

**Pentosans $(C_5H_8O_4)_x$.** Polysaccharides present with cellulose in plant tissue including straw and sawdust. They can be utilized by HYDROLYSIS which converts them to pentoses (general formula $C_5H_{10}O_5$) for the production of FURFURAL.

**Percentile.** When a set of measurements is made the difference between the smallest (lowest) value and the largest (highest) is the range in which there are $N$ observations. Within the range the observations may be grouped, e.g. into 100 groups to obtain percentile values. As an illustration of use: the top 1% of the observed values will all be greater than the concentration represented by the 99th percentile; similarly 10% will be greater than the 90th percentile.

**Percolating filter.** ⇨ SEWAGE TREATMENT.

**Percolation.** The movement of GROUNDWATER through an AQUIFER under the influence and direction of the HYDRAULIC GRADIENT. (⇨ INFILTRATION)

**Periodic table.** All elements can be classified in terms of the periodic table, which arranges them in order of ATOMIC NUMBER (number of PROTONS in the NUCLEUS) from left to right. Most naturally occurring elements up to URANIUM exist as mixtures of ISOTOPES (⇨ RADIOACTIVE ISOTOPE), so their ATOMIC MASS is a non-integral number. Man-made elements beyond uranium are highly radioactive and toxic (⇨ PLUTONIUM). The vertical columns in the periodic table are known as Groups and are often named for their striking similarity in physical and chemical properties:

Group I:     Li, Na, K, Rb, Cs – alkali metals
Group II:    Be, Mg, Ca, Sr, Ba – alkaline earth metals
Group VI:    O, S, Se – chalcogenides
Group VII:   F, Cl, Br, I – halogens
Group 0:     He, Ne, Ar, Kr, Xe – noble gases

The elements can also be classified in terms of their properties. Most of the elements are metallic (⇨ METALS) and lie to the left-hand side of the table. The unreactive noble gases lie at the extreme right-hand side of the table, immediately adjacent to the non-metals. These elements are generally poor conductors of heat and electricity and usually occur as compounds. An intermediate group of elements (B, Si, Ge, As, Sb, Te) show intermediate physical properties and are hence termed semi-metals or semiconductors. (⇨ APPENDIX II)

**Permanent hardness.** ⇨ HARDNESS.

**Permanganate value (PV).** Used as a rapid effluent treatment test. It is the amount of oxygen absorbed from a standard potassium permanganate solution during 4 hours at 27°C, and is a crude measure of the oxygen demand of an effluent.

**Permeability.** ⇨ HYDRAULIC CONDUCTIVITY.

**Permeation tube.** Used for calibrating instruments and analytical methods used for measuring concentrations of pollutants in air. It consists of a sealed polymer tube containing a liquefied sample of the gas to be measured. At a fixed temperature, this diffuses through the walls of the tube at a constant rate, so that by allowing a stream of air to flow past the tube at a known rate, mixtures of this gas with air at very low, but accurately known, concentrations can be prepared.

**Peroxyacetylnitrate (PAN).** One of a number of complex compounds present in PHOTOCHEMICAL SMOG. It causes irritation to eyes and is toxic to plants.

**Pest control.** The term for the control of pests (e.g. tsetse flies, mosquitoes, cotton boll weevils) which affect public health, or attack resources of use to man.

The techniques used other than PESTICIDES are:

1. Sterilization, i.e. use of irradiated males which controls or stops reproduction of the species. (⇨ IRRADIATION)
2. Administration of juvenile hormones so that the species cannot metamorphose and therefore cannot reproduce.
3. Exposure to sex attractants: pheromones, specific chemical produced by the female attracts the male into an insect trap and hence the male can be destroyed.
4. Plant breeding to develop insect-resistant plants.
5. Modified planting practices: (a) a favourite food is planted nearby, therefore drawing the insect to it; (b) crop rotation.
6. Encouragement of predatory insects or animals, e.g. the use of small fish which prey on mosquito larvae.
7. Infection by viruses, bacteria or parasites, e.g. a strain of nematode (minute thread worm), which is a parasite of mosquitoes, is being introduced for mosquito control.
8. Deprivation of brooding ground, e.g. drainage of swamps to deny mosquitoes their breeding grounds, controlled tipping of DOMESTIC REFUSE to deny breeding grounds to flies.

**Pesticide.** A product or substance used in the control of pests such as vermin, mosquitoes, moulds, and weeds, all of which may affect public health or attack resources of use to man. (The term subsumes insecticides with which it is often used interchangeably.) There are three main classes:

1. Chlorinated hydrocarbons (e.g. DDT) which are long-lived and capable of being concentrated biologically.
2. ORGANOPHOSPHATES which are short-lived and degrade to 'harmless' end-products.
3. Artificial pyrethrins originally based on natural sources from pyrethrum flower heads but now being synthesized in very large amounts.

Widely used pesticides are:

*Chlorinated hydrocarbons (persistent in the environment):* DDT and its metabolites, e.g. DDE. Used widely to control malarial mosquitoes and houseflies, but strains have evolved with resistance. Acute oral toxicity to mammals low, but it becomes concentrated in fatty tissues, especially around and in vital organs.

Aldrin and dieldrin: formerly widely used as seed dressing, but use suspended because of death of birds and other wildlife. Their use is restricted in the UK because of effects on the FOOD CHAIN.

HCH and lindane (its gamma isomer): used for seed treatment. At one time widely used in gardening products.

*Organophosphorus insecticides (short-lived in the environment):* These are chemically related to nerve gases and some, such as parathion, can be dangerous to use. Others, such as malathion, are much less toxic and are in widespread use in agriculture, gardening and public health. The carbamate grouping of insecticides, which includes aldicarb and carbaryl, is similar in its effects to organophosphorus compounds.

*Pyrethrins* (⇨ PYRETHROIDS): Pyrethrin I, allethrin and bioallethrin are natural products or synthetic chemicals closely related to natural products. Often a mixture, making build-up of resistance in houseflies, for example, more difficult. They are relatively safe.

The chlorinated hydrocarbon class concentrates in the fatty tissues and vital organs of birds. In time of stress, body fat is mobilized and death can result.

It is postulated that pesticides have synergistic effects (⇨ SYNERGISM) in combination with air pollution and that dietary deficiency can markedly increase any hazard, particularly with the ORGANOCHLORINE and ORGANOPHOSPHATE groups.

Another class, not now often used, is the inorganics, which are preparations of zinc, copper, arsenic or mercury. All are extremely toxic to human and animal life, but *may* be of use should resistance to organics develop.

Pesticides have undoubtedly contributed greatly to human health and increased food yields and will continue to do so – but unless caution is exercised, man will be in a race to develop new and/or more powerful insecticides as resistant strains of pest develop. There are dangers of over-zealous insecticide use. ($\Rightarrow$ ECOSYSTEM; CARCINOGEN; TERATOGENS; BIOLOGICAL CONCENTRATION)

**PET (polyethylene terephthalate).** Used in the manufacture of lightweight beverage container bottles. ($\Rightarrow$ PLASTICS; RECYCLING)

**Petrochemical.** An intermediate chemical derived from petroleum, hydrocarbon liquids or natural gas, e.g. ethylene, benzene, propylene, toluene, xylene.

**Petroleum.** A range of distillate and residual liquid fuels derived from OIL. The principal residual fuels in order of increasing boiling point are gasoline, kerosine and gas-oil, which includes diesel fuel. Residual oil is the refinery remainder after distillation and may be blended with gas-oil as a fuel in large industrial furnace and for large slow-moving reciprocating engines. The average CALORIFIC VALUE (gross) is 45 MJ/kg.

**pH.** A measure of the alkaline or acid strength of a substance. The pH value of any solution in water is expressed on a logarithmic scale to the base 10. It is defined and calculated as the logarithm of the reciprocal of the hydrogen-ion concentration of a solution and may be expressed in symbolic form as:

$$pH - \log_{10}\left[\frac{1}{H^+}\right]$$

where $H^+$ is the concentration of hydrogen ions. ($\Rightarrow$ LOGARITHMS)

What this means in practice is that the pH scale ranges from 0 to 14 with the mid-point 7 indicating neutrality. If acid is added to water, the $H^+$ value increases and the pH decreases. Thus a pH value less than 7 is acidic. If greater than 7, it is alkaline (the opposite of acidity). Each unit increase in pH value expresses a change of state of 10 times the preceding state (because of the logarithmic scale). Thus, pH 5 is 10 times more acidic than pH 6;

and pH 9 is 10 times more alkaline than pH 8.

pH measurements can be made by observing colour changes in special indicator chemicals or with indicator impregnated paper (litmus paper), and also by pH electrodes. The pH value of effluents, toxic liquid wastes, etc. is a crucial parameter in effluent treatment. Acidic solutions, for example, would require neutralization with an alkali prior to disposal so that the pH is 7, and treatment can be undertaken without the added complications of acidity (or alkalinity).

The pH of soils is an important factor in soil management – acidic soils require neutralization with limestone (calcium carbonate) which replaces one hydrogen ion in the soil with one calcium ion. However, as the calcium is leached, it must be replaced or the acidity returns. Other techniques, such as the use of CHELATING AGENTS, may also be used in soil management to remove or sequester unwanted ions. (⇨ LEACHING)

**Phenols.** A group of aromatic organic compounds which are highly toxic to living organisms. They can poison sewage treatment systems and taint water in very small concentrations.

**Phosdrin.** ⇨ ORGANOPHOSPHORUS COMPOUNDS.

**Phosphates (as pollutants).** Essential inorganic substances for normal plant metabolism and applied as fertilizer to rectify a deficiency of phosphorus in the soil. Pollution can occur by too much being applied (or too little taken up by plants) and the surplus is leached into rivers, where it may contribute to EUTROPHICATION, and into groundwater.

**Phosphorus (P).** An element that plays an essential role in the growth and development of both plants and animals. In plants it is the energy exchange between adenosine diphosphate and adenosine triphosphate that provides energy, derived as a by-product of photosynthesis, at sites removed from the green parts in which photosynthesis occurs. In animals and plants, phosphorus is an essential component of DNA. It is impossible to conceive therefore of any substitute for phosphorus as a plant nutrient, and a phosphorus deficiency will retard growth and seriously affect yields.

In FERTILIZERS, phosphates may be associated with excessive quantities of fluorine, a cumulative plant poison.

**Phosphorus cycle.** The phosphorus cycle is sedimentary as are those of the elements calcium, iron, potassium, manganese, sodium and sulphur. In essence, the cycle operates as follows: compounds such as calcium phosphate, sodium phosphate, etc.

are leached from the rocks into the soil and water, where they are taken up by plant roots. The plants are subsequently consumed by herbivores, which in turn are eaten by carnivores. On the death of either herbivores or carnivores, decomposition takes place and the compounds are returned to the soil through the action of water – hence sedimentary cycle.

In the sea there is a similar cycle in which phosphate compounds from sediments pass through the aquatic food chain, some fish being eaten by sea birds, whose droppings (guano) are rich in these compounds, and recycling eventually takes place.

**Photochemical smog.** Photochemical smog appears to be initiated by nitrogen dioxide. Absorbing the visible or ultraviolet energy of sunlight, it forms nitric oxide to free atoms of oxygen (O), which then combine with molecular oxygen ($O_2$) to form OZONE ($O_3$). In the presence of hydrocarbons (other than methane) and certain other organic compounds, a variety of chemical reactions takes place. Some 80 separate reactions have been identified or postulated, but two stages have been identified:

*Stage 1:* Smog concentration is linked to both the amount of sunlight and HYDROCARBONS respectively.
*Stage 2:* the amount is dependent on the initial concentration of NITROGEN OXIDES.

Many different substances are formed in sequence including formaldehyde, acrolein, PAN, etc. The low-volatility organic compounds formed, condense to the characteristic haze of minute droplets which is called photochemical smog. The organics irritate the eye, and also, together with ozone, can cause severe damage to leafy plants such as tobacco and endive. Photochemical smog tends to be most intense in the early afternoon, when sunlight intensity is greatest. In this respect it differs from traditional SMOG, which is most intense in the early morning and is dispersed by solar radiation.

This type of smog was first recognized in the Los Angeles area and its persistence has resulted in the passing of legislation to curb drastically automobile emissions. Control can be effected by controlling either hydrocarbon or industrial nitrogen oxide emissions. Photochemical activity of the type involved in smog formation also occurs in the UK as indicated by measurement of OZONE, the usual indicator pollutant of photochemical reactions. ($\Rightarrow$ AUTOMOBILE EMISSIONS; SMOG; NITROGEN

OXIDES; VOLATILE ORGANIC COMPOUNDS)

**Photodegradation.** The process whereby ultraviolet radiation in sunlight attacks a chemical bond or link in a polymer or chemical structure, e.g. plastic.

**Photosynthesis.** The means by which CHLOROPHYLL enables radiant energy to be used to accomplish the chemical conversion of elements in the atmosphere into organic matter. Chlorophyll is contained in organisms such as green and purple bacteria, blue-green algae (in fresh water), phytoplankton (at sea) and green plants (on land). The organisms live in those areas that receive sunlight such as the top few centimetres of soil and rivers and lakes. In the seas, sunlight can penetrate over 100 metres and this is the province of phytoplankton. On land the green plants account for most photosynthesis.

Photosynthesis can be summarized as:

$$nCO_2 + 2nH_2A + \text{energy} \rightarrow (CH_2O)_n + nA_2 + nH_2O$$

or carbon dioxide plus hydrogen donor ($H_2A$) plus energy gives organic compounds (carbohydrates) plus a free, i.e. gaseous, compound plus water. Photosynthesis, as carried out by green plants and phytoplankton, uses carbon dioxide plus water for the hydrogen donor and the equation for such a reaction is:

$$nCO_2 + 2nH_2O + \text{energy} \xrightarrow{\text{green plant}} (CH_2O)_n + nO_2 + nH_2O$$

In this way the oxygen content of the atmosphere is maintained, carbon dioxide is fixed, and a carbohydrate source, $(CH_2O)_n$, is available for incorporation in cell structure or as a source of energy directly or indirectly for all plants and animals. ($\Rightarrow$ CARBON CYCLE; OXYGEN CYCLE)

The 'purple' and 'green' bacteria can use hydrogen sulphide ($H_2S$) as the hydrogen donor with sulphur as a by-product.

**Photosynthetic efficiency.** The percentage of total energy falling on the earth this is fixed by plants. It is approximately 6 per cent.

**Phthalates.** Organic compounds used as PLASTICIZERS in PVC manufacture. As they are not polymerized, they can migrate or volatilize and enter food from packaging or drinking water from PVC pipes. They are persistent and bioaccumulative, and are accumulated in the fatty tissues of oily fish such as herrings and mackerel. Concern has been expressed that the mechanism of accumulation is akin to that of DDT with attention being focused on dioctyladipate (DOA, a plasticizer used in food wrap films). The European Chemical Industry Federation (CEFIC) states that the acute toxicity of plasticizers is extremely low and puts this into perspective, by publishing data (shown below) showing a range of substances and their poison class. Plastic film producers are now using even less DOA.

Acute toxicity of various substances

| Category* | Examples of toxic substances | Lethal dose ($LD_{50}$) in mg/kg body weight (for oral application) |
|---|---|---|
| Very toxic (less than 25 mg/kg body weight) | Clostridium botulinum toxin | 0.00000003 |
| | Hydrocyanic acid | 0.7–1.0 |
| | Arsenic (arsenic oxide) | 1.4–4.3 |
| Toxic (25–200 mg/kg body weight) | Sodium nitrite | 57–86 |
| | Barbiturates | 47–143 |
| Harmful (200–2000 mg/kg bodyweight) | Oxalic acid | 375 |
| | Carbon tetrachloride | 457–686 |
| Not classified as harmful (more than 2000 mg/kg body weight) | Ethanol | 3300 |
| | Common salt | 7150–14300 |
| | Plasticizers (e.g. dioctylphthalate (DOP)) | More than 30000 |

*Based on the EEC classification levels for acute toxicity – CEFIC Publication *Plasticizers*, January 1990.

**Physico-chemical effluent treatment.** The treatment of effluents by non-biological means, e.g. PRECIPITATION and settling.

**Phytoplankton.** Free floating minute plants in sea, lake and river surface waters where sufficient sunlight is available for PHOTO SYNTHESIS. Phytoplankton are said to be responsible for up to one quarter of Europe's ACID RAIN emissions due to their production of dimethyl sulphide which converts to SULPHUR DIOXIDE and hence to acid rain. This process occurs in spring and summer. (⇨ ZOOPLANKTON)

**Pica.** The compulsive habit of some children to eat non-food matter such as paint and soil. While not of widespread importance, this has caused serious problems when the material has contained substances such as lead from the pigment in old paintwork.

**Pickling.** The removal of scale from iron and steel usually by means

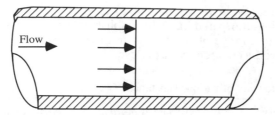

**Figure 113**   'Ideal' fluid velocity profile.

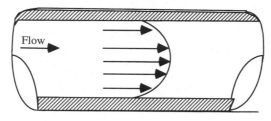

**Figure 114**   Real fluid velocity profile.

of immersion in a hot hydrochloric or sulphuric acid bath. WASTES include spent pickling liquor, sludges and rinse water. (⇨ POLLUTION PREVENTION PAYS)

**Pilot plant.** A small treatment plant which is built to obtain basic design reliability and cost data before the full scale plant is designed.

**Pipe flow.** The factors which affect the flow of a fluid in a pipe are:

- Velocity of the fluid (mean velocity).
- Viscosity of the fluid.
- Density of the fluid.
- Diameter of the pipe.
- Friction where the fluid is in contact with the pipe.

If the effects of viscosity and pipe friction are ignored, a fluid would travel through a pipe in a uniform velocity across the diameter of the pipe. The velocity profile would be as in Figure 113. In practice, viscosity affects the flow rate of the fluid and works together with the pipe friction to further decrease the flow rate of the fluid near the pipe wall (see Figure 114).

   The most important of the factors affecting fluid flow in pipes can be pulled together in one dimensionless quantity to express

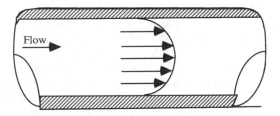

**Figure 115**   Laminar flow.

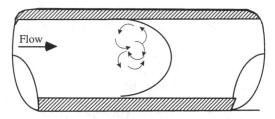

**Figure 116**   Turbulent flow velocity profile.

the characteristics of flow. This is known as the REYNOLDS NUMBER.

$$\text{The pipe Reynolds number} = R_e = \frac{\rho V D}{\mu}$$

where $\rho$ = density (kg/m³)
   $V$ = mean velocity in the pipe (m/s)
   $d$ = internal pipe diameter (m)
   $\mu$ = dynamic viscosity (kg/ms).

It should be noted that the Reynolds number ($R_e$) is dimensionless. In simple terms:

$$R_e = \frac{\text{dynamic force}}{\text{viscous force}}$$

So at very low velocities, the dynamic force will be low and, therefore, the Reynolds number will be small. Similarly, a high viscosity fluid will result in a low Reynolds number. This will result in viscous forces holding back the fluid flow at the pipe walls with the highest fluid velocity at the centre of the pipe. This is similar to the first figure except that the velocity profile is a parabola. It is known as LAMINAR FLOW and normally occurs at Reynolds numbers below 2000 (see Figure 115).

When Reynolds numbers are above 2000, i.e. high velocities and/or low viscosities, the flow breaks up and TURBULENT FLOW occurs with a much flatter velocity profile (see Figure 116).

In the process world, unless very viscous fluids are being piped, turbulent flow is the norm. This is certainly the case for STEAM and compressed air where Reynolds numbers well in excess of 2000 are encountered and intimate mixing between particles of the fluid can be assumed. For mud slides and lava flows laminar flow is usual. (⇨ VISCOSITY)

**Plastics.** Any substance that is capable of plastic flow or deformation under certain conditions or at some stage of its manufacture and thus can be moulded into shapes by heat and/or pressure. The common definition of plastics relates to those products of the chemical industry called 'polymers' which fall into two groups, thermoplastics and thermosetting.

Thermoplastics retain their potential plasticity after manufacture and can be re-formed by heating. The main types in this group are polyethylene, polypropylene, polystyrene and polyvinyl chloride (PVC) which embrace the whole range of domestic use. This group can be recycled to produce similar artefacts or be incorporated in lower grade ones depending on the degree of separation obtained and product purity.

The thermosetting group includes mainly resins such as the Bakelite or epoxy varieties, which are made into light switches, etc., and cannot be reused as there is permanent and irreversible change in the chemistry on setting.

It takes about seven times as much energy to make a cast iron pipe as it does to make the same pipe in PVC. Plastics save energy, first in producing and processing then in their transport and again in their durability. The energy saved by plastics is reflected in the price people pay for products made from plastics.

Plastics save energy because of their intrinsic properties. These include:

- thermal insulation
- electrical insulation
- high performance to weight ratios
- durability and chemical resistance

(⇨ PLASTICS PACKAGING, RECYCLING)

**Plastics, degradation.** The very durability of plastics which is an excellent property for many purposes makes them virtually indestructible when discarded, and as RECYCLING is difficult on both cost and quality grounds, means have been sought to make them degrade when their useful life is over. There are two postulated routes – photodegradation using ultraviolet radiation (solar radiation) and biodegradation.

- *Photodegradation* makes use of the fact that window glass *removes* ultraviolet radiation. Therefore plastic goods kept indoors are not exposed to ultraviolet rays, but when discarded on tips, etc. they would be exposed. Thus, the incorporation into the polymer of ultraviolet-sensitive groups would cause degradation on rejection.

- *Biodegration* is currently a long shot, but possible, as ICI's Agricultural Division has come up with a truly biodegradable plastic (one that disappears completely leaving no residue). It is made from a natural polymer called poly-3-hydroxybutyrate (PHB) which is found in the cells of certain bacteria. The bacteria can be grown on sugar glucose substrates and the PHB extracted. This biodegradable polymer should have an important future. (There are still a number of practical problems to overcome, such as its brittleness.) This could become the plastic of the future. The main problem at the moment is the artificially low price of oil against the high price of sugar (necessary to produce the bacteria). So, currently, it is cheaper to produce plastics from the finite resources of oil rather than the renewable resource of glucose. It is extremely short-sighted to continue to regard plastics as cheap and disposable materials and the development of biodegradable replacements from renewable resources is very welcome news.

**Plastic, disposal** The last phase in the life of a product is its disposal when it is no longer of any use. Disposable objects generally end up, with a very few exceptions, in DOMESTIC REFUSE. In Europe, 70 per cent (UK 95 per cent) of domestic refuse is land filled, 30 per cent is incinerated (UK 5 per cent), and the rest is composted or recycled as raw material. Plastic objects, then, are disposed of by land-filling, by INCINERATION with or without energy recycling (energy recovery), or by recycling the raw material. The problem is that while plastics occupy 7 per cent by weight, they occupy 25 per cent by volume in municipal waste and recycling options are under urgent consideration. (⇨ LANDFILL)

**Plastics packaging (UK statistics).** Tables below give a breakdown of plastics packaging in use in the UK and the assessed condition of plastics packaging waste, respectively.

Plastics packaging materials used in the UK

|  | Films | | Rigids | | Total |
| --- | --- | --- | --- | --- | --- |
|  | Quantity (kt) | Percentage of total plastic use | Quantity (kt) | Percentage of total plastic use | Quantity (kt) |
| Polyethylene | 753 | 50.8 | 179 | 12.1 | 932 |
| Polypropylene | 61 | 4.1 | 116 | 7.8 | 177 |
| Polystyrene |  |  | 124 | 8.4 | 124 |
| PVC | 15 | 1.0 | 120 | 8.1 | 135 |
| Polyester | 3 | 0.2 | 87 | 5.9 | 90 |
| Other | 8 | 0.5 | 17 | 1.1 | 25 |
| Total | 840 | 56.6 | 643 | 43.4 | 1483 |

It should be noted that much is contaminated and therefore unsuitable for materials recycling but excellent as a FUEL in INCINERATION with energy recovery.

The conclusion from the table on page 423 is that to merely set a material recovery target whilst ignoring the degree of contamination and, therefore, the major problems (in both cost and quality) that can be incurred in re-processing would be to impose unnecessary burdens on industry and the environment. This is especially so when an environmentally sound and financially viable option, energy recovery, exists to effect 'valorization' as well.

This point was accepted by the House of Commons Environment Committee in its Second Report on Recycling when it stated 'We agree that energy recovery is appropriate for parts of the household waste plastics stream'. It is clear then that individual material appraisals need to be made.

*Sources:* Packaging and Industrial Films Association, Oriented Polypropylene Film Manufacturers Association and the Flexible Packaging Association, *Management of Waste and Plastic Packaging Films,* January 1993.
The Consortium of the Packaging Chain (COPAC), *Action Plan to Address UK Integrated Solid Waste Management,* 26 October 1992.
German Federal Bulletin, 'Survey of Sales Packaging Consumption 1991', 27 August 1992.

Assessed condition of used flexible plastics packaging in the UK

| | Total (×1000 tonnes) | 'Clean' (×1000 tonnes) | Product contaminated (×1000 tonnes) | Ink contaminated (×1000 tonnes) | Mixed materials (×1000 tonnes) |
|---|---|---|---|---|---|
| **Domestic use of polyethylene** | | | | | |
| Refuse sacks | 66 | | 50 | 16 | |
| **Food contact packs** | | | | | |
| Frozen/chilled food | 29 | 10 | | 19 | |
| Moist products (i) | 44 | 5 | 33 | 35 | |
| Dry products (ii) | 157 | 40 | | 70 | 41 |
| Bread bags | 12 | | | 12 | 47 |
| **Non-food contact packs** | | | | | |
| Carrier bags | 30 | | 10 | 70 | |
| Counter bags | 50 | 37 | | 13 | |
| Dry cleaning | 11 | 10 | | 1 | |
| Mail envelopes | 14 | | | 4 | 10 |
| Subtotal | 533 | 102 | 93 | 240 | 98 |
| **Industrial use of polyethylene** | | | | | |
| Shrink/stretch film | 129 | 78 | 38 | 13 | |
| Sacks | 31 | 2 | | 23 | 6 |
| Liners | 25 | | 25 | | |
| Bubble pack | 7 | 3 | | | 4 |
| Others | 28 | 1 | 11 | 12 | 4 |
| Subtotal | 220 | 84 | 74 | 48 | 14 |
| Total polyethylene | 753 | 186 | 167 | 288 | 112 |
| Polypropylene | 61 | 8 | 15 | 23 | 30 |
| PVC | 15 | | | | |
| Polyester | 3 | | | | 3 |
| Cellulose film | 12 | 2 | | 6 | 4 |

House of Commons Environment Committee, Recycling Report, July 1994.

**Plastics, recycling.** Approximately 33 per cent of the 20 million tonnes per year plastics production in Europe has a very short life – the rest are virtually in captive use. The bulk of this 33 per cent is thermoplastics which can in theory be recycled. (⇨ RECYCLING)

Direct recycling to obtain virgin plastics is difficult due to contamination, mixtures of grades, and the type of original product. To overcome this difficulty, selective collections are organized, as well as sorting by hand (PVC bottles, for example) at sorting centres. Thus, discarded plastic materials collected in this way are of a more uniform consistency and can be recycled more easily.

There are also extrusion and injection processes capable of utilizing mixtures of plastics as raw materials for making simple objects such as posts, mats, etc. The melting temperature for working with these mixtures must remain below 220°C, at which temperature PVC decomposes. This is not a problem for the other mass-produced polymers, except for polyethylene terephthalate (PET), the working temperature of which is of the order of 260°C.

An indirect recycling route is to take virtually as-received plastics waste and to grind it, mix in fillers and put it through a high-energy extruder and use the resultant material for 'low-quality' purposes such as fence posts, pallets and roof tiles, where finish is relatively unimportant but the plastics property of durability and resistance to decay is. This method has substantial promise and is expected to have a rapid growth in the USA where recycling is becoming a way of life. However, unless waste plastics can be collected at source, separation from DOMESTIC REFUSE is not economic as it comprises ca. 7 per cent by weight (but 25 per cent by volume). Thus the normal domestic refuse disposal processes will prevail, but with the proviso that PYROLYSIS or INCINERATION would be the most suitable energy recovery process for plastics because of their hydrocarbon content and high volatility. However, West Germany has introduced mandatory deposits of up to 50 pfennigs (15 pence) on plastics beverage containers which has resulted in PET bottles being withdrawn from the market or recycled. Restrictions have also been placed on many non-refillable drinks containers. An alternative approach is being tried in France where special public receptacles are provided for the deposit of PET bottled-water containers. These are ground up and re-used for high-grade plastics manufacture and are eminently suitable for non-food grade container manufacture.

The throw-away era of plastic bottles and non-refillable drinks containers is coming to an end in some European countries, with West Germany and Italy leading the way. Draft EC measures on plastic wastes are:

- Substantial VAT reductions for recycled products which fully meet user standards for competing products.
- Improvement of these standards in terms of the properties required of products and not in terms of materials.
- Reserving public procurement contracts entirely for recycled products which fully meet the standards set in the specifications.
- Establishment of Information Centres dealing with available quantities of plastic wastes, which are mixed with each other to varying extents and will be collected separately from other wastes. The Centres will operate as plastic-waste exchanges to disseminate all the necessary information to recyclers. They must have available information on quantities, grades and prices on where the waste is located. They will also centralize information on recycling businesses and available recycling technologies. Community-wide co-ordination will be provided by an organization concerned with recycling. The same organization will develop a classification scheme for these secondary plastic materials (mixed to varying extents) to help to stabilize the markets for them and their prices, so as to make the latter independent of fluctuations in resin prices.
- Financial assistance to encourage research and development into technologies for the selective collection, automatic sorting and recycling of plastics.
- Indelible marking, by producers, of all plastic products to show each plastic used, according to a pre-established code. One code will be provided for complex or laminated plastics and another for biodegradable plastics.

Where a producer has been able to ascertain that a collection and recycling system is available for these products in the countries of destination, he will also be permitted to place a recyclability mark on them.

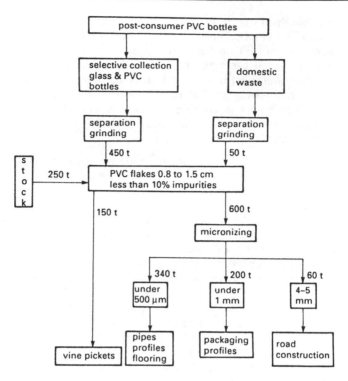

**Figure 117**   Material flow (tonnes) created by post-consumer PVC bottles in France during the first half of 1986.

These accord with EC waste management principles of:

- the prevention and reduction of waste arisings
- the increase of recycling and re-use
- and the safe disposal of unavoidable waste.

What can be stated is that with EC (1989) consumption of plastics numbering at least 20 million tonnes per year, a lot more thought will be given to plastics recycling, and/or materials substitution, e.g. refillable glass for non-refillable plastic bottles, if recycling is not implemented. One such scheme is being assessed in France, as shown in Figure 117. This relies primarily on source separation of PVC bottles. Another scheme launched in the UK aims to recycle 50 per cent of all domestic plastic

containers by 1995. This anticipated the EC measures above.
(⇨ POLYVINYL CHLORIDE, SPI MARKING SYSTEMS)

**Plasticizer.** A substance incorporated in a material, usually a plastic or elastomer, to improve its flexibility and processability. Examples are the addition of water to clay for the production of pottery and PHTHALATES to PVC to produce the desired degree of flexibility in the finished product.

**Plate count.** The number of bacterial colonies which grow on a plate of nutrient agar from a diluted water sample incubated for a specified period and temperature.

**Plume.** The stream of gases issuing from a stack which retains its identity and is not completely dispersed in the surrounding air. Near the stack the plume is often visible owing to water droplets, dust, or smoke that it contains, but it often persists downwind long after it has become invisible to the eye.

**Plume rise.** The rise of plume from stacks is a function of atmospheric conditions and the plume discharge (efflux) velocity, temperature and density. In still atmospheric conditions the plume will rise vertically; in strong WIND conditions it can be carried away horizontally or drawn downwards in the low pressure area behind the chimney. Further complications arise if the chimney is in a built-up area or in countryside of varying contours as the wind patterns can then cause downdraughts much more frequently.

To avoid downdraught, the exit velocity of the plume should be at least one-and-a-half times the wind velocity. Thus, where noxious fumes are concerned, a meteorological survey is required to determine the annual wind velocity profile so that the plume will not be drawn down for, say, at least 95 per cent of the year. The stack gas temperature is also very important as this determines the gas buoyancy. The higher the temperature, the greater the buoyancy. The gas eventually cools to ambient and then the plume disperses downwind in an ever-widening cone.

One form of plume dispersion that can cause severe pollution under certain circumstances is a fanning plume, i.e. one that fans out horizontally into a thin layer under stable atmospheric conditions that restrict its spread vertically.

**Plutonium (Pu).** Plutonium-239 is an artificial radioisotope with a HALF-LIFE of 24 400 years. It is made by bombarding uranium-238 with neutrons. It is reactive and emits ALPHA RADIATION. It is a bone-seeking poison similar to radium but several times as toxic, and is one of the most hazardous substances known. Its most hazardous form may be the respirable plutonium dioxide particles produced by combustion – of the order of 10 thousand million particles per gram of metal – with each particle suspected of carrying a substantial risk (perhaps 1 per cent or more) of lung cancer. The plutonium in breeder reactor fuel is already oxidized into a refractory ceramic which, it is claimed, cannot produce respirable particles. However, the sodium coolant in a breeder may be reactive enough in an accident to reduce the plutonium dioxide fuel back to plutonium metal. (⇨ NUCLEAR POWER; PLUTONIUM, THE HOT-SPOT CONTROVERSY)

**Plutonium, the hot-spot controversy.** The magnitude of a dose of ionizing radiation is defined as the energy absorbed from the radiation per unit mass of tissue. If a specified amount of a radioactive material undergoes decay in an organ, the energy released can be calculated from a knowledge of the physics of the isotope in question. The fraction of this energy absorbed within the organ can be estimated from a knowledge of the penetration capacity of the relevant radiation: for example, the energy of $\alpha$-particles is entirely absorbed within less than a millimetre from the source, whereas X-rays are much more penetrating and only a small fraction of their energy will be absorbed within a metre or so of their source. Such considerations provide an estimate of the total energy absorbed within the organ. The dose to the organ is then estimated by dividing that amount of energy by the mass of the organ. In general, the risk of cancer is considered to increase in proportion to the dose of radiation received by any tissue.

The hot-spot controversy is about the potential for lung-cancer induction by very small particles of plutonium that may be inhaled and lodge in the lung. Plutonium emits $\alpha$-particles and it is contended by critics of present recommendations of 'acceptable' doses that the conventional form of calculation may lead to gross underestimates of the dangers of such particles. They point out that all the energy of the $\alpha$-particles emitted by the plutonium is absorbed in a small sphere of tissue surrounding the 'hot-spot'. This sphere is no more than a millimetre or so in diameter and the dose within is therefore thousands of times higher than is calculated by averaging the energy over the entire mass of the lung. Cells immediately adjacent to the 'hot particle' will simply be killed, which is of little importance, but doses of *all* lower levels down to zero will be delivered at increasing distances from the particle. Thus, any particular dose critical in causing cancer must occur somewhere in the vicinity of the hot spot.

**PM10.** Particles with a diameter less than $10\,\mu m$ which can be inhaled beyond the larynx. Those with a diameter less than $2.5\,\mu m$ are called 'respirable particles' and can penetrate to the inner lungs. DIESEL vehicle (truck, bus and taxi) emissions are responsible for most PM10s, with 87 per cent of UK black smoke emissions attributable to them, in London. Such emissions may be carcinogenic.

**PNdB** (perceived noise decibels). A frequency-weighted noise unit used for AIRCRAFT NOISE measurement. ($\Rightarrow$ NOISE; DECIBEL)

**Point source.** $\Rightarrow$ EMISSION SOURCE.

**Poison, nuclear.** Material in a reactor that absorbs neutrons strongly, thus reducing the reactivity. Some of the fission products (e.g. xenon) are nuclear poisons.

**Pollutant.** A substance or effect which adversely alters the environment by changing the growth rate of species, interferes with the food chain, is toxic, or interferes with health, comfort, amenities, or property values of people. Generally pollutants are introduced into the environment in significant amounts in the form of sewage, waste, accidental discharge, or as a by-product of a manufacturing process or other human activity. A polluting substance can be a solid, semi-solid, liquid, gas or sub-molecular particle. A polluting effect is normally some kind of waste energy such as heat, noise or vibration.

Pollutants may be classified by various criteria:

1. *Natural or synthetic:* sulphur dioxide is an example of a natural pollutant; the class of CHLORINATED HYDROCARBONS is an example of a synthetic pollutant. Natural substances can be assimilated into biological cycles; they may often undergo BIODEGRADATION. Some synthetic substances such as DDT, are not biodegradable; they are often toxic and accumulate in biological systems.

2. *Effect:* a pollutant may affect man, a complete ecosystem, a single individual member or component of an ecosystem, an organ within the individual member, a biochemical or cellular sub-system (e.g. crop growth rates).

3. *Properties:* e.g. toxicity, persistence, mobility, biological properties.

4. *Controllability:* the ease with which a pollutant can be removed from air or water is a very important factor. For

example, most grit can be readily removed from flue gases, whereas sulphur dioxide cannot without a great deal of expense.

In addition, the environmental attributes of the system into which a pollutant is to be discharged must be taken into account. If a watercourse is to be used for sewage discharge, the BIOCHEMICAL OXYGEN DEMAND imposed by the sewage must be such that it does not swamp the DISSOLVED OXYGEN of the stream.

The maintenance of biological processes is fundamental to our continued existence and health, and must not be grossly overloaded or resources can be irretrievably lost. The gross pollution of the Great Lakes is one example of how man has severely diminished a major natural resource. (⇨ PRIMARY POLLUTANT; SECONDARY POLLUTANT)

**Polluter-pays principle.** After the Second World War, when problems of pollution began to attract public attention and provoke protests, it was clear that the installation of plant and processes for pollution abatement would require financing from some source or other. Firms were evidently willing to install such plant if funds were provided from public sources, but such a policy did not have much appeal to taxpayers. Accordingly, much was made of the slogan 'the polluter must pay', interpreted as meaning that public money would not be used to subsidize pollution abatement for private industry. In practice, capital grants, subsidies, payments for reduced crop yields through using less fertilizers and tax allowances are all part of incentives to encourage industry and farmers to control pollution – in addition to increasing prices to consumers.

**Pollution.** Pollution is the introduction into the environment of substances or effects that are potentially harmful or interfere with man's use of his environment or interfere with species or habitats. (⇨ ENVIRONMENTAL PROTECTION ACT 1990)

**Pollution control.** The term for administrative mechanisms for control *and* the various technical processes and devices available for reducing emissions of waste streams.

In the UK, the administrative control is effected by legislation (e.g. Alkali Act 1906, Clean Air Act 1956 and 1968, Control of Pollution Act 1974, Health and Safety at Work Act 1974, and Environmental Protection Act 1990) and its enforcement and implementation through statutory bodies such as the Waste Disposal Authority for a region; the National Rivers Authority,

Pollution Inspectorate, and the Health and Safety Executive.

The actual control of pollution is through process selection and plant construction. For example, in stack gas emission control, is it better to filter out particulates and scrub and neutralize the gases, or specify chimney height and efflux velocity and rely on the atmosphere for dilution an dispersion? These are typical considerations for only one problem – DILUTE AND DISPERSE versus CONCENTRATE AND CONTAIN. (⇨ BEST PRACTICABLE ENVIRONMENTAL OPTION)

**Pollution conversion.** In the elimination of one or more sources of pollution, it is important that new ones are not created. If solid waste is incinerated, air pollution may occur instead, which may be more serious than the original problem. Similarly, disposal of sewage SLUDGE too liberally on agricultural land could result in the build-up of toxic metals. Washing or scrubbing of exhaust gases can lead to a water-pollution problem. The disposal of urban solid waste can pollute groundwater and produce METHANE. (⇨ INTEGRATED POLLUTION CONTROL)

**Pollution Hazard Appraisal (PHA).** For each process regulated by Her Majesty's Inspectorate of Pollution (HMIP), in addition to an Operator Risk Appraisal, a PHA must also be carried out against the following characteristics:

- Hazardous substances
- Techniques for prevention and minimization
- Techniques for abatement
- Scale of process
- Location
- Frequency of operation
- Offensive substances in the process.

Each is assigned a factor on a scale from 1–5 to reflect HMIP's view of its relative importance, where 1 is low hazard potential and 5 is high hazard potential.

Each result is then multiplied by a weighting factor which reflects HMIP's view of the relative importance of each factor. From this a Pollution Hazard Value is derived for the process as a whole.

*Examples of PHA*
Location: remote, e.g. coastal and/or rural site, PHA rating 1.
Location: close proximity to areas of high population and/or where physical factors would exacerbate effects of any releases, PHA rating 5.

*Source:* Draft Report, *Operator and Pollution Appraisal*, (OPRA), consultative document, April 1995.

**Pollution Index.** Used in chimney height calculation to determine limiting pollutant.

$$PI \ (m^3 \, s^{-1}) = \frac{1000 \times \text{pollutant emission} \ (g \, s^{-1})}{(\text{guideline concentration} - \text{background concentration}) \ (mg \, m^{-3})}$$

**Pollution indicators, natural.** ⇨ BIOTIC INDEX; MOSSES; GLADIOLI; LICHENS.

**Pollution of the Environment (Environmental Protection Act 1990).** Pollution of the environment due to the RELEASE (into any environmental medium, from any process) of substances which are capable of causing harm to man or any other living organisms supported by the environment.

**Pollution permit.** A system of permitting companies with excess pollution discharge 'allowances' (consents) to trade them with those who need to discharge more than they are allowed. The net pollutant total is constant but the distributions vary according to market forces. Pollution permits are already in use in the US.

The pollution regulator sets a maximum level for emissions or discharges and companies can then buy and sell permits among themselves.

A sulphur dioxide scheme could be easiest as the UK has a national 'bubble' of sulphur dioxide which can be allocated among all power stations and oil refineries as long as the net allowable total is achieved.

**Pollution Prevention Pays.** Two examples are given below. The first from a compendium of *3P Success Stories* compiled by the Environment Engineering and Pollution Control Department, 3M, St Paul, Minnesota. The second on pickling waste minimization is from the Royal Commission on Environmental Pollution, 11*th Report* (example submitted by The University of Aston). It is also worthwhile recording the 3M (UK) plc Corporate Environment Policy Statement (November 1989) as an example enlightened corporate self and public interest: Under its worldwide Environmental Policy, 3M will continue to recognize and exercise its responsibility to:

● Solve its own environmental pollution and conservation problems
● Prevent pollution at the source wherever and whenever possible

- Develop products that will have a minimum effect on the environment
- Conserve natural resources through the use of reclamation and other appropriate methods
- Assure that its facilities and products meet and sustain the regulations of all federal, state and local environmental agencies
- Assist, wherever possible, governmental agencies and other official organizations engaged in environmental activities.

Waste stopper: pumice on copper 3M Company, St. Paul, Minnesota

| | |
|---|---|
| **Problem** | 3M's electronic product plant in Columbia, Mo., makes flexible electronic circuits from copper sheeting. Before sheeting can be used in the production process, it has to be cleaned. |
| | Formerly, the metal was sprayed with ammonium persulphate, phosphoric acid and sulphuric acid. This created a hazardous waste that required special handling and disposal. |
| **Solution** | Cleaning by chemical spraying was replaced by a specially designed new machine with rotating brushes that scrubbed the copper with pumice. |
| | The fine abrasive pumice material leaves a sludge that is not hazardous and can be disposed of in a conventional sanitary landfill. |
| **Payoff** | 40 000 pounds a year of hazardous waste liquid prevented. |
| | $15 000 first year savings in raw materials and in disposal and labour costs. |
| | In the third year of use, the new cleaning machine had saved enough to recover the $59 000 it cost. Because of increased production each year, costs saved and volumes of pollution prevented continue to rise. |

Descaling of hot rolled steel RCEP 11th report

| **The old technology** | Acid pickling |
|---|---|
| Wastes produced | Depleted hydrochloric and sulphuric acid |
| | Acidified rinse water |
| Reasons for change | To achieve greater control over waste disposal |
| | To reduce rising waste disposal costs |

**Options for on-site waste reduction**

Neutralizing acid liquors       Requires additional chemicals
                                Residual sludges need disposal
Recovery of acid for re-use     Possible especially with modern
                                ion exchange systems to recover
                                both acids
Shot blasting                   Physical rather than chemical
                                process
                                Leaves smaller volumes of inert
                                wastes
                                Allows savings of up to 50 per
                                cent of original descaling costs

**Options adopted**
Shot blasting is now the preferred method for descaling drawn steel. Acid recovery still used in plants producing steel bars in a variety of shapes and sizes, for which shot blasting is not suitable.

---

**Polyelectrolyte.** Long-chain organic compounds used to cause FLOCCULATION of dispersed non-settling matter in water which can then be removed by sedimentation.

**Polyethylene (polyethene).** A THERMOPLASTIC polymer of ethylene (ethene) $(C_2H_4)$. It has good flexibility and is used very widely in packaging. ($\Rightarrow$ PLASTICS, RECYCLING; PVC)

**Polymer.** A chemical compound made by the repeated joining of MONOMER molecules. ($\Rightarrow$ POLYMERIZATION)

**Polymerization.** The joining together of MONOMER molecules by 'addition polymerization' in which case the POLYMER is a simple multiple of the monomer molecule, or by 'condensation polymerization', where the resulting polymer does not have the same empirical formula as the basic monomer constituent. The term is also used to cover the process of copolymerization, in which the polymer is built up from two or more different kinds of monomer molecules. Many plastics and textile fibres are made from natural or synthetic polymeric substances.

**Polyvinyl chloride (PVC).** PVC is a thermoplastic polymer, that is it can be softened for shaping by raising its temperature and then can be hardened by cooling without any chemical change taking place.

The monomer formula is $CH_2{=}CHCl$ which is chloroethene though it is commonly called vinyl chloride monomer (VCM). This monomer is usually produced in a three-stage process. Stage 1 is the production of ethylene dichloride by reacting ethylene dichloride (EDC) (1.2 dichloroethane) which has the formula $CH_2Cl$. The EDC is then catalytically cracked to produce VCM, hydrogen chloride and some chlorinated hydrocarbon by-products. The hydrogen chloride is usually reacted with more ethylene and oxygen in the oxychlorination process to produce more VCM. The oxychlorination process also gives rise to chlorinated by-products. The VCM is distilled, cooled and liquefied under pressure. It is transported and used as a liquefied gas. The raw materials for VCM manufacture are often transported as EDC rather than as ethylene and chlorine.

Though ethylene and VCM differ in only one atom, the polymers produced from them are very different. Unlike polyethylene (PE), PVC is never processed alone because its decomposition temperature is lower than its softening temperature. It is possible to process 'pure' PE for use for example in very low dielectric loss cable insulation. Most PE, like most other plastics including PVC, are processed in a physical blend with other additives or ingredients. The chlorine in the PVC molecule makes it much more polar than polyethylene. That polarity makes PVC compatible with a wider range of types and quantities of additive than any other commodity polymer.

While the necessity of processing PVC with other ingredients may seem a weakness, the wide range of properties these additives make possible is the key strength of PVC. By a choice of additives PVC formulation can be hard or soft, brittle or tough, flammable or flame resistant, matt or glossy, conducting or insulating, opaque or transparent and they are available in a very wide range of colours. A PVC formulation is simply a physical mixture and no new chemicals are formed in the mixture. The other ingredients include:

- heat/light stabilizers – organic or inorganic metal compounds;
- impact modifiers – ethylene vinyl acetate (EVA), chlorinated polyethylene (CPE) or acrylic polymers;
- fillers – calcium carbonate;
- lubricants – metal compounds, waxes, oxidized polyethylene;
- plasticizers – phthalates, phosphates, adipates;
- pigments – organic or inorganic;
- fire retardants – antimony oxide, aluminium trihydrate.

The formulation ingredients are chosen to produce the physical and processing properties required in a particular application and shaping process. The ingredients are also chosen to meet national or international standards for safety critical applications such as toys, food contact or medical devices.

While PVC is widely used in short-life applications, some 50–60 per cent of applications are long-life, for example in window frames, pipes and other construction applications. End-of-life PVC components may readily be recycled into second life applications though the economics of recycling are presently doubtful. The medium-life application of PVC in computer enclosures is already being profitably recycled and long-life applications will be readily recyclable when larger quantities become available.

*Raw materials*
The production of 1 tonne of polyethylene requires the fractionation of 18.7 tonnes of crude oil. Because 57 per cent of the basic PVC molecule is chlorine derived from common salt (sodium chloride) only 8 tonnes of oil are needed for each tonne of PVC polymer. Some PVC formulations include oil-based additives, such as the phthalates, so their dependence on non-renewable fossil fuels is greater, but never at the level needed for polyethylene or polypropylene.

Since salt is neither a source of energy nor a scarce raw material, PVC has a resource advantage over other polymers. Chlorine is produced from salt by the electrolysis of a solution of salt in water. This produces not only chlorine but two other very valuable raw materials, caustic soda and hydrogen. Caustic soda is essential to the production of numerous materials such as paper, soap, aluminium and viscose rayon. Hydrogen is used as a raw material, for example in margarine manufacture, or as a fuel. Chlorine gas is toxic but once it is part of the PVC molecule it becomes an inert component. The PVC industry uses some 30 per cent of the chlorine made in western Europe and the rest finds very important applications throughout the rest of the chemical industry.

The energy cost of several polymers, including the energy content of the plastic themselves, have been calculated in a LIFE CYCLE ANALYSIS (LCA) project organized by the Association of Plastics Manufacturers in Europe. It should be noted that these are values for the polymers alone. For a ready-to-mould or extrude compound these values must be added in proportion to those of the other ingredients. Few of the polymer additive suppliers have yet produced LCA data.

| Polymer | Energy cost (MJ/kg) |
|---------|---------------------|
| PVC     | 53                  |
| PE      | 69                  |
| PP      | 73                  |
| PS      | 80                  |
| PET     | 84                  |

### Health aspects

Various materials used in the production of PVC and PVC formulations have been under the spotlight of health concerns in the past 20 years or so.

*Vinyl chloride monomer.* Until the mid 1970s was known only as a low acute toxicity, anaesthetic gas which on long exposure at high levels could cause Reynards disease (restriction of finger arteries) and acreolysis of finger bones. Research into these phenomena led to an understanding that long term worker exposure could cause angiosarcoma of the liver. About 175 cases have been diagnosed since the 1970s as being associated with PVC manufacture in the western world. Radical changes in working practices, agreed by industry, the unions and government have reduced worker exposure to well below the threshold of hazard. Residual VCM levels in PVC have never been high enough to represent a significant hazard to the public. Present levels are accepted by, for example, the MAFF or the European Pharmacopoeia, as representing an insignificant hazard in the most sensitive approach such as food-contact goods and medical devices.

*Ethylene dichloride.* This too is a carcinogen but it is made and processed in chemical plant designed to minimize emissions and so is acceptable to the regulatory authorities.

*PVC dust.* The health effects of PVC polymer dust remain a matter for controversy. It has been suggested that very fine, respirable dust might cause the miners' disease, pneumoconiosis. This is not accepted by all the regulatory authorities but all accept that exposure to dust should be limited. There are moves in Europe to limit total PVC dust to $2.5 \, mg/m^3$ which should give respirable dust levels of about $0.1 \, mg/m^3$. The Association of Plastic Manufacturers (Europe) is investigating this problem and has (1994–95) a medical working party studying the problem.

*Heavy metal stabilizers.* Governments in Europe are keen to reduce the quantity of heavy metal compounds used and released to the environment. The regulations concerning sensitive applications of plastics forbid their use but there are long-life applications of PVC where they are still the additive of choice.

CADMIUM compounds are used as heat and light stabilizers in PVC exposed to the weather. The industry has significantly reduced the use of cadmium stabilizers by moving to lead or lead/barium/cadmium blends. During 1993 the EC enacted a Cadmium Directive which limited the use of cadmium stabilizers to a short list of long-life applications including window-frame material. Cadmium pigments are only allowed in high melting point plastics such as PP and PET. During 1995 the Cadmium Directive is to be reviewed. Some countries, such as Norway and Sweden, have banned cadmium completely.

Lead compounds have been more widely used than cadmium compounds and to date they are only nationally regulated. They still find wide use in PVC pipes, windows and conducts. There have been worries about their use in potable water pipes but the DoE Drinking Water Inspectorate reauthorized their use in 1995.

Organo-tin compounds have been widely used in PVC pipe and packaging compounds. No hazard in use is known but concern about the effects of related organo-tin anti-fouling paint on shellfish has brought questions about their use as stabilizers.

*Plasticizers.* Here the major group of compounds used are the PHTHALATES, the most used being di(2 ethyl hexyl)phthalate. Over the past 20 years the phthalates have been accused of being CARCINOGENS. They are carcinogenic for rodents at very heavy doses but no cancer effects have been proved for humans.

During 1994–95 phthalates have been grouped with a wide range of other chemicals as endocrine disrupters, also called oestrogen mimics. While the pressure groups claim specific but long-term subtle effects on immune and reproductive systems, medical experts still regard oestrogen mimics as 'not proven' and they are asking for research funds to develop the topic. Certainly no significant effects of this kind have been so far reported from PVC formulation plants.

*End-of-life disposal*
As indicated earlier, PVC formulation can readily be recycled,

though such processes are not always economic. Alternatively PVC can be a minor component in feedstock recycling processes where polymer molecules are broken down into oil feedstock hydrocarbons. Alternatively PVC waste can be burnt, with heat and power recovery, in modern municipal solid waste (MSW) INCINERATORS equipped to extract hydrogen chloride from stack gases for second life application in steel cleansing. MSW incineration is not a significant contributor to acid rain, most of which (98 per cent) comes from power stations or motor transport. Only 0.3 per cent of acid rain can be attributed to MSW incineration and of that only 0.15 per cent can be linked to hydrogen chloride from incinerated PVC.

**Ponding.** The filling up of the VOID spaces in a biological filter, or in a landfill site where liquids are co-disposed in the deposited refuse and further throughput is not possible due to an accumulation of solids. Usually denoted by pools of liquid which will not drain away.

**Pool price.** The normal price paid by regional electricity companies for marginal supplies of electricity. Computed on a half-hourly basis. Prices can range from 1.5 to 11 pence/kWh, depending on the cost of bringing in additional generating capacity.

Pool prices hit a new record on 23 January 1995, as the graph in Figure 118 shows.

At the afternoon peak half-hour the pool purchase price – the 'wholesale price' of electricity that the pool pays to generators – reached £633 per/MW hour, equivalent to 63.3 pence a unit, about eight times the 'retail price' paid by domestic consumers.

The system marginal price (SMP), the price bid by the least economic plant expected to be called into use to meet demand, rose at times of peak demand.

By far the biggest element in the pool price is the capacity payment. This is calculated from the probability of loss of supply and the notional value that customers would put on that loss of supply.

Capacity payments are intended to cover the costs the generators face in making available for generation plant which is seldom needed but which is essential to maintain an adequate level of security of supply.

**Population.** Number of individuals of a certain species that live in a particular area at a particular time.

The tables below give the population growth statistics and selected world fertility rates.

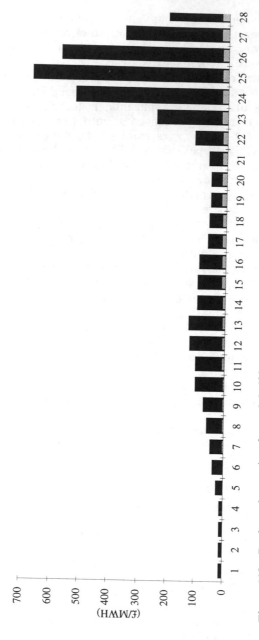

**Figure 118**  Pool purchase prices for each half hour.
*Source: National Power News*, March 1995.

World population growth

| Date | Growth |
| --- | --- |
| 1 AD | 300 000 000 |
| 1750 | 790 000 000 |
| 1975 | 4 100 000 000 |
| 1995 | ca. 6 000 000 |

World fertility rates by region for the early 1980s

| | Total population (millions) | Crude birth rate $(CBR)^2$ (per 1000) | Total fertility rate $(TFR)^3$ |
| --- | --- | --- | --- |
| North Africa | 124 | 41 | 5.9 |
| West Africa | 161 | 49 | 6.8 |
| East Africa | 150 | 47 | 6.6 |
| Middle Africa | 60 | 45 | 6.2 |
| Southern Africa | 36 | 36 | 5.2 |
| Southwest Asia | 110 | 37 | 5.5 |
| Middle South Asia | 1036 | 37 | 5.3 |
| Southeast Asia | 393 | 33 | 4.5 |
| East Asia | 1243 | 20 | 2.6 |
| Middle America | 192 | 34 | 4.9 |
| Caribbean | 31 | 26 | 3.4 |
| Tropical South America | 220 | 32 | 4.3 |
| Temperate South America | 44 | 24 | 2.7 |
| North America | 262 | 15 | 1.8 |
| North and west Europe | 236 | 12 | 1.7 |
| Eastern and southern Europe | 253 | 15 | 2.0 |
| USSR | 274 | 20 | 2.5 |
| Oceania[1] | 24 | 21 | 2.5 |

[1] Australasia, Polynesia, Micronesia and Melanesia.
[2] This is a very simple measure and useful for comparison purposes.
[3] TFR measures the average number of children that would be born to a woman in childbearing age range.

**Pores.** Small VOID spaces in rocks or aggregates. The volume of the pore space is an important parameter in estimating the volume of water, or oil held in storage in relation to the total volume of the aquifer or oil field respectively.

**Poverty.** The UK Government does not have a national definition of poverty; however, the European Council of Ministers defines poverty as the condition of 'persons whose resources (material, cultural and social) are so limited as to exclude them from the minimum acceptable way of life in the Member State in which they live'.

**Precautionary principle.** The reduction of risks to the environment by taking avoiding action before any serious problem arises.

**Precipitation:**

1. General term for the release of water from the atmosphere. It can be in the form of rain, snow, hail, dew or hoar frost. In all cases, condensation is initiated by nuclei of water droplets or ice crystals forming in clouds of moist air cooled below the DEW-POINT. The nuclei then grow and coalesce in the clouds and then, provided the droplets are large enough to overcome any rising air currents, precipitation may take place.

2. The formation of an insoluble substance by chemical reaction which occurs in solution; a method used in treating some liquid hazardous wastes.

3. The process of removing suspended matter in water, sewage and industrial effluents by the addition of suitable chemicals. The use of coagulants such as aluminium sulphate, $Al_2(SO_4)_3$, is widespread in water treatment.

**Precision.** The closeness of agreement between the results obtained by applying the experimental procedure several times under prescribed conditions.

**Prescribed process.** The successor to SCHEDULED PROCESSES under the UK Environmental Protection Act. A process subject to INTEGRATED POLLUTION CONTROL (e.g. boiler $> 50$ MWth output), waste incinerators ($> 1$ tonne per hour), cement manufacture, iron and steel smelting, paper manufacture.

**Prescribed substances.** Releases to air, water and land which are regulated by HER MAJESTY'S INSPECTORATE OF POLLUTION (HMIP) and/or the NATIONAL RIVERS AUTHORITY (NRA), such as SULPHUR OXIDES, NITROGEN OXIDES, MERCURY, ALDRIN, ENDRIN, SOLVENTS, DIOXINS, PESTICIDES.

**Pressure.** The force exerted per unit area. Atmospheric pressure is the amount of pressure exerted by the atmosphere above absolute zero pressure and which changes with elevation above sea level (atmospheric pressure decreases as altitude increases). Most pressure gauges indicate a pressure that is referenced to atmospheric pressure, i.e. atmospheric pressure = 0 bar gauge (more commonly shown as bar g) at sea level.

Pressure is measured in the SI system in newtons per square metre. *Note:* the BAR (1 bar $= 10^5$ N/m$^2$) and millibar (1 mbar $= 10^2$ N/m$^2$) are in very common use, the former for high pressures and the latter for variations in atmospheric pressure.

To obtain absolute pressure the atmospheric pressure must be added to the gauge pressure (i.e. that which is measured on pressure gauges).

At sea level, the mean atmospheric absolute pressure $= 1.013$ bar. So at 0 bar g, the absolute pressure $= 0$ bar g $+ 1.013 = 1.013$ bar abs. 2 bar g $= 2 + 1.013 = 3.013$ bar abs., etc.

**Pressure, partial.** The pressure exerted by one of the gaseous components of a mixture of gases on the assumption that it alone exists in the same volume as the mixture.

**Pressure vessel.** In NUCLEAR REACTOR DESIGNS, this is the reactor containment vessel built to withstand high internal pressure, constructed from steel or concrete.

**Pressurized water reactor (PWR).** ⇨ NUCLEAR REACTOR DESIGNS; SIZEWELL.

**Presumptive limit.** These are standards of emissions to atmosphere which represent the result which may reasonably be expected if the requirements of best practicable means (BPM) are applied satisfactorily to a works controlled by HMIP. Presumptive limits are not statutory limits, but they are supported by the force of law because works which fail to meet them are presumed not to be using BPM and may be prosecuted accordingly. One strength of these limits is that they are not statutory limits and so may be changed easily to meet the changing pollution control needs. The limits are set by HM Inspectorate of Pollution after consultation with the industry concerned, and while they are uniform within an industry across the country, they can be varied to meet particular circumstances at individual works. Presumptive limits provide a guide to works management in the design and running of plant where a simple measurement of emission indicates the degree of control exercised by the operators. The limits also assist the Inspectorate in the execution of its duties.

**Primary air.** The air supplied to a FUEL in its early stages of combustion, i.e. at or on the grate.

**Primary efficiency.** ⇨ PHOTOSYNTHETIC EFFICIENCY.

**Primary pollutant.** A pollutant emitted directly into the environment such as SO$_2$, CO. (⇨ SECONDARY POLLUTANT)

**Primary sludge.** The sludge consisting of settled solids from the primary stage of a SEWAGE TREATMENT plant. (⇨ SLUDGE, SEWAGE)

**Priority list.** ⇨ RED LIST.

**Probable reserves.** Underdeveloped oil or gas reserves which are deemed to be recoverable from known formations but not yet proven due to lack of data. (⇨ PROVEN RESERVES)

**Producer Responsibility Group (PRG).** ⇨ PACKAGING.

**2-Propenal.** A colourless liquid ALDEHYDE with a choking odour which occurs in PHOTOCHEMICAL SMOG.

**Proteins.** A group of nitrogenous organic compounds of high molecular weight (up to 10 000 000) which are essential components of all living matter. ENZYMES are an important class of proteins. (⇨ AMINO ACID)

**Proton.** A stable elementary particle with an electrical charge equal and opposite to that of an ELECTRON.

**Protozoa.** Small unicellular animals having a well-defined nucleus (in contrast to bacteria). They are found in a variety of habitats, e.g. fresh and salt water and soil, and occupy an important position in these ECOSYSTEMS where they normally consume dead organic matter and wastes, although they can also be parasitic. (⇨ ACTIVATED SLUDGE)

**Proven reserves.** The estimated quantities of oil or gas reserves (or other natural resource) which geological and engineering data demonstrate with reasonable certainty to be recoverable in future years from known reservoirs under existing technological and economic conditions. (⇨ PROBABLE RESERVES)

**Provisional tolerable weekly intake.** ⇨ MERCURY.

**Proximate analysis.** A simple method of analysis of solid fuels for measuring the percentages of free moisture, volatile matter, ash and FIXED CARBON. The last named is obtained by difference from 100 per cent. Proximate analysis is used to check quality of bulk deliveries and for daily works control. (⇨ ULTIMATE ANALYSIS)

**Proximity principle.** The requirement to treat wastes close to where they arise, e.g. within the boundary of the plant or community in which they are generated. The problem is not to be exported to someone else's back yard. This poses major problems for cities if they do not wish to export their wastes and implies reduction, recycling and the adoption of waste to energy practices, e.g. it is envisaged for London by 2005 that almost all its Municipal Solid Wastes will be treated in this fashion to avoid reliance on distant landfills in other administrative regions, such as Bedfordshire.

**Psychrometry.** The measurement of atmospheric humidity. Usually done by the use of two thermometers: (a) dry bulb; (b) wet bulb. Readings are converted to a humidity value by means of charts or tables.

**Public health.** Public health embraces the health of both the individual and the community. In its widest sense it means the mental and physical health of the people, which in turn ensures the well-being of future generations – a resource to be husbanded just as much as the resources of land, air and water. All environmental factors are involved: food (type, quantity, wholesomeness), condition of work, home and play, pollutants, noise, etc.

Originally public health was primarily concerned with the prevention of diseases, e.g. CHOLERA or other sewage-borne diseases that used to abound in the UK and still do in many countries. We are still dependent on these early concepts and practices to prevent the return of such diseases, but also the wider spheres outlined above have great importance once basic health is established.

Public health embraces an awareness of new hazards, risks or pollutants introduced by technology in the name of progress. The hazards or risks must be controlled so that the benefit is maximized to all. We also learn how pollutants or substances (e.g. VINYL CHLORIDE) are more dangerous than first realized. Thus, the field is dynamic and only through greater awareness of the risks of new substances of developments can progress be made. The aim of many of the entries in this book is to bring this awareness to the fore, so that we may use our new-found chemicals, process and practices wisely for the individual, the community and future generations.

The basic definition of public health has been given by C. E. A. Winslow, *The Cost of Sickness and the Price of Health*, World Health Organization, Monograph Series No. 7, 1951.

**Pulmonary irritants.** A group of air pollutants which affect the mucous lining of the respiratory tract which comprises the nasopharynx, the tracheobronchial area and the lung tissue or alveoli.

The nasopharynx defence mechanisms include the hairs at the nasal entrance, which filter out large particles, and the mucous glands, which wash out many of those particles that escape the hairs. The bronchial tree contracts to prevent dust entry (e.g. coal dust) and hence reduce the amount of particulate matter. Coughing also ejects mucus in which the particulates are trapped. While bronchial spasms lessen the amount of dust, they also reduce the amount of air received. The alveoli consist of minute air sacs (around 300 million) filled with capillaries through which oxygen enters the blood and carbon dioxide is removed.

The main respiratory diseases are:

1. Emphysema, where the alveoli lose their oxygen/carbon dioxide exchange properties caused by smoking and chronic exposure to air pollutants.
2. Pulmonary fibrosis. Scarring of the lung tissue.
3. Pulmonary oedema. A drowning of the lung tissue in fluid caused by exposure to highly concentrated amounts of irritant or corrosive pollutants.

Severe pulmonary irritants include NITROGEN OXIDES, SULPHUR OXIDES and CHLORINE which can cause severe bronchial problems.

*Source:* G. L. Waldbott, *Health Effects of Environmental Pollutants,* 2nd edn, C. V. Mosby, St Louis, 1978.

**Pulp.** The raw material for PAPER making. Pulp is usually obtained from trees, esparto grass and other long-fibred, CELLULOSE-containing materials. There are two main classes of pulp:

1. Chemical pulp, obtained from wood by chemical means, i.e. by sulphite, sulphate or soda processes which dissolve the LIGNIN and release the long cellulose fibres.
2. Groundwood or mechanical pulp, made by grinding the wood so that the fibre are separated. Groundwood is mainly used for newsprint because of its low quality, due to short fibre length.

The effluent from pulpmills has a high BIOCHEMICAL OXYGEN DEMAND and SUSPENDED SOLIDS and requires treatment before discharge. One recycling and pollution abatement method in certain pulpmill effluents is the manufacture of SINGLE-CELL PROTEIN by growing and harvesting yeasts on the dilute sugar solutions in the effluent.

**Pure oxygen activated sludge process.** The use of pure oxygen instead of air in the activated sludge process to treat wastewaters. Treatment is usually carried out in covered, completely mixed

tanks positioned in series (see Figure 119). In each stage
mechanical aerators ensure that the oxygen is dispersed
adequately within the tank contents. The reason for using pure
oxygen in place of air is to take advantage of the higher
concentration driving forces, primarily within the liquid phase,
to give higher oxygen transfer rates per unit reactor volume.
This allows for higher organic loadings (0.4–1.0 kg $BOD_5$/kg
MLVSS/day) compared to conventional air activated sludge
systems (0.2–0.6 kg $BOD_5$/kg MLVSS/day). Thus using pure
oxygen a significant increase in plant capacity can be achieved
without a corresponding increase in reactor volume. This
feature is attractive where severe space limitations are present.

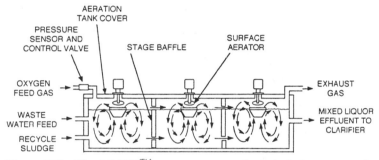

**Figure 119** The UNOX™ pure oxygen activated sludge process in
schematic form. Reproduced by permission of Wimpey Construction Ltd.

**Putrefaction.** Uncontrolled ANAEROBIC decomposition of organic
wastes, e.g. in refuse or in sewage effluent.

**Putrescible wastes.** Wastes that can undergo PUTREFACTION such
as food wastes, sewage SLUDGE, slaughterhouse residues.

**PVC.** ⇨ POLYVINYL CHLORIDE.

**Pyrethroids.** These are the active insecticidal constituents of
pyrethrum flowers (*Chrysanthemum cinerariaefolium* and *Chrysanthemum coccineium*). The insecticidal properties are due to
five differing compounds: pyrethrins I and II, cinerins I and II,
and jasmolin II. They are present in the achenes (single-seeded
fruits) of the flowers in concentrations of 0.7 to 3 per cent, and
may be extracted by organic SOLVENTS. As the cost of this
extraction from the flowers is high, synthetic derivatives called
allethrins are in common use.

Pyrethroids are highly unstable to the action of air, moisture
and alkalis. The residues deteriorate rapidly after application.

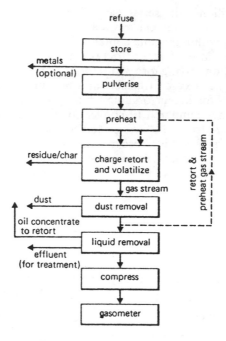

**Figure 120**  Flow diagram for pyrolysis of refuse for gas production.

The characteristic insecticidal action is a very rapid knock-down of insect life followed by a substantial recovery due to detoxication enzymes present in the organisms.

Pyrethroids have low mammalian toxicity because of rapid detoxification by enzymes in the body.

**Pyrolysis.** The heating of organic matter of wastes containing organic matter such as DOMESTIC REFUSE in a closed retort in the absence of air (see Figure 120). The subsequent volatilization produces combustible gases, a low-calorific value combustible char, a mixture of oils and liquid effluent.

The gas has a CALORIFIC VALUE half that of natural gas and requires modified appliances for its combustion. The oil may not be present in sufficient quantities to justify a refinery stream in the UK and is therefore sent back for gasification in the pyrolysis reactor. The char can be upgraded to a fuel equivalent to low-grade COAL. The liquid effluent is treated to prevent water pollution. The process requires around 50 per cent of the fuel produced – the remainder is available for sale.

Pyrolysis is an embryonic waste disposal process which is receiving much attention in the USA as a means of conserving energy resources and recycling organic wastes. It is already in use for energy recovery from tyres and waste PLASTICS.

# Q

**$Q_{10}$.** The amount by which the growth or activity of an organism or an enzymic reaction increases per 10°C rise in temperature.

**Quality assurance.** The guarantee that the quality of a product or service is actually what is claimed on the basis of the quality control applied in creating the product or providing the service. Quality assurance is there to protect against lapses in QUALITY CONTROL.

**Quality control.** The maintenance of the quality of a product or service above a minimum standard based on known and accepted criteria.

**Quality factor (Q).** This ranges from 1 to 20 and is used to weight the biological effects that IONIZING RADIATION has on human tissue.

- for X-rays, gamma rays ($\gamma$), and beta rays ($\beta$), Q = 1
- for NEUTRONS                                    Q = 10
- for alpha ($\alpha$) particles                  Q = 20

(⇨ IONIZING RADIATION; DOSE MEASUREMENT)

**Quartz.** Crystalline form of silica ($SiO_2$). Usually white coloured but can be colourless. Is a major constituent of granite, schist and sandstones. GLASS making requires quartz sands as a raw material.

**Quicklime.** Alternative description, burnt lime, chemical description, calcium oxide, CaO.

Quicklime is one of the most versatile chemical products available and is indispensable to many industries: steelmaking, construction products, industrial chemicals, oil, pharmaceuticals and many environmental treatment processes.

# R

---

**Rad.** A measure of absorbed radiation dose = 0.01 joules per kg, now replaced by the gray (Gy) = 100 rads. ($\Rightarrow$ IONIZING RADIATION (DOSE MEASUREMENT))

**Radial flow tank.** A circular tank with a central inlet used in sewage works and water treatment processes. The outlet consists of a weir around the circumference. The floor usually slopes to the centre where settled sludge can be drawn off.

**Radiation.** Figure 121 shows the sources of radiation to someone living in the UK. This is now 2.5 millisieverts (mSv) per year. The actual dose will depend on life-style and geographical location.

**Radiation sickness.** The severe effects caused by a large dose of ionizing radiation to the whole body, e.g. from a nuclear reactor accident or nuclear weapon fallout. (Known as a non-stochastic effect by the nuclear professionals.) ($\Rightarrow$ STOCHASTIC EFFECTS)

**Radical (free radical).** A reactive atom or portion of a stable molecule which contains an odd number of bonding electrons. Characterized by very short atmospheric lifetimes in the lower atmosphere but longer lifetimes in the upper atmosphere. Typical atmospheric free radicals are $HO$, $HO_2$, $Cl$, $NO_3$, $NH_4$. They are responsible for most atmospheric chemical reactions.

**Radioactive decay.** $\Rightarrow$ RADIONUCLIDE.

**Radioactive half-life.** $\Rightarrow$ HALF-LIFE.

**Radioactive isotope.** The nucleus of an atom is made up of NEUTRONS and PROTONS. The number of protons in the nucleus is called the ATOMIC NUMBER and characterizes the chemical element. The total number of neutrons ($n$) and protons ($p$) determines the mass of the nucleus. If two atoms have nuclei with the same number of protons but differing numbers of neutrons, they are said to be atoms of different isotopes of the same chemical element, and are shown with their respective mass number ($n + p$), e.g. $^{235}U$ and $^{238}U$ are isotopes of uranium. The isotopes of an element are identical in chemical

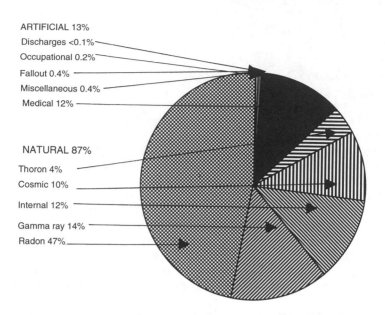

ARTIFICIAL 13%
Discharges <0.1%
Occupational 0.2%
Fallout 0.4%
Miscellaneous 0.4%
Medical 12%

NATURAL 87%
Thoron 4%
Cosmic 10%
Internal 12%
Gamma ray 14%
Radon 47%

**Figure 121** Composition of the total radiation exposure to the UK population.
*Source:* Environment in Trust, *Radioactive Waste Management – A Safe Solution*, Department of the Environment.

properties and in all physical properties except those dependent on atomic mass.

Nuclear changes may result from neutron bombardment which transmutes an atom of one element to either an atom of another element or an atom of a different ISOTOPE of the same element.

For the use of the isotopes of uranium, ⇨ HALF-LIFE; IONIZING RADIATION; NUCLEAR ENERGY.

**Radionuclide.** An atom that has an unstable nucleus which spontaneously disintegrates and emits ALPHA and BETA PARTICLES or gamma radiation, or both. Radium, uranium and strontium are examples of radionuclides, although radioactive isotopes of all common elements – carbon, potassium and hydrogen, for example – occur naturally. The process of spontaneous transmutation is called radioactive decay. (⇨ HALF-LIFE; IONIZING RADIATION; RADIOACTIVE ISOTOPE)

**Radon.** A radioactive gaseous element emitted naturally from rocks and minerals where radioactive elements are present. It is released in non-coal mines, e.g. tin, iron, fluorspar, uranium. Radon is an ALPHA PARTICLE emitter as are its daughter or decay products and has been indicted as a cause of excessive occurrences of lung cancer in uranium miners. New exposure standards have been adopted in the USA, Sweden and the UK based on a standardized dose measure called a 'working level month' or WLM. The new standards stipulate that a miner should not be exposed to more than 4 WLM per year.

Concern has been expressed at radon levels in some UK housing, usually adjacent to granite rocks or old tin mining regions. Levels of radon up to 300 BECQUERELS per cubic metre have been found. These can be reduced substantially by simply installed ventilation systems. An action level of 200 Becquerels per cubic metre would warrant action to reduce the concentration. This level is under review.

**Rain.** Precipitation of water from clouds. In the UK this can fluctuate greatly. For example, at the Welsh hydro-power plant at Rheidol, more than 178 mm (seven inches) of rain fell on the mountains above Nant y Moch reservoir between 10.00 am on the 26th and 10.00 pm on the 28th December 1994 – 10 per cent of the average annual rainfall in just two-and-a-half days.

It was more than a third of the total rainfall for the month, in what was already the third wettest December since the station opened in 1963.

In just seven hours on the 27th, 3 000 000 cubic metres (675 million gallons) of water flowed into Nant y Moch. The reservoir has a capacity of 24 million cubic metres.

Downstream, water cascaded down the Rheidol at more than 40 cubic metres a second. Normal winter flow is three to four cubic metres per second.

The incidence of such heavy fluctuations in precipitation means that landfilling of wastes in high rainfall areas is an extremely difficult task which can lead to severe ground and surface water pollution. (⇨ LANDFILL, HYDROLOGICAL CYCLE)

*Source: The GEN, February 1995, p. 7.*

**Real-time measurement.** A measurement that is made simultaneously with the event that is measured. In pollution studies it is to be contrasted with integrating, or time-averaging, methods in which the average concentration of a pollutant over a given period of time is determined. Real-time measurements are of particular importance when the pollutants involved are hazardous to health.

**Recessive genes.** ⇨ IONIZING RADIATION EFFECTS.

**Recharge.** The replenishment of AQUIFERS by natural or artificial means. Natural recharge includes precipitation, seepage from streams and lakes or by underground leakage. Artificial recharge eusually involves the construction of recharge basins on suitable porous strata with water purified from near at hand rivers or canals.

**Recombinant DNA technology.** A range of modern techniques in molelcular biology which enable sections of DNA to be transferred between individuals of the same or different species. This enables beneficial characteristics to be introduced into organisms which may be crop plants, sources of chemicals or microbes to be used for destroying toxic wastes. In this way tomatoes which ripen on demand, bacteria which produce human insulin, crops resistant to herbicides and cocktails of organisms for decontaminating polluted land have been produced. Concern has been expressed about the escape of such 'synthetic' organisms into the environment, particularly when bacteria resistant to antibiotics are concerned. (⇨ GENETIC POLLUTION)

**Record keeping.** For pollution control purposes, statutory log of all monitoring and inspection required by HMIP or other regulatory bodies.

**Recycling.** Reusing a material not necessarily in its original form. The natural recycling of the substances required for life are the keystone of our existence on earth. In the CARBON CYCLE the reuse of the same material is obtained with the aid of solar energy.

We cannot by-pass the conservation laws of mass or energy and in a world of increasing material scarcity it is important to make the best use of all resources. The intelligent adoption of recycling techniques allied with good design practice which allows for materials to be reclaimed after their useful life is over can do much to conserve raw materials and energy, minimize pollution and save money. It would be too simplistic to expect people voluntarily to cut back their standard of living so that major energy and materials savings can be made, but WASTE MINIMIZATION followed by recycling does offer the potential for considerable savings without major sacrifices on the part of the consumer.

Recycling falls into three classes.

*Reuse:* this is typified by the returnable bottle which makes several trips from bottler to consumer and back again where it is cleaned and refilled. Reuse may be allocated the highest availability in the recycling spectrum in that least energy and process complexity is normally expended in getting the material or article back into use. Typical reuse items are compressed gas cylinders, 45 gallon drum (a worldwide standard item).

*Direct recycling:* using the returnable bottle as our example, once it is unfit for reuse it may be cleaned and broken down to CULLET at the glassworks and used to make more bottles. Direct recycling depends on the quality of the recycled material and its cost, which should not exceed that of the raw material. Currently most direct recycling occurs at the factory where the product is made, e.g. misshapen or broken bottles formed during glass manufacture are in fact fed back to the melting chamber. Industry calls this recycling and it is not to be confused with material reclaimed from waste or point of use. Thus paper with a 20 per cent recycled content may in fact be paper where surplus pulp fibres, mill offcuts and spoiled rolls have been internally re-routed back through the pulping process. Direct recycling has an intermediate availability in that both energy expenditure and process complexity may be required in getting the material back into use.

*Indirect recycling:* this practice often makes no pretence at reclaiming the material for use as such but rather gets a second bite at the cherry. Continuing with our glass bottle, it is quite probably that it will eventually end up in domestic refuse where it can be extracted by screening and separation in conjunction with other bottles. These bottles will probably be of different colours and varying degrees of cleanliness and are unsuitable for cullet use unless costly optical sorting is used. The bottles may, however, be ground up and used for a highly skid-resistant and durable road-surfacing material. Similarly, waste plastic containers which *en masse* are unsuitable for direct conversion to new containers may be ground up and used for plastic fence-posts, pallets and chipboards, where appearance and structure are not primary considerations. Other forms of indirect recycling embrace the conversion of refuse to combustible gases, or the use of heat from the combustion of refuse for district heating by means of INCINERATION with heat recovery. Indirect recycling is the lowest form of recycling. Normally once processed in this phase, the material is no longer available for

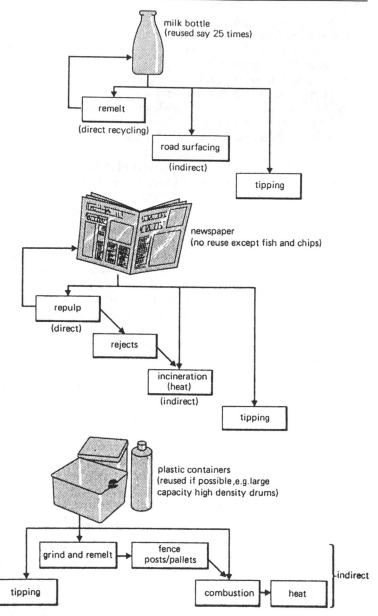

**Figure 122**   Recycling routes for various consumer products.

use except for landfill or incineration. The downgrading in use of several typical products is shown in Figure 122.

The use of LIFE CYCLE ANALYSIS can help to decide whether recycling saves resources or if it is better to incinerate with energy recovery as is the case with many PLASTIC materials in household waste.

LANDFILL tipping is not a form of recycling, although many claims may be made for it to be thought of in this way; it is the sink for discarded materials, just as our surroundings form the last resting place for degraded energy. (⇨ ENERGY; LAWS OF THERMODYNAMICS)

**Recycling, financial incentives for.** A variety of taxes and levies could be imposed as incentives for recycling materials, and hence conserving resources, or to cover the costs of disposal of used materials.

*Direct subsidies to reclaimers.* Grants and price-support systems would ensure that recycled materials had the financial edge over original materials.

*Disposal tax.* A tax levied on a product according to the cost of disposal, e.g. a levy on a new car to allow for its eventual disposal. Suitable allowances (i.e. deductions or avoidance of tax) can be included for recyclable products. Another version is a LANDFILL or MINIMIZATION surcharge as every tonne of material sent for disposal (implemented by Sweden in 1993).

*General pollution tax.* A version of the POLLUTER-PAYS PRINCIPLE. If one assumes that recycling processes give rise to less pollution, this would favour recycling.

*Packaging tax.* This to some extent overlaps with the disposal tax. Products would be taxed on the packaging value added or the amount of packaging used over a certain limit adjudged sufficient for the purpose. This could work in conjunction with container deposits to encourage the return of used containers for recycling.

*Virgin materials tax.* A tax applied to limit the use of particular materials by raising the price artificially. This would reflect their lifetime at current rates of use, e.g. for oil based on the RESERVES–PRODUCTION RATIO.

*Virgin materials levy.* Levies imposed on primary materials by producer countries to restrict consumption. Britain is likely to be on the receiving end of such a measure. (⇨ COLLECTION TAX)

*Recycling levies.* The table below shows Incpen's calculation of the fees payable by packers and fillers in four countries for recovering value from used packaging. One reason for the disparity is that the criteria on which levies are based and the method of calculation are not the same in all cases. Nonetheless, the figures are a good indication and represent a fairly realistic picture of the situation at present. French and Belgian levies may well rise in the future as their systems are not yet operating at full capacity. However, there is evidence to suggest that their fees will still end up considerably lower than those in Germany and Austria.

|           | *Levy (£/tonne)* | | | |
|-----------|---------|--------|--------|---------|
|           | *Belgium* | *France* | *Austria* | *Germany* |
| Paper     | 9       | 37     | 161    | 170     |
| Plastic   | 224     | 30     | 922    | 1255    |
| Glass     | 4       | 2      | 45     | 64      |
| Steel     | 30      | 16     | 265    | 240     |
| Aluminium | 50      | 33     | 394    | 638     |

Germany has very high recycling targets and places substantial hurdles in the path of waste to energy. France, on the other hand, is much more pragmatic, preferring VALORIZATION instead, which admits a spectrum of value recovery methods.

*Source: Incpen Newsletter,* Autumn 1994.

Figure 123 shows the effects of various levels of recycling on the fuel value or CALORIFIC VALUE (CV) of MUNICIPAL SOLID WASTE.

It can be seen that the CV is noticeably unaffected by recycling and would still remain within the WASTE TO ENERGY (WTE) plant operating range. (⇨ BOTTLES, PAPER)

It may reasonably be asked if the recycling of some of the principal components of MSW, especially bottles and paper, actually save energy or resources when ENERGY ANALYSIS is performed. An optimum needs to be struck which permits the recovery of the energy content of MSW too. (⇨ VALORIZATION)

**Recycling, product specification, legislation for.** This idea is based on the inherent wastefulness of the one-trip bottle and the throw-away-after-use syndrome. Legislation should be able to outlaw (in most cases) this wastefulness, and insist on a reusable or recyclable design. For example, the State of Minnesota has

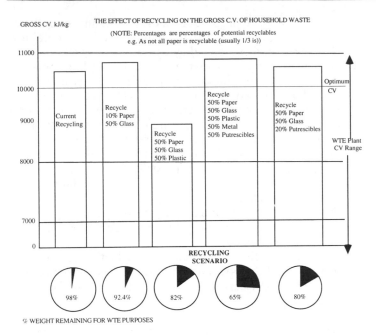

**Figure 123** Recycling effect on calorific value.
*Source:* Hampshire County Council.

legislated for packaging controls which require all new packages and any changes in the packaging of existing products to be examined to see if they constitute a solid waste disposal problem or are inconsistent with the State's environmental policies.

Such legislation must be applied fairly; the glass, one-trip bottle should not be made the only banned form of container. Aluminium and plastics are much more energy intensive and thus they and other packaging materials must also be considered.

Product specification at the design stage can also achieve many of the desired ends of recycling legislation. The following criteria are suggested as being of use:

1. Increase the product's lifetime where possible, eliminating planned obsolescence. This should not only include increased durability but ease of repair and maintenance.
2. Design products for reuse or for multiple use where suitable.

This would mean in some cases the redesigning of bottles to standard sizes and shapes with no non-standard moulding or closure methods so that they can be used by any bottler.

3. Design for ease of reclamation and recycling. Where a product presents unnecessary problems in reclamation, it should be replaced by a more easily reclaimed product. This will depend on many factors, not least the local conditions for reclamation. However, a number of products can be identified as difficult to reclaim and can be dealt with.

4. Design for disposal. If disposal is a problem, the product should be replaced by a more easily disposable product.

5. Use the least energy-intensive material that will do the job. Also, take into account the relative scarcity of materials and the lifetime of the product if it is to be reused.

6. Consideration should be given to the polluting effects of a material's manufacture.

7. Fully inform the public of the product's suitability for recycling and the raw materials and energy consumed in its production. Also certification of the environmental impact of its disposal.

**Recycling case study – paper.** In paper recycling, the environmental benefits of recycled fibres are all too often counteracted by the negative effects of the largely untreated residual products emitted from the recycling plants.

Stora Papyrus Dalum in Denmark is the first fine paper manufacturer to resolve this environmental dilemma. All the raw material (waste paper) is utilized and reused in different ways. No residues leave the mill to end up on the landfill or pollute the water or air.

The raw material for both Cyclus and Cyclusprint (both 100 per cent recycled fines) is waste paper from printing works and offices, mostly in Denmark, northern Germany, southern Sweden and Norway (the Oslo area).

The waste paper is converted into pulp at the Stora Papyrus de-inking plant at Naestved, south of Copenhagen. The waste paper is de-inked and the filler and coating are separated from the paper fibres.

The stock (mixture of fibre and water) is then pressed into thick sheets and packed in pulp bales, ready for delivery by rail to the papermill in Odense.

The actual de-inking plant itself is similar to the plants in most other mills making pulp from waste paper. What makes

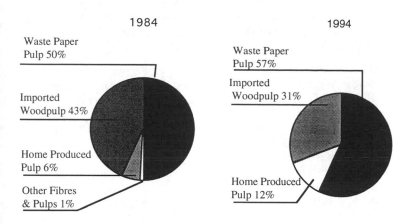

**Figure 124**
*Source:* Paper Federation of Great Britain Industry Facts, published 1995.

Naestved unique is the fact that the residual products from de-inking are 100 per cent recycled.

From 100 000 tonnes per year (tpy) waste paper, 75 000 tpy fibres are recovered, leaving 15 000 tpy clay and chalk which is sent to a nearby cement factory, and 10 000 tonnes of sludge, plastics, etc. The sludge is used as a land conditioner, and any plastics and combustibles are sent to the local district heating plant.

The company has put a lot of effort into making the whole process virtually closed, and has halved the water consumption in three years. Water which is released is pumped to Odense Council Sewage Works for final purification. The Council Technical Department has compared the quality of the water upstream of the mill with the water after purification. Their tests show that the water released is cleaner than the water taken in.

*Source: Sweden Today,* January 1995.

The UK paper recycling scene is shown in Figure 124. It can be seen that waste paper consumption as percentage of usage has increased from 50 per cent (1984) to 57 per cent (1994). It should be noted that many statistics are often based on total consumption of paper and board (in 1994, 11.6 million tonnes per year) when a better comparison is in the UK production of paper and board which shows that 57 per cent of all furnish comes from recycled sources (1994).

Some imports will always be necessary into the UK and it is to the great credit of the British paper industry that it achieves such a high level of recycling already.

**Red List.** A list of 23 dangerous substances, designated by the UK, whose discharges to water should be minimized under the BATNEEC principle and whose DISCHARGE CONSENTS are required to ensure that strict environmental quality standards are met and maintained in the receiving waters. The NATIONAL RIVERS AUTHORITY would have to be satisfied that this can be done before consent is granted by HMIP.

In addition, the *Third North Sea Conference*, March 1990 agreed a priority list of 39 substances whose discharge to rivers and estuaries should be reduced by 50 per cent by 1995 (over 1985 levels). The list includes the Red List substances, plus those below:

| *Red list substances* | *Additional substances on priority list* |
|---|---|
| Mercury and its compounds | Copper |
| Cadmium and its compounds | Zinc |
| Gamma-hexachlorocyclohexane | Lead |
| DDT | Arsenic |
| Pentachlorophenol | Chromium |
| Hexachlorobenzene | Nickel |
| Hexachlorobutadiene | Chloroform |
| Aldrin | Carbon tetrachloride |
| Dieldrin | Azinphos-ethyl |
| Endrin | Fenthion |
| Polychlorinated biphenyls | Parathion |
| Dichlorvos | Parathion-methyl |
| 1,2-Dichlorethane | Trichloroethylene |
| Trichlorobenzene | Tetrachloroethylene |
| Atrazine | Trichloroethane |
| Simazine | Dioxins |
| Tributyltin compounds | |
| Triphenyltin compounds | |
| Trifluralin | |
| Fenitrothion | |
| Azinphos-methyl | |
| Malathion | |
| Endosulfan | |

($\Rightarrow$ BLACK LIST; GREY LIST)

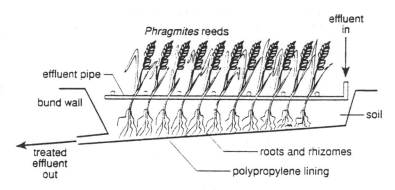

**Figure 125** Cross-section through a reed bed.

**Reducing agent.** Chemical which combines with oxygen, i.e. removes if from a substance. Term also used for the addition of hydrogen to a substance. The opposite of an OXIDIZING AGENT.

**Reed beds.** These are beds of soil or gravel media in a sealed pit containing reeds (usually the species *Phragmites australis*) for treatment of effluent (see Figure 125). They were first used for the treatment of sewage in Germany in the 1960s. The effluent to be treated is distributed through pipes and nozzles onto the bed. The rhizomes (underground stems) of the reeds grow vertically and horizontally, opening up the bed to provide a passage for effluent. OXYGEN is passed via the leaves and stems of the reeds, through the hollow rhizomes, and out through the roots to the rhizosphere (the area surrounding the rhizomes). Here large populations of aerobic bacteria effect the biodegradation of the organic matter in the effluent. Anaerobic bacteria in anoxic areas of the soil also contribute to breakdown of pollutants.

In reed bed systems for sewage treatment, screened, settled sewage is passed through horizontal flow beds. BOD removal is in the range of 80–90 per cent.

Owing to the great diversity of microbiol species in soil, flexibility is possible with reed beds; for instance, the fungal species *Actinomyces, Streptomyces* and *Basidiomyces*, which are capable of biodegrading many synthetic chemicals such as the common pesticides and chlorinated hydrocarbons, are found in soil but not normally in effluent treatment plants. Fully mature reed beds have been used to remove traces of phenol, methanol, acetone and amines from industrial effluents.

Unlike mechanical treatment plants, reed beds are not noisy or unsightly. They do, however, require a greater land area for equivalent treatment levels.

**Refrigerants.** A fluid medium (gas or liquid) for removing heat rapidly in refrigerating equipment.

*Refrigerant choices:*

*CFCs* (mainly CFC 11 and CFC 12). These substances are highly damaging to the ozone layer, and also contribute to global warming. Further use of these materials will rely on existing stocks and recycling; inevitably, prices are rising fast. HCFCa (mainly HCFC 22 and HCFC 123). Are substantially less damaging to the ozone layer, and were not controlled under the original protocol. However, European legislation now restricts production and use after the year 2000 and the timescale for total phase-out is expected to tighten.

*Ammonia.* This was the workhorse of the refrigeration industry until about 30 years ago, when CFCs were introduced as the safer options. Many locations have continued to use ammonia (without serious incident). It is cheap and efficient, does not contribute to ozone depletion or global warming, but needs careful handling because it is toxic and mildly flammable. Often the best option as it is a natural substance and is cheap and effective.

*Other refrigerants.* These mainly consist of the HFCs, a new class of materials which contain no chlorine and therefore do not damage the ozone layer. If released into the atmosphere they do, however, contribute to global warming. They have been commercially available only for the last 2–3 years. HFC 134a is by far the most widely used of the HFCs, and is generally considered to be non-toxic and non-flammable. There are also many HFC-blend refrigerants tailored for specific jobs, mainly for small to medium size commercial chillers, rather than large process cooling units. (⇨ CFCs)

*Source: Courtald's Environment Matters, March 1995.*

**Refuse.** ⇨ DOMESTIC REFUSE.

**Refuse derived fuels.** ⇨ WASTE DERIVED FUELS.

**Reinluft process.** A process for removing (and recovering) oxides of nitrogen and sulphur, from stack gases. The gases are mixed with oxygen and converted to a higher oxidation state (a CATALYST may be used to increase the efficiency of conversion) and passed upward, in countercurrent fashion, over a bed of carbon in a multi-chambered adsorber. The oxides are adsorbed at different temperatures. Finally, they are swept out of the carbon bed at high temperature in a stream of nitrogen or carbon dioxide and reduced. The process may be carried out in the presence of water vapour, in which case the corresponding acids, rather than the gaseous oxides, are recovered.

**Relative biological effectiveness (RBE).** ⇨ IONIZING RADIATION (DOSE MEASUREMENT).

**Relative humidity.** The ratio of the actual amount of moisture in the air to the amount needed for saturation at the same temperature.

**Release (Environmental Protection Act 1990).** Release includes any emission of a substance into the air in relation to water, any entry including any discharge of the substance into water/sea and in relation to land, any deposit, keeping or disposal of the substance in or on land.

**Rem (Roentgen Equivalent Man).** A measure of the effective radiation dose absorbed by human tissue. It is the product of the dose in RADS and the QUALITY FACTOR. Now replaced by the Sievert (Sv) = 100 rem. (⇨ IONIZING RADIATION (DOSE MEASUREMENT))

**Renewable energy.** Replenishable power source (⇨ SOLAR ENERGY; TIDAL POWER; WAVE POWER). A means of attaining SUSTAINABLE DEVELOPMENT.

The UK has a target of 1500 MWe of new electricity generating capacity by the year 2000. This is assisted by the NON-FOSSIL FUEL OBLIGATION ORDER. Figures 126 and 127 give the 1993 renewable energy breakdown and status, respectively of UK renewable energy technologies.

The table overleaf gives a summary of 1991 and 1994 Non-Fossil Fuel Obligation (NFFO) contracts for England and Wales.

*Source:* Digest of UK Energy Statistics, 1994, 'The Government's Policy for Renewables and the Non-Fossil Fuel Obligation', *Renewable Energy Bulletin* No. 5, Department of Trade, October 1993, p. 5.

**Repeatability.** The closeness of agreement between successive results obtained with the same method, test material and under the same conditions (e.g. same laboratory, same apparatus).

**Reprocessing.** The reclamation of in-house arisings and those purchased from within the relevant industry for the subsequent manufacture of finished products.

|                        | 1991   | 1994 (*NFFO* 3) |
|------------------------|--------|-----------------|
| Waste                  |        |                 |
|   MW DNC*     | 271.48 | 241.87          |
|   No. of schemes | 10 | 20              |
|   Price p/kWh | 6.55  | 4               |
| 'Other'                |        |                 |
|   MW DNC      | 30.15  | 122.06          |
|   No. of schemes | 4  | 9               |
|   Price p/kWh | 5.9   | $8.75^1/5.23^2$ |
| Hydro                  |        |                 |
|   MW DNC      | 10.86  | 14.48           |
|   No. of schemes | 12 | 15              |
|   Price p/kWh | 6.00  | 4.85            |
| Landfill gas           |        |                 |
|   MW DNC      | 48.45  | 82.07           |
|   No. of schemes | 28 | 42              |
|   Price p/kWh | 5.7   | 4               |
| Sewage gas             |        |                 |
|   MW DNC      | 26.86  |                 |
|   No. of schemes | 19 |                 |
|   Price p/kWh | 5.9   |                 |
| Wind                   |        |                 |
|   MW DNC      | 84.43  | $145.9^3/19.7$  |
|   No. of schemes | 49 | 31/24           |
|   Price p/kWh | 11.0  | 4.8/5.99        |
| Cumulative capacity[4] |        |                 |
|   MW DNC      | 472.23 | 626.9           |
|   No. of schemes | 122 | 141            |

*DNC = Declared NeH Capacity
[1] Energy crops.
[2] Residuals.
[3] Wind has two bands, $> 1.6$ MW and $< 1.6$ MW, respectively.
[4] 340 MW is already operational from NFFO 1 and 2 contracts. It is anticipated that NFFO 3 will double this capacity in practice.

**Reproducibility.** The closeness of agreement between individual results obtained with the same method on identical test material but under different conditions (e.g. different laboratories or apparatus).

**Reserves.** The amount of substance, e.g. oil, that still remains available to be exploited at a cost below a certain level. They are normally referred to as recoverable reserves, i.e. the total quantities of oil that are likely to be brought to the surface for commercial use within a certain time-span and level of technology.

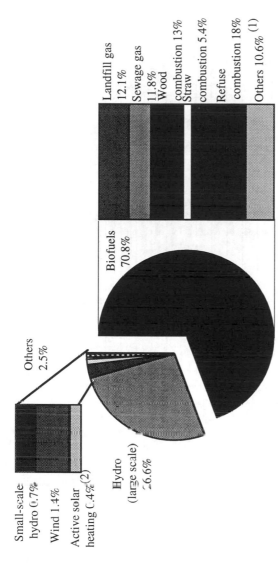

**Figure 126**   Renewable energy use in UK in 1993.
(1) 'Others' includes farm waste digestion and chicken litter, industrial and hospital waste combustion.
(2) Excludes all passive use of solar energy.
*Source:* Dukes, 1994.

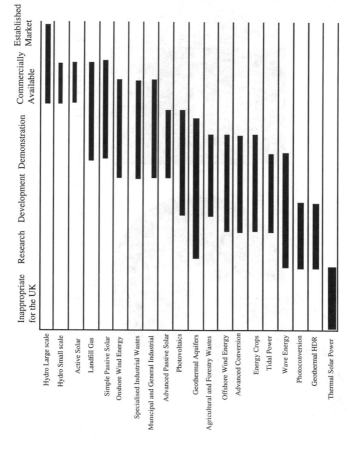

**Figure 127** Current status of renewable technologies in UK (1993).

**Reserves–production ratio (R/P ratio).** The ratio between the annual production rate and the proved recoverable reserves remaining in the ground, i.e. the number of years of production remaining at the rate of production in the year in question. Despite the impressive discoveries of oil, the R/P ratio has recently been falling steadily due to a continually increasing demand.

The current (1988) global R/P ratio for oil is 41. The UK's is estimated at 5, but this shifts as new ways are found to extend the life of oil fields.

It should be noted that a low R/P ratio does not allow much time to develop alternative energy sources and life-styles, or contain economic growth and stabilize populations and ENERGY CONSERVATION is one sure way of extending R/P ratios.

*Source: B.P. Statistical Review of World Energy, July 1989.*

**Reservoirs.** The provision of storage capacity to remove fluctuations in the flow of a material or component. A controlled reservoir has a means of discharging flow control. An uncontrolled reservoir only has a spillway which allows discharge when capacity is attained.

**Residence time.** In air pollution studies, the length of time during which a given molecule of an air pollutant remains in the atmosphere (it may, at the end of that time, be replaced by another molecule, so the residence time is not the duration of air pollution). In chemical engineering, the term is applied to the length of time that a given material remains in a vessel through which it is flowing.

**Residual fuel oil.** The residue from the distillation of petroleum which is not economical to process further. Can be blended with 'heavy' distillates as fuel for low-speed reciprocating engines or used blended as a boiler fuel.

**Resource recovery.** Recovery of materials, fuel or energy from waste. (⇨ PACKAGING)

**Resources.** Anything that is of use to man. (⇨ RESOURCES, RENEWABLE; RESOURCES, NON-RENEWABLE)

**Resources, non-renewable.** Substances which have been built up or evolved in a geological time-span and cannot be replaced except over a similar time-scale. Examples are copper, tin, COAL and OIL. It is often (erroneously) stated that when, for example, high-grade copper ore runs out, low-grade copper ore will become economically workable. However, this view neglects the facts of energy resources depletion and increasing pollution with lower grade burdens. Even if there were unlimited supplies of energy, the limitations imposed by the LAWS OF THER-MODYNAMICS and climatic stability mean that there are limits to how much energy man may use in working low-grade ore deposits. Furthermore, ore sources do not necessarily become more plentiful with lower grades. (⇨ ARITHMETIC–GEOMETRIC RATIO)

RECYCLING is one method of conserving finite resources. Some resources such as land, water and air have definite limits on the amount of exploitation they can sustain. Water is a renewable resource, but it is finite in its rate of supply as dictated by the HYDROLOGICAL CYCLE.

**Resources, renewable.** Resources that derive from solar energy such as fish, trees, wind, rain. Plant-life such as timber or grass should be managed for MAXIMUM SUSTAINABLE YIELD. If the yield exceeds this rate, the system gives ever-diminishing returns. In a fishery pushed past its limit, the catch is maintained by collecting greater and greater numbers of younger fish until there is extinction. Resource management is now a necessity due to the pressures of population growth and affluence. (⇨ WHALES)

**Reverse osmosis.** A means of separating solutions such as sugar in water or desalting by means of a suitable semi-permeable membrane and the application of pressure to the solution. (⇨ DESALINATION)

**Revolving screen.** Separation device, used in gravel extraction, rock-crushing and domestic refuse RECYCLING, consisting of a perforated drum which allows materials of varying sizes to be graded and removed. (⇨ AIR CLASSIFIER)

**Reynolds number.** A dimensionless parameter (i.e. a pure number which has no units) which is one of several such numbers used in the study of fluid flow. It is the ratio of inertial forces to viscous forces and is defined as $RE = \rho v l / \eta$, where $\rho$ is the density of a fluid, $\eta$ its viscosity and $v$ its velocity, and where $l$ is a linear dimension that depends on the problem under study (for flow in a pipe, for example, it is the diameter of the pipe). The value of the Reynolds number helps to determine whether flow will be laminar or turbulent, i.e. 'streamlined' or 'mixed up'. The determination of a flow regime is of crucial importance in sewer and water pipeline design. (⇨ LAMINAR FLOW; TURBULENCE)

**Richter scale.** A logarithmic scale developed in 1935 that measures how much energy is released by ground movement from the centre of a seismic shock. A 6.0 earthquake is 10 times more powerful than a 5.0 earthquake and a 7.0 earthquake is 100 times more powerful and so on. The October 1989 California earthquake measured 6.9 on the Richter scale. The greatest ever recorded is 8.9. (⇨ LOGARITHMS)

**Right to discharge.** An alternative to the principle that the polluter must pay is the proposal that any firm wishing to discharge a pollutant to the environment would have to pay for the right to discharge such pollutants. The interpretation is essentially a change in emphasis, from curative to preventive legislation, and would have the advantage of ensuring that firms maximize their benefits after payment. (⇨ POLLUTER-PAYS PRINCIPLE; BEST PRACTICABLE MEANS)

**Ringelmann charts.** A series of four charts of graduated shades from white to black for assessing the darkness of a plume of smoke by visual comparison. The standard chart (BS 2742) is a set of numbered grids differing from one another in the width and spacing of black lines printed on a white background.

Ringelmann **1** is equivalent to 20 per cent black, **2** is 40 per cent black, **3** is 60 per cent black, **4** is 80 per cent black. The scale may be extended to shade **0** (white) and shade **5** (black). The charts must be used at a certain specified distance so that the grid forms a grey scale. Micro and miniature smoke charts are also available and are more convenient to use.

In law 'dark smoke' is smoke which is as dark or darker than Ringelmann **2**, and 'black smoke' is as dark or darker than Ringelmann **4**.

**Risk.** The chance of a particular adverse effect occurring in a given period of time, e.g. the chance of causing illness or death per year. Low risk can only be defined by comparison with every day life, e.g. heart disease (UK) accounts for approximately 50 per cent of male and female deaths; 50 children per day (UK) are admitted to hospital with cigarette smoke related medical conditions.

Annual accidents UK (1993):

| Road deaths | 3470 | injuries | 300 000 |
| Industrial deaths | 430 | injuries | 168 000 |

*Risk assessment*

When risk is assessed it is important to take on board that linear extrapolation of effects from high to lower doses is often not valid. In a third or more of instances in which a maximum tolerable dose elicited extra tumours in rodents, one-half that dose did not. Many scientists have pointed out that huge doses of non-genotoxic substances are accompanied by toxicity, cell death, and cell replacements. This creates conditions favourable for growth of tumours. At doses in which cellular death does not occur, tumours would not be produced by non-genotoxic substances. The majority of chemicals are not genotoxic, nor does metabolism of them give rise to genotoxic intermediates. Thus the linear extrapolation is not applicable to the majority of chemicals. However, the extreme care taken is shown by the following example of dioxins.

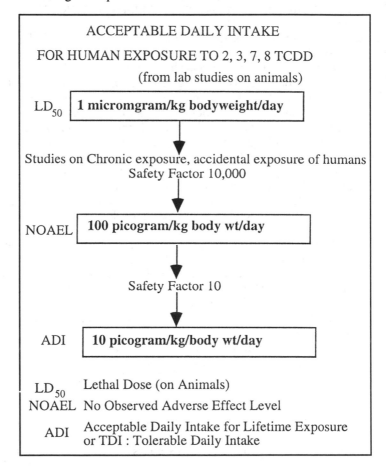

ACCEPTABLE DAILY INTAKE

FOR HUMAN EXPOSURE TO 2, 3, 7, 8 TCDD

(from lab studies on animals)

$LD_{50}$  **1 micromgram/kg bodyweight/day**

Studies on Chronic exposure, accidental exposure of humans
Safety Factor 10,000

NOAEL  **100 picogram/kg body wt/day**

Safety Factor 10

ADI  **10 picogram/kg/body wt/day**

$LD_{50}$  Lethal Dose (on Animals)
NOAEL  No Observed Adverse Effect Level
ADI  Acceptable Daily Intake for Lifetime Exposure
or TDI : Tolerable Daily Intake

i.e A Factor of $\dfrac{1}{100,000\ \text{th}}$ of the $LD_{50}$ is used in practice.

Total daily intakes are of the order $\dfrac{1}{100\ \text{th}}$ of the ADI or less (incineration plants may contribute a negligible 0.006% of dioxin ADI for a maximally exposed individual).

> The current mode of extrapolating high-dose to low-dose effects is erroneous for both chemicals and radiation. Safe levels of exposure exist. The public has been needlessly frightened and deceived, and hundreds of billions of dollars wasted. A hard-hearted, rapid examination of phenomena occurring at low exposures should have a high priority.

*Source:* P. H. Abelson, 'Risk assessments of low-level exposures', *Science*, vol. 2, 9 September 1994.

**River Ecosystem Classification.** The Surface Waters (River Ecosystem) (Classification) Regulations 1994 (SI 1994 No. 1057), prescribe a system for classifying the quality of rivers and canals, to provide the basis for setting statutory water quality objectives (WQOs) under section 83 of the Water Resources Act 1991 in respect of individual stretches of water.

The River Ecosystem Classification comprises five hierarchical classes, in order of decreasing quality: RE1, RE2, RE3, RE4 and RE5. The criteria which samples of water are required to satisfy are set out, for ease of reference, in the table overleaf.

**River Purification Board.** An authority (Scotland) responsible for the prevention of pollution of streams in its area. Set up in 1975 under the Local Government (Scotland) Act 1973.

**River Quality Objectives (RQO).** A series of measures which ensure that river quality is checked more directly against all the quality standards needed to support the following uses

- Abstraction for public water supply.
- Salmonid fishery.
- Cyprinid fishery.
- Amenity and conservation.
- Abstraction for industrial water supply.
- Spray irrigation of field crops.
- Livestock watering.

The determinands most often involved in the decision-making process are DISSOLVED OXYGEN, BIOCHEMICAL OXYGEN

Criteria for River Ecosystem Classification

| Class | Dissolved oxygen (% saturation) 10 percentile | BOD (ATU) (mg/l) 90 percentile | Total ammonia (mg N/l) 90 percentile | Un-ionized ammonia (mg N/l) 95 percentile | pH lower limit as 5 percentile; upper limit as 95 percentile | Hardness (mg/l $CaCO_3$) | Dissolved copper (µg/l) 95 percentile | Total zinc (µg/l) 95 percentile |
|---|---|---|---|---|---|---|---|---|
| RE1 | 80 | 2.5 | 0.25 | 0.021 | 6.0–9.0 | ≤10 | 5 | 30 |
| | | | | | | >10 and ≤50 | 22 | 200 |
| | | | | | | >50 and ≤100 | 40 | 300 |
| | | | | | | >100 | 112 | 500 |
| RE2 | 70 | 4.0 | 0.6 | 0.021 | 6.0–9.0 | ≤10 | 5 | 30 |
| | | | | | | >10 and ≤50 | 22 | 200 |
| | | | | | | >50 and ≤100 | 40 | 300 |
| | | | | | | >100 | 112 | 500 |
| RE3 | 60 | 6.0 | 1.3 | 0.021 | 6.0–9.0 | ≤10 | 5 | 300 |
| | | | | | | >10 and ≤50 | 22 | 700 |
| | | | | | | >50 amd ≤100 | 40 | 1000 |
| | | | | | | >100 | 112 | 2000 |
| RE4 | 50 | 8.0 | 2.5 | — | 6.0–9.0 | ≤10 | 5 | 300 |
| | | | | | | >10 and ≤50 | 22 | 700 |
| | | | | | | >50 and ≤100 | 40 | 1000 |
| | | | | | | >100 | 112 | 2000 |
| RE5 | 20 | 15.0 | 9.0 | — | — | — | — | — |

DEMAND, and AMMONIA. The impact of other substances, for example metals and pesticides, is also assessed against the standards set down in the River Quality Objectives. These substances are also prominent in several of the directives issued by the European Community. (⇨ WATER QUALITY STANDARDS; ⇨ RIVER ECOSYSTEM CLASSIFICATION)

**River regulation.** Upland reservoirs in the upper reaches of the river are used to regulate the flow and especially to increase it at times of low run-off. Abstraction is carried out in the lower reaches close to the main areas of demand – the yield can be much greater than it water were taken from the reservoir directly. River regulation is a common means of augmenting water supplies.

Occasionally AQUIFERS are used as extra sources of water to augment the flow of a river in dry weather. The conjunctive use of aquifers and rivers has resulted in increased yields of water from the Thames and is also under survey for the Great Ouse. (⇨ WATER SUPPLY)

**RME (rape methyl ester)** ⇨ BIODIESEL.

**Road traffic noise.** The disturbing features of traffic noise are its general level and its variability with time. The latter refers to short-term variations due to the passage of individual vehicles, and to longer period variations at different times of day due to the general changes in traffic flow. It has been found that the dissatisfaction expressed by occupants of dwellings varies in accordance with the peak noise levels. Hence an index known as the '10 per cent level' ($L_{10}$) is used. $L_{10}$ is the level of noise in dB(A) exceeded for just 10 per cent of the time. For the measurement and prediction of noise from traffic, the average of $L_{10}$ values for each hour between 6 a.m. and 12 midnight on a normal weekday has been recommended in the UK as giving

| $dB(A)$ | Situation |
|---|---|
| 80 | At 60 feet from the edge of a busy motorway carrying many heavy vehicles, average traffic speed 60 mph, intervening ground grassed. |
| 70 | At 60 feet from the edge of a busy main road through a residential area, average traffic speed 30 mph, intervening ground paved. |
| 60 | On a residential road parallel to a busy main road and screened by the houses from the main road traffic. |

satisfactory correlation with dissatisfaction. This is known as $L_{10}$ (18 hours), and permits accurate predictions for design purposes. Some values of $L_{10}$ (18 hours) and typical conditions in which they are experienced are shown in the table above.

($\Rightarrow$ NOISE; DECIBEL; HEARING; NOISE INDICES; AIRCRAFT NOISE; INDUSTRIAL NOISE MEASUREMENT)

**Roentgen.** A unit of exposure to radiation based on the capacity to cause ionization. It is equal to $2.58 \times 10^4$ coulomb per kg in air. Generally an exposure of 1 roentgen will result in an absorbed dose in tissue of about 1 rad. ($\Rightarrow$ REM; IONIZING RADIATION (DOSE MEASUREMENT))

**Rotary kiln.** A slowly revolving drum, lined with refractory material and fired by gas or oil, used in the processing of ores and cement manufacture. Rotary kilns can also be used to incinerate sewage sludge and industrial wastes to obtain a high degree of BURN OUT. A recent recycling innovation is to use domestic refuse as part of the fuel supply for cement manufacture in rotary kilns as the needs of refuse combustion and cement manufacture complement each other.

ENEAL incineration process guaranteed pollutant emissions

| Emission | $mg/m^3$ |
|---|---|
| Dust | $<30$ |
| $NO_x$ | $<35$ |
| $SO_2$ | $<100$ |
| CO | $<30$ |
| HCN | $<0.05$ |
| HCl | $<2$ |
| Total aromatic hydrocarbons | $<30$ |
| Heavy metals | $<50$ |
| TCDD + PCDD + TCDF + PCDF | $<0.006$ |
| PCB | $<0.1$ |
| $O_2$ | 10% |

*Concentrations to standard temperature and pressure conditions.

One innovative use of rotary kilns adopted is the 'ENEAL' incineration process where whole scrap tyres (up to 1.2 m diameter) are used as the feed. The overall efficiency (based on the energy value of the scrap tyres) is claimed to be 90 per cent with air pollutant emission levels as set out in the table above. (⇨ EFFICIENCY; INCINERATION)

**Rotating biological contactor (RBC).** A series of slowly rotating parallel disks on a common shaft whose lower halves are immersed in liquid effluent which requires biological treatment. The biological film is aerated on the disks as they rotate alternately in the effluent reservoir and out, in contact with the air.

**Royal Commission on Environmental Pollution (RCEP).** A standing Royal Commission established in 1970 to advise with matters of both national and international importance, to prevent the pollution of the environment and by specific research in this field, any future possible dangers to the environment. The Royal Commission has now published 18 reports on different aspects of pollution. The 17th Report (1993) is on incineration and strongly recommends its adoption (with energy recovery) for MSW management in preference to LANDFILL.

The 18th Report (1994) is on transport and the environment and contains key recommendations for a sustainable transport policy. This implies major emissions reduction from road transport sources.

**Royal Commission Standard.** The treated sewage effluent standards proposed by the UK Royal Commission on Sewage Disposal (1898–1915) were no more than 30 mg/l SUSPENDED SOLIDS and 20 mg/l BOD, i.e. a 30/20 effluent. The Commission expected that the effluent would be diluted with eight volumes of clean river water of BOD 2 mg/l and hence no undue pollution of the receiving stream would result. Other discharge restrictions can be placed on heavy metals, cyanides, phenols, ammonia, etc. Where ammonia is specified, this is usually set at 10 mg/l. Hence, the Royal Commission Standard is often quoted as 30/20/10.

Effluent standards should be determined on the basis of the particular receiving river, but 30/20 is often the norm which is applied. However, this was not met by up to 20 per cent of UK sewage effluent treatment plants in 1989. One extreme example of this occurred in 1989 when treated sewage discharge standards were 'relaxed' on the River Ouse in Yorkshire whose water is abstracted by Drax Power Station for cooling purposes. The result was that sewage particles and bacteria were spread in the vicinity of the power station by the cooling towers; a novel if unwelcome form of POLLUTION CONVERSION.

**Run-off.** The volume derived from snow or rain falling on a surface and which does not permeate into the soil.

**Rylands v. Fletcher (1968).** Classic law case which established that the liability for the consequences of non-natural or special operations on land resides with the owner of the land, i.e. if the use to which the land is put carries an increased danger to others, then the owner is liable for the consequences. In *Rylands v. Fletcher*, an escape of water from a reservoir flooded an adjacent mine. The ruling made stated

> In particular the rule states that anyone who brings or collects and keeps on his land anything likely to do mischief if it escapes must keep it at his peril and if he does not do so is prima-facie strictly liable for all that damage which is the natural consequence of its escape, and the defendant's use of the land must be non-natural.

This ruling could have consequences for all who pollute knowing that they do so or who could have foreseen that their operations could do so.

*Cambridge Water Company v Eastern Counties Leather*
The decision in the case of *Cambridge Water Company v Eastern Counties Leather* (water supplies pollution from solvent spillage) is an important one in relation to potential liability at common law for environmental damage. The use of premises as a tannery was held to be a 'non-natural user of land' within the rule in *Rylands v Fletcher*. As such, it was generally thought that liability would be 'strict', ie. that it would not be necessary to prove fault, but only a causal link and that the damage suffered was not too remote. In the event the House of Lords treated the claim under the general heading of nuisance and held that it was necessary also to show that the damage suffered was forseeable in order to establish liability. Since at the time of the spillages of solvent in the Eastern Counties Leather premises it was not foreseeable that the level of contamination in Cambridge Water's borehole would be in breach of some standards yet to be formulated, Eastern Counties Leather was held not to be liable at common law. However, it should be noted that Eastern Counties Leather subsequently found themselves in trouble with the National Rivers Authority acting under their statutory powers contained

in the Water Resources Act 1991 and agreed to carry out remediation works.

*Source:* David Cuckson, Head of Environmental Law Group, London.

(⇨ DUTY OF CARE; LANDFILL GAS; NUISANCE)

# S

Safe yield[1]. In water supply, the long-term rate at which water can be extracted from an AQUIFER without a continuing progressive decline in its water level or other adverse effects.

Safe yield[2]. In fisheries, the allowable catch which will not endanger the breeding stock. (⇨ MAXIMUM SUSTAINABLE YIELD)

Salinity. Total amount of dissolved material expressed in terms of kilograms of material per million kilograms of feedwater, i.e. parts per million (ppm) of total dissolved solids.

Typical sea water has a salinity of 35 000 ppm of which 30 000 ppm is common salt (NaCl). Accepted potable water standards are a maximum of 250 ppm as salt and a total dissolved solids content of 500 ppm and these figures should preferably be lower. Bearing this in mind, if a desalting process can sustain only a 90 per cent separation of total dissolved solids from a feed of 35 000 ppm, then the product water will have a dissolved solids content of 3500 ppm. Water of this salinity is obviously not potable. This restriction applies to both electrodialysis and reverse osmosis but not to distillation, as product purity can range from 1 to 100 ppm from a feed of 35 000 ppm (⇨ DESALINATION).

The salinity of sea and brackish waters is:

| Source | Salinity – total dissolved solids (ppm) |
|---|---|
| Potable well water | 300–500 |
| Approximate limit for irrigation | 1000 |
| Brackish well water | 1500 6000 |
| Baltic Sea | 2000–3000 |
| Arabian Gulf | 44 000 |
| Typical sea water | 35 000 |

Salt. A compound which results from the replacement of the hydrogen ATOMS of an ACID, usually by metal atoms to form, for example, chlorides (NaCl), sulphates ($CaSO_4$), and carbonates ($CaCO_3$).

**Sample.** A part of a population selected with the object of estimating some characteristic of the whole population. Can be random or spot. *Random sample*, a sample selected in such a way that all possible samples of the same size have the same chance of being chosen. *Spot sample* ('grab sample'), a sample of air, effluent, etc. collected over a short period of time and usually taken to a central laboratory for analysis.

**Sampler, personal.** A device attached to a person that samples air in the immediate vicinity so that personal exposure to pollutants may be determined.

**Sampling, isokinetic.** The taking of a sample of flowing gas (particularly gas flowing through a duct) in such a way that the sample does not undergo any change in either velocity or direction at the inlet of the probe. Used in taking samples of stack gases for measurement of their dust concentration.

**Sampling, particulate.** Particulate sampling is dominated by two well-established methods. The reference test method (BS 3405) is to undertake isokinetic sampling and to determine the particulate burden in the gas by weighing filters. The most common continuous method is to shine a light beam across the duct or chimney and to measure the reduction in light intensity (opacity measurement). An alternative technique is to insert a probe into the gas stream and to monitor the particles by the charge transfer that occurs between particles and probe (the triboelectric effect); see Figure 128. ($\Rightarrow$ SAMPLING, ISOKINETIC) Figure 129 gives the range of suitability for the different technologies.

Many gases, such as carbon dioxide, carbon monoxide and oxygen, present no particular problems for either sampling or analysis. Infra-red and other optical or electrochemical techniques of measurement have been widely used for many years. However there are other gases which for such reasons as having a rather high DEW POINT (which causes condensation), being highly reactive or because they strongly absorb onto the surfaces of the sampling system, present great difficulties. HYDROGEN FLUORIDE (a pollutant of particular concern in the ceramic and brick industries) is such a gas which because of its high chemical reactivity is especially difficult to sample and measure. An alternative route has been developed based on the pollutant removal from the gas stream into a liquid as

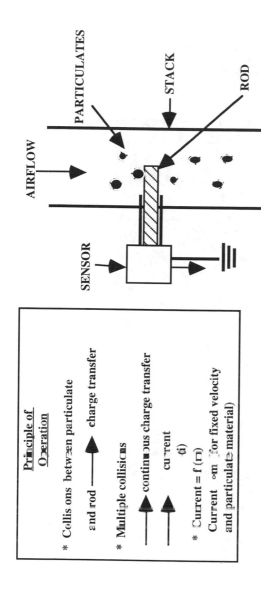

**Figure 128** Particle impingement, principle of operation.

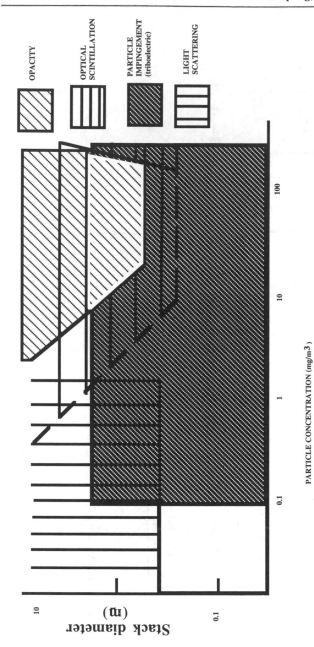

**Figure 129**   Suitability of dust emission technologies for different applications.

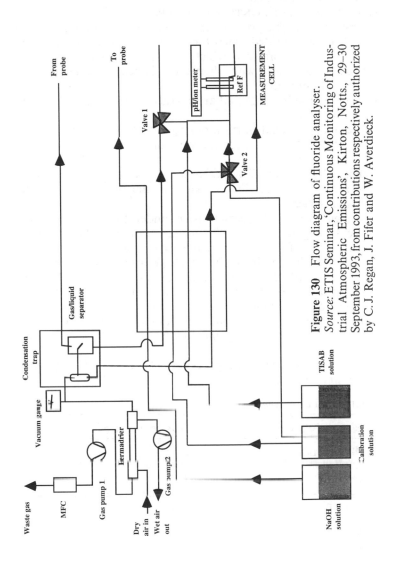

**Figure 130** Flow diagram of fluoride analyser.
*Source*: ETIS Seminar, 'Continuous Monitoring of Industrial Atmospheric Emissions', Kirton, Notts., 29–30 September 1993, from contributions respectively authorized by C.J. Regan, J. Fifer and W. Averdieck.

early as possible in the sampling system and subsequently analysed using the well established technique of selective ion electrode. Figure 130 gives the flow diagram of the ETIS project fluoride analyser.

**Sanitary landfill.** ⇨ CONTROLLED TIPPING.

**Saturation (air).** At a given temperature and (total) pressure air is said to be saturated when the water contained in it is at the water vapour pressure which would occur when the water vapour is in contact with a free water surface at that temperature and total pressure.

**Saturation (in soil).** Soils that are charged with water to the fullest extent possible.

**Saturation zone.** That portion of an AQUIFER which exists below the WATER TABLE, i.e. groundwater fills all the pore spaces and voids.

**Scaling.** The peeling off of oxides/corrosion products in the form of scales thus allowing a fresh metallic or mineral face for further attack.

**Scheduled Process.** Processes listed under the UK Alkali etc. Works Regulation Act 1906 (termed 'works' in the Act) along with a list of 'noxious and offensive gases'. Later additions were consolidated in the Alkali etc. Works Order 1966. These were subscribed in the Control of Industrial Air Pollution (Registration of Works) Regulations 1989. Her Majesty's Inspectorate of Pollution now has oversight over Air, Water and Waste and permission embraces the former Alkali Inspectorate's duties of policing emissions from the Scheduled Process. (Renamed PRESCRIBED PROCESS)

A listing of scheduled processes (effective 1986) is given in the table below.

Scheduled processes

| Scheduled Process | Substance controlled |
|---|---|
| Alkali works | HCl |
| Aluminium | Total fluoride |
| Amines | Total amine |
|  | Trimethylamine |
| Ammonia | $NH_3$ |
| Arsenic | $As_2O_3$ |
| Beryllium | Be |
| Bisulphite | Total acidity |
| Cadmium | Cd |

| Scheduled Process | Substance controlled |
| --- | --- |
| Cement | Particulate matter |
| | $H_2S$ |
| Ceramics | Particulate matter |
| Chemical fertilizer | Total acidity |
| | HCl |
| Chemical incineration | HCl |
| | $H_2S$ |
| | Particulate matter |
| Chlorine | Hg |
| | Chlorine |
| | HCl |
| Copper | Particulate matter |
| | Fume |
| | Cu |
| | Zn |
| | Pb |
| Diisocyanate | Diisocyanate |
| Electricity | Particulate matter |
| Gas and coke | Particulate matter |
| Hydrochloric acid | HCl |
| Hydrofluoric acid | Total gaseous F |
| | Total acidity |
| | Particulate matter |
| Iron and steel | |
| Lead | Pb |
| | Particulate matter |
| Lime | Particulate matter |
| Metal recovery | HCl |
| | Particulate matter |
| Mineral | Particulate matter |
| Nitric acid | Total acidity |
| | $NO_x$ |
| Petroleum | S |
| | $H_2S$ |
| | Particulate matter |
| Phosphorus | $P_2O_5$ |
| Sulphide | $H_2S$ |
| Sulphuric acid | Sulphur |
| Tar | Tar |
| | $H_2S$ |
| | Particulate matter |
| Vinyl chloride | Vinyl chloride |

Discharge Authorizations are now used for all scheduled processes. They contain the following items:

1. The name and address of the owner of the works.
2. The location of the works and the local authority.
3. The category of works and the process description.
4. Details of the type and capacity of the abatement measures and chimneys.
5. Sources, types and volumes and presumptive limits of process emissions.
6. Details of continuous, manual or environmental monitoring.

With the implementation of the UK Environment Protection Act, INTEGRATED POLLUTION CONTROL is applied to PRE-SCRIBED PROCESSES by HMIP, and by local authorities to less complex processes. The likely numbers are 5000 complex processes (HMIP) and 27 000 others (LAs).

**Schistosomiasis (bilharzia).** The disease caused by the worms *Schistosoma haematobium* and *Schistosoma mansoni*, which use human beings as a primary host and snails as secondary hosts. Irrigation ditches make ideal transmission networks for the diffusion of these snails. ($\Rightarrow$ DAM PROJECTS)

**Scintillation counter.** A radiation detector in which the radiations cause individual flashes of light in a solid (or liquid) 'scintillator' material. Their intensity is related to the energy of the radiation. The flashes are amplified and measured electrically and displayed or recorded digitally as individual 'counts'.

**Scrap.** Discarded material from manufacturing, processing or remnants after an article's useful life has run out. ( $\Rightarrow$ RECYCLING; DOMESTIC REFUSE)

**Screen[1].** A device for size separation, e.g. screening of DOMESTIC REFUSE to + 200 mm and − 200 mm particle size.

**Screen[2].** An array of fixed or moving bars used to remove large solids from sewage effluent.

**Screenings.** The product from SCREENS. In sewage effluent treatment, screenings range from 0.01 to 0.03 m$^3$/d per 1000 population.

**Scrubber.** Device for flue gas cleaning, such as spray towers, packed scrubbers and jet scrubbers. The gas is passed through wetted packing or a spray to ensure intimate contact with the scrubbing water. This removes particles down to 1 $\mu$m in diameter. The gas flow is usually countercurrent to the water flow. Scrubbers for particulate control produce SLUDGE that requires dewatering and disposal as well as contaminated effluent which may be recirculated or treated before discharge. Scrubbers may also serve to control gaseous pollutants, in which case an alkaline solution is used. The data below give a comparison of emissions from a semi-dry scrubbing system and a wet scrubbing system, installed on German municipal solid waste incinerators, using $Ca(OH)_2$ solution as the neutralizing agent for gases.

| Component | Semi-dry ($mg/m^3$ normal dry 11% $O_2$) | Wet ($mg/m^3$ normal dry 11% $O_2$) |
|---|---|---|
| HCl | 50 | 10 |
| $SO_2$ | 100 | 50 |
| HF | 0.5 | 0.5 |
| Particulate | 10 | 10 |
| Heavy metals | 1 | 1 |
| Hg vapour | 0.2 | 0.05 |

**Scum.** A layer of fats, oils, soaps, etc. which can collect on the surface of polluted lakes, rivers and the SEDIMENTATION tanks of sewage effluent treatment plants.

**Sea water.** ⇨ SALINITY.

**Second law of thermodynamics.** ⇨ LAWS OF THERMODYNAMICS.

**Secondary air.** The air introduced above (and beyond) the bed of burning fuel or waste to promote the complete combustion of volatile materials which are released after the first stage of combustion.

**Secondary liquid fuel (SLF).** A blend of organic chemicals no longer suitable for recovery/original use. Substantial commercial interests are now pushing for its use in CEMENT KILNS which are substantial consumers of FUEL, e.g. a kiln input requirement of 127 MW. Hence, a substantial replacement fuel market exists if the price is right and environmental standards are met comparable to those of HAZARDOUS WASTE INCINERATION. Typical UK proposals are for up to 25 per cent of the heat input to be met by SLF, one brand of which is marketed as CEMFUEL™.

**Secondary materials.** Materials which have fulfilled their primary function and which cannot be used further in their present form. Also, materials which occur as by-products from the manufacturing or conversion of primary products.

**Secondary pollutant.** Formed from a PRIMARY POLLUTANT as a result of chemical changes such as photochemical and other reactions. Examples are OZONE and $NO_2$. (⇨ PRIMARY POLLUTANT)

**Sediment.** The deposit of silt and accumulated organic and/or inorganic materials at the bottom of rivers, lakes, seas, etc.

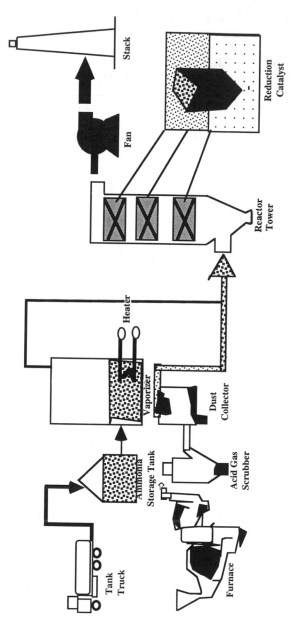

**Figure 131**   Selective catalytic NO$_x$ reduction system.

Sediments act as sinks for PESTICIDES and can contain concentrations as much as 800 times that of water in the case of DIELDRIN. They are therefore a source of secondary contamination as well as allowing the bottom feeders (e.g. invertebrates) to accumulate the pollutants at much greater concentrations than water analysis would indicate, thereby contaminating the FOOD CHAIN. (⇨ MINAMATA DISEASE)

**Sedimentation.** The settling out of 'SETTLEABLE' SOLIDS from sewage effluent in sedimentation tanks which allow very slow rates of flow so that settling takes places.

**Selective catalytic $NO_x$ reduction system.** This system achieves greater $NO_x$ reductions than selective non-catalytic $NO_x$ reduction and can meet the requirements of more stringent environmental regulations when used in conjunction with a non-catalytic de-$NO_x$ system (Figure 131).

**Selective non-catalytic $NO_x$ reduction system (SNC).** This system injects ammonia into the furnace (Figure 132). The $NO_x$ content is reduced by the resulting reducing reaction.

Features:

(1) Simple and maintenance-free installation.
(2) Favourable cost/removal ratio.

**Selectivity[1].** The characteristic of radiation, toxic chemicals or heavy metals (if ingested) to affect certain organs of the body to a much greater degree than the whole body dose would indicate. (⇨ MINAMATA DISEASE; HALF-LIFE)

**Selectivity[2].** The characteristic of plants and INSECTS to adapt to their environment by means of genetic selection of the most favourable strains.

**Selenium (Se).** Often referred to as a toxic metal, selenium is not, in fact, a metal but has certain metallic properties. It is a member of the sulphur group. It is produced as a by-product of the refining of copper, nickel, gold and silver ores. It is used extensively in the electronics industry, and also in paints and rubber compounds. It is a micro-nutrient at levels of 0.02 to 1 part per million. Maximum allowable concentration in drinking water is 0.01 part per million. It can act as a systemic poison. The $LD_{50}$ for one selenium compound is as low as 4 micrograms per kilogram body weight. Known cases of poisoning are rare. (⇨ LETHAL DOSE)

**Settleable solids.** Suspended solids which will settle out of an effluent (e.g. sewage) in two hours when stationary.

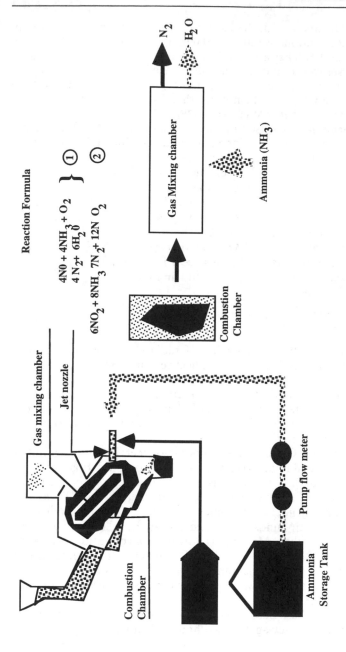

**Figure 132**  Selective non-catalytic $NO_x$ reduction system.

**Settling chamber.** A chamber inserted between a furnace and its stack in which coarse particulate matter settles out of the gas stream. Also used in SEWAGE TREATMENT for the separation of grit and other solids before further treatment.

**Seveso.** ⇨ DIOXIN.

**Sewage.** The liquid wastes from a community. Domestic sewage is from housing. Industrial sewage is normally from mixed industrial and residential areas. (⇨ SEWAGE TREATMENT)

**Sewage treatment.** The reduction of the organic loading that raw sewage would impose on discharge to streams and water-courses. It is carried out by oxidation of the sewage, i.e. contact with air, which oxidizes most of the wastes to allow discharge. Several steps are required:

1. Preliminary screening to remove large suspended solids, metal and rags.
2. Grit removal.
3. Sedimentation to allow as much suspended organic solids as possible to settle.
4. Biological oxidation in either of two plants:

    (a) percolating filter, i.e. a packed bed of clinker to stones 2 metres deep, through which the sewage trickles, and in which the surface area is maximized; *or*
    (b) activated sludge, in which the sewage is aerated in tanks by agitators which maintain the level of dissolved oxygen as high as possible. The sludge is recycled to seed the raw sewage and allow treatment to proceed faster.

In both cases the aerobic bacteria are able to grow and the end result is sludge and an effluent which should be very low in biochemical oxygen demand (20 milligrams per litre or less) unless the plant is overloaded. Discharge of the effluent, provided there is at least an eight-fold dilution in the river, should not normally cause any problems following Step 5 below (and if necessary Step 6 in cases where there are strict discharge requirements).

5. Sedimentation in chambers called humus tanks where the discharge from biological oxidation has most of its residual suspended solids removed. The discharged effluent is made to flow upwards through the tank at very low velocities (1 to 2 metres per hour) so allowing the suspended solids to be removed as sludge, which is then disposed of. For the

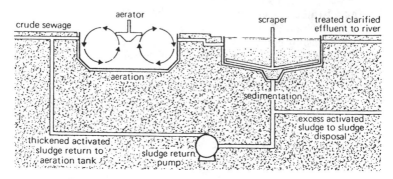

**Figure 133**   Activated sludge effluent treatment process.

activated sludge process, a portion is recycled as shown in
Figure 133.

6. Tertiary treatment. If the treated effluent is still of inadequate
quality (e.g. too high in suspended solids) it can be micro-
strained in special filters. Tertiary treatment also consists of
other filters, lagoons and the use of REVERSE OSMOSIS.
Tertiary treatment will grow in use as higher DISCHARGE
STANDARDS are applied to reduce or contain the pollution of
inland rivers from sewage treatment plant effluent and in
particular achieve a reduction in nitrate levels.

(⇨ AEROBIC PROCESSES; BIOCHEMICAL OXYGEN DEMANDS;
ROYAL COMMISSION STANDARD; SLUDGE, SEWAGE)

**Sewage treatment, deepshaft process.** This technique has been
developed by ICI and is claimed to cut the costs of sewage
treatment by up to 50 per cent. It relies on a deep shaft
(approximately 135 metres) into which sewage and biodegradable
effluents are admitted as shown in Figure 134. The mixture is
circulated by air injection and the mixing processes together
with the increased pressure, enable the aerobic bacteria to
reduce the biochemical oxygen demand of the wastes at rates of
up to 10 times that of conventional sewage treatment plants. A
baffle arrangement keeps untreated and treated effluent separate.
The method still relies on aerobic bacteria but has increased the
intensity of operation, thus allowing plants to be greatly
reduced in area.

This plant is an offshoot of aerobic fermentation research
where the tower or shaft is much shorter as the bacteria require

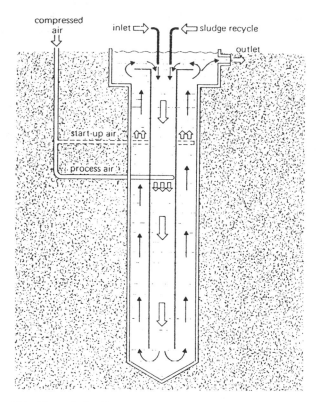

**Figure 134** Deepshaft effluent treatment process (as developed by ICI).

less contact time with the substrate to be fermented. (⇨ BIODEGRADATION; BIOREACTOR)

**Sewerage.** A network of pipes and associated appliances for the collection of transport of domestic and industrial SEWAGE.

**Shadow flicker.** Under certain combinations of geographical position and time of day, the sun may pass behind the blades of a wind turbine and cast a shadow. When the blades rotate the shadow flicks on and off. The effect only occurs inside buildings where the flicker appears through a window opening. The seasonal duration of this effect can be calculated from the geometry of the machine and the latitude of the site. (⇨ WIND ENERGY)

**Shear.** A machine used to reduce scrap metal materials to a small size for handling, transportation or smelting purposes.

**Short-term exposure limit (STEL).** ⇨ TIME-WEIGHTED AVERAGE.

**Shredder.** A machine used to break up materials into smaller pieces by tearing and/or impact.

**Sick building syndrome.** 'Sick building' syndrome is used to describe a building in which a significant number (more than 20 per cent) of building occupants report illness perceived as being building-related. The complaints are characterized by a range of symptoms including, but not limited to, eye, nose, and throat irritation, dryness of mucous membranes and skin, nosebleeds, skin rash, mental fatigue, headache, cough, hoarseness, wheezing, nausea and dizziness. The introduction of new building materials, decreased ventilation and decreased air leakage have all contributed to the problem.

**Sievert (Sv).** ⇨ IONIZING RADIATION.

**Silage.** Green leaf cattle food which has had molasses added to promote fermentation and preservation. The liquors from silage production can have BODs up to 300 times that of domestic sewage effluent. They are highly polluting and can be a seasonal cause of fish deaths in small streams.

**Silicosis.** ⇨ FIBROSIS.

**Silt.** Normally a wet mixture of particles between 4 and 60 $\mu$m diameter often found in the base of sewers and streams. Intermediate between CLAY and MUD.

**Single-cell protein (SCP).** Protein manufactured by microbial means from organic substances such as oil, natural gas, cellulose or sewage. Microbial conversion offers the means of converting wastes into useful protein for both animals and man, because of the rapid rate of metabolism of micro-organisms and their subsequent growth and reproduction. Under suitable conditions many microbes, especially YEASTS, may be maintained in an exponential growth situation in continuous culture with a doubling time of 15–20 minutes. (⇨ EXPONENTIAL CURVE)

The production technology uses aerobic FERMENTATION with copious supplies of air to obtain maximum growth rate. A single strain of organism is chosen because this allows the optimum production of protein from the substrate, e.g. the yeast, *Candida utilis* is used for protein production from molasses and a similar strain is used on oil. The disadvantage of single strains is that sterility must be maintained.

One major SCP process is that of ICI which uses a bacterium (*Methylophilus methylotrophus*) which can live on methanol, synthesized from 'cheap' North Sea gas. The bacterium is grown in a $15\,000\,m^3$ fermenter which is kept supplied with methanol, ammonia and nutrients and a sterile air supply.

Cell concentration is maintained at 3 per cent in the fermenter by removal of the product. This is followed by separation and drying. The product is marketed as Pruteen which contains 70 per cent protein and $3600\,kcal/kg$ metabolizable energy. This makes it a high protein animal feedstuff which has to compete in the market place with fishmeal (and to some extent soy protein). The process is particularly applicable to those areas with large cheap natural gas resources.

SCP is a major growth area and is seen by many as the only route to alleviating shortages of feedstocks for both cattle and humans. ($\Rightarrow$ SYMBA)

**Sink[1].** The source of cooling water or air for power stations or other fuel-conversion devices where reject thermal energy is discharged.

**Sink[2].** In air pollution, the receiving area for material removed from the atmosphere, e.g. in PHOTOSYNTHESIS, plants are sinks for carbon dioxide.

**Site licence[1].** Licence issued by the Nuclear Installations Inspectorate covering safety of design, construction, operation and maintenance of facilities at a nuclear site.

**Site licence[2].** A waste disposal site licence issued by a Waste Disposal Authority. ($\Rightarrow$ WASTES)

**Sizewell Nuclear Power Station.** Site of UK's only PRESSURIZED WATER REACTOR (PWR) Sizewell 'B'. The public inquiry hinged on the economic case that was made for its construction. The proponents claimed that power would be produced at comparable or better costs than those of a coal-fired power station. This argument was exposed as fatally flawed after building permission was obtained. In particular, the following factors were either not examined properly or glossed over: the much lower cost of coal in the future; under-estimation of the capital costs; alleged massive under-estimates of decommissioning costs; the prospects for much increased imports of electricity from Scotland; the ready supply of British sector North Sea gas for fuelling high-efficiency and low cost COMBINED HEAT AND POWER stations.

A fitting commentary on the Sizewell 'B' public inquiry is provided by Professor P. Odell's letter to the *Financial Times* (21 November 1989): 'The inspector and his economic assessor simply failed to get to grips with the prospects for nuclear power economies, and produced a report with wrong conclusions'.

He further stated that on the *Financial Times*' own figure of expected full generating costs (8p–10p per kWh), even a successfully completed (and completed on time) Sizewell station will cost taxpayers and/or electricity consumers some £400 million a year for every year of its life. Odell claims it would be cheaper to abandon construction now (November 1989).

An escalating cost analysis for the proposed Hinkley Point C PWR station has been provided by Lord Marshall (in retrospect) after abandonment of the UK's PWR construction programme. The tables show the 'public sector price' and the 'private sector price' for Hinkley Point C electricity.

'Public sector price' for Hinkley Point C electricity

| Price component | pence/kWh |
|---|---|
| Price quoted at public inquiry | 3.09 |
| Inflation | +0.33 |
| Additional building costs | +0.10 |
| Capitalized company overheads | +0.11 |
| Operating cost increases | +0.68 |
| Interest during construction to be recovered | −1.09 |
| Total | 3.22 |

'Private sector price' for Hinkley Point C electricity

| Price component | pence/kWh |
|---|---|
| Public sector price | 3.22 |
| Adjustment from 8% to 10% rate of return | +0.71 |
| Uncertainties: | |
| 70% availability (instead of 75%) | |
| Fuel reprocessing and | +0.54 |
| decommissioning uncertainties | |
| New basis for calculating profit | +1.03 |
| Increase for 20 year contract (instead of | |
| lifetime contract) | +0.75 |
| Total | 6.25 |

*Source:* 'What went wrong in the UK – Lord Marshall's view', *Nuclear Engineering International*, January 1990.

The fluctuating fortunes of UK nuclear power are charted in the table below (note the exclusion of interest during construction and the assumption of 40 year amortization).

So, it all depends on:

(a) what subsidies are available
(b) how optimistic or pessimistic the analysts are
(c) what is left out (e.g. full decommissioning costs).

Lifetime generating costs of Sizewell C twin PWR Station (1993 money values)

| | |
|---|---|
| Net output | $1310 \times 2$ MWe |
| Capital cost (excluding interest during construction) | £3520 million |
| Unit costs | |
|     Capital costs including interest during construction | 1.9 pence/kWh |
|     Fuel cost | 0.4 pence/kWh |
|     Operations and maintenance costs | 0.6 pence/kWh |
|     Decommission cost | 0.01 pence/kWh |
| Total unit cost | 2.9 pence/kWh |

*Principal assumptions:* the capital cost of £3520 million excludes £110 million initial fuel charge; 85 per cent lifetime factor; 40 year amortization period/station lifetime; 8 per cent real rate of return on capital; 60 and 54 month construction times for the first and second reactors from first structural concrete to first fuel load; 2 per cent real rate of return on funded provisions.
*Source: The Prospects for Nuclear Power in the UK*, Department of Trade and Industry, HMSO, 1995.

Despite all the controversy, Sizewell is now (July 1995) on full power.

*Sources:* A. Cottrell, Letter to the *Financial Times*, 7 January 1974.
J. T. Edsall, *Environmental Conservation*, vol. 1, no. 1, 1974, p. 32.
J. Collier (M.D. Nuclear Electric plc) extract from *Journal of Power and Energy*, vol. 207, 1993; and *Nuclear Electric Briefing Note*, 6 October 1994, Davies, D. R., Bsc, CEng FICE, 'Planning and acquiring the UK's first PWR power station', *Proc. Instn., Civ. Engrs., Civ. Engng.*, Sizewell B Power Station, 1995.
ENDPIECE: British Energy has abandoned plans for any new nuclear powerstations on economic grounds, 11 December 1995.

**Slag.** The non-gaseous waste material formed in a metallurgical furnace; it is largely non-metallic but usually contains some metal. Slag contains as much of the undesirable constituents of the ore as possible and is withdrawn from the furnace in the molten form. Alkaline slags retain much of the sulphur in the fuel and the ore, and the gases issuing from the furnace may be virtually sulphur-free.

Slag is also formed in high-temperature incineration processes where refuse is converted to a molten slag which is removed from the bottom of a special cupola. Steelworks slag is often processed and recycled because of its high ferrous content. Other uses are road foundations and in sintering plants for the manufacture of lightweight aggregate building blocks.

**Slaked lime.** A common name for CALCIUM OXIDE to which water has been added. It is used in INCINERATION plants and FLUE GAS DESULPHURIZATION to remove acid gases.

For example, the 400 000 tonnes per year South East London Combined Heat and Power Plant (SELCHP) uses 8000 tonnes per year of calcium oxide which is slaked on site and used in the gas SCRUBBERS to remove $SO_x$ and HCl.

**Slick.** A floating patch of oil or SCUM on the surface of a body of water.

**Sludge, sewage.** The term for the solids that settle out during primary settlement before the aerobic treatment of sewage. It is a slimy, offensive material with 95 per cent or greater water content. Its disposal is a major problem and four main methods are employed:

1. Conversion into METHANE gas by digestion, and subsequent use as a soil dressing.
2. Dewatering followed by incineration.
3. Spreading on land (accounts for 40 per cent of UK sludge disposal).
4. Dumping at sea from specially constructed sludge vessels (to be phased out by 1998).

Digestion is an ecologically sound method and is practised in large sewage works where the methane evolved is used to run the plant and heat the digesters. (Surplus gas may be available for sale.) The process renders the sludge free from pathogenic organisms and offensive odours and reduces by 30 per cent the weight of solid matter to be disposed of. Subsequent solids disposal – land spreading or dumping – is very much simplified. Incineration can be costly and involve the use of fuel, as opposed to the generation of methane by the digestion process.

It is often adopted where the sludge is produced in large volumes as in inland cities where metal content may be high. Land spreading is sometimes used in areas where the sludge is relatively free from HEAVY METAL contamination. This may restrict its use on the same land to once in 30 years to prevent accumulation of toxic heavy metals in excessive quantities. The average heavy-metal concentration in UK sewage sludge is:

| Metal | Concentration (mg/kg dry solids) |
| --- | --- |
| Zinc | 1820 |
| Copper | 613 |
| Nickel | 188 |
| Cadmium | 29 |
| Lead | 550 |
| Chromium | 744 |

These values accord with EC permissible levels, but of course individual analyses are needed before application to land.

Ocean dumping is used by large coastal cities; the sludge is dumped in deep trenches outside the tidal flow to be diluted by the sea, but this is rapidly being phased out due to EC pressure. Incineration in autothermic (self-sustaining) incinerators is being canvassed as the most likely solution once ocean dumping is terminated. FLUIDIZED BED combustion is also an up-and-coming method with units designed to burn ca. 15 000 tonnes of solids per year extracted from 240 000 tonnes of wet sludge, which is dewatered to 30 per cent MOISTURE CONTENT. This permits autothermic combustion once the combustion process is started. The combustion temperature is 850°C which destroys any toxic compounds present. Alkaline scrubbing of flue gases is employed. (⇨ ANAEROBIC PROCESS; SEWAGE TREATMENT, HUMUS)

**Slurry.** Fine particles suspended in water so as to form a paste with the properties of a viscous fluid, e.g. wet coal fines from a wet coal preparation plant. These can constitute a low grade fuel.

**Smog.** Collective name for a FOG containing man-made air pollutants usually trapped near the ground by a TEMPERATURE INVERSION. Constituents can be smoke, sulphur dioxide, unburnt hydrocarbons and nitrogen oxides. Smog should not be confused with PHOTOCHEMICAL SMOG.

**Smoke.** Gas-borne solids resulting usually from incomplete combustion or chemical reaction. The particles are usually less than 2 micrometres in diameter. Other solid particles in smoke are compounds of silica, fluoride, aluminium, lead, acids, bases, and organic compounds such as phenols.

Smoke density is measured by photometric methods of estimated visually. (⇨ RINGELMANN CHART)

Smoke is synergistic (⇨ SYNERGISM) with SULPHUR OXIDES and can have significant adverse health effects at concentrations as low as 200 micrograms per cubic metre.

**Smoke control area.** In the United Kingdom, an area designated under the Clean Air Acts as one in which smoke emissions from domestic chimneys and industrial stacks are substantially reduced (only authorized fuels may be burned and approved furnaces used).

**Smoke shade.** The measurement of the darkness of a plume of smoke by means of a shaded card such as the RINGELMANN CHART in which the observer finds the shade which provides the closet match to the plume.

**Smoke stain.** This is a patch of dark material accumulating on a white filter paper after a measured volume of air has been drawn through it. The darkness or change in reflectivity of light from the white paper can be used as a measure of the amount of dark particulate material in the air.

**Soil.** The topmost layer of decomposed rock and organic matter which usually contains air, moisture and nutrients, and can therefore support life. Soil types include sand, clay, loam (a sand–clay mixture) and peat (which contains a large proportion of decaying plant matter). In tropical zones it may be lateritic, i.e. clay formed by the weathering of igneous rocks, which is notoriously difficult to cultivate, and once cultivated can set into a rock-like mass making further cultivation impossible.

It is a highly complex and variable material. About 50–56 per cent by weight is mineral matter derived from underlying rock, consisting of a variety of inorganic compounds of different forms and composition; 25–35 per cent is water; 15–25 per cent soil gases and 5–10 per cent organic material; however, these proportions can vary enormously. The organic component comprises material derived from living and dead organisms. Each gram of soil can contain 100 million bacteria, 500 000 fungi, 100 000 algae and 50 000 protozoa.

Soils are predominantly CATION exchangers, but some soils have also developed the capacity to adsorb ANIONS such as sulphate. The adsorption sites are usually oxides of iron and aluminium, predominantly in the lower horizons of the soil. Sulphate adsorption is very important in determining the response of soils to ACID RAIN.

**Soil moisture deficit (SMD).** The difference between the moisture remaining in the soil at any time and the FIELD MOISTURE CAPACITY. The SMD is important in agriculture and determines whether irrigation is required or not. In the UK the mean annual SMD ranges from less than 12 millimetres (west coast) to more than 150 millimetres (East Anglia) where irrigation is now extensively practised.

**Solar constant.** The amount of solar radiation incident, per unit area and time, on a surface which is perpendicular to the radiation and is situated at the outer limit of the atmosphere (the earth being at its mean distance from the sun). Its value is approximately $1400 \, J/s/m^2$.

**Solar energy.** The prospect of a low maintenance cost, low environmental impact, 'free' energy source is obviously enormously attractive, and a considerable amount of research into practical methods of converting the sun's rays directly into usable energy has been mounted in the last few years. Such techniques would have the further advantage of helping to redress the energy imbalance between the developed and underdeveloped countries, since a majority of the latter are situated in tropical zones.

Engineering studies of large centralized solar electrical power systems suggest that capital costs per unit of energy output are within striking distance of traditional sources of energy, assuming the usual improvements of technique that would inevitably be associated with a serious commitment to such a system. Such schemes, however, would be in many ways a misuse of this ultimate, and in many ways ideal, resource, the principal advantage of which is that it is freely available in sufficient intensity between latitudes $\pm 30°$. There would appear to be little point in designing vast centralized systems of energy conversion which then have to overcome distribution problems, when there is no technological reason why small individual units, capable of providing single dwellings with heating, lighting and cooking facilities could not be developed. Rudimentary forms of such devices are already in use in many countries and would almost certainly become competitive with conventional energy systems if properly developed on a commercial basis.

Such diffuse solar technology would obviously be ideal in

developing countries where capital is not available for the development of large centralized power systems. Furthermore, such devices are not limited to tropical areas. Modern 'selective black' surfaces, which are highly absorbent through the visible spectrum but poor radiators in the far infra-red, can attain very high working temperatures even on a cloudy day and bring solar energy use to the northern hemisphere too.

**Solar heating.** A common means of putting the sun's energy to use is to cover a black water-filled metal panel with glass or plastic. The water is then heated by the GREENHOUSE EFFECT. Similar techniques apply to solar distillation where the sea or brackish water to be distilled is not enclosed in a metal and can therefore be evaporated. The glass sheet is cooler than the vapour evolved and therefore condensation takes place and the result is pure distilled water. Solar devices are essentially area intensive, because of the low solar radiation density. Production rates of 3–6 litres/m$^2$ are possible depending on location.

Other methods of solar heating include the use of black-coated pool bases, thus allowing the body of water to warm up very cheaply. Reflectors, 'concentrating' lenses, etc. have been tried mainly to generate high temperatures in special situations. (⇨ SOLAR ENERGY)

**Solid waste.** Any refuse, certain sludges and other discarded materials, including solid and semi-solid materials resulting from industrial, commercial, mining, agricultural operations and domestic activities.

**Solution mining.** As the reserves of high-grade ores diminish, the prospects of recovering metals from low-grade ores and spoil heaps is receiving great attention. The means are varied but solution mining, whereby the derived metal is leached *in situ* and then recovered from the leachate, is one of the most promising. (⇨ LEACHING)

This technique enables extremely low-grade ores to be mined. For example, copper ores with 0.5 per cent copper would require, by conventional means, extraction of the ore, crushing and milling, and at least 200 tonnes of rock would be removed per tonne of copper recovered. Solution mining would leach the copper from the rock *in situ* using a dilute sulphuric acid solution, after it had been fractured, say, by explosion, or crushed into large lumps (not milled which is extremely expensive) and piled into mounds. The copper ion solution can then be ponded, where it is placed in contact with scrap iron and ION EXCHANGE takes place. The reaction is as follows:

$$Fe + Cu^{++} \rightarrow Fe^{++} + Cu$$

| iron | copper | ferrous | copper |
|------|--------|---------|--------|
|      | ions   | ions    | deposit |
| (in solution) | | (in solution) | |

The iron goes into solution and the copper is deposited on the bed of the pond. The pond is drained and the copper deposit removed. By this means low-value scrap iron is substituted for high-value copper. Now, the process can stop there *but* a more useful technique is to regenerate the leach solution through the action of bacteria *Thiobacillus ferroxidans*, which can oxidize the ferrous sulphate to ferric (a form of iron) and release sulphuric acid in the process. The leach liquor is thus regenerated. This method is used in the extraction of uranium from low-grade ores in spoil heaps, old mine workings, etc.

The bacterial re-oxidation process is the key to the development of this work and the leaching techniques are now being applied not only to low-grade copper but also nickel ores, and the possibility of recovering aluminium from non-bauxite sources is also under consideration. It also has obvious environmental benefits where mining is done *in situ*. This technique may grow as high-grade ore reserves decline. (⇨ ARITHMETIC–GEOMETRIC RATIO)

**Solvent extraction.** A chemical separation method (used in nuclear fuel processing etc.). Two immiscible liquids, one being a mixture of dissolved substances and the other a good solvent for the material to be extracted, are agitated together. The material required passes into the solvent from which it can be recovered when the two liquids are allowed to separate out. The method is the basis for the purification of uranium and the extraction of plutonium.

**Solvents.** Omnibus term for liquids which are used to dissolve other substances. The table gives a list of those in common use.

For water pollution purposes, the ORGANOCHLORINE solvents such as trichloroethylene, which is widely used for industrial cleaning purposes, are of concern as their presence in trace quantities can render groundwater non-potable on grounds of taste. Several boreholes in the UK have been closed because of penetration of industrial solvents into groundwater from either careless or improper disposal or leakages from industrial wastes. Solvent recovery is a widely practised form of RECYCLING where spent solvents are distilled and reused. However, the cheaper solvents are often incinerated or dumped in HAZARDOUS

List of solvents in everyday use

| Solvent | Boiling point (°C) |
| --- | --- |
| Acetic acid | 118 |
| Acetone | 56.5 |
| Benzene | 80 |
| Cyclohexanol | 160 |
| Ethanol | 78 |
| Ethylene glycol | 197 |
| Furfural | 162 |
| Glycerol | 290 |
| Hexane (normal) | 69 |
| Isopropanol | 82.5 |
| Methanol | 65 |
| Methylethylketone (MEK) | 80 |
| n-Propanol | 97 |
| Toluene | 111 |
| Trichlorethylene | 87 |
| White spirit | 155–195 |
| plus, of course, water | |

WASTE landfill sites (not recommended but done nevertheless if the site licence permits it).

One of the WASTE MINIMIZATION targets in the chemical industry is the use of much reduced solvent quantities.

Waste solvent, previously disposed of through an external agency, is now being reclaimed and reused on site by Aerospace Advanced Materials at Runcorn, UK.

This example of recycling and waste minimization is achieved through investment in a £4500 low-maintenance distillation unit which, although unable to handle all the large quantities of waste solvent and resin produced at the plant, is making a valuable contribution to the company's improved environmental performance.

The recovered solvent is worth 56.6p per litre and the cost of running the recovery unit is 6.5p per litre. Capital recovery period has been estimated at:

| Litres daily | Payback period |
| --- | --- |
| 50 | 36 weeks |
| 100 | 18 weeks |
| 150 | 12 weeks |
| 200 | 9 weeks |

*Source: Courtald's Environmental Matters*, March 1995.

**Somatic mutation.** A mutation arising in a non-reproductive cell. (⇨ IONIZING RADIATION)

**Soot.** Finely-divided carbon particles which adhere together. Soot is often left in flues when fossil fuels are incompletely burned.

**Sootblowing.** The use of jets of steam or compressed air to remove soot deposits from boilers.

**Sorption.** Process of ADSORPTION or ABSORPTION of a substance on or in another substance. (Sorption covers both processes.)

**Sound.** Periodic wave-like fluctuations of air pressure. The amount by which the pressure changes is known as the sound pressure (which is the 'mean' of pressure peaks generated by the waves), and the rate at which the fluctuations occur is the frequency. High-frequency sound is characterized by screeches and whistles; low frequency sound by rumbles or booms.

The sound pressure level is a measure of the sound pressure differences using the logarithmic decibel (dB) scale. The nature of the decibel scale is illustrated in Figure 135. It will be seen that any tenfold increase of sound pressure on a linear scale corresponds to a rise of sound pressure level of 20 on the decibel scale. Thus taking the pressure of a just-audible sound and ascribing to this a value of 0 dB, a sound of ten times that pressure has a level of 20 dB. A sound of a little more than three times the pressure of the just-audible one has a level of 10 dB. The sound pressure at which a sensation of pain begins in the ear is about one million times greater than that of the quietest sound that can be heard, and this has a level of 120 dB. Decibels therefore give a manageable way of measuring sound pressure, as well as having a close conformity to the ear's scale of response. The expression of a sound pressure level is always relative to the reference level of (in this instance) the quietest sound and is further discussed in the DECIBEL entry. (⇨ DECIBEL; HEARING, NOISE; ROAD TRAFFIC NOISE; AIRCRAFT NOISE; INDUSTRIAL NOISE MEASUREMENT; NOISE INDICES)

**Source.** The place, places or areas from where a pollutant is released into the atmosphere or water, or where noise is generated. A source can be classified as *point source*, i.e. a large individual generator of pollution, an *area source*, or a *line source*, e.g. vehicle emissions and noise.

**Special waste.** ⇨ WASTES.

**Species[1].** In chemistry a given kind of ATOM, MOLECULE or RADICAL which has a characteristic chemical structure and composition.

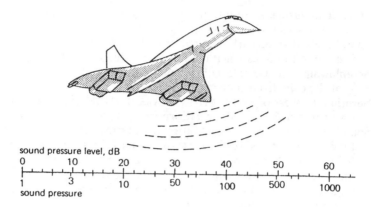

**Figure 135**   The decibel scale for sound pressures.

**Species².** In botany or zoology, a group of closely-related individuals showing constant differences from allied groups.

**Specific capacity.** The specific capacity of a well is its yield per unit of drawdown, usually expressed as cubic metres per hour (i.e. difference in level between the pumping water level and the natural water level).

**Specific gravity.** The ratio of an object's or substance's weight to that of an equal volume of water. The property is useful in, for example, sink float separation of minerals or wastes.

**Specific volume.** The volume of unit mass, e.g. the specific volume of STEAM (saturated) at 0.1 BAR (pressure) is $14.56\,mg^3/kg$; at 1 bar it is $1.673\,m^3/kg$.

**SPI (Society of Plastics Industry) marking systems.** A system of labelling plastic bottles for ease of recycling as set out below:

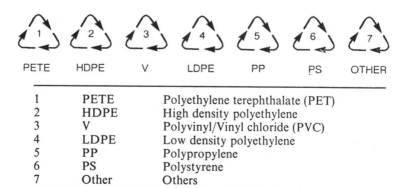

| 1 | PETE  | Polyethylene terephthalate (PET) |
|---|-------|----------------------------------|
| 2 | HDPE  | High density polyethylene        |
| 3 | V     | Polyvinyl/Vinyl chloride (PVC)   |
| 4 | LDPE  | Low density polyethylene         |
| 5 | PP    | Polypropylene                    |
| 6 | PS    | Polystyrene                      |
| 7 | Other | Others                           |

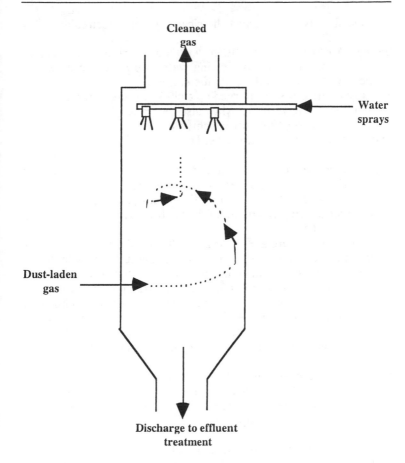

**Figure 136**  Spray tower scrubber

**Spoil bank.** A deposit of colliery or other mine or quarry waste. In addition to stone, spoil banks may contain sufficient fine and low-grade coal to enable spontaneous combustion to occur. If this happens, a very serious source of air pollution can occur. Acid water run-off is also a frequent companion which creates severe local stream pollution.

**Spray tower.** Air pollution control device for PARTICULATE removal from DUST laden gases or for scrubbing out of acid gases using an alkaline solution in which the gases are cleaned by contacting with a water spray (Figure 136). The effluent discharge is subsequently treated and the cleaning water recirculated. (⇨ SCRUBBER)

**Springs.** A discharge of GROUNDWATER appearing at the ground surface as continuously flowing water. Springs appear where the WATER TABLE intersects the surface.

**Stack gas.** The gases discharged up a chimney stack for dispersion into the ATMOSPHERE. May also be called FLUE GASES or EXHAUST GASES (for motor vehicles).

**Standard.** Broadly, something used as a basis of comparison, often a unit of reference.

**Standard deviation.** The most common measure of spread or deviation in a set of observations. It is the square root of the average of the squares of the differences (VARIANCE) of each observation from the mean of those observations. (⇨ NORMAL DISTRIBUTION)

**Standard temperature and pressure (STP).** As the density of gases depends on temperature and pressure, it is customary to define the pressure and temperature against which the volume of gases are measured. The normal reference point is standard temperature and pressure – 0°C at a standard atmosphere of 760 millimetres of mercury (approximately 100 000 pascal, the SI unit of pressure). All gas volumes are referred to these standard conditions. (⇨ GASES, PROPERTIES OF; NORMAL TEMPERATURE AND PRESSURE)

**Statutory Consultees.** These are the Health and Safety Executive and, where water is concerned, the National Rivers Authority (or Scotland River Purification Board). The following may be consulted where appropriate:

(a) The Minister of Agriculture Fishes and Food.
(b) The Secretary of State for Wales.
(c) The sewerage undertaker for any process which may involve discharge to a sewer.
(d) The Nature Conservancy Council for England, for processes located in England which may affect a site of special scientific interest; the Countryside Council for Wales for similar processes located in Wales; and the Scottish equivalent.
(e) The Harbour Authority for processes which may involve the release of a prescribed substance into a harbour under its control. For Scotland the National Consultees are also used.

**Statutory limit.** A specific upper limit, which by law cannot be exceeded.

**Steam.** Water in the vapour state. It can be in the wet condition (i.e. containing water in the vapour), dry (i.e. free from water, usually called dry saturated where it is both dry and at the same temperature as the boiler water) or superheated (i.e. at a higher temperature than the boiler water). All three steam conditions are encountered in steam engineering and denote differing energy contents per unit mass as well as densities.

Steam tables are a convenient method of showing the various related properties of saturated steam, for example see the table on the following page.

The steam tables show the properties of what is usually known as 'dry saturated steam'. This is steam which has been completely evaporated, so that it contains no droplets of liquid water.

In practice, steam often carries tiny droplets of water with it and cannot be described as dry saturated steam. Nevertheless, it is important that the steam used for process or heating is as dry as possible and correct STEAM TRAPPING and separation can improve steam quality.

Steam quality is described by its 'dryness fraction' – the proportion of completely dry steam present in the steam being considered. It is usually expressed as a decimal value less than 1, i.e. 0.95 represents 95 per cent dry steam. The volume of 1 kg of steam at any given pressure is termed its specific volume and the volume occupied by a unit mass of steam decreases as its pressure rises. The small droplets of water in wet steam have mass but occupy negligible volume.

**Steam tables.** ⇨ STEAM.

**Steam trap.** A device for removing moisture from STEAM in order to supply dry process steam. Their correct operation is vital in the high EFFICIENCY running of the SYSTEM as the example in Figure 137 illustrates.

| Trap size (mm) | Orifice size (mm) |
|---|---|
| 15 | 3 |
| 20 | 5 |
| 25 | 7.5 |
| 40 | 10 |
| 50 | 12.5 |

Related properties of saturated steam

| Gauge pressure (bar) | Absolute pressure (bar) | Temperature (°C) | Specific enthalpy | | | Specific volume steam (m³/kg) |
| | | | Water (kJ/kg) | Evaporation (kJ/kg) | Steam (kJ/kg) | |
| --- | --- | --- | --- | --- | --- | --- |
| 3 | 4.013 | 143.75 | 605.3 | 2133.4 | 2738.7 | 0.461 |
| 5 | 6.013 | 158.92 | 670.9 | 2086.0 | 2756.9 | 0.315 |
| 10 | 11.013 | 184.13 | 781.6 | 2000.1 | 2781.7 | 0.177 |
| 15 | 16.013 | 210.45 | 859.0 | 1935.0 | 2794.0 | 0.124 |

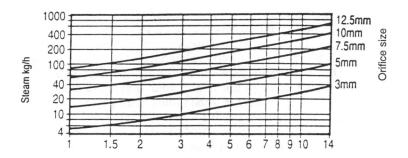

**Steam pressure bar g**

**Figure 137** Typical wastage through steam trap leaks.

*Example*
A process plant has 200 steam traps of which 10 per cent fail annually.

The plant operates with steam at 6 bar g and runs 12 hours per day, 6 days per week for 50 weeks per year = 3600 hours per year.

Thus, the average number of failed traps per annum = 20.

From the chart, energy loss via sharp edged orifice (for traps in use) = 50 kg/h.

Allowing an actual energy loss of 50 per cent of sharp-edged orifice = 25 kg/h per failed steam trap.

So, 20 × 25 × 3600 = 1 800 000 kg per year, i.e. 1800 tonnes of steam are wasted each year. Good maintenance pays off!

*Source:* Spirax Sarco Technology, Cheltenham.

**Stochastic effects.** As applied to radiation exposure, this refers to effects where the probability of an event (e.g. cancer) taking place is proportional to the dose received, i.e. no threshold is assumed. Non-stochastic effects occur only after a threshold has been exceeded, i.e. a 'large' dose has been received. (⇨ IONIZING RADIATION)

**Stoichiometric.** The exact or fixed proportions of elements in a chemical compound or of reactants to produce a compound. Thus in CARBON DIOXIDE the stoichiometric ratio of carbon atoms to OXYGEN is 1:2. Stoichiometric amounts satisfy a balanced chemical reaction with no excess of reactants or products.

**Stoker.** A mechanical feeding mechanism to control the rate of solid fuel admission into a boiler. Usually, for coal, a continuous moving chain (chain grate) or ram feed. For DOMESTIC REFUSE, can be rotating cylindrical grates or various designs of rocking bars to ensure both feed control and agitation of the fire-bed.

**Stokes' law.** A mathematical expression for the drag of a small sphere falling through an infinite fluid: $D = 6\pi\mu r u$, where $\mu$ is the viscosity of the fluid, $r$ is the radius of the sphere, and $u$ the velocity of the sphere. It is valid only for restricted conditions (laminar flow and low REYNOLDS NUMBER). Stokes' law is widely used in the study of the settling of particulate matter both out of the atmosphere and in water treatment.

**Stomata.** The small apertures of pores in a leaf epidermis (skin) or young stem of a plant and also of externally secreting gland in animals. In plants they allow the passage of gases and vapours into and out of leaves.

**Storage reservoir.** A reservoir, normally constructed by damming a valley in an upland catchment area, for the provision of water for the public supply.

**Storm overflow.** A device used in storm and sewage flow sewerage systems to prevent overloading. Flows in excess of a predetermined quantity are discharged untreated (with the possible exception of screening) direct to a nearby receiving water course, usually a stream. Severe cases of local stream pollution have been caused by this.

**Stratification.** The separation of a lake or sea into distinct layers of strata. These layers are characterized by 'warm' water on top and 'cold' on the bottom with an intermediate transitional band or 'thermocline' which is stagnant and seals off the bottom laver.

Figure 138 shows the stratification for Grasmere in the Lake District. Note that the amount of DISSOLVED OXYGEN has dropped to zero in the bottom layer and thus this anaerobic zone will not support aquatic aerobic life. Stratification can also occur in estuaries when freshwater floats over salt water.

**Stratosphere.** The 'upper' portion of the earth's atmosphere above the TROPOSPHERE extending to a height of about 80 kilometres. The temperature increases with height in this region.

**Straw.** Plant stems remaining after cereal crops have been harvested. In the UK the varieties are wheat, barley and oats; each one has its own characteristics. Barley and oat straw can be chopped, milled and fed to ruminants (sheep and cattle) as a carbohydrate source which requires protein supplementation for a balanced diet. As straw is produced in abundance in the UK (annual

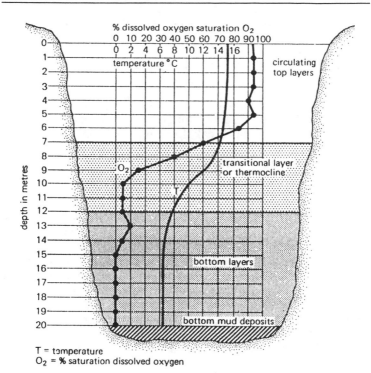

**Figure 138** A stratified lake based on measurements of Grasmere, 2 September 1974. Temperature and dissolved oxygen are measured at metre intervals.

production ca. 9 million tonnes), many uses have been proposed including papermaking, enzymatic decomposition to produce glucose and/or high protein feedstock. Chemical delignification by boiling with caustic soda to produce an easily digestible feedstock is one possible use being explored. (⇨ LIGNIN)

As a fuel, straw has a CALORIFIC VALUE (GROSS) of 14.7 MJ/kg which compares with 28 MJ/kg for coal. However, it may be a cheap fuel, but it does require transporting and possibly briquetting or pelleting before combustion. It is normally burnt in a cyclone furnace with an associated waste heat boiler. The overall EFFICIENCY of this arrangement is ca. 75 per cent which is comparable with many coal fired boilers of similar outputs (4 MW or less).

A microbiological method for breaking down straw for

feedstuffs is to use selected strains of FUNGI. This technique can be carried out simply on the farm, so avoiding high transport and processing costs.

The process involves the following steps:

Straw
↓
Inoculation ← add cellulosic fungus + a scavenger organism to help break down mixed carbohydrates
↓
Incubation ← control temperature/moisture
↓
straw/protein feed

This technique can provide an upgraded ruminant feedstock that should require no protein supplement. (⇨ CELLULOSE)

**Strontium-90.** Radioactive isotope of strontium produced in nuclear reactions as a fission product. It is chemically similar to calcium and has a half-life of 28 years. Like calcium it tends to be concentrated in milk. Thus, if milk contaminated with strontium-90 is drunk, the strontium-90 will be concentrated in the bones.

**Structure plan.** A written statement, approved by the Secretary of State for the Environment, of the County Planning Authority's general policies and main proposals for change over a period of up to 15 years.

**Sublimation.** Sublimation is the process by which a solid is converted directly to the vapour phase by application of heat. Condensation of this vapour to a solid, without an intermediate liquid phase, is known as desublimation.

Unlike DISTILLATION, which is applied to substances in the liquid phase, sublimation can be used to purify substances or mixtures which tend to decompose or polymerize at temperatures about their melting points.

For many gases this approach works well.

**Substantial change.** The meaning of substantial change is given in section 10/7 of the Environmental Protection Act 1990. Guidance of what constitutes a substantial change is also available from HMIP. As a general principle, any change or increase in the release of any prescribed substance to a relevant medium should be considered as substantial. For IPC processes, any increase in rate of overall throughput beyond 5 per cent of the design capacity or authorization conditions, or changes in the process operating parameters, e.g. temperature or feed stock, or changes in process equipment are likely to be considered substantial changes.

**Substitute liquid fuel (SLF).** Also known as SECONDARY LIQUID FUEL.

SLFs and their use in CEMENT kilns are causing serious concern to the established HAZARDOUS WASTE disposal industry. For example, Cleanaway have urged that

1. 'Fuels' derived from wastes should remain classified as wastes through to final destruction.
2. Until the single ENVIRONMENT AGENCY is established the Department of the Environment should require a single authority to regulate both blenders and kilns to ensure that inappropriate wastes are excluded from waste-derived fuels.
3. The same stringent emission limits and reporting parameters should apply to all facilities burning hazardous waste.

*Source: The Responsible Incineration of Hazardous Waste*, Cleanaway Limited.

**Subsurface water.** That part of rainfall which is not evaporated or which does not flow away as surface RUN-OFF, it penetrates into the ground and thereby becomes subsurface water.

**Sugars.** Carbohydrates, i.e. compounds of carbon, hydrogen and oxygen, which are usually crystalline and dissolve in water to give a sweet-tasting solution. They may be classified by molecular structure, e.g. as mono-, di-, tri- or poly-saccharides. ($\Rightarrow$ CELLULOSE)

Monosaccharides $\Big\langle$ $C_6H_{12}O_6$, hexoses (6 carbons), e.g. glucose, fructose
$C_5H_{10}O_5$, pentoses (5 carbons), e.g. xylose

Disaccharides $\quad C_{12}H_{22}O_{11}$, e.g. sucrose, maltose, lactose, cellobiose

Trisaccharides – $C_{18}H_{32}O_{16}$, e.g. raffinose

**Sulphate.** Salts of sulphuric acid containing the $SO_4^{2-}$ group. Sulphur dioxide emitted from the burning of sulphur in fuels is oxidized slowly in the atmosphere to sulphur trioxide ($SO_3$) which forms sulphuric acid with moisture and sulphates with basic materials such as ammonia or metals and their oxides. Sulphates so formed are particulates. Sea spray is a substantial natural source of airborne sulphate particles. The chemical reactions for acid formation (as well as acid attack on limestone) are given in equations [1], [2] and [3]:

$$SO_2 + \tfrac{1}{2}O_2 \rightarrow SO_3$$
(Sulphur  (Oxygen)  (Sulphur
dioxide)  trioxide) [1]

$$SO_3 + H_2O \rightarrow H_2SO_4$$
(Sulphur  (Water)  (Sulphuric acid)
trioxide) [2]

$$H_2SO_4 + CaCO_3 \rightarrow CaSO_4 + CO_2 + H_2O$$
(Sulphuric  (limestone)  (Calcium  (Carbon  (Water)
acid)  sulphate)  dioxide) [3]

Hence, buildings can be subjected to corrosive acid attack by the combustion of sulphur-containing fuels. (⇨ ACID RAIN)

**Sulphur.** A non-metallic element, atomic number 16, relative atomic mass 32.6, symbol S. A constituent of all living matter, which occurs in all fossil fuels and is emitted (in the form of sulphur oxides) on combustion. Attention has been given to the possibility of removing the sulphur from such fuels.

**Sulphur cycle.** The three principal forms of sulphur present in the lower atmosphere, sulphur dioxide, hydrogen sulphide and sulphates, have been formed and emitted by both natural and industrial processes. Other mechanisms return these compounds to the earth's surface. For example atmospheric sulphur dioxide is oxidized to the trioxide, which combines with water and is washed out on to the earth's surface as sulphuric acid or sulphates; bacterial action converts the sulphates into hydrogen sulphide, which is then oxidized to sulphur dioxide. This cycle maintains the global atmospheric concentration of sulphur dioxide at a roughly constant level.

**Sulphur dioxide.** ⇨ SULPHUR OXIDES.

**Sulphur oxides.** Sulphur dioxide ($SO_2$) and sulphur trioxide ($SO_3$). Of the two, sulphur dioxide predominates and in the presence of particular catalysts conversion to sulphur trioxide can take place.

Sulphur oxides occur naturally from sources such as volcanoes, sulphur springs, decaying organic matter PHYTOPLANKTON activity. Annual global production by man is about 100 million tonnes per year with over 90 per cent produced in the northern hemisphere. Man-made sources are principally combustion of fuels which contain sulphur, brickworks and spontaneous combustion in coal mine spoil heaps. The man-made sulphur dioxide contribution is usually concentrated in industrial and

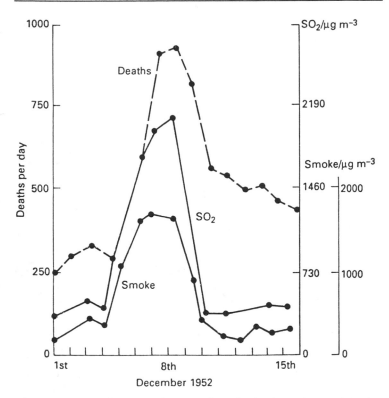

**Figure 139**   Health effects of $SO_2$ and smoke in December 1952 during London smog.

domestic areas and can severely affect health during SMOG conditions. It is synergistic ($\Rightarrow$ SYNERGISM) in combination with smoke. Together they affect the respiratory tracts and about 1 per cent of the population encounter bronchial spasms at concentrations of between 300 and 500 micrograms per cubic metre ($\mu$g/m³). Above $57\,000\,\mu$g/m³, waterlogging of the lungs takes place and eventually respiratory paralysis. During the 1952 London smog the extra deaths of 4000 people in one week and another 8000 in the following three months were attributed to the combination of sulphur dioxide and smoke (see Figure 139).

The effects of sulphur dioxide and smoke, respectively, on man begin at concentrations of 300–500 $\mu$g/m³ for sulphur

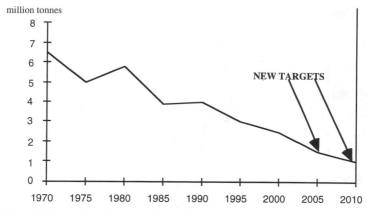

**Figure 140** UK emissions of sulphur dioxide trends (HMIP).

dioxide and $250\,\mu g/m^3$ for smoke. But the effect of sulphur dioxide on LICHENS begins at $40\,\mu g/m^3$. Thus the damage to lichens does not correlate with the health effects on man. But where should the permissible level of exposure be set?

It is clear that society puts up with different levels of effect, according to what is being affected – in this case plants or people – and it is apparent that sulphur dioxide levels are set for people. This may or may not be a wise judgement. The effects of lower sulphur dioxide levels may be increased crop yields, but the costs would be the costs of desulphurizing fuel before combustion, or the flue gases after combustion and before emission. COST-BENEFIT ANALYSIS could possibly be helpful in this area. This example illustrates how much is yet to be done to determine what is an acceptable level of any particular pollutant.

Removal of sulphur oxides from the atmosphere is usually accomplished in the form of ACID RAIN with serious effects on property through corrosion and in some cases on aquatic ecosystems due to the alteration of pH.

The emissions trend and new targets are shown in Figure 140.

**Sulphur trioxide.** ⇨ SULPHUR OXIDES.

**Sulphuric acid.** A dense oily liquid, colourless when pure; formula $H_2SO_4$. It is highly corrosive and poisonous. Sulphuric acid is the most widely used of all industrial chemicals. As a pollutant it occurs in the atmosphere in the form of an aerosol, called sulphuric acid mist, produced by the oxidation of atmospheric sulphur dioxide as well as by direct emissions from stacks. These fine droplets are more difficult to remove from the air than gaseous sulphur dioxide, their life in the atmosphere is longer, and they can travel great distances with the wind. They can reach the alveoli in the lungs without being absorbed in the wider bronchial passages, or in the nose and throat; they can therefore, be potentially very harmful.

**Sunshine recorder (or heliograph).** An instrument which records the time interval during which solar radiation reaches sufficient intensity to cast distinct shadows.

**Supercritical oxidation.** The heating of water beyond 374°C to pressures greater than 217.7 atmospheres enables it to break down complex organic compounds into simpler chemical building blocks.

A test plant has run smoothly on sewage sludge in the US (Modell Environmental Corporation) which is heated with oxygen in a long, three-stage reactor tube. The supercritical water breaks the raw material down into simple oxides. After heating and cooling, the by-products are clean water, clean carbon dioxide gas and, in the case of sewage sludge, an odour-free brown powder. As the system is completely sealed, there are no other emissions.

*Source:* S. Haig, 'Alchemy of sludge', *Financial Times*, 22 February, 1995.

**Supersonic flight.** If the speed of an aeroplane is greater than the speed of sound in air then this is termed supersonic flight, and the plane is said to be travelling at greater than Mach 1 (the speed of sound in air). When this is the case, the plane creates a so-called shock wave and a noise or 'boom' accompanies the flight path. The boom intensity is a function of the speed, altitude and plane design. (⇨ CONCORDE)

**Supply curve.** A supply curve shows, for a range of prices, the quantities of a commodity that will be offered for sale. Such a curve may be drawn for a single seller or for a number of sellers in a market. (⇨ DEMAND)

**Surface inversion.** A surface or ground inversion is a temperature inversion based at the earth's surface, i.e. an increase of air temperature with height beginning at the ground level. (⇨ TEMPERATURE INVERSION)

**Suspended solids.** The dry weight of solids captured by filtering a known volume of untreated sewage effluent, other effluents or river water. Usually expressed as mg/l.

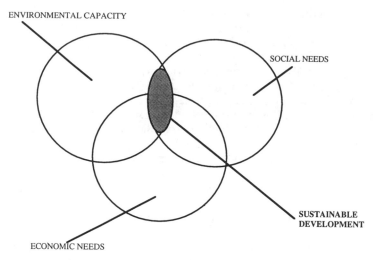

**Figure 141**   Measuring sustainable development.

**Sustainability.** *Sustainable development*: formally it is 'development which meets the needs of the present generation without compromising the ability of future generations to meet their own needs'. This should make us think about how we currently use our natural resources at the future expense of our children.

The question of equity between the generations, industrialized and industrializing countries, and sectors in society is now firmly on inter-governmental agendas because, in reality, the pace and type of development for some is at the expense of others.

There is now a growing recognition that social, environmental and economic needs must be fully integrated if sustainability is to be achieved (see Figure 141).

*Measuring sustainable development*: statistical indicators on a range of themes can be chosen which together give as good an indication as possible of progress. The themes are shown in the table below:

---

*Environmental capacity*
- Resource use and waste   water, materials, land take, energy,
                                                sources of land, water, air, food, etc.
- Pollution
- Biodiversity

*Quality of life*
- Basic needs           food, water, shelter, energy
- Information/training/ education
- Leisure and culture
- Freedom – political      participation
          – personal      freedom from fear, crime or persecution
- Access/transport       to goods, services and people
- Income                    adequate and fair
- Work                     opportunity to, including voluntary
- Health                   physical and mental
- Beauty/aesthetics       of places, spaces, objects

*Source:* Avon County Council Newsletter, 'The Environment in Avon', Issue 6, Spring 1995.

*Sustainability encouragement through resource pricing*

In the short term the use of energy and other resources responds only slightly to increases in price. But this should not be misinterpreted as suggesting that this is an inefficient instrument. Over the long term price increases can be expected to have a very great effect, even on the consumption of car fuels, which is often said to be particularly unresponsive. Figure 142 shows a surprisingly strong relationship between per capita fuel consumption and fuel prices that were consistently kept at different levels regardless of world market prices: the higher the price, the less energy was consumed.

*Sustainability examples*

Sustainable landfill: a sustainable landfill will

- be located in an area that is not hydrogeologically sensitive, i.e. not close to water resources;
- be located in strata that do not facilitate rapid movement of materials from the site;
- be located in strata that are conducive to the attenuation of contaminants;
- contain only those materials that will degrade or will remain immobile;
- exclude mobile persistent toxic species;
- be managed to promote degradation processes;
- be managed to collect and utilize landfill gas;
- be designed to facilitate, at the appropriate stage of degradation, slow migration from the site of the mobile components in the site.

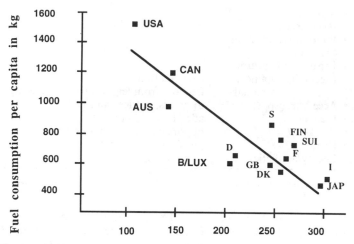

**Figure 142**   Fuel prices and consumption per capita.
*Source:* E. U. von Weizäcker, 'Let prices tell the ecological truth', *Our Planet*, (UNEP), Vol. 7, No. 1, 1995.

Fuel prices (index USA 1988 = 100)

Sustainable landfill should result in the re-introduction in a controlled manner to the environment of the raw materials of society that have been exploited and ceased to have a useful function. The sustainable development policy of the UK government recognizes the inevitable need for landfill facilities and that it is the BEST PRACTICABLE ENVIRONMENTAL OPTION (BPEO) for certain waste materials and is the only alternative for wastes which cannot be treated.

While in theory, sustainable landfill is possible, whether it is practically achievable is not certain. Even if the ultimate sustainable landfill proves unachievable, its pursuit will improve landfill practices. This in itself is a valid justification for the pursuit of sustainability.

M. J. Carter, M. J. Carter Associates, 'Towards sustainable landfill', *Scottish Envirotec*, February 1995, p. 22.

*Sustainability and CHP*
In 1993, just as in 1983 and 1973, the UK throws away more thermal energy from power stations than it obtains in the form of natural gas from the North Sea. This energy flow (100 GW)

provides more than enough energy to heat all the buildings in
England, Scotland and Wales.

*Source:* D. Green, 'Sustainable energy policy demands efficiency',
*Energy in Buildings and Industry,* March 1995.

One point is of pivotal significance. No long-term strategy of
poverty alleviation can succeed in the face of environmental forces
that promote persistent erosion of the natural resources upon which
we all depend. And no environmental protection programme can
make headway without removing the day-to-day pressures of
poverty that leave people little choice but to discount the future so
deeply that they fail to protect the resource base necessary for their
own survival and their children's well-being.

To meet the challenges of 'Attacking Poverty, Building Solidarity,
Creating Jobs', the global community must devise new and innovative
strategies that attack the vicious cycle of poverty and environmental
degradation. This will be just the first step.

*Source:* Elizabeth Dowdeswell, United Nations Under-Secretary General
and Executive Director, UNEP, commenting on the World Summit for
Social Development 6–12 March 1995, Copenhagen, *Our Planet,* Vol. 7
No. 2.

There are no magic solutions. In the long run there is no alternative
to restoring equilibrium between resources and population and the
environment, working for a relatively steady state society. The first
step to wisdom is to recognise the problem. The next is to do all
possible to prepare for it. Governments must work in the first
instance to manage it within their countries. They will need help
from the international community, especially from those who have
contributed most to the creation of the problem. Action to
accommodate refugees across frontiers would be immensely more
difficult.

*Source:* Sir Crispin Tickell, former United Kingdom Ambassador to
the United Nations, *Our Planet,* Vol. 7, No. 3, 1995

(⇨ COMMON INHERITANCE)

The strongest ecological effect of ecological tax reform comes from
making it more profitable to lay off kilowatt hours than to lay off
people.

*Source:* E. U. von Weizäcker, 'Let prices tell the ecological truth', *Our
Planet,* (UNEP), Vol. 7, No. 1, 1995.

**Symba.** A process, developed by the Swedish Sugar Company,
which uses starch as a substrate for the eventual production of
SINGLE-CELL PROTEIN using the *Candida utilis* or *C. torula*

yeast strain. For the purpose *Candida utilis* is propagated with an organism, *Endomycopsin*, which produces an ENZYME (amylase) which can split starch into SUGARS. The organisms grow in symbiosis, the amylase activity of *Endomycopsin* converting the starch to sugars which the *Candida utilis* – the faster grower of the two – uses as a fermentation substrate, i.e.

Starch $\xrightarrow{\text{\textit{Endomycopsin}}}$ Glucose $\xrightarrow{\text{\textit{Candida utilis}}}$ Yeast cell substance (single-cell protein)

The Symba process has been used in effluent treatment from potato processing plants and has reduced the BIOCHEMICAL OXYGEN DEMAND by 85 per cent in 10 hours. It reduces pollution and obtains a useful product at the same time.

The process, or its counterparts, is under active investigation for single-cell protein production from waste carbohydrates (cellulose, starch, 'simple' sugars), and from the organic substances in chemical industry waste streams. The waste fibres and bark from pulp and board production are also fruitful sources for production of single-cell protein. (➪ SINGLE-CELL PROTEIN)

**Symbiosis.** A compatible association between dissimilar organisms to their mutual advantage. The classic case is the association of nitrogen-fixing bacteria with plants of the clover family. The bacteria occupy nodules on the roots of the plants. The bacteria fix the nitrogen from the air into nitrates for the plants, who in their turn supply the bacteria with carbohydrates as an energy source.

Symbiosis can be put to commercial use as in the SYMBA yeast process. (➪ SINGLE-CELL PROTEIN)

**Symmetrical distribution.** A set of values or observations which are distributed evenly about the MEAN. Often met as a bell-shaped curve or NORMAL DISTRIBUTION.

**Synergism.** A state in which the combined effect of two or more substances is greater than the sum of the separate effects, e.g. smoke and sulphur dioxides. It is the opposite of ANTAGONISM. (➪ PARTICULATES; SULPHUR OXIDES)

**Systemic agent.** An agent that affects the body as a whole and not a particular part or organ. (➪ MERCURY)

# T

**Tailings.** The residual fine-grained waste rejected after mining and processing of ore, usually after washing.

**Tar sands.** The Alberta tar sands (oil sands) constitute the largest known reserve of petroleum in the world (approximately 900 000 million barrels of 'in place' heavy oil). It will cost considerably more to extract this oil than from conventional oil fields. Efforts are being made to improve existing processing technology.

Essentially, the extraction process is one of open-cast mining, in which the 'overburden' of vegetation and earth must first be removed in order to expose the bitumen-soaked sand deposits. These deposits are extremely difficult to handle, being sticky and corrosive in summer and rock-hard in winter. The capital investment required for the large-scale extraction of such material is huge, and present techniques are probably only capable of recovering about 10 per cent of the synthetic crude which is potentially available in such deposits. ($\Rightarrow$ OIL SHALES)

**TCDD.** $\Rightarrow$ DIOXIN, TOLERABLE DAILY INTAKE.

**Temperature.** A property of a body or substance and a measure of how 'hot' it is, or how much thermal energy it contains. Temperature is measured on several scales, e.g. the centigrade or Celsius and Fahrenheit scales are both measured from a reference point – the freezing point of water  which is taken as 0°C or 32°F. The boiling point of water is taken as 100°C or 212°F respectively. For thermodynamic devices, it is usual to work in terms of absolute or thermodynamic temperature where the reference point is absolute zero, which is the lowest possible temperature attainable. For absolute temperature measurement the thermodynamic or KELVIN (K) scale which uses centigrade divisions is used.

Temperature is a fundamental measurement in most pollution work. The temperature of a stack gas plume, for example, determines its buoyancy and how far the plume of effluent will rise before attaining the temperature of its surroundings. This in turn determines how much it will be diluted before traces of the pollutant reach ground level.

**Temperature inversion.** The temperature of the air normally decreases at increasing heights above the ground. A variety of meteorological conditions can occur to reverse this trend and cause a layer of warmer air to overlie a cooler layer. The cooler air cannot then rise because it is heavier and so any air pollutants emitted below the inversion layer are trapped. Inversion of the normal temperature profile can begin at any height above the ground but the lowest are more noticeable in their effect as they trap the smoke and fumes from domestic chimneys. When they occur between 150 and 900 metres above the ground, they can trap the discharges from all chimneys except those of the largest power stations.

Inversions commonly (but not always) occur in valleys or

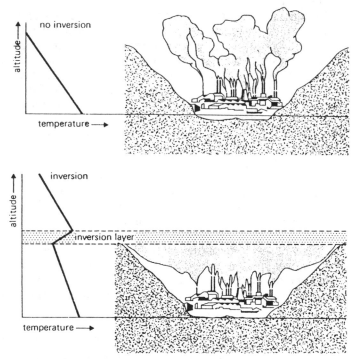

**Figure 143**  The effects of a temperature inversion.

basins due to radiation cooling of the ground at night. This cools the air near the ground, and being cooler, it can drain into the low-lying areas as a katabatic flow.

Figure 143 shows the temperature profile with height for typical non-inversion and inversion conditions. In extreme circumstances a halt may be required on industrial processes until the inversion lifts. In practice this temperature profile can assume a wide variety of forms.

A temperature inversion can also be associated with an area of high pressure or anticyclone. In this, air drifts downwards very slowly, warming as it does so.

Near the surface, however, the warm ground or sea sets off more vigorous upward motion.

These two air movements meet about one kilometre above the surface and a temperature inversion forms at their boundary.

A related phenomenon associated with industrialized cities is the *heat island*. This is most noticeable at night and early mornings and is caused by hot air layers forming at building or chimney height level which are warmer than both ground conditions and the air above the layer. This heat island can be 5–7°C warmer than ground conditions and, or course, can trap any pollutants emitted within it. The heat islands are dome-shaped and can disperse towards midday when the temperature increases, but in turn they may be replaced by a higher level inversion.

**Temporary hardness.** ⇨ HARDNESS.

**Teratogen.** An agent that causes birth defects. A strong teratogen is DIOXIN found in herbicides. The $LD_{50}$ for guinea pigs fed with dioxin is as low as 600 parts per million million. Fish in Vietnamese waters (1970–71) had concentrations on average of 540 parts per million million as a result of the use of herbicides as defoliants (1962–70). The use of defoliants in that country has been suggested as being responsible for rises in stillbirths and birth defects.

**Terminal velocity.** The maximum velocity at which a particle settles out in air or water.

**Tertiary recovery.** Recovery of oil or gas from a reservoir over and above that which can be obtained by primary and secondary recovery. It requires methods such as heating the reservoir to reduce the oil VISCOSITY.

**Tetraethyl lead (TEL).** Tetraethyl lead and tetramethyl lead (TML) are the principal additives to petrol to raise the octane

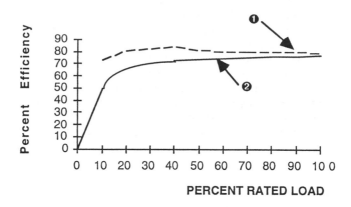

**Figure 144**   Specimen boiler efficiency vs per cent load (courtesy Clayton Boilers, Cheshire).

rating and thus reduce the tendency to 'knock' in spark-ignition internal combustion engines. Now being phased out in EC and USA due both to suspected links with increased blood lead levels associated with particulate lead emissions and to the use of CATALYTIC REACTORS which require lead-free petrol to avoid poisoning of the catalyst. (⇨ OCTANE NUMBER)

**Therm.** A measure of heat (thermal energy) which is used as the common basis for charging for heat supplied by gas or steam in the UK. It is $100\,000\,(10^5)$ British Thermal Units or $1.055 \times 10^8$ joules. It will eventually be superseded by megajoules (MJ). (⇨ BTU)

**Thermal efficiency.** The efficiency of a thermal energy conversion device. It is defined as the ratio of energy value of fuel supplied to useful energy output from the device. The thermal efficiency of modern coal fired power station boiler is ca. 90 per cent; that of an internal combustion engine ca. 14 per cent; that of a coal-fired power station 38 per cent. (⇨ CARNOT EFFICIENCY)

Figure 144 shows the effect of percentage loading on thermal (input–output) efficiency for two boiler designs ❶ and ❷ efficiencies. Clearly, the load cycle needs to be appraised before selection of the boiler system but ❶ has good low loading compared to ❷.

Since boilers operate most of the time at less than 100 per cent load rating, fuel costs cannot readily be compared unless this information is available and in comparable terms.

It is important to remember when comparing thermal efficiency claims that the percentage increase in fuel costs will be greater than that nominal difference in efficiency. For example, 80 per cent versus 75 per cent efficiency at partial load, a 5 per cent difference in efficiency, translates to a 6.25 per cent saving in fuel usage.

$$1 - \frac{75}{80} \times 100 = 6.25 \text{ per cent savings in fuel usage.}$$

Figure 145 shows how useful heat can be lost through soot build-up on boiler tubes and lead to a drop in efficiency manifested as an increase in fuel consumption.

**Thermal oxide reprocessing plant (THORP).** The plant being built by British Nuclear Fuels Limited at Windscale, Cumbria, for reprocessing oxide fuel from thermal reactors.

**Thermal pollution.** The heat released from the combustion of fossil fuels or the dissipation of energy from prime movers, such as electric motors, which is eventually converted to heat. All such releases end up as waste heat in the sinks of air and water.

Direct thermal pollution of water usually occurs at power stations where 60 per cent or more of the heat content of a fuel ends up as waste heat which must be removed by cooling water which is then discharged to rivers or coastal waters. This form of thermal pollution depletes DISSOLVED OXYGEN and can change aquatic ecosystems. It also increases any existing BIOCHEMICAL OXYGEN DEMAND and therefore *reduces* the capacity of a stream to assimilate organic wastes. The waste heat from power stations could be used for glasshouse or DISTRICT HEATING, but the heating load and power station capacity and duty require careful matching. In 1989, low river conditions in the River Ouse in Yorkshire, which is used for cooling the Drax power station, were such that diluted sewage effluent passed into the cooling towers with subsequent deposition of dried particulate matter in the vicinity.

Thermal pollution of the atmosphere can cause local instabilities and in North America there is reason to believe that the industrialized Atlantic Seaboard has its own man-made climate induced by the energy of air and pollutants released.

**Thermal processing.** A collective term for the disposal or conversion of domestic refuse or wastes by INCINERATION or PYROLYSIS.

**Thermal wheel.** A slowly rotating brick-lined heat exchanger which

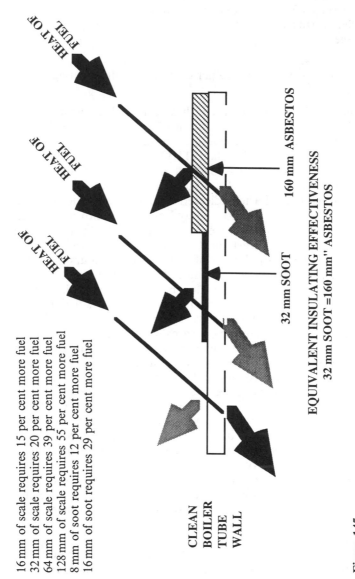

16 mm of scale requires 15 per cent more fuel
32 mm of scale requires 20 per cent more fuel
64 mm of scale requires 39 per cent more fuel
128 mm of scale requires 55 per cent more fuel
8 mm of soot requires 12 per cent more fuel
16 mm of soot requires 29 per cent more fuel

HEAT OF FUEL

HEAT OF FUEL

HEAT OF FUEL

160 mm ASBESTOS

32 mm SOOT

EQUIVALENT INSULATING EFFECTIVENESS
32 mm SOOT =160 mm'' ASBESTOS

CLEAN
BOILER
TUBE
WALL

**Figure 145**

*Source: The Clayton Report,* Clayton Boilers of Belgium, NV, 1995 (UK, Woolstoon, Warrington).

recovers heat from (say) furnace exhaust gases on one half, while the other half is giving up its recovered heat to incoming gases such as preheating air for combustion in boilers or blast furnaces. Recovery of 30–50 per cent of the heat in exhaust gases can be obtained.

**Thermionic converter.** A device for DIRECT ENERGY CONVERSION which gives an electrical current from the electrons emitted by the heating of a suitable metal. This thermionic emission produces a net flow of electrons from a high-temperature electrode (the cathode) to the lower temperature electrode (the anode), both being enclosed in an evacuated enclosure.

However, the efficiency of the device is low unless the gap between the electrodes is very small (less than $25\,\mu m$). In addition, as radiation takes place from the hot surface to the cold surface, large temperature differences cannot be used.

**Thermodynamic heating.** If a COMBINED CYCLE/COMBINED HEAT AND POWER plant is used in conjunction with HEAT PUMPS for DISTRICT HEATING then it is possible to obtain from 100% FUEL ENERGY, heat totalling 220%, where the heat pump extracts heat at ambient temperature.

The relevant energy balance is given in the Figure 146 for a COEFFICIENT OF PERFORMANCE of 3.5

Even if 10 per cent transmission losses are assumed for the electricity and the heat, the useful heat is still 198 per cent, which is more than double the heat produced by a good boiler. It should be noted that this figure refers to the maximum heat load; averaged over the year, conditions are even more favourable.

**Thermodynamic temperature scale.** ⇨ TEMPERATURE, KELVIN.

**Thermoelectric converter.** A device for DIRECT ENERGY CONVERSION which gives an electric current when the two junctions of a loop of two wires of different metals are kept at different temperatures. This is the principle of the thermocouple for temperature measurement. However, for ordinary metals the voltage produced per unit of temperature difference is extremely low and it is only with the advent of semiconductors that the voltage obtainable rose from microvolts per degree kelvin to millivolts per degree kelvin (a factor of 1000). Semi-conductor technology also allows vast numbers to be connected in series, so that useful outputs can be obtained. The electrical charge carriers can be electrons as in metals (or in semiconductor terminology 'n-type' materials) or they can be positive (or 'p-type' materials). Thus an 'n-pair' with low resistance to heat flow and low electrical resistance

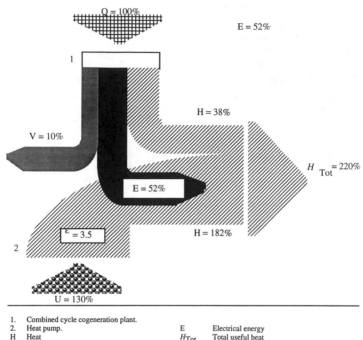

1. Combined cycle cogeneration plant.
2. Heat pump.
H   Heat
Q   Fuel
V   Losses
ε   Coefficient of performance

E      Electrical energy
$H_{Tot}$    Total useful heat
U      Ambient heat

**Figure 146**   Principle of thermodynamic heating. The combination cycle cogeneration plant with sequential combustion and heat pumps yields useful heat equivalent to 220 per cent of the primary fuel energy (100 per cent).
*Source: ABB Review*, March 1995.

would make the ideal unit for thermoelectric converter construction. The semiconductor is indeed such a device and just as for the THERMIONIC CONVERTER the 'working fluid' is the flow of charge carriers, so that the Carnot restrictions also apply. (⇨ CARNOT EFFICIENCY)

The thermoelectric device operates at lower temperatures than the thermionic converter and is therefore suited for waste-heat recovery applications as in, say, the exhaust gas stream of a gas turbine. Practical efficiencies are 10 per cent or less due to the temperature restrictions but this can be a useful addition in a total energy situation. The USSR manufactures a thermoelectric generator powered by a paraffin heater for radio receivers in rural areas.

**Thermoplastic.** A plastic which deforms on heating and which can be heated repeatedly and reformed. Hence, source separated plastics can be recycled, e.g. POLYVINYL CHLORIDE and POLYETHYLENE.

**Thermoselect process.** A proprietary process for gasification and direct smelting of MUNICIPAL SOLID WASTE (MSW) in which it is stated that organic components are totally destroyed and inorganic compounds smelted at 2000°C.

The products are a combustible synthesis gas, mineral products and metals. The sequence of operations is given in Figure 147.

An energy balance is given in Figure 148 for two production lines totalling 20 tonnes per hour. This implies that 1 tonne of MSW provides 350 kWh electricity.

**Thorium cycle.** A nuclear fuel cycle in which fertile thorium-232 is converted to fissile uranium-233. It is regarded as a possible alternative to the uranium-238/plutonium-239 cycle.

**Three Mile Island.** A major (1979) US nuclear reactor incident in which a circulating pump failure plus a stuck valve on the pressurizer allowed the pressurized water coolant to escape. The reactor core overheated in minutes and core meltdown was narrowly avoided. Total cost of clean-up was in excess of $US 1 billion. The US nuclear power industry was severely set back by this avoidable event which was compounded by operator error and mechanical failures.

The lesson is, nothing is certain except mishap and the most careful planning and emergency procedures must be openly agreed with the communities who host these facilities.

**Three-minute mean concentration.** The maximum permissible concentration of pollutant at ground level from a stationary source, i.e. the concentration of pollutant averaged over three minutes. For example, sulphur dioxide from a factory chimney is often specified as a three-minute mean value. Yearly averages are likely to be between 5 and 15 per cent of this value, and monthly averages up to 25 per cent owing to wind changes and plume-dispersal effects.

In addition to the three-minute mean concentration an absolute mass emission limit may also be placed on the total

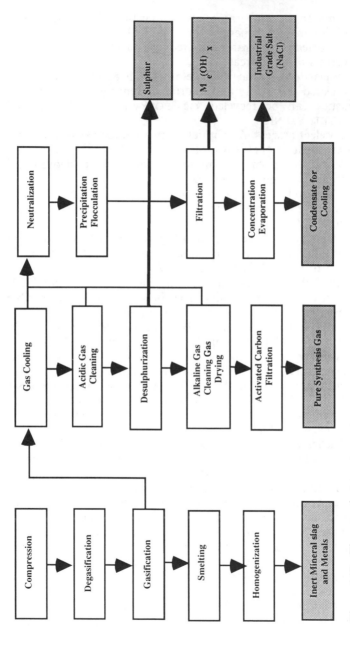

**Figure 147**  Thermoselect sequence steps.

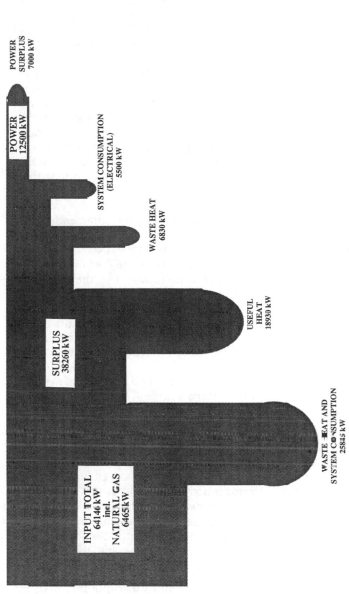

**Figure 148** Energy balance for thermoselect process (two production lines, throughput of 10 te/h each).
*Source:* R. Stahlberg, 'Thermoselect recovery of energy and raw materials from waste', *Recycling 95 Congress Proceedings*, Chalmers University, 1–3 February 1995.

amount of a pollutant that may be emitted from a stationary source per hour. Thus, the mass emission limit for a smelter discharging less than 85 cubic metres per minute of gases up the stack is set at 45 kilograms of lead per 168 hours which may not approach anywhere near the three-minute mean value, but is required to prevent despoilation of land through build-up of lead with its dangers to animals. (⇨ EMISSION STANDARD)

**Threshold.** In environmental terms, the dividing line between exposure to, say, a noise level above which physical damage may take place and below which damage will probably not take place. The threshold is a concept that is applied in many areas of industrial exposure to airborne pollutants. However, the basis is that below the threshold none but the most sensitive or infirm would suffer. There are substances for which many believe there is no threshold effect whatsoever, e.g. VINYL CHLORIDE and ASBESTOS. IONIZING RADIATION is another area where a threshold may in effect be postulated by a recommended dose limit but the only safe exposure is no exposure at all. (⇨ CARCINOGEN; MUTAGEN; THRESHOLD LIMITING VALUE)

**Threshold limiting value (TLV).** Formerly in the UK, the threshold limiting value of an airborne pollutant was the maximum concentration of that pollutant (or mixture of pollutants) to which it was believed 'healthy' workers in industry could be repeatedly exposed day after day without adverse health effects, on an eight-hour day. Now superseded by occupational control limits. (⇨ CONTROL LIMITS, OCCUPATIONAL; TIME-WEIGHTED AVERAGE)

**Tidal power.** Generation of power by the ebb and flow of the tides. This is an example of an 'income' source of energy and takes two forms:

1. BARRAGES with conventional hydroelectric turbines installed in special sluices, so that the incoming and outgoing tides generate power. The availability of suitable tidal locations is severely restricted and, as with WATER POWER, the use of this form of energy source is not expected to grow significantly.

2. Floating power stations which rise and fall with the tide and have wave-powered floats or paddles which can drive water turbines for power generation. This method is not restricted to the availability of sites as for barrages, and, though still in embryonic form, may have some potential in the long term.

**Time lag.** The time needed for an event or disruption to manifest itself. There are many examples in environmental matters of a time-lag effect. Once such example is the use of nitrogen fertilizers on agricultural land from which water filters into the London Chalk Basin. For many years, water from the aquifer in this area has been within World Health Organization limits for nitrates; now after 20 years or more of fertilizer application, some wells are showing higher than normal nitrate levels.

The latency period for the development of cancer is well known. This can take tens of years dating from exposure to the carcinogenic agent in the first instant. It is for this reason that before the introduction of new health products, food additives, hair dyes, etc., the most exhaustive tests must be carried out for mutagenic and carcinogenic properties. Hair dyes are a good case in point; they have been implicated as being carcinogenic but confirmation will have to wait for at least a decade in all probability.

The time-lag effect often obscures the cause of the problem encountered and leads to treatment of the symptoms only. ($\Rightarrow$ CARCINOGEN; MUTAGEN)

**Time-weighted average (TWA).** Used in the control of occupational exposure to hazardous substances or harmful emissions such as noise or IONIZING RADIATION. This is done by controlling the DOSE that the body is permitted to receive over a period of time. A long-term exposure limit (LTEL) is the sum of the concentrations averaged over 8 hours. If, say, for a compound this is 100 ppm, then exposure to a concentration of 200 ppm would only be permitted to 4 hours, which is the same as a TWA of 100 ppm for 8 hours, i.e. time-weighted.

Note that there is often an STEL or short-term exposure limit (concentration averaged over 10 min) for chemical or radiation exposure. In the above example, if the STEL were 600 ppm then exposure to 400 ppm for 4 hours is permissible. If the STEL were 300 ppm it would not be allowed. ($\Rightarrow$ CONTROL LIMITS, OCCUPATIONAL)

**Titanium (Ti).** The ninth commonest element in the earth's crust. Over 90 per cent of titanium produced is used in the form of the oxide, titania, as a white pigment or mineral filler in the paint, paper and plastic industries. Titanium metal and its alloys have a favourable weight-to-strength ratio combined with excellent resistance to corrosion.

**Titration.** A method for the quantitative determination of a substance in solution by the addition in measured amounts of a reagent that reacts with the substance until the reaction is complete. This is indicated by a colour change in the solution, by precipitation, by the colour change of an added indicator or by electrical measurement which indicates the end point of the reaction.

**Tolerable daily intake (TDI).** A threshold value set by the EC Scientific Committee on Food for individual additives or trace chemicals below which no adverse health effects may be expected, e.g. the value for DOA is 30 mg/kg body weight. (⇨ PHTHALATES)

**Ton/tonne.** The British (UK) ton is an old unit of weight equivalent to 20 hundredweights or 2240 pounds, avoirdupois measure. In the USA it is called the long ton or gross ton; the American equivalent is the short ton or net ton, 0.893 × UK ton. The metric equivalent, the tonne (tonneau) or metric ton, is equal to 1000 kilograms.

**Total dissolved solids (TDS).** The solids residue after evaporating a sample of water or effluent expressed in mg/litre.

**Total energy.** The integrated use of all or most of the heat involved in the combustion or fossil or nuclear fuels. Thus, instead of just generating electricity, a total energy scheme would generate electricity and sell heat for both DISTRICT HEATING and factory processes. (⇨ COMBINED HEAT AND POWER)

**Toxic action of pollutants.** There are three main mechanisms by which the human organism is affected by toxic pollutants.

1. They influence enzymatic action by, for example, combining with the enzyme so that it cannot function (⇨ ENZYMES)
2. They can combine chemically with the constituents of cells, as, for example, carbon monoxide combining with blood haemoglobin so that oxygen transport to the brain is affected.
3. Secondary action because of their presence. Hay fever is brought about by pollen and the system reacts to produce histamine.

The factors of importance are the concentration of the pollutant, the length of exposure, the age, the activity – whether slight or heavy exertions – and the health of exposed person/population.

**Toxic wastes.** ⇨ WASTES, HAZARDOUS (i); WASTES, OPTIONS (iv); DOSE.

**Toxicity characteristics leaching procedure.** US test used as an indication of the leaching elements in incinerator ash. The acidic solutions used exaggerate the leachability of the ash. (US ash practice is to use monofills where the ash cannot come in contact with acids in co-disposal sites.) Hence, there is minimal LEACHATE production from such sites and great confidence can be placed in ash monofills.

The alternative of LANDFILL of crude MSW produces highly polluting leachates and LANDFILL GAS, both of which can require decades of after-treatment following deposition.

**Trace elements.** Elements which occur in minute quantities as natural constituents of living organisms and tissues. They are necessary for the maintenance of growth and development; the shortage of any one may result in reduced growth, physiological troubles and eventually death. However, in large quantities they are generally harmful.

Trace elements include copper, silicon, cobalt, iron, zinc, iodine and manganese.

**Tracer.** Commonly used to determine the movement of ground water through AQUIFERS in order to determine both the direction and velocity of ground water flow. They are also used to study dispersion in the atmosphere.

Tracers may be classified by method of detection, namely, colorimetry, chemical determination, radioactivity, electrical conductivity, etc. Certain radioisotopic tracers, e.g. tritium, can be used as field tracers without danger of contamination, but others must be carefully controlled because of dangerous radiation levels.

**Trade effluent.** Any waste waters from an industrial process. For direct discharge into a water course, strict consent conditions have to be met and expensive treatment plant may have to be installed. Under the Water Act 1989 the NATIONAL RIVERS AUTHORITY generally determines consent conditions and maintains a register. It is often cheaper to discharge to a sewer and have the waste water treated at the local sewage works. This can pose a considerable treatment problem and the object is to ensure that the trader pays a charge for the services rendered. The control of trade effluents to sewers is controlled by the Control of Pollution Act (1974).

**Tramp elements.** ⇨ METALS.

**Transfer station.** A depot from where local municipal waste collection vehicles are unloaded for transfer/shipment into large road vehicles, rail wagons or barges to a treatment or disposal site.

**Transferable drug resistance.** ⇨ ANTIBIOTIC.

**Transmissivity.** The transmissivity of an aquifer is the product of the HYDRAULIC CONDUCTIVITY (permeability) and the aquifer thickness. It is an essential measure of the movement of ground water through an aquifer, particularly in the vicinity of production wells. Measured in gallons per day per foot or square metres per day.

**Transpiration.** Water is transferred from the soil to the leaves of plants by capillary action and osmosis. At the leaf surface, the water transpires or evaporates and the vapour diffuses to the atmosphere.

As evaporation also takes place from the surfaces of lakes and rivers, it is common to use the term evapo-transpiration to account for the land-based water vapour component of the HYDROLOGICAL CYCLE.

**Tributyltin (TBT).** A marine anti-fouling agent which is on the RED LIST. It is extremely toxic, concentrations as low as two parts per trillion can affect marine life. Banned in UK and France for use on boats greater than 25 m in length. EC ban awaited

**Trichlorofluoromethane (CFCl$_3$) (Freon 11).** A member of the group known as halogenated fluorocarbons, used as an AEROSOL PROPELLANT. Roughly 2 million tonnes (1989) are liberated per year. It breaks down chemically, releasing free chlorine which combines with ozone to deplete the earth's OZONE SHIELD. However, due to TIME LAG, the effects are only now becoming apparent. If all CFCl$_3$ production were stopped immediately, the decomposition will continue for decades. The EC has now set mid 1997 as the date for a total ban on all CFCs. ($\Rightarrow$ CHLOROFLUOROCARBONS)

*Trichoderma viride.* A fungus which is capable of breaking down crystalline cellulose by the production of an ENZYME of the cellulase group which can accomplish the decomposition in a few days. Mutant strains of *Trichoderma viride* have been developed which accelerate the decomposition. *Trichoderma viride* is the basis of the Natick process for the production of glucose from cellulose. ($\Rightarrow$ ENZYME TECHNOLOGY)

**Trippage.** The number of trips that a returnable bottle makes in its lifetime. In the UK, average trippage is 25 times for a milk bottle. In the USA, 10 times for a soft drink bottle. ($\Rightarrow$ RECYCLING)

**Tritium (H$_3$).** An ISOTOPE of hydrogen, atomic mass number 3. It is used as a fuel in fusion power research.

**Trophic levels.** $\Rightarrow$ FOOD CHAIN.

**Tropopause.** The upper limit of the TROPOSPHERE.

**Troposphere.** The 'lower' portion of the atmosphere about 8

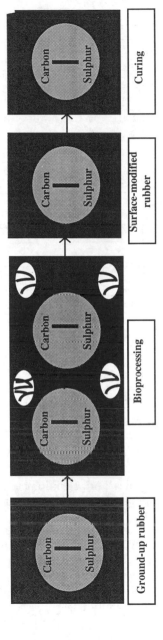

**Figure 149** Bioprocessing of waste tyre rubber.
*Source*: Bioprocessing helps make waste tyres recyclable, *World Wastes*, March 1995.

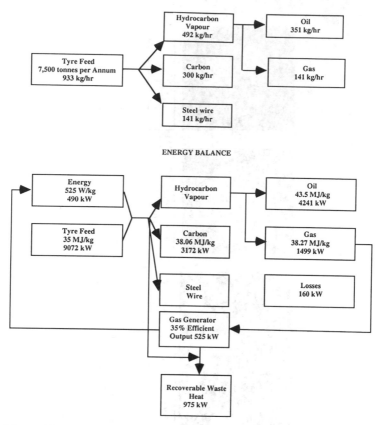

**Figure 150** Tyre pyrolysis mass and energy balances (courtesy of BRC Environmental Services Limited, February 1994).

kilometres high at the poles and 16 kilometres high at the equator. In the troposphere the earth's temperature usually decreases with height. (⇨ STRATOSPHERE)

**Turbidity.** Reduced transparency of the atmosphere, caused by absorption and scattering or radiation by solid or liquid particles, other than clouds, held there in suspension. In a liquid, scattering of light due to suspended particulate matter.

**Turbulence.** Movements which are uncoordinated and in a state of continuous change in liquids and gases.

**Turndown ratio.** The ratio of maximum output to minimum output of a plant or boiler based on continuous operation.

**TWA.** ⇨ TIME-WEIGHTED AVERAGE.

**Typhoid.** Water-borne disease caused by drinking sewage-contaminated water. (⇨ PUBLIC HEALTH)

**Tyres.** There are approximately 0.4 million tonnes per year (tpy) of scrap tyres produced in the UK. These have a gross energy value of 0.5 tpy of coal equivalent and burning them for energy recovery is one way of both effecting their disposal and recovering something of value. More commonly, they are tipped in holes in the ground, often illegally, or on old, industrial sites. Uncontrolled combustion can and does take place causing severe levels of water and air pollution.

Other methods of reuse include rubber crumb manufacture for shock-resistant surfaces in playgrounds or as an asphalt additive, but both of these uses are costly.

One prototype US approach is bioprocessing to remove the sulphur in the tyres by grinding the rubber to MICRON size at a cost of $1.4/kg which compares with virgin rubber compounds at $2.2/kg.

The Pacific Northwest Laboratory has successfully demonstrated the technology as shown schematically in Figure 149, using micro-organisms which selectively attach to the carbon sulphur cross-links.

Another approach is to use PYROLYSIS to recover the ENERGY content, plus the steel reinforcement in the tyre casings. The mass and energy balance for one such scheme is given in Figure 150.

# *U*

**Ultimate analysis.** The proportion by weight of the chemical and inert constituents of a fuel. Usually done as an elemental analysis by percentage of carbon, oxygen, hydrogen, nitrogen, sulphur, water and inerts (ash).

**Ultrafiltration.** A membrane separation process which uses larger pore size membranes than those of REVERSE OSMOSIS. Used in the tertiary treatment of biologically treated sewage effluent.

**Ultraviolet disinfection.** In sewage effluents, even after secondary treatment there are large populations of micro-organisms present of enteric origin. These comprise bacteria and viruses, many capable of causing disease. Macrobial cells contain nucleic acids which store genetic data and to survive and reproduce a cell must be able to replicate the biochemical data in these nucleic acids. As these acids can absorb light of different wavelengths, it has been shown that ultraviolet irradiation can be absorbed by them, thereby damaging or rearranging their genetic information. That damage renders a cell unable to replicate and results in the death of that cell. It is therefore possible to arrange for secondary sewage effluent to be irradiated by a suitable array of ultraviolet lamps in order to render the final effluent disinfected. The table on page 550 shows the results before and after UV treatment of the Bellozanne disinfection plant operated by the States of Jersey.

**Ultraviolet radiation (UV).** Radiation which falls between visible light waves and X-rays. The longest UV waves have wavelengths slightly less than those of violet light (the limits of the human eye) UV radiation stimulates the body to produce vitamin D.

**Unburnt hydrocarbons.** Airborne particles of hydrocarbon fuels not consumed in combustion and which are emitted in exhaust or flue gases particularly from an INTERNAL COMBUSTION ENGINE. Main components are CARBON particles, PARAFFINS, olefins, and aromatic compounds. (⇨ AUTOMOBILE EMISSIONS; PHOTOCHEMICAL SMOG)

Example of results from small-scale trial

| Intensity (mW/cm²) | T% | S.S. (mg/l) | Total coliforms/100 ml | | | Faecal coliforms/100 ml | | |
|---|---|---|---|---|---|---|---|---|
| | | | Before | After | Reduction | Before | After | Reduction |
| 4.8 | 56 | — | 56000 | 30 | 99.95 | 5000 | <10 | 99.80 |
| 1.8 | 10 | 71 | 14000000 | 290000 | 97.93 | 2200000 | 85000 | 94.14 |
| | | | 10000000 | 290000 | 97.10 | 2000000 | 69000 | 96.5 |
| 4.8 | 50 | 7 | 90000 | 50 | 99.94 | 4000 | <10 | 99.75 |
| | | | 79000 | 40 | 99.95 | 2000 | <10 | 99.50 |

SS = Suspended Solids

T = Transmissivity

*Sources:* D. O. Lloyd and J. Clark, 'Waste Management in Jersey', Paper 5, States of Jersey Seminar, March 1993.

G. W. Shepherd, Paper 20, States of Jersey Seminar, March 1993.

**Unit cost.** The cost per unit of production or service, e.g. the cost per tonne of waste incinerated.

**Unsaturated zone.** That part of an aquifer where water only partially fills the interstices, i.e. the unsaturated zone exists everywhere above the WATER TABLE.

**Uranium (U).** The principal radioactive element used in the production of nuclear (or radioactive) energy. Uranium has several isotopes but only the natural isotope known as $^{235}U$ is 'radioactive' or unstable. Uranium-235 comprises only about 0.7 per cent of all natural uranium, the rest being largely $^{238}U$ which can only be used as a nuclear fuel after treatment.

Uranium-235 breaks down into lighter elements, neutrons and heat during nuclear fission. The 238 form can be converted into the active element plutonium-239 by the process known as breeding and the form of the element thorium, known as thorium-232, can be converted into fissionable $^{233}U$ in a breeder reactor.

1 kilogram of uranium produces the thermal energy equivalent to about 11 800 barrels of crude oil. Conversion efficiencies are less than fossil-fuelled power stations as the maximum steam temperature of nuclear reactors is lower. ($\Rightarrow$ NUCLEAR POWER; NUCLEAR REACTOR DESIGNS)

**Urea.** A nitrogenous component for FERTILIZER manufacture. Also used as an animal feed additive and in plastics manufacture (urea formaldehyde). Manufactured from CARBON DIOXIDE and AMMONIA reacted at high temperature and pressure.

Formula $H_2N \cdot Co \cdot NH_2$

# V

**Vacuum filter.** Used in SLUDGE dewatering. Usually consists of a cloth covered drum or belt which is partially submerged in the sludge. The vacuum applied to the other side of the cloth effects dewatering.

*Valdez.* Name of a supertanker owned by US Exxon group which ran aground on a reef in Prince William Sound, Alaska (March 1989) spilling 11 million gallons of crude oil and causing one of the largest man-made ecological disasters. The oil slick covered over 3000 square miles.

**Valorization.** The recovery of value from waste. This can (and does) embrace recycling, composting and energy recovery. It embodies a pragmatism which, in the case of PACKAGING wastes after RECYCLING, has been carried out (or judged inappropriate in the case of contaminated plastics); INCINERATION with energy recovery can be practised. The name of the game is added value and not dogmatic adherence to one solution only.

**Valpak.** The body set up by the UK packaging industry to recover real value from PACKAGING waste.

Its functions are as follows:

- Valpak will agree with material-specific organizations (MOs) acting on behalf of businesses in that material sector what funds will be required to execute the plan. Valpak will also agree with waste management companies and local authorities their funding needs.
- Valpak will register appropriate companies (including importers of packaged goods) and inform them of the formula to be applied in calculating the packaging levy that they should pay to Valpak. The Producer Responsibility Group (PRG) together with the Departments of the Environment and Trade and Industry has agreed to appoint an independent expert to recommend the point(s) within the packaging chain at which the levy should be raised. Companies

who ship packaged goods directly to industrial or commercial organizations will be treated like packers.

- Valpak will manage the collection and disbursement of funds.
- Valpak will co-ordinate the setting up of additional domestic collection and sorting schemes to meet re-processing and end-use market demand.
- Valpak will monitor value recovery achievements and regularly report on progress.

The Valpak value recovery targets are given below:

|  | Quantities by year ($\times$ 1000 tonnes) | |
|---|---|---|
|  | 1993 | 2000 |
| Total packaging waste | 7292 | 8051 |
| Domestic | 3600 | 3757 |
| Commercial and industrial | 3692 | 4294 |
| Total quantity recycled | 2199 | 4003 |
| From domestic sources | 513 | 1306 |
| From commercial and industrial sources | 1686 | 2697 |
| Total recovered by waste to energy | 150 | 650 |
| Total diverted from landfill | 2349 | 4653 |
| Value recovery (%) | 32 | 58 |
| Residual packaging waste to landfill | 4943 | 3398 |

The UK packaging body of Producer Responsibility Group (PRG) has commented:

> Statistics on the size of the packaging market and on waste handling are notoriously imprecise. This poses significant problems even in determining current value recovery rates. Projecting to 2000 is difficult.

PRG are also concerned that a value recovery rate, as requested by the Minister and used as a basis in the proposed PACKAGING WASTE DIRECTIVE, fails to provide any impetus for source reduction or re-use.

PRG therefore considers that the best single measure of the success of the plan will be the reduction in the tonnage of packaging waste going to landfill. This is expected to reduce from 4.9 million tonnes today (out of over 100 million tonnes of waste landfilled and 7.3 million tonnes packaging on the market) to about 3.4 million tonnes in 2000 (relative to a projected packaging market of 8.1 million tonnes).

Landfill diversion and value recovery are greater for commercial (and industrial) packaging waste than for domestic waste. This makes environmental and economic sense because it is easier to collect and re-process commercial material. (⇨ VALORIZATION)

*Source: Report on Public Consultation*, Producer Responsibility Group, Committee Response to the PRG 'Real Value from Packaging Waste', June 1994.

**Vanadium (V).** Hard, white metal, used as a steel alloying element and as a chemical industry catalyst. The THRESHOLD LIMITING VALUE is 500 micrograms per cubic metre for dusts and 50 micrograms per cubic metre for fumes of vanadium pentoxide ($V_2O_5$).

Vanadium affects most metabolic processes in the human organism. The lethal dose is between 60 and 120 milligrams. Chronic exposure to environmental air concentrations of vanadium can lead to bronchitis.

**Vapour pressure.** The pressure exerted by a vapour. Usually saturated vapour pressure, i.e. the pressure of the vapour in contact with the parent liquid, e.g. water vapour pressure. Saturated vapour pressure increases with temperature hence steam can be generated if the parent liquid (water) is contained at high pressure in the boiler tubes.

**Variance.** A statistical term – the square of the standard deviation.

**VCM (vinyl chloride monomer).** ⇨ POLYVINYL CHLORIDE.

**Vector.** An organism (e.g. animal, fungus) which transmits or acts as a carrier of parasites, e.g. the *Anopheles* mosquito is the vector for the malaria parasite, the *Aedes* mosquito the vector for yellow fever, the rat flea the vector for the plague, and the tsetse fly the vector for sleeping sickness.

The use of insecticides to eliminate such vectors, together with generally improved standards of medical practice, have resulted in significant reductions in death rates of the underdeveloped countries.

**Vehicle emissions.**

*The problem*
*Holiday smog.* Britain suffered its worst winter smog for years over Christmas 1994 but the Government failed to issue a health warning.

High levels of nitrogen dioxide, which brings on respiratory disease, and is one of the main factors in the asthma epidemic, settled over four of the country's biggest cities for many hours on 23 December and Christmas Eve.

The holiday smog, which followed the most polluted summer this decade, reached dangerous levels when nitrogen dioxide emitted by a rush of cars on last-minute Christmas errands failed to disperse because of a temperature 'inversion', when cold air near ground level is trapped by warmer air higher up.

The gas reached 198 ppb (parts per billion) in east Birmingham on 23 December, nearly treble the previous record for the year, and 257 ppb in Walsall, about two and a half times as bad as previously.

High levels were also recorded in Leeds and Manchester. London suffered worst, with two days of pollution twice as bad as anything that had been encountered all year. Over 23 December and Christmas Eve there were 27 hours of serious nitrogen dioxide pollution over much of the centre of the city. The highest level of all, 288 ppb, was recorded near Victoria.

The crisis passed on Christmas Day as the traffic died down and winds dispersed the pollution.

At no time during the crisis did the Government issue health warnings to advise people with respiratory trouble to stay indoors or ask drivers not to take out their cars, even though it had issued them during the polluted summer.

The Department of the Environment said that it had not warned the public because it thought that the weather was going to change earlier than it did.

*The amounts*

The table below gives the percentage of airborne pollutants from road transport, by class of vehicle, as a percentage of total road transport emissions (1990).

|  | Cars | Light goods vehicles | Heavy goods vehicles | Public service vehicles | Motor cycles |
|---|---|---|---|---|---|
| Carbon monoxide | 88 | 7 | 3 | 1 | 1 |
| Nitrogen oxides | 72 | 7 | 19 | 3 | — |
| Volatile organic compounds | 84 | 7 | 6 | 1 | 2 |
| Particulates | 6 | 7 | 77 | 10 | — |
| Sulphur dioxide | 37 | 9 | 47 | 6 | — |

*Source:* Royal Commission on Environmental Pollution, 18th Report, 1994.

*How gaseous emissions are measured:* Gaseous emissions can be measured directly from either undiluted hot exhaust or from cool exhaust after dilution with air.

| | |
|---|---|
| HC | Heated flame ionization detector |
| NO$_x$ | chemiluminescence analyser |
| CO | non-dispersive infra-red |
| Particulates | These are usually measured after diluting the exhaust with air to simulate the condition at the outlet of the exhaust system. A sample is drawn through a filter and weighed. Particulates are sampled from a dilution tunnel, either a full flow constant volume sampling system, or a mini-dilution tunnel in which only a sample of the exhaust is diluted. |
| Smoke | Can be measured in the undiluted exhaust either by passing a sample through a white filter and measuring the darkening effect, or by passing a beam of light through the exhaust and measuring the opacity. |

*Source: Perkins Power News*, Vol. 5, No. 2, June 1995.

*Reduction of vehicle emissions*
Volvo Trucks has cut emissions from its transport operations in and around its manufacturing bases in Gothenburg, Sweden. Traffic-generated pollution has been cut by 50 per cent in just 5 years, by using cleaner fuels, improving vehicle scheduling and productivity and from developments in truck design.

Particulate emissions have been reduced by 25 per cent and the level of hydrocarbons and nitrogen oxide has decreased by approximately 10 per cent each by using diesel fuel with lower levels of sulphur and aromatics both in suppliers' vehicles and in Volvo's own trucks.

Relocating loading and unloading terminals for goods and finished products, improving road signs, driver information points and equipment, has produced reductions in regional transport of around 2800 km/day. And by co-ordinating incoming deliveries to Volvo and making more use of information technology, long haul transport volume has decreased by around 1600 km/day. Further reductions of up to 300 km/day are envisaged if plans for combined road/rail transport are implemented.

The advanced technology in Volvo's new 12-litre engine

Passenger car exhaust emissions legislation (EEC Directives 91/441/EEC, 94/12/EEC etc.)

| | Effective date | | | | | | | | | | | |
|---|---|---|---|---|---|---|---|---|---|---|---|---|
| | 1992 Stage I | | | | 1994 | | 1996 Stage II* | | | | 1999 Stage III*** | |
| | IDI | | DI | | | DI | IDI | | DI | | IDI & DI | |
| Applicability | TA | CoP | TA | CoP | TA | CoP | TA | CoP | TA | CoP | TA | CoP |
| HC & NOX (g/km) | 0.97 | 1.13 | 1.36 | 1.58 | 0.97 | 1.13 | | 0.7 | | 0.9 | | 0.5 |
| CO (g/km) | 2.72 | 3.16 | 2.72 | 4.42 | 2.72 | 3.16 | | 1 | | 1 | | 0.5 |
| Pm (g/km) | 0.14 | 0.18 | 0.2 | 0.25 | 1.14 | 0.18 | | 0.08 | | 0.1 | | 0.04 |

*Notes:* *, Proposal; ***, Stage III at discussion stage; TA, type approval; CoP, conformity of production; DI, direct injection; IDI, indirect injection.

*Source: Automotive Diesel Engineers and the Future*, Ricardo Consulting Engineers Ltd, Shoreham-by-Sea, West Sussex, 1994.

Heavy duty engine exhaust emissions legislation (EEC Directive 91/542/EEC)

| Applicability | Until June 1992 | | July 1992 Stage 1 | | October 1995 Stage II | | 1999? Stage III (Discussion) | |
|---|---|---|---|---|---|---|---|---|
| | TA | CoP | TA | CoP | TA | CoP | TA | CoP |
| HC | 2.4 | 2.6 | 1.1 | 1.23 | 1.1 | | 0.7 | |
| CO | 11.2 | 12.3 | 4.5 | 4.9 | 4.0 | | 2.5 | |
| NO$_x$ (g/kWh) | 14.4 | 15.8 | 8.0 | 9.0 | 7.0 | | <5.0 | |
| PT | | | | | | | | |
| <85 kW | | | 0.61 | 0.68 | 0.15 | | <0.12 | |
| <85 kW | | | 0.36 | 0.4 | | | | |

*Source: Automotive Diesel Engines and the Future*, Ricardo Consulting Engineers Ltd, Shoreham-by-Sea, West Sussex, 1994.

Air quality standards: WHO guidelines, EC standards (bold type) and EC guide values (italics)

| | Less than 1 hour | 1 hour | 8 hours | 24 hours | Units |
|---|---|---|---|---|---|
| Carbon monoxide | 100 (15 mins) 60 (30 mins) | 30 | 10 | 1 | mg/m$^3$ |
| Nitrogen dioxide | | 400 **200** | | 150 | µg/m$^3$ |
| Sulphur dioxide | 500 (10 mins) | 350 | | *100–150* *40–60* | µg/m$^3$ |
| Combined exposure to sulphur dioxide and suspended particulates (black smoke) | | | | 125 SO$_2$ plus one of the following: 125 black smoke 120 total suspended particulates 70 thoracic particles | 50 SO$_2$ and 50 black smoke |
| Suspended particulates (black smoke) | | | | **80** **130** **250** *100–150* *40–60* | |
| Ozone (health) | | 150–200 | 100–120 **110** | | µg/m$^3$ |

| | | |
|---|---|---|
| Ozone (vegetation) | 200 | 65 | **60** (April to September, over the growing season) |
| Lead | **200** | **65** | 0.5–1.0 **2.0** |
| Formaldehyde | 10 (30 mins) | | |

*Source:* Royal Commission on Environmental Pollution, 18th Report, 1994.

Freight transport modes: energy use and emissions

| | Rail | Water transport | Road | Pipeline | Air |
|---|---|---|---|---|---|
| Specific primary energy consumption (kJ/tonne-km) | 677 | 423 | 2890 | 168 | 15839 |
| Specific total emissions (g/tonne-km) | | | | | |
| Carbon dioxide | 41 | 30 | 207 | 10 | 1206 |
| Methane | 0.06 | 0.04 | 0.3 | 0.02 | 2.0 |
| Volatile organic compounds | 0.08 | 0.1 | 1.1 | 0.02 | 3.0 |
| Nitrogen oxides | 0.2 | 0.4 | 3.6 | 0.02 | 5.5 |
| Carbon monoxide | 0.05 | 0.12 | 2.4 | 0.0 | 1.4 |

*Source:* J. Whitelegg, *Traffic Congestion – Is there a way out?*, Leading Edge Press, 1992.

introduced in its FH range in the autumn of 1993 brought further benefits, albeit at a late stage in the programme. As Volvo's older vehicles are replaced, the D12A engine (already able to meet Euro II emission regulations due to come into force in 1996) in the FH series will make a significant contribution towards reducing emission levels still further.

*Source:* Issued by Volvo Truck and Bus Ltd, December 1994.

*Legislation.* The tables above set out the emission legislation in the European Community.

Another option is to switch to rail or water transport.

*Electric cars.* One should be aware that for electric vehicles, generating electricity for recharging batteries can cause considerable environmental harm.

Analyses have been done on the environmental effects of gasoline as compared with those of electricity generation. In response to the electric vehicle mandate, automakers have proposed ultra-low emission vehicles.

For vehicles that are to be mass-produced in late 1997, lead-acid batteries are likely to be the only practical technology. Smelting and recycling the lead for these batteries will result in substantial releases of lead to the environment. Lead is a neurotoxin, causing reduced cognitive function and behavioural problems, even at low levels in the blood.

Using 4 per cent losses from virgin production, 2 per cent losses from recycling and reprocessing, and 1 per cent losses from battery manufacturing, we calculated the amount of lead discharges to be from 1340 mg of lead per kilometre (for the existing technology battery that has the lowest energy density and shortest lifetime distance and uses virgin lead) to about 117 mg of lead per kilometre (for a goal technology battery that has high energy density and long lifetime driving distance and uses scrap lead). If a large number of electric cars are produced, the demand for lead for batteries will surge, requiring more lead to be mined.

A 1998 model electric car is estimated to release 60 times more lead per kilometre of use relative to a comparable car burning leaded gasoline. The United States banned TEL in large part for health reasons. Electric vehicles would introduce lead releases to reduce urban ozone, a lesser problem. This argues for the urgent commercialization of alternatives to lead-acid batteries.

*Source:* B. L. Lave, C. T. Hendrickson and F. C. McMichael, *Science*, Vol. 268, 19 May 1995.

*Reduction of noise nuisance from transport*
The Royal Commission on Environmental Pollution 18th Report proposed the following:

To reduce daytime exposure to road and rail noise to not more than 65 dB $L_{Aeq. 16h}$ at the external walls of housing;
To reduce night-time exposure to road and rail noise to not more than 59 dB $L_{Aeq.8h}$ at the external walls of housing.

*Noise measurements relating to transport.* In assessing human exposure to noise we also need to take account of variation in noise levels over time. This can be expressed in a number of different ways, for example:

| dB $L_{Aeq}$ | the mean level of the sound |
| dB $L_{A90}$ | the level of sound exceeded for 90 per cent of the time (background noise) |
| dB $L_{A10}$ | the level of sound exceeded for 10 per cent of the time. |

Other suffixes are used to represent the level of exposure over all or part of a 24-hour period, for example:

dB $L_{Aeq.8h}$
dB $L_{Aeq.16h}$
dB $L_{Aeq.18h}$

**Venturi effect.** A local decrease of pressure, e.g. caused when the wind blows through a narrow mountain pass or between buildings. The effect can be put to use in pipes by inserting a contracting length, a throat and then an expanding length and measuring the pressure drop between the throat and upstream. This can then be correlated with the flow rate of the fluid in the pipe. Also used in a high efficiency SCRUBBER for air pollution control. (⇨ VENTURI TUBE)

**Venturi tube.** A tube whose internal diameter gradually decreases to a throat and then gradually increases again to its original value. Such tubes are used in flowmeters and also in venturi scrubbers.

**Vinyl chloride (CH₂CHCl).** A colourless gas used in the manufacture of POLYVINYL CHLORIDE (PVC), a PLASTIC. The vapour has been found to be highly carcinogenic and has been linked with the deaths of 20 workers globally from angioma, a rare form of liver cancer. In the UK there is evidence that the vapour has caused impotence, stiffening of the joints, bad circulation and shortness of breath. In the USA very strong standards set exposures at a maximum of 5 parts per million and TIME-WEIGHTED AVERAGED exposures as low as 1 part per million. This has also been adopted by Sweden.

There is a great concern at the potential health effects of this substance and no threshold effect is postulated for it, as for other pollutants, i.e. exposure to any concentration may cause damage. (⇨ CARCINOGEN; MUTAGEN; THRESHOLD LIMITING VALUE)

**Viscosity.** A measure of the internal flow resistance of a liquid or gas. Viscosity decreases as temperature rises for liquids.

**Visual impact appraisal.** ZVIs, or zones of visual intervisibility (also intrusion or influence), are maps which display the areas from which an object or a number of objects may be expected to be seen. (An area under study is usually modelled by the computer).

Other techniques used include photomontage, created by either photographing a perspective which has been very accurately modelled into a photographic background, or by the direct manipulation of the photo image in the computer. The technique is extremely useful for wind turbine locations in the areas of scientific interest and tall chimney stack siting.

**Vitrification.** The process of making a glassy non-crystalline solid from materials capable of vitrifying when heated into their vitrification range, e.g. clays vitrify when heated to high temperatures, hence ceramics and brickmaking. Vitrification is claimed to be one of the 'ultimate solutions' for the safe containment of NUCLEAR REACTOR WASTES but has not been put to the test for the hundreds (or thousands) of years needed for it to be shown as being totally reliable.

**VOC.** ⇨ VOLATILE ORGANIC COMPOUNDS.

**Voest-Alpine process.** An Austrian high-temperature gasification process which can use contaminated scrap plastic (e.g. car scrap) as a feedstock. It is particularly useful for thermosetting plastics which cannot be recycled like a THERMOPLASTIC. The process comprises a reactor supplied with preheated air and a primary fuel to attain a temperature of 1600°C which decomposes the plastics feedstock. The resulting combustible gas is purified and used as an industrial fuel. The residues consist of glassy granules

VOC emissions

| Chemical | Emission (Tonnes per year) | | Maximum annual average (site)[a] | 2-Year network average (ppbv)[b] | Annual ESL (ppbv) | ACGIH TLV-TWA 420 (ppbv) |
|---|---|---|---|---|---|---|
| | 1987 | 1988 | | | | |
| Ethylene | 2146 | 2715 | 20.9 (7) | 13.2 | _[c] | _[d] |
| Propylene | 2322 | 1379 | 19.2 (8) | 11.2 | _[c] | _[d] |
| Toluene | 1350 | 992 | 7.7 (15) | 5.7 | 98 | 238 |
| Benzene | 489 | 804 | 5.1 (8) | 3.4 | 1 | 24 |
| Xylene (mixed isomers) | 551 | 622 | 5.3 (15) | 2.5 | 100 | 238 |
| Styrene | 710 | 600 | 1.8 (3) | 0.3 | 50 | 119 |
| Acetone | 450 | 497 | 8.9 (8) | 6.2 | 245 | 1786 |
| Chlorobenzene | 379 | 477 | 0.4 (11) | <0.2 | 75 | 179 (24)[e] |
| Dichloromethane | 465 | 371 | 2.9 (11 & 13) | 3.0 | 7 | 119 |
| Methyl tert-butyl ether | 285 | 312 | 1.7 (15) | 0.4 | _[f] | _[g] |

[a] The maximum annual (average) concentration measured during the two-year period at the site is indicated in parentheses.
[b] Average concentration for all sites during the two-year monitoring period (parts per billion volume).
[c] 30-minute average screening level is less than the annual average screening level.
[d] Simple asphyxiant. No TLV-TWA has been established.
[e] On TLV-TWA Notice of Intended Changes list (Threshold Limit Value-Time Weighted Average).
[f] No effects screening level has been established.
[g] No TLV-TWA has been established.
ESL = Health Effect Screening Level.
ACGIH = American Congress of Government and Industrial Hygienists.

which can be landfilled or, it is claimed, used as a construction material.

**Void ratio.** The ratio of VOID volume in a filter medium to the total volume occupied. (⇨ SEWAGE TREATMENT)

**Voids.** The open spaces between solid material in a porous medium. Voids, PORES and interstices are terms which are closely interrelated and in part synonymous.

**Volatile acids.** Mainly acetic, propionic and butyric acids which are produced during anaerobic decomposition of sewage sludge and organic wastes.

**Volatile matter[1].** The ratio of the weight of dry matter lost on heating a sludge sample to 600°C to the initial weight of sludge. This is often used as an approximation of its organic content.

**Volatile matter[2].** In fuel analysis, the loss of weight (corrected for moisture) when a solid fuel is heated to 900°C in the absence of air. The greater the volatile matter, the more secondary air is needed to avoid smoke products during combustion.

**Volatile organic compounds (VOCs).** Organic compounds (e.g. ethylene, propylene, benzene, styrene, acetone) which evaporate readily and contribute to air pollution directly or through chemical or photochemical reactions to produce secondary air pollutants, principally OZONE and PEROXYACETYL NITRATE. VOCs embrace both HYDROCARBONS and compounds of carbon and hydrogen containing other elements such as oxygen, nitrogen or chlorine. The degree to which a VOC contributes to ozone formation depends on:

(a) total mass of the VOC emitted
(b) molecular mass of the VOC
(c) reactivity of the hydrocarbon with OH RADICALS
(d) the VOC chemical structure.

VOCs that are not methane are often designated NMOCs, i.e. non-methane VOCs.

The table on page 565 opposite gives a comparison of 1987 and 1988 VOC emissions to measured concentrations in Harris and Chambers Counties, Texas (courtesy of UNEP Industry and Environment, March 1991).

Substantial reductions in VOC emissions from the Western industrialized nations will shortly be required. A target reduction of 30 per cent had been mooted from early 1990 levels.

**Volatilization.** The driving off of VOLATILE MATTER as a vapour solid or liquid by heating.

**Volumetric measurements.** Methods of measuring gaseous pollutants (usually in air) that make use of accurately measured volumes of liquid reagents and volume(s) of air reacted with them.

**Voluntary separation.** The separation from domestic refuse of glass bottles, food and beverage cans or newspaper by individuals or groups, at home or in local collection centres.

**Vortex separator.** A vertical cylindrical tank into which air- or liquid-borne particulate matter is introduced tangentially. Separation is effected by centrifugal force with the particulate matter falling downwards for removal from the conical base.

**Waste Collection Authority (WCA).** Outside Greater London, the council of a district charged with the responsibility for the collection of household waste.

The Environmental Protection Act 1990 confers the following duties and powers on the District Councils as Waste Collection Authorities:

- A duty to arrange for the collection of household waste and, if requested, from commercial and industrial premises.
- A duty to deliver for disposal all waste collected by the Authority, other than waste for which arrangements for recycling have been made, to places directed by the Waste Disposal Authority.
- A duty to inform the Waste Disposal Authority of any new arrangements it proposes to make for recycling.
- Powers to provide plant and equipment for the sorting and baling of waste retained by the Authority for recycling.
- Powers to require household waste to be placed in receptacles of a specified type and number, including for the separation of waste which is to be recycled and that which is not.
- A duty to carry out investigations and prepare a recycling plan in respect of household and commercial waste arising in its area.
- Powers to make payments based on net savings in collection costs (collection credits) to third parties who collect waste, which would not otherwise be collected by the Authority, for recycling.
- Powers to buy or otherwise acquire waste with a view to recycling it and use, sell or otherwise dispose of waste (or anything produced from it) belonging to the Authority.

The above statement still holds true in 1995. There is, unfortunately, too much hype and hysteria over wastes

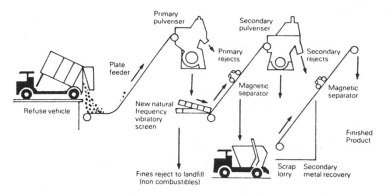

**Figure 151**    Blue Circle refuse processing flow chart.

management. Only when this dies down can national debate be fostered.

**Waste derived fuel.** Fuel made from wastes which can be loose, e.g. shredded paper and plastics from MUNICIPAL SOLID WASTE (MSW) or compressed ('densified') into pellets. Fuel from MSW can have up to half the energy value of UK industrial coal. It is typically used as a solid fuel supplement in ratios of up to 50:50 coal and waste derived fuel on a heat input basis, or it can be blown in the loose state into specially adapted boilers or cement kilns. The Blue Circle process is shown in Figure 151.

Other waste derived fuels can based on agricultural residues such as rice hulls, sawdust, logging residues and STRAW. Most are characterized by energy values roughly 30 per cent that of an industrial coal. They can also pose handling and storage problems.

Waste oil and spent SOLVENTS can also be used as fuels, preferably in specially adapted boilers. However, contamination with PCBs can lead to DIOXIN emissions. (⇨ INCINERATION; PELLETED FUEL)

**Waste Disposal Authority (WDA).** Authorities established by the Local Government Act (England & Wales) 1972 and in 1973 for Scotland.

In England, they are basically the County Councils and statutory authorities such as the London Waste Regulatory Authority. In Wales and Scotland, they are the District Councils. Their functions are to be changed under the Environmental Protection Act, which specifies Waste Regulation Authorities in their place.

**Waste factor.** A term used in ENERGY ANALYSIS to measure the departure from ideal energy needed to effect a transformation. It is defined as:

$$\frac{\text{(actual energy required to effect transformation)} - \text{(ideal energy)}}{\text{(actual energy)}}$$

Ideal energy is obtained from physical or chemical calculations which show the theoretical energy consumption required to manufacture a given component or product.

A waste factor close to 1 represents a very wasteful and inefficient process. Thus, it can be used to show scope for improvement in energy utilization.

| Product | Actual energy (1968) (MJ/kg) | Ideal energy (MJ/kg) | Waste factor |
|---------|------------------------------|----------------------|--------------|
| Iron | 25 | 6 | 0.76 |
| Petrol | 4.2 | 0.4 | 0.90 |
| Paper | 38 | 0.2 | 0.99 |
| Aluminium | 190 | 25 | 0.87 |
| Cement | 7.8 | 0.8 | 0.90 |

From E. Gyftopoulos et al., Free Energy Use, Actual and Ideal, Thermo-Electron Corp., Waltham, Mass., 1974.

**Waste heat.** Term used to denote heat that is normally rejected to the environment, e.g. the hot water discharge from power-station cooling circuits which may be 10–15°C above ambient temperature. Other examples include the jacket cooling water from large industrial residual oil fuelled engines which may be at a temperature of ca. 90–100°C. The higher the temperature of the waste heat source, the greater the prospect of using it for an additional source of energy for heating purposes such as DISTRICT HEATING where a modified steam power station cooling circuit can provide both power and high-temperature hot water.

**Waste management.** The management of wastes at all stages from production, handling, storage, transport, processing and ultimate disposal which includes the Duty of Care. (⇨ WASTES – DUTY OF CARE; WASTES, (vii) WASTE MINIMIZATION

**Wastes – Duty of Care.** The Royal Commission on Environmental Pollution 11th Report, 1985, recommended in para 13.74:

(a) that producers of controlled wastes have a duty of care to take all reasonable steps, having regard to the hazards presented by their wastes, to ensure that their wastes are subsequently managed and disposed of without harm to the environment;

(b) that the steps that it is reasonable for the waste producer to take in different circumstances to discharge his duty of care shall be contained in a code of practice issued by the Secretary of State; and

(c) that producers of commercial, industrial and certain domestic wastes who engage contractors to transport their wastes remain liable for the proper disposal of those wastes unless they use a contractor registered as a waste transporter in accordance with our later recommendation and provide him in writing with an unambiguous indication of the nature of the wastes and clear instructions for their disposal.

*EC Framework Directive on Waste*
This has four main mandatory elements:

(a) each member state must have a competent authority or authorities with responsibility for waste;

(b) the competent authorities must prepare waste management plans;

(c) undertakings handling waste must have a permit from the competent authority;

(d) the polluter-pays principle must be applied.

Waste is defined as any substance or object which the holder disposes of or is required to dispose of pursuant to the provisions of national law in force. (⇨ DUTY OF CARE; ENVIRONMENTAL AUDIT)

**Wastes (solid and hazardous).** This extended entry is provided to outline most matters relevant to both public concern and the environment resulting from solid and hazardous waste disposal practices:

**(i) Hazardous wastes, causes for concern.** Hazardous wastes disposal in the UK has been a long running cause for concern. Starting with *The Report of the Technical Committee on the Disposal of Solid Toxic Wastes*, March 1970 (the Committee was appointed in July 1964), through The House of Lords Select Committee on Science Technology *Hazardous Wastes Disposal*, July 1981, The Royal Commission on Environmental Pollution (RCEP) *Managing Wastes: The Duty of Care*, December 1985, *Three Hazardous*

*Wastes Inspectorate Reports* (HWI), 1985–1988, The House of Commons Environment Committee *Toxic Waste*, February 1989, finishing with another House of Lords Report, *Hazardous Waste Disposal* April 1989.

The RCEP 11th Report, *Managing Wastes: The Duty of Care* (1985), stated that the main areas of public concern about wastes were:

- Toxic wastes which have mutagenic, teratogenic or carcinogenic effects, or other latent, slow effects on health;
- The tipping of hazardous wastes and the risk of acute poisoning to man and other living organisms;
- The spread of toxic or noxious materials placed on the ground to gardens, homes, play areas, etc;
- Irreversible damage to the environment by irretrievable disposal of persistent substances to land or the marine environment;
- Contamination of ground water;
- The lowering of the amenity value of an area through the presence of wastes – especially litter, but also the noise, smell and other nuisances associated with waste collection and disposal;
- The apparent waste of resources which could be recovered from discarded materials.

The 1989 House of Lords Report said in paragraphs 3 and 7:

3. Eight years after making their first report on Hazardous Waste Disposal, the committee are dismayed at how little has changed. Legislation to improve the control of waste disposal has still not been introduced; the standards of control by local authorities still vary widely; many of those authorities have not drawn up the plans for waste disposal required of them by the Control of Pollution Act 1974; incidents of pollution by waste continue.

7. In 1981 the Committee considered 'the present degree of control exercised by waste disposal authorities to be defective and urgently in need of strengthening . . . It must be made possible for the public to feel confident that real control is taking place and that disposal is geared to the best practicable, not the cheapest tolerable, means.' The Committee still believe this to be the key to acceptable waste disposal.

*Source:* Hazardous Waste Disposal, House of Lords 1st Report, Session 1980–81.

This extract gives a flavour of the concerns felt by the House of Lords. It is fair to say that most major disposal contractors wish to see standards raised but they have to compete with those whose practices leave much to be desired. Also, thanks to a deadline issued by the Department of the Environment most WASTE DISPOSAL PLANS (V) have been filed – 15 years after the passage of the Control of Pollution Act 1974. Another example of what can happen is given below from the second *Hazardous Waste Inspectorate* (HWI) *Report*, appropriately titled *Hazardous Waste Management – Ramshackle and Antediluvian?*, July 1986:

2.11.2  The second potential incident involved tonneage quantities of solid waste materials which react violently with water to produce a poisonous gas which, under the appropriate conditions of humidity, is spontaneously flammable in air. The disposal of these materials had been the subject of an invitation to tender, and one of the proposals, fortunately intercepted by the HWI, had been for a landfill disposal route in sealed drums. Investigations showed that this did not reflect the disposal philosophy of the environmental adviser of the producing company, but that the invitation to tender had originated from a particular manufacturing site, with no reference to HQ. Again, a proposal of potentially disastrous consequences. Again, without intervention by the Inspectorate this landfill disposal could have gone ahead and without, we believe, breaching the landfill site licence.

2.12  This is an example (and there are others) of what can go wrong at co-disposal sites. The HWI commented – 'It is small wonder that the UK's championship of co-disposal landfill is looked at askance by other nations if these examples were to be typical of current practice.'

The HWI 2nd Report gave further examples of malpractice including:

2.16  The HWI is also aware of two cases of the deliberate landfill disposal of low flash-point solvent wastes at sites in two different WDA areas in England. These wastes, which were water-miscible solvents and which arose in bulk, were deliberately diluted with water (and in one case, with leachate from the site at which their disposal was intended) in order to bring the aqueous solvent mixture within the landfill site licence conditions in terms of flash point or fire point. This is entirely contrary to DOE advice which recommends that bulk flammable solvents should be re-covered, used as a fuel or incinerated, and that large volumes of aqueous waste should not, in general, be landfilled.

The HWI stated that the preceding examples were evidence of a continuing trend by waste generators to seek the cheapest rather than the environmentally most suitable disposal option.

The Inspectorate gave further examples in the report. The RCEP 11th Report has commented:

> NIMBY (not in my back yard) cannot be dismissed as irrational: it is the expression of concern about a risk as perceived by the local inhabitants. But the fact there is public concern does not necessarily mean that there is a real environmental hazard. A proper evaluation of the risk requires access to the relevant information and its interpretation, and the public will not be reassured by the interpretations provided by the putative polluter, who has an interest. In the absence of public confidence in the role of bodies that are both authoritative and independent, interpretations by the press or pressure groups are often accepted, even if they go beyond what an informed expert would regard as justified by the evidence.

**(ii) Waste classifications.** The common definition for waste is that for which there is no further use, the implication being that there is nothing that can be done, hence they must be disposed of. However, the UK Environmental Protection Act (EPA) 1990 (Section 75) defines waste as:

(a) any substance which constitutes a scrap material or an effluent or other unwanted surplus substance arising from the application of any process; and

(b) any substance or article which requires to be disposed of as being broken, worn out, contaminated or otherwise spoiled, but does not include a substance which is an explosive within the meaning of the Explosives Act 1875.

There are innumerable classifications for waste. However, they can be grouped by: (i) *origin*, e.g. clinical wastes, household or urban solid wastes, industrial wastes, nuclear wastes, agriculture; (ii) *form*, e.g. liquid, solid, gaseous, slurries, powders; (iii) *properties*, e.g. toxic, reactive, acidic, alkaline, inert, volatile, carcinogenic; (iv) *legal definition*, e.g. special, controlled, household and industrial, where specific definitions or criteria are employed as follows:

*Controlled wastes:* are defined by the UK EPA 1990 Section 75 as household; industrial and commercial waste and any such waste.

*Commercial wastes:* are defined in the EPA 1990 Section 75(7). Waste from premises used wholly or mainly for the purposes of a trade or business or the purposes of sport, recreation or entertainment excluding (i) household and industrial waste; (ii) mining, quarrying and agricultural waste; (iii) waste of any other description prescribed by regulations made by the Secretary of State for the purposes of the above.

*Demolition wastes:* Masonry and rubble wastes arising from the demolition or reconstruction of buildings or other civil engineering structures.

*Difficult wastes:* Another UK term (not legally defined) is that of Difficult Wastes which includes wastes that could be harmful to the environment or whose physical properties present handling problems. A list of potentially 'difficult' wastes is given below. Note these are not necessarily *special wastes*, but rather fall loosely into 'hazardous' wastes definitions or classes used in other countries. 'Hazard' relates to the situation and circumstances as well as the properties of the waste materials. Toxicity is also not conducive to rigorous definition unless target organisms and levels and duration of exposure are stated. Hence special wastes is a sub-set of hazardous wastes that has been more precisely defined leaving authorities free to devise their own classification and criteria for hazardous wastes which has led (in the UK) to variability of standards between authorities.

Classification of Difficult Wastes include:

(a) Inorganic acids
(b) Organic acids and related compounds
(c) Alkalis
(d) Toxic metal compounds
(e) Non-toxic compounds
(f) Metal (elemental)
(g) Metal oxides
(h) Inorganic compounds
(j) Other inorganic materials
(k) Organic compounds
(l) Polymetallic materials and precursors
(m) Fuel, oils and greases
(n) Fine chemicals and biocides
(p) Miscellaneous chemical waste
(q) Filter materials, treatment sludge and contaminated rubbish
(r) Interceptor wastes, tars, paint, dyes and pigments
(s) Miscellaneous wastes
(t) Animal and food wastes

*Hazardous wastes – definition:* There are major problems in finding an acceptable definition of hazardous wastes, because of the following:

(i) The hazards created depend on the waste disposal method used.

(ii) The quantity of toxic material present, e.g. household waste may contain small quantities of toxic compounds and be innocuous. On the other hand, larger concentrations of the same toxic material could pose very serious problems. However, small disposable concentrations of some toxic compounds can contaminate large bodies of water, e.g. phenol. (The taste threshold for phenol in water can be as low as 0.001 ppm.)

(iii) The hazards posed by other industrial materials so that hazardous wastes are not unjustly singled out for opprobrium.

The 1981 House of Lords Report considered these matters and recommended that the WHO definition should be used. Namely:

(a) short-term hazards, such as acute toxicity by ingestion, inhalation or skin absorption, corrosivity or other skin or eye contact hazards or the risk of fire or explosion; or

(b) long-term environmental hazards including chronic toxicity upon repeated exposure, carcinogenicity (which may in some cases result from acute exposure but with a long latent period), resistance to detoxification processes such as biodegradation, the potential to pollute underground or surface waters or aesthetically objectionable properties such as offensive smells.

This recommendation does not have official acceptance in the UK.

*Household wastes:* Defined in the UK EPA 1990 Section 75(5) as waste from:

(a) domestic property, that is to say, a building or self-contained part of a building which is used wholly for the purposes of living accommodation;

(b) a caravan (as defined in section 29(1) of the Caravan Sites and Control of Development Act 1960) which usually and for the time being is situated on a caravan site (within the meaning of that Act);

(c) a residential home;

(d) premises forming part of a university or school or other educational establishment;

(e) premises forming part of a hospital or nursing home.

*Industrial wastes:* 'Industrial waste' is defined in the UK EPA 1990 Section 75(6) as waste from:

(a) any factory (within the meaning of the Factories Act 1961);

(b) any premises used for the purposes of, or in connection with, the provision to the public of transport services by land, water or air;

(c) any premises used for the purposes of, or in connection with, the supply to the public or gas, water or electricity or the provision of sewerage services; or

(d) any premises used for the purposes of, or in connection with, the provision to the public of postal or telecommunications services.

Generally taken to include waste from any industrial undertaking or organization.

*Inert wastes:* Wastes that will not react physically, chemically or biologically and are non-polluting under normal conditions, e.g. demolition waste, although this could contain organic matter and give rise to LEACHATE generation.

*Poisonous wastes (UK):* This was used under the now defunct UK Deposit of Poisonous Waste Act 1972 (passed incidentally by Parliament in 10 days after several well-publicized incidents of drums of industrial cyanide wastes being dumped in children's playgrounds, ditches etc. by unscrupulous transport companies). The scope of this Act was very broad, in effect requiring 'the giving of notices in connection with the removal and deposit of waste and for connected purposes'.

The definition of poisonous was given as

> to subject persons or animals to material risk of death, injury or impairment of health or as to threaten the pollution or contamination (whether on the surface or underground) of any water supply; and where waste is deposited in containers, this shall not of itself be taken to exclude any risk which might be expected to arise if the waste were not in containers.

This law had the effect of ensuring very strict controls on all manner of poisonous, noxious or polluting waste whose presence on the land is liable to give rise to an environmental hazard. However, the Deposit of Poisonous Waste Act was repealed (being adjudged to be too all-embracing and liable to misinterpretation) and replaced by the Control of Pollution (Special waste) Regulations 1980.

The Deposit of Poisonous Waste Act used an exclusive list approach whereby wastes which presented no significant short-term personal hazards or long-term environmental hazards were excluded from the Act and all others were to be treated as poisonous. This was criticized by some interested parties as being too wide ranging and imprecise. However, the Act did achieve very tight controls on poisonous wastes.

The special waste definition is much narrower using an inclusive list approach, i.e. where certain named substances are to be deemed to be 'special' and subject to the UK Special Wastes Regulations 1980. The effect has been to lose strict control of a large percentage of formerly notifiable wastes under the Deposit of Poisonous Waste Act.

*Special wastes (UK):* are principally defined (under the UK Special Waste Regulations 1980) by the criterion 'is ingestion of up to 5 cubic centimetres of waste likely to cause death or serious tissue damage to a 20 kg child?' This begs many questions and is widely recognized as offering loopholes. Also, environmental effects are specifically excluded from the special wastes category which has led to very narrow categories of wastes in this classification.

The special wastes classification also embraces wastes which contain or are contaminated with materials provided in a schedule to the regulations (e.g. PCBs) or have a FLASH POINT of 21°C or less, or are medicinal products only available on prescription.

These wastes require pre-notification to the relevant authorities of their transport and deposit under the UK Special Wastes Regulations 1980.

The UK Government is reviewing the Special Wastes Regulations 1980 to embrace all controlled (i.e. controlled under the 1974 Control of Pollution Act and subsequently the EPA 1990) waste which is or may be dangerous or difficult to dispose of. The wastes are to be defined by reference to:

(a) a list of wastes which by their nature are considered likely to have characteristic properties which makes them dangerous or difficult to dispose of. These wastes are set out in a list. Some of these wastes are specifically identified (e.g. A090 –

waste oil, or A100 – PCBs) but for the most part they are categorized by the particular source or process from which they arise;

(b) a list of substances which, if they are present in wastes, may cause the waste to have characteristic properties which make it dangerous or difficult to dispose of. These substances are set out in a list, e.g. B502 – barium compounds, B521 – zinc compounds;

(c) a list of the characteristic properties which cause waste to be dangerous or difficult to dispose of. These are given below:

Explosive
Flammable (liquid)
Flammable (solid)
Emission of flammable gas
Oxidizing
Toxic (poisonous harmful)
Infectious
Corrosive/chemical irritant
Reactive
Toxic (delayed)
Ecotoxic

(d) a list of readily identifiable household wastes which are considered likely to have properties which make them dangerous or difficult to dispose of, e.g. solvent-based adhesives in quantities in excess of 250 ml or 250 grams; bleach, creosote, wood preservatives, in excess of 1 litre.

The inclusion of ecotoxicity remedies the important omission in the original 1980 regulations. It is defined as follows:

(a) A substance which may present an immediate, delayed or accumulative risk for one or more sectors of the environment; and

(b) in determining whether a substance presents a risk to the environment particular account shall be taken of:

(i) its effect on animals and other living organisms, including aquatic organisms;

(ii) its effect on plants, water and land; and

(iii) the likelihood of its entering the food chain and its persistence in the food chain.

**(iii) Wastes, clinical.** Clinical waste is defined in regulation 1 (2) of

The Controlled Waste Regulations 1992 (SI 1992/588) as meaning:

(a) any waste which consists wholly of partly of human or animal tissue, blood, other body fluids, excretions, drugs or other pharmaceutical products, swabs, or dressings, or syringes needles or other sharp instruments, being waste which unless rendered safe may prove hazardous to any person coming into contact with it; and

(b) any other waste arisings from medical, nursing, dental, veterinary, pharmaceutical or similar practice, investigation, treatment, care, teaching or research, or the collection of blood for transfusion, being waste which may cause infection to any person coming into contact with it.

The need for very strict control and secure hygienic disposal of clinical wastes cannot be over-emphasized and is discussed in depth in Waste Management Paper No. 25 *Clinical Wastes*. These papers are a series numbering from 1 to 28 on important waste management topics published in the UK by HMSO and regularly updated.)

*Group A Wastes:*
(a) All human tissue, including blood (whether infected or not), animal carcasses and tissue from veterinary centres, hospitals and laboratories, and all related swabs and dressings.
(b) Waste materials where the assessment indicates a risk arising from, for example, infectious disease cases.
(c) Soiled surgical dressings, swabs and other soiled waste from treatment areas.

*Group B Wastes:* Discarded syringe needles, cartridges, broken glass and any other contaminated disposable sharp instruments or items

*Group C Wastes:* Microbiological cultures and potentially affected waste from pathology departments (laboratory and post-mortem rooms) and other clinical or research laboratories.

*Group D Wastes:* Certain pharmaceutical products and chemical wastes.

*Group E Wastes:* Items used to dispose of urine, faeces and other bodily secretions or excretions assessed as not falling within Group A. This includes used disposable bed pans, incontinence pads, stoma bags and urine containers.

The shape of careful clinical wastes management is exemplified by the City of Copenhagen which has established seven individual collection schemes for clinical wastes:

- Collection of special hospital waste from all hospitals and from other sources generating more than 50 kg of special hospital waste per week.
- Collection of special hospital waste from doctors' and dentists' practices, nursing homes, visiting nurses, etc., generating less than 50 kg of special waste per week.
- Collection of chemical waste from doctors' and dentists' practices, nursing homes, visiting nurses, etc.
- Collection of dead animals from veterinary hospitals, laboratories, etc.
- Collection from pharmacies of discharged medicine, disposable syringes, etc.
- Emptying of syringe boxes in public places.
- Clean-up of syringes, etc., from certain troubled back-yards, playgrounds, etc.

ALL THE ABOVE MUST BE SEPARATED FROM HOUSE-HOLD WASTE.

The collection and transportation system is designed with one main objective: safe handling of the waste from the user via collection to its ultimate disposal.
    This ensures two purposes:

(a) rational and safe handling which protects the staff in all steps of the collection and treatment system from direct contact to the waste, and
(b) rendering the collection system visible by the waste handlers as they must never be unaware of which waste product they are handling.

The above precautions are essential in today's society.

*Sources:* The UK National Association of Waste Disposal has also published *Clinical Waste Guidelines*, November 1994.
C. Vennicke, 'Incineration of Hospital Waste', presented at Wastes Management Meeting WM95, Copenhagen, 1995.

(iv) **Waste disposal options for industrial wastes.** There is a wide range of options available and given that a strategy for WASTE MINIMIZATION (vii) is followed, and that recovery, re-use or recycling options are exhausted or not feasible, then the choice boils down to:

(a) *Disposal on site.* This is used by many companies who LANDFILL their waste arisings as the cheapest option where permissible. Such sites must have a WASTE MANAGEMENT LICENCE (vi) and are of course subject to the Special Waste Regulations (UK).

(b) *Landfill.* This is used for around 80 per cent of hazardous waste disposal. It may be by *co-disposal* whereby the liquid hazardous wastes are run into beds of settled domestic wastes, where a variety of absorption and filtering processes take place. The substances that may be deposited and their manner of deposition is laid down in the site licence. Landfill disposal has been viewed as a cheap option and is often not a totally environmentally secure means of disposal. But, in strictly controlled sealed landfill sites with thorough checks on the WATER BALANCE of the site, backed up by rigorous LEACHATE treatment and LANDFILL GAS controls, and sound aftercare programmes, landfill disposal can be very acceptable. (⇨ LANDFILL)

However, the scope for abuse is considerable and it should not be looked on as a cheap option.

The 1981 House of Lords Report on *Hazardous Waste Disposal*, identified the following requirements for the 'sensible' landfill of hazardous waste:

1. Proper selection of landfill site;
2. sampling, analysis and specification of the waste;
3. skilled operation of the selected site for the disposal of the waste specified (including and necessary treatment); and
4. external monitoring of the site both during operation and after the site has been closed for disposal.

Unfortunately, very few UK landfill sites are equipped to analyse wastes received 'at the gate' and reliance has to be placed on the manufacturer's description. Reputable companies will sample but standards and practices can vary as paragraph 2.17.1, from the HWI 2nd Report of 1986 illustrates:

2.17.1 The HWI is aware of a major waste disposal contractor who was forced either to lose a longstanding and very substantial disposal contract by a treatment route to a landfill competitor, or himself to offer a cheaper landfill route. The contractor chose the latter course and some 2,500 tonnes per annum of acid at pH 1 and alkali at pH 8 have now been diverted from treatment to a landfill disposal route.

(c) *Treatment.* This falls into four categories: biological, physical, chemical and fixation.

*Biological* treatment can only be used on organic aqueous wastes and may be either an AEROBIC or an ANAEROBIC process.

*Chemical* treatment embraces a spectrum of techniques to, for example, make soluble wastes insoluble, destroy toxicity (oxidation of cyanides) or neutralization of ACIDS or ALKALIS.

*Physical* treatment techniques include settling, sedimentation, filtration, flotation, evaporation, distillation, e.g. the separation of oil and water using settlement lagoons or the recovery of SOLVENTS by distillation are widely used.

*Fixation* is the encapsulation or sealing of the hazardous wastes in a variety of matrices such as organic polymers, resins, fly ash/cement mixtures. The matrices are unreactive and are claimed to encapsulate the wastes although controlled release might be a better term. The long-term integrity of some of these processes has been questioned, but if properly conducted, they are undoubtedly a much better environmental option than co-disposal or straight landfill, which is still practised in the UK, of for example, acid wastes whose pH can be as low as 3 at some sites.

(d) *Deep underground disposal.* This is practised in the UK in a few deep underground mineshafts which discharge the liquid low or non-toxic ('nuisance') wastes into old mine workings where there is minimal risk to water supplies. Worked out salt mines or brine deposits can also be used. This is used in West Germany for *'difficult wastes'* which are very securely packaged.

(e) *Dumping at sea.* This is controlled by the Oslo and London Conventions implemented through the UK Dumping at Sea Act 1974. The licensing authorities are the Minister of Agriculture, Fisheries and Food (MAFF), the Secretaries of State for Scotland and Wales, and the Northern Ireland Department of the Environment. MAFF reported that, in 1979, 74.42 million tonnes of waste were licensed for dumping at sea by the UK, of which only 1.27 million tonnes were industrial waste (excluding colliery waste and fly ash).

In fact 45.21 million tonnes were actually dumped, including 0.60 million tonnes of industrial waste (House of Lords 1980–81 Report).

Both Conventions prohibit the dumping at sea of certain BLACK LIST substances, such as organohalogen and organosilicon compounds, carcinogens, mercury and its compounds, etc. Most of the industrial wastes dumped at sea from the UK are liquids or sludges of low or medium toxicity; wastes containing more than traces of Black List materials are not licensed for dumping (House of Lords 1st Report 1981).

Much tighter restrictions on dumping at sea are already in train. This means that the UK will shortly have to find land-based options for sewage SLUDGE disposal, one-third of which is currently dumped in the North Sea. Industrial waste disposal in the North Sea is expected to end completely by 1993.

(f) *Incineration.* This is really the only environmentally secure option if properly conducted, for pathogenic materials, inflammable liquids, carcinogenic substances such as PCBs or materials containing DIOXINS. Organic wastes are destroyed and there is the possibility of energy recovery from spent SOLVENTS. The highly aggressive operating conditions and intermittent loading of waste incinerators means high maintenance costs. (Marine waste incineration was banned in 1990.) (⇨ INCINERATION)

(v) **Waste disposal plan.** Required under Section (50) of the Environmental Protection Act 1990. The plan of the Waste Regulation Authority (WRA) must include information on:

(a) the kinds and quantities of controlled waste which the authority expects to be situated in its area during the period specified in the plan;

(b) the kinds and quantities of controlled waste which the authority expects to be brought into or taken for disposal out of its area during that period;

(c) the kinds and quantities of controlled waste which the authority expects to be disposed of within its area during that period;

(d) the methods and the respective priorities for the methods by which in the opinion of the authority controlled waste in its area should be disposed of or treated during that period;

(e) the policy of the authority as respects the discharge of its functions in relation to licences and any relevant guidance issued by the Secretary of State;

(f) the sites and equipment which persons are providing and which during that period are expected to provide for disposing of controlled waste; and

(g) the estimated costs of the methods of disposal or treatment provided for in the plan.

A waste recycling plan must also be produced.

In preparing the plan the WRA is also required to consult any relevant water authorities, collection authorities (District Councils in England), any WDAs affected by the export of controlled wastes, relevant other persons engaged in the disposal of controlled waste. Members of the public can make representations but are not statutory consultees, as the WDA may make alterations which it 'considers appropriate in consequence of the representations'.

**(vi) Waste management licence.** A waste management licence is defined under Section 35 of the Environmental Protection Act 1990:

> A waste management licence is a licence granted by a waste regulation authority authorising the treatment, keeping or disposal of any specified description of controlled waste in or on specified land or the treatment or disposal of any specified description of controlled waste by means of specified mobile plant.

The licence is granted to the following persons, that is to say:

(a) in the case of a licence relating to the treatment, keeping or disposal of waste in or on land, to the person who is in occupation of the land; and

(b) in the case of a licence relating to the treatment or disposal of waste by means of mobile plant, to the person who operates the plant.

on the terms and conditions as appear to the waste regulation authority to be appropriate; the conditions may relate:

(a) to the activities which the licence authorizes, and

(b) to the precautions to be taken and works to be carried out in connection with or in consequence of those activities;

and accordingly requirements may be imposed in the licence which are to be complied with before the activities which the licence authorizes have begun or after the activities which the licence authorizes have ceased. Under Section 36, the authority must also refer the application to the National Rivers Authority and the Health & Safety Executive (HSE), (in Scotland the relevant River Purification Authority and the HSE).

Under the EPA, the suitability of a person to be a licence holder will be taken into account when an application is made. The authority must be satisfied that the applicant is a 'fit and proper person'. Whether a person is or is not a fit and proper person to hold a licence will be determined based on whether he or another relevant person has been convicted of a relevant offence, whether he is technically competent to manage the licensed activities and whether he is able to make adequate financial provision to discharge the obligations arising from the licence. This is an important check, as too often in the past unsuitable licensees have either not discharged their responsibilities or have simply walked away from them by handing their licence back to the then Waste Disposal Authority.

The grounds for refusing a licence include prevention of pollution of the environment, harm to human health or severe detriment to amenities of the locality. The environment is defined in Section 29 as:

> all, or any, of the following media, namely land, water and the air. Pollution of the environment means pollution of the environment due to the release or escape (into any environmental medium) from:
> (a) the land on which controlled waste is treated,
> (b) the land on which controlled waste is kept,
> (c) the land in or on which controlled waste is deposited,
> (d) fixed plant by means of which controlled waste is treated, kept or disposed of,
> of substances or articles constituting or resulting from the waste and capable (by reason of the quantity or concentrations involved) of causing harm to man or any other living organisms supported by the environment.

A licence can no longer be handed back at any time. It may only be surrendered if the Waste Regulation Authority accepts the surrender. On an application to surrender a licence the Authority will inspect the site and may request additional information. The Authority shall determine whether it is likely or unlikely that the condition of the land will cause pollution of the environment or harm to human health. If the Authority is so satisfied, it must then refer the matter to the National Rivers Authority, who may appeal to the Secretary of State if it disagrees with the surrender of the licence. Where the surrender of a licence is accepted the Authority will issue the applicant with a Certificate of Completion. This is a major advance on the Control of Pollution Act 1974, where the conditions of the licence ceased to be enforceable once the licence was handed back regardless of the state in which the land was left. This loophole has left several local authorities with massive problems due to past site mismanagement.

Contravention of the conditions of the waste management licence is an offence under the Act. The current situation regarding conviction for contravention of licence conditions depends on the offence being committed while waste is being deposited. This has proved very difficult to establish when action has been taken for contravention of licence conditions under the Control of Pollution Act. A licence may be revoked by the Waste Regulation Authority, under Section 38, if it appears to the authority:

(a) that the holder of the licence has ceased to be a fit and proper person by reason of his having been convicted of a relevant offence; or
(b) that the continuation of the activities authorized by the licence would cause pollution of the environment or harm to human health or would be seriously detrimental to the amenities of the locality affected; and
(c) that the pollution, harm or detriment cannot be avoided by modifying the conditions of the licence.

Licence conditions may, for example, include:

(a) duration of licence;
(b) supervision by the licence holder of the licensed activities;
(c) the kind and quantities of waste which may be dealt with;
(d) precautions to be taken on land to which the licence relates;
(e) compliance with planning permission conditions;
(f) hours during which waste may be dealt with;
(g) the works to be carried out relating to the land, plant or equipment to which the licence relates before the activities authorized by the licence are begun or after the activities which the licence authorizes have ceased.

Site licences are available for public inspection at the relevant Waste Regulation Authority office.

**(vii) Waste minimization.** There is a hierarchy of waste management which begins with waste minimization before proceeding to actual disposal:

1. Do not create the waste product in the first place.
2. If there is a waste product, re-use it.
3. If it cannot be re-used, recover or reclaim the primary material for new manufactured products.
4. If primary materials recovery is not practicable, recover it for secondary materials or if combustible use it for fuel.
5. If none of these is practicable proceed to the various waste disposal options choosing the one that has the least environmental impact.

(⇨ iv, WASTE DISPOSAL OPTIONS).

As an example of a waste minimization strategy for the chemical industry, the following stages are suggested:

1. Design for maximum chemical conversion.
2. Design for maximum energy efficiency.
3. Use low hazard solvents.
4. Make minimum use of solvents.
5. Make minimum use of process water.
6. Ensure minimum dilution of any carrier liquid (otherwise there will be greater clean up costs).
7. Ensure low inventories of materials and product.

*Source:* Discussion at Institution of Chemical Engineers meeting on Pollution Control, I.C.I. Pharmaceuticals, Alderley Park, Cheshire, 2 November 1989.

In addition, through applying the POLLUTION PREVENTION PAYS approach, many wastes can be reduced in volume or altered to be less environmentally damaging.

The House of Commons Environment Committee in its 1989 Report had this to say on waste minimization in paragraphs 34 and 35:

34. We conclude that waste minimisation has many environmental and economic benefits and much more could be achieved in the UK using existing technologies. We feel that as well as environmental disadvantages, British industry could be prejudiced in the future if it fails to keep pace with European and US

developments in waste minimisation and clean technology. The Government must take a stronger lead in encouraging waste minimisation.

35. We recommend:
   (i)   that waste minimisation should be given primary importance in the whole waste production/waste regulation process;
   (ii)  that a statement of principles for waste minimisation should be incorporated in the Code of Practice on the duty of care;
   (iii) that waste minimisation should be specifically included in the environmental impact assessment of new plant;
   (iv)  that HMIP through IPC (integrated pollution control) and through publicising examples of good practice, should give a greater lead on this issue; and
   (v)   that WDAs should encourage the application of waste minimisation techniques in their own areas.

Much too often, when wastes are created they are LANDFILLED or 'dumped' according to whether there is engineered disposal or just plain sticking them in a hole in the ground as cheaply as possible. It is then that POLLUTION arises which, as F. Fraser Darling (BBC, Reith Lectures, 1969) stated, 'comes from getting rid of wastes at the least possible cost'.

**(viii) Waste quantities.** The Department of Environment Report, December 1985, estimated that there is a total of 500 million tonnes of wastes provided annually in the UK.

The report also gives the estimated 1994 UK annual industrial and agricultural waste arisings:

Estimated total annual waste arisings in the UK, by sector.

| Sector | Annual arisings (million tonnes) | % of total arisings |
|---|---|---|
| Households | 20 | 5 |
| Commercial waste | 15 | 3 |
| Construction and demolition | 70 | 16 |
| Sewage sludge | 35 | 8 |
| Dredged spoils | 35 | 8 |
| Controlled wastes total | 245 | 56 |
| Mining and quarrying waste | 110 | 25 |
| Agricultural waste | 80 | 25 |
| Total of all waste | 435 | 100 |

*Source: Sustainable Waste Management, A Waste Strategy For England and Wales*, Department of the Environment Welsh Office, Consultation Draft, January 1995.

**(ix) Waste recycling plan.** Required under the Environmental Protection Act 1990 (Section 49). Each waste collection authority, as respects household and commercial waste arising in its area is required:

(a) to carry out an investigation with a view to deciding what arrangements are appropriate for dealing with the waste by separating, baling or otherwise packaging it for the purpose of recycling it;

(b) to decide what arrangements are in the opinion of the authority needed for that purpose;

(c) to prepare a statement ('the plan') of the arrangements made and proposed to be made by the authority and other persons for dealing with waste in those ways;

(d) to carry out from time to time further investigations with a view to deciding what changes in the plan are needed; and

(e) to make any modification of the plan which the authority thinks appropriate in consequence of any such further investigation.

The information in the plan should include:

(a) the kinds and quantities of controlled waste which the authority expects to collect during the period specified in the plan;

(b) the kinds and quantities of controlled waste which the authority expects to purchase during that period;

(c) the kinds and quantities of controlled waste which the authority expects to deal with in the ways specified in subsection (a), previous page, during that period;

(d) the arrangements which the authority expects to make during that period with waste disposal contractors or, in Scotland, waste disposal authorities and waste disposal contractors for them to deal with waste in those ways,

(e) the plant and equipment which the authority expects to provide under section 48(6) above or 53 below; and

(f) the estimated costs or savings attributable to the methods of dealing with the waste in the ways provided for in the plan.

**Wasteplex.** A concept of a highly integrated centre for the reception of domestic and industrial wastes from population catchment areas of around 500 000 people. This allows economic sorting of the wastes into directly and indirectly reusable fractions. (⇨ RECYCLING)

**Wastes separation.** The use of mechanical, manual or other means

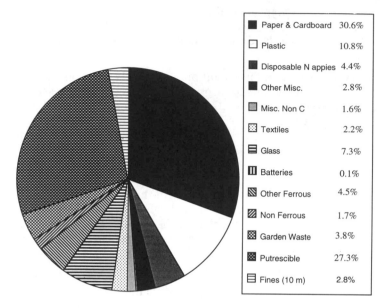

**Figure 152**   City of Coventry household waste composition for 1995.

of segregating specific materials from a mixed waste stream, normally a municipal or domestic waste stream but commercial wastes that are not of uniform composition and origin may also be separated.

**Water.** One of the prime resources for life which must be preserved from contamination for the public WATER SUPPLY. It has many industrial uses: a conveying medium for wastes, slurries, woodpulp, etc.; a heat-exchange medium; it is used in steam raising and also as a solvent. Its basic chemical formula is $H_2O$, yet it has other molecular combinations such as $H_8O_4$, which gives it many unique properties such as expansion on freezing.

For the purpose of this entry the chemical, physical and biological attributes are considered.

1. *Chemical properties.* These have a great influence on the use for drinking or industrial purposes or the support of aquatic life. The main parameters are:
   (i)   *pH* – a measure of the acidity or alkalinity. Natural

water supplies are usually in the pH range 6–8, whereas industrially contaminated waters can have any value in the range pH 1–13 and require neutralization to pH 7 before discharge. (⇨ pH)

(ii) *Hardness* – this prevents soap lathering properly. Also when the water is heated, scale may be deposited on the heating surfaces, which can ruin the efficiency of heat transfer and in extreme instances, cause burn out of boiler tubes or heat exchanges. The hardness is measured by the concentration of calcium ions ($Ca^{2+}$) and magnesium ions ($Mg^{2+}$), present usually as calcium carbonate ($CaCo_3$) and magnesium sulphate ($MgSO_4$).

(iii) DISSOLVED OXYGEN and, related to this, the chemical and biological oxygen demand. These values are of great importance to biological systems. (⇨ BIOCHEMICAL OXYGEN DEMAND; CHEMICAL OXYGEN DEMAND)

2. *Physical characteristics.* The colour, taste and appearance of water are of great importance in its suitability for consumption. The solids in solution and in suspension play an important part. The following characteristics are measured:

(i) *Turbidity* – an indication of the presence of colloidal ('gluey' suspensions of fine particles) particles such as silt or bacteria from, say, sewage treatment.

(ii) *Colour* – usually due to dissolved substances such as peat acids which give highland streams their brown appearance.

(iii) *Taste* and *odour* due to the presence of dissolved solids or gases, e.g. 0.001 milligrams per litre (1 part in 100 million) of phenolic liquors can taint water and render it unpalatable. The gases can be of biological origin from algae or from bacterial action on wastes or impurities.

(iv) *Temperature* – this has a direct bearing on the saturation concentration of dissolved oxygen in the water; the greater the temperature, the less oxygen the water can hold in solution and therefore temperature has important biological consequences.

3. *Biological properties.* The variety and type of water organisms and aquatic life give a very good indication of its biological state of health and the BIOTIC INDEX is used as such a measure. Other tests are made to indicate the presence of

various classes of micro-organism algae and bacteria, and especially the bacterium, *Escherichia coli*, which indicates the presence of faecal matter.

($\Rightarrow$ COLIFORM COUNT; WATER SUPPLY; WATER QUALITY STANDARDS; PUBLIC HEALTH)

**Water balance.** The method used to estimate the quantities of LEACHATE produced in a LANDFILL site and/or the capacity of the site to accept additional liquid waste inputs.

This is best understood by considering the landfill hydrological cycle (which can be regarded as a local sub-system of the hydrological cycle). This local sub-system may be complex, with important liquid inputs to, and outputs from, the landfill storage system, not normally encountered or considered in regional hydrologic cycles. An idealized picture of the main components is shown in Figure 153. The water balance for the site is determined by the difference between these rates of inflow and outflow, viz.

$$(\text{Rate of inflow}) - (\text{Rate of outflow}) = \Delta S$$

where $\Delta S$ is the rate of change in the quantity of liquid, $S$, stored within the landfill. Thus, the rate of change in storage is given by

$$\Delta S = (P + R_1 + G + W + L) - (E + T + R_2 + R_3 + I + Q)$$

Of these, the liquid abstracted from storage ($Q$), the surface springs and seepages ($R_3$), and any recharge to groundwater ($I$), can be considered as leachate and require to be managed as such. In some cases, surface run-off $R_2$, which would normally be clean, becomes mixed with leachate and increases the volume requiring attention.

The equation above reflects the reality that most of the component factors vary continuously with time. However, in order to calculate a water balance for a particular period, say one year, it is normally only feasible to divide the period into a limited number of discrete smaller periods (e.g. days, weeks or months) and to compute values for each period. The size of the sub-period chosen can make a difference to the calculated values for the overall rate and pattern of leachate production.

**Water consumption.** The quantity of water abstracted and used for any purpose, irrespective of the state or time in which it is returned to the source or to the atmosphere. Per capita

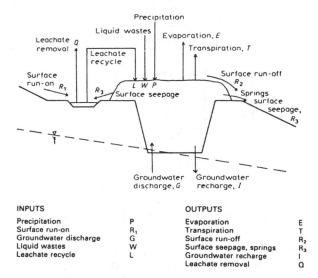

**INPUTS**

| | |
|---|---|
| Precipitation | P |
| Surface run-on | $R_1$ |
| Groundwater discharge | G |
| Liquid wastes | W |
| Leachate recycle | L |

**OUTPUTS**

| | |
|---|---|
| Evaporation | E |
| Transpiration | T |
| Surface run-off | $R_2$ |
| Surface seepage, springs | $R_3$ |
| Groundwater recharge | I |
| Leachate removal | Q |

**Figure 153**   Hydrological cycle at a landfill.

consumption is the amount of water supplied to an area divided by the size of the resident population. The UK average per capita consumption is 45 000 litres or 10 000 gallons per year.

**Water pollution.** Can be caused by a variety of means, e.g. farm pollution from animal wastes and SILAGE liquor, LEACHATE from LANDFILL sites, and spoil heaps, SOLVENT discharge to sewers or to land and inadequate SEWAGE TREATMENT works. Pollution incidents are shown overleaf.

Her Majesty's Inspectorate of Pollution, First Annual Report (1987–88) stated that in 1986, there were 4333 water authority sewage treatment works (in England and Wales) with numerical consents, of which 1000 (23%) failed to comply with their DISCHARGE CONSENTS. The principal reasons for this state of affairs were given as:

(a) Old and overloaded equipment.
(b) Poor maintenance.
(c) Possible overloading because of trade effluent discharges to sewer.
(d) Under-investment.

Total pollution incidents

| Year | Total | % | Serious | % | Prosecutions | % |
|---|---|---|---|---|---|---|
| *1987* | | | | | | |
| Farm | 3890 | 19 | 990 | 51 | 225 | 67 |
| Industrial | 7575 | 37 | 617 | 32 | 98 | 29 |
| Sewage | 4177 | 20 | 221 | 11 | 6 | 2 |
| Other | 5017 | 24 | 117 | 6 | 8 | 2 |
| Total | 20659 | 100 | 1945 | 100 | 337 | 100 |
| *1980* | | | | | | |
| Farm | 1671 | 15 | | | | |
| Total | 10797 | 100 | | | | |

*Source:* Water Authorities Association, *Water Pollution from Farm Waste,* 1988 *England & Wales,* 1989.

**Water power.** The free renewable source of energy provided by falling water has been put to use for centuries. Hydroelectricity is practised in many countries, yet in the USA there are only five plants with capacities in excess of 1000 megawatts. In the UK it supplies less than 2 per cent of the nation's power needs and there is little scope for further development as almost all sites have been used. It is capital intensive, requiring costly dams and earthworks which may have a finite lifetime due to silting up behind the dam. This will restrict its use at any one site to 100–200 years unless silt-laden floodwater is diverted to lengthen the life of the reservoir. Globally there are locations that can, in total, supply energy comparable to the present rate of energy consumption. South America, Africa, Asia and the USSR account for 80 per cent of the potential capacity. Because of the high capital investment per megawatt capacity, its contribution to future energy supplies is not expected to be significant in the West. (➪ BARRAGE)

**Water quality objectives (WQOs).** WQOs specify formal minimum quality standards, and have already been set to give effect to the EC Dangerous Substances legislation arising from the Directive 76/464/EEC, and subsequent daughter Directives (via the Surface Waters (Dangerous Substances) (Classification) Regulations 1989 (LS 2286) and 1992 (SI 337)), and the EC Bathing Waters legislation arising from Directive 76/160/EEC (via the Bathing Waters (Classification) Regulations 1991 (SI 1597)).

The requirements of these Directives, and the methodologies necessary for compliance assessment, are contained within the Directives and the domestic Regulations and Notices arising from them.

Details of WQOs assigned to river stretches, compliance with WQOs, and the monitoring data upon which compliance assessment is based, together with any of the 'exceptional circumstances' which have been identified by the NRA as applying, will be included on the public register.

**Water Quality Standards.** Water for domestic use should comply with purity standards which are internationally acceptable. The World Health Organization has published the *WHO Guidelines for Drinking Water Quality*, while in Europe the water supplied to the public is controlled by the EC Directive on *The Quality of Water Intended for Human Consumption* 80/778/EEC. OJ No. L229. This Directive covers all water supplied whether it be surface or groundwater.

The standards to be met relate to:

1. Organoleptic factors, e.g. colour, turbidity, taste, odour, pH, conductivity, etc.
2. Physico-chemical factors, e.g. chloride, sulphate, calcium, magnesium, aluminium, dissolved oxygen, etc.
3. Substances undesirable in excessive amounts, e.g. nitrate, ammonia, fluoride, iron, manganese, etc.
4. Toxic substances, e.g. arsenic, mercury, lead, cadmium, etc.
5. Microbiological parameters.

Toxic substances have been given Maximum Admissible Concentrations (MAC) values, while for other parameters Guide Levels (GL) values and MAC values are quoted. The following tables show the main parameters:

(a) Organoleptic factors

| Parameter | Expressions of result | GL | MAC |
|-----------|----------------------|-----|-----|
| Colour | Pt/Co scale (mg/l) | 1 | 20 |
| Turbidity | $SiO_2$ (mg/l) | 1 | 10 |

(b) Physico-chemical parameters

| Parameter | Expression of result | GL | MAC |
|---|---|---|---|
| Hydrogen ion concentration | pH unit | 6.5–8.5 | – |
| Conductivity | $\mu$S/cm at 20 °C | 400 | – |
| Chlorides | Cl (mg/l) | 25 | – |
| Calcium | Ca (mg/l) | 100 | – |
| Magnesium | Mg (mg/l) | 30 | 50 |
| Aluminium | Al (mg/l) | 0.05 | 0.2 |

(c) Substances undesirable in excessive amounts

| Parameter | Expressions of result | GL | MAC |
|---|---|---|---|
| Nitrate | $NO_3$ (mg/l) | 25 | 50 |
| Ammonium | $NH_4$ (mg/l) | 0.05 | 0.5 |
| Phenols | $C_6H_5OH$ ($\mu$g/l) | – | 0.5 |
| Iron | Fe ($\mu$g/l) | 50 | 200 |
| Manganese | Mn ($\mu$g/l) | 20 | 50 |
| Copper | Cu ($\mu$g/l) | 100 | – |
| Zinc | Zn ($\mu$g/l) | 100 | – |

(d) Toxic substances

| Parameter | Expression of result | GL | MAC |
|---|---|---|---|
| Arsenic | As ($\mu$g/l) | – | 50 |
| Cadmium | Cd ($\mu$g/l) | – | 5.0 |
| Mercury | Hg ($\mu$g/l) | – | 1.0 |
| Nickel | Ni ($\mu$g/l) | – | 50 |
| Lead | Pb ($\mu$g/l) | – | 50 |
| Antimony | Sb ($\mu$g/l) | – | 10 |
| Pesticides and related products | $\mu$g/l | – | 0.5* |
| | | – | 0.1† |
| PAHs | $\mu$g/l | – | 0.2 |

*Total MAC of all pesticides present
†MAC allowable for any one pesticide

(e) Microbiological factors

| Bacteria | Sample volume (ml) | GL | MAC | |
|---|---|---|---|---|
| | | | Membrane filter | Multiple tube |
| Total coliforms | 100 | – | 0 | < 1 |
| Faecal coliform | 100 | – | 0 | < 1 |
| Faecal streptococci | 100 | – | 0 | < 1 |

The EEC now has 13 Directives (1990) on water quality. Discharges to the aquatic environment are controlled by Directives which protect surface and groundwater against pollution. Each of these directives has two lists and the aim is to eliminate pollution by substances quoted in List 1 and reduce pollution by those in List 2. These Directives are 'enabling' Directives which will come fully into force as daughter Directives are enacted. Directives on cadmium, mercury and hexachlorocyclohexane have been passed, while there are proposed directives for aldrin, dieldrin, endrin, DDT, carbon tetrachloride and pentachlorophenol. (⇨ DIRECTIVES; EC)

The Directive on the *Quality of Surface Water Intended for Drinking Water* proposes standards for surface water to be used in the public supply and defines three categories of water treatment (A1, A2, A3) from simple physical treatment and disinfection to intensive physical and chemical treatment. The treatment to be used depends on the quality of the water abstracted. The Directive uses I (Imperative) values for parameters known to have an adverse effect on health and also G (Guide) values for those which are less adverse. There is also a Directive which complements the 'surface water abstraction' Directive by indicating the methods of measurement and the frequency of sampling and analysis required.

The quality of coastal and estuarial waters is considered in Directives on the quality of bathing water, the quality required for shellfish water and the quality of water for freshwater fish.

The bathing water Directive sets out quality requirements for bathing water, defines a monitoring programme and gives reference methods of analysis.

The freshwater fish and shellfish Directives apply to designated waters which require protection to make them suitable for fish life or to ensure that the quality of shellfish is suitable for human consumption.

The Directive on waste from the titanium oxide industry has as its aim the prevention and progressive reduction of pollution from the industry until eventually the pollution is eliminated. This Directive applies to all sections of the environment and not just the aquatic environment.

There has been control on water pollution in the UK since 1951. This control has been enacted in The Rivers (Prevention of Pollution) Acts 1951 and 1961 along with the Control of Pollution Act 1974. Section 31 of the latter act controls pollution to the 'three-mile off shore limit' and also certain underground waters.

**Water supply.** A complex subject illustrated in Figure 154 with the processes used to supply London with 'wholesome' water from the River Thames.

The presence of the coliform bacteria, *Escherichia coli*, is monitored in the source as an indicator of faecal pollution. In Figure 154 the count is originally 6680 per 284 cubic centimetres. The abstracted water is pumped to a reservoir for storage and settlement where faecal bacteria are reduced through natural processes and a large portion of the suspended solids settle out. (⇨ COLIFORM COUNT)

The water is taken from the storage reservoir via the draw-off arrangement under the dam base to an aeration basin so that it is fully oxygenated, to the primary filter and micro-strainers where most plankton and algae are removed. (Prior to this stage, coagulation is often practised whereby colloidal particles which carry negative charges and thus will not naturally come together, are coagulated by special agents and removed by sedimentation.)

Sand or other filtration follows and this is a most important stage as the sand is very much more than a fine strainer. The sandfilter is a large shallow basin on which there is 1 metre of sand on a gravel base. The water is introduced at the top of the bed and drawn off at the bottom through collector pipes and channels. Filtration in this case takes place under gravity although pressure filters are also available.

Two separate layers develop in the bed: the top 2 millimetres comprise micro-organisms which decompose organic matter and consume nitrates and phosphates and release oxygen. This is the autotrophic layer which has a strong purifying action, on

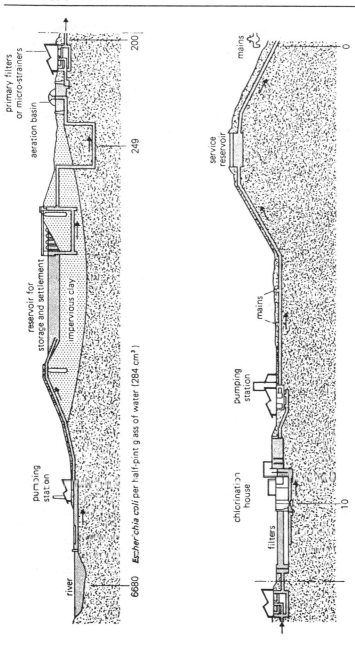

**Figure 154** *Escherichia coli* count in London's water supply.

whose surface there is a dirt film which can impede the flow of water considerably. Below the autotrophic layer is the heterotrophic zone (30 centimetres) where non-pathogenic bacteria continue the decomposition of the organic matter and render the water in a very pure condition. In the London example the *E. coli* count is 200 before filtering and 10 after. The sand filter thus performs a complex range of duties. Its performance depends on the physical and chemical characteristics of the suspended solids and the chemical composition of the water.

Filtration can be by either the *slow sand filter* or the *rapid sand filter* method. The slow method requires a lot of space and is not now often used as the dirt layer previously referred to must be skimmed off by draining the filter, and new sand spread on top every two or three months. The flow of water per unit area is low. The rapid gravity sand filter is 20 times faster and this is accomplished by backwashing with compressed air. The rapid filter is not nearly so efficient as the slow filter so coagulation is required as a pretreatment process. The rapid filter does not oxidize organic matter and nitrogen compounds so these may require removal at a later stage or anthracite sand filters can be used. These are a layer of large anthracite grains on top of smaller sand grains. These filters clog less rapidly as the particles adhere to the larger grains and thus larger filtration rates are possible.

The water with an *E. coli* count of 10 is now sterilized by chlorine injection at a rate proportional to the water flow. The *E. coli* count is now zero and the presumption is that the water is now wholesome. It is then pumped to the mains where it is distributed to the consumers. Note that service reservoirs and/or water towers are required to maintain adequate pressure in the supply mains and to iron out fluctuations in demand.

Another form of sterilization (not shown in Figure 154) is the use of OZONE, which is a powerful oxidizing agent and which also reduces colour, taste and odour in water. A dose as low as 0.0001 per cent or 1 part per million destroys all bacteria within 10 minutes. The ozone must be manufactured by electric-arc discharge on site and is thus more expensive than chlorination. No matter what method is adopted, extensive testing and analysis is carried out by all water authorities.

**Water table.** The upper surface of the SATURATION ZONE below which all void spaces are filled with water.

**Watt (W).** Unit of power defined as 1 JOULE per second. Conversion factor 1 horse power = 745.7 W.

A megawatt (MW) is equal to 1000 kilowatts (kW) or 1 000 000 watts (W). It is used as a measurement of electrical generating capacity. (⇨ KILOWATT-HOUR)

**Watt-hour (Wh).** Unit of energy equal to 1 watt acting for 1 hour, i.e. 3600 joules.

**Wave power.** The harnessing of the energy in the waves to generate electricity. Engineering assessments show that to get the equivalent amount of energy from 1 kilogram of coal requires 1 tonne of sea-water to fall through 3 kilometres. Thus, the capital costs for the harnessing of wave power may be high.

**Weathering.** The chemical and mechanical breakdown of rocks and minerals under the action of atmospheric agencies.

**Wells and boreholes.** A hole drilled into the ground to tap an aquifer for water supplies (or an oil or gas field for oil or gas). Shallow wells in soft formations can be dug by hand or with power augers. Deeper and/or resistant rock formations are drilled by machine. Once the well has been drilled it must be completed, i.e. the hole cased if necessary to prevent collapse with a slotted casing to allow the water to enter. The well is then developed by explosives, chemicals or compressed air to shatter the rock and increase the yield.

Once pumping commences, the safe yield is established. This can be restricted by salt-water INFILTRATION in some areas. This means that a well's yield is not necessarily the maximum rate at which water can be extracted, but the maximum rate at which the quality of the water or the supply of other nearby wells is not affected. *Note:* Wells can have other uses, e.g. as observation wells to monitor the abstraction of water from an aquifer or the migration of LEACHATE or LANDFILL GAS from a LANDFILL SITE.

**Whale harvests.** Whales fisheries serve as an archetypal model of over-exploitation. As the large whales (blue and fin) were hunted to near or total extinction, the industry shifted to harvesting not only the young of the larger species but also smaller species.

After the Second World War the International Whaling Commission (IWC) was established, with the brief of regulating harvests and protecting certain endangered species. In practice the IWC has turned out to be a relatively impotent organization with inadequate power of inspection and non-existent powers of enforcement. The IWC made an early fundamental error in establishing quotas on the basis of 'blue whale units' (bwu), a unit being one blue whale, or its equivalent (2 fin whales, $2\frac{1}{2}$ hump-back whales, or 6 sei whales). The total permitted quota was initially set at 16000 units with the blue whale the preferred species. In the 1950s the number of blue whales caught declined sharply and the whalers turned their attention to the fin and hump-back whales. During the 1960s the blue whales and fin whale catch continued to decline, and in 1963 a Commission committee report warned that blues and hump-backs were in serious danger of extinction. It recommended that these species be totally protected and the fin whale catch limited. It also urged that the 'blue whale unit' be dropped in favour of limits on individual species. Predictably, these warnings and recommendations were ignored and the total catch quota was limited to 10000 bwu. By this time the fin whale harvest was estimated to be three times the MAXIMUM SUSTAINABLE YIELD. Subsequently the committee recommended a bwu total for 1964–5 of 4000; for 1965–6, 3000; and only 2000 for 1966–7, in order to allow recovery of the whale stocks. Again it was ignored. All four countries then engaged in Antarctic whaling – Japan, the Netherlands, Norway and Russia – voted against accepting the recommendation and continued to practise their trade according to the sound economic principle of obtaining a maximum return on invested capital. (⇨ ECONOMICS)

Norwegian and Japanese whalers continue to kill minke whales despite a 1986 moratorium.

In 1988 only 22 blue and fin whales were recorded, where once the blue whales numbered 250000.

There is now, hopefully, some evidence that a long process of recovery is under way and that population growth rates of 5–10 per cent per year are now taking place from a very low base. The fear is that the practitioners of whaling will use this increase as justification for the continuous extension of their activities. One positive aspect of the whale population increase is that the effects of toxic compounds in the marine environment may not be as severe as was feared.

Fish removed are also over-exploited and have remained static globally at around 100 million tonnes per year. The FAO has stated that: '70 per cent of fish stocks are over-fished, depleted or rebuilding from earlier over-fishing. The most exploited areas are around the North Atlantic, around the

American and European land masses, the North Sea, the central Baltic, the Mediterranean and Black Seas and the central-west Pacific'.

Fish farming measures to stop over-fishing and a reduction in the waste of fish discarded at sea are all urged.

**Whey.** Organic liquid left from cheese manufacture which is a very strong, highly polluting material, one cubic metre of whey being equivalent to the daily pollution production from approximately 600 people. In wastewater terminology, the composition of a typical whey is:

| | |
|---|---|
| Total solids | 66 000 mg/l |
| Volatile solids | 59 400 mg/l |
| Chemical oxygen demand | 65 000 g/l |
| Biological oxygen demand | 45 000 mg/l |
| Total nitrogen | 1 500 mg/l |

A waste of this strength is uneconomic to treat by conventional AEROBIC PROCESSES, due to the high power requirements for aeration, and for this reason whey has not traditionally been treated in aerobic biological systems.

In anaerobic digestion however, whey is broken down to produce a gas rich in METHANE, giving the treatment process a large positive energy balance. (⇨ ANAEROBIC DIGESTION)

**Will Secure Test.** A requirement for NON-FOSSIL FUEL OBLIGATION funding.

It 'requires' that a guaranteed security of supply of the fuel (e.g. municipal solid waste) is available and the technology to be used is reliable. (⇨ INCINERATION, FLUIDIZED BED COMBUSTION)

**Wind.** The movement of air parallel to the earth's surface caused by differences in atmospheric pressure and the earth's rotation. Above a certain height the wind blows at constant velocity in a direction parallel to the isobars – lines of constant pressure on the weather chart. The velocity is proportional to the isobar spacing the pressure gradient – and is known as the *gradient wind*. This occurs above the gradient height, approximately 600 metres above ground. Below the gradient height, the wind is slowed down by ground effects, mainly friction, and is also stirred near ground level by obstructions so that eddies are set up which are very useful in diluting and dispersing air pollutants emitted at or near ground level.

**Wind energy.** Energy (normally electrical power) obtained from the wind using wind turbines. The exploitation of wind energy should be regarded in the context of UK and EU policy on SUSTAINABLE DEVELOPMENT. Wind energy as with other forms of renewable energy occurs naturally and repeatedly in the environment. For the UK, commercial development of wind energy is still relatively new. It began in 1990 with the introduction of the NON-FOSSIL FUEL OBLIGATION (NFFO) and by November 1994 over 35 wind energy projects had been built including several with individual turbines and 26 wind farms comprising either a large number of turbines or a cluster of a few.

In order to provide some numbers, the power available from the Finnish Phyätunturi Arctic Test Wind Turbine site are given in the table below:

| | |
|---|---|
| Rated power | 220 kW |
|    at wind speed | 13 m/s |
| Rotor diameter | 25 m |
| Rotor swept area | 490 m$^2$ |
| Weights | |
|    machinery | 6800 kg |
|    rotor | 5700 kg |
|    tower | 13 500 kg |
| Manufacturer | Wind World AS, Denmark |
| Estimated annual production on | |
|    Phyätunturi | 600 MWh |

*Source: CADET Renewable Energy Newsletter, January 1995.*

However there can be legitimate concern over inappropriate siting of wind turbines. A development chart is given in Figure 155 to help responsible development.

**Wind profile.** A graphical representation of the variation of wind speed as a function of height or distance.

**Wind rose.** A star-shaped diagram indicating the relative frequencies of different wind directions for a given location and period of time. The most common form consists of a circle from whose centre are drawn a number of radii to indicate different points of the compass, the length of each radius being proportional to the number of times during the given period that the wind blew from that direction.

**Wind speed.** The wind speed of a potential WIND ENERGY site is a crucial factor in determining the economic viability of a project.

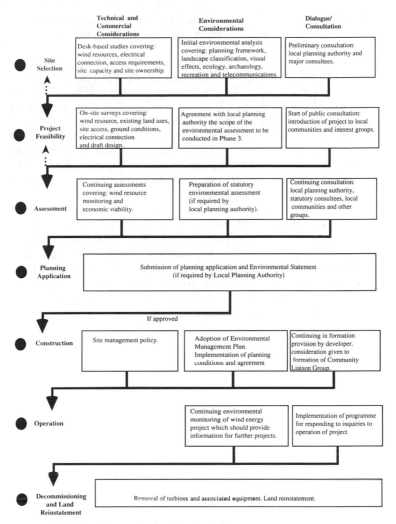

**Figure 155** Development chart for wind energy.
*Source:* British Wind Energy Association, November 1994.

Since energy yield is a function of wind speed (proportional to the cube of the wind speed), the higher the wind speed, the greater the energy yield.

**Wobbe Index.** The Wobbe Index is the amount of the gas delivered to a burner via an injector. The energy input is a linear function of the Wobbe Index. Two gases of differing composition, but having the same Wobbe Index will deliver the same amount of energy for any given injector under the same injector pressure.

**Woodmark label.** This label tells buyers that their wood product has come from well-managed forests, and that by purchasing the product they are helping to support responsible timber production, rather than poor forestry practice or destructive logging.

**Wood alcohol.** ⇨ METHANOL.

**Working plan.** Document submitted in support of waste disposal licence and planning application detailing the engineering and restoration proposals and the conduct of operations. It should include a technical drawing or drawings for each state of the development and a detailed statement of the way the operations are to be carried out, including provisions for landfill gas control and any aftercare programme.

# *X*

---

**Xenon effect.** The rapid but temporary poisoning of a reactor by the build-up of xenon-135 from the radioactive decay of the fission product iodine-135. (Xenon-135 is a strong absorber of neutrons and until it (and its parent) have largely decayed away reactor start-up can be difficult.) ($\Rightarrow$ POISON, NUCLEAR)

**X-ray.** Ionizing or electromagnetic radiations of the same type as light but with much shorter wavelength. The absorption of the rays depends on the density of the material; as bone is denser than flesh this allows X-ray photographs to be taken. All radiation exposure carries risks. However, medical irradiation is voluntary in the sense that the benefits incurred by the diagnostic X-rays far outweigh the risks. Having said that, pregnant women and/or fertile women are not normally exposed to X-rays as the foetus is particularly susceptible in the first instance and the germ cells in the second. ($\Rightarrow$ IONIZING RADIATION)

# Y

**Yeast.** Unicellular fungi which are able to multiply asexually by a budding process. Yeasts are of great importance in FERMENTATION, the manufacture of pharmaceuticals and SINGLE-CELL PROTEIN. One major strain used for single-cell protein from molasses and the sugar in the spent liquors from sulphite pulp manufacture is *Candida utilis*, which has an approximate 50–60 per cent protein content.

**Yellow-cake.** Concentrated crude uranium oxide, the form in which most uranium is shipped from the mining areas to the fuel manufacturers.

# Z

**Zinc (Zn).** Metallic element, atomic number 30, atomic mass 65.38. Principal ore, zinc sulphide (ZnS).

**Zinc equivalent.** The sum of the copper, zinc and nickel contents of a sludge, in milligrams per kilogram, after multiplying the copper content by 2 and the nickel content by 8. Used to indicate the rate of addition of metals and the cumulative concentration in the soil with reference to the effect on the growth of crops. A maximum safe limit of 250 mg/kg zinc equivalent in dry top soil has been suggested.

**Zircaloy.** An alloy of zirconium and tin used for fuel cladding in water reactors.

**Zone of visual influence.** A zone of visual influence provides a representation (usually presented as a map with markings or colourings) of the area over which a site and/or a proposed development, such as a wind turbine or chimney stack, may be visible.

**Zoonoses.** Infections that can be transmitted from animals to humans such as cryptosporidiosis, a diarrhoeal disease which can be life threatening in individuals with suppressed immune systems. Other zoonoses with the potential to cause ill health include chlamydia (also known as enzootic abortion, which can cause abortion in pregnant women and is a risk during lambing and open days), leptospirosis, ort and ringworm.

**Zooplankton.** Floating and drifting aquatic animal life. (⇨ PHYTOPLANKTON)

# *Pollution and the environment – organizations*

**General**

*Departments of state and other public bodies*

AEA Technology, National Technology Centre, Culham, Cubington OX14 3DE (01235 463040)

Countryside Commission for Scotland, Battleby, Redgorton, Perth PH1 3EW (01738 627921)

Countryside Commission for Wales, Plas Penthos, Ifordd Penthosd, Bangor, Gwynedd LL57 2LQ (01248 370444)

Countryside Commission, John Dower House, Crescent Place, Cheltenham, Gloucestershire, GL50 3RA (01242 521381)

Department of the Environment, 2 Marsham Street, London SW1P 3EB (0171 276 3000)

> (Government Offices for Regions with environmental responsibilities are located in Bedford, Birmingham, Bristol, Leeds, Liverpool, London (south-east), Manchester, Newcastle-upon-Tyne and Nottingham)
>
> Central Directorate of Environmental Protection, Romney House, 43 Marsham Street, London SW1P 3PY (0171 276 8731)
>
> Her Majesty's Inspectorate of Pollution, Romney House, 43 Marsham Street, London SW1P 3PY (0171 276 8142) (Regional Offices in Bedford, Bristol, Cardiff, Fleet, Lancaster, Leeds and Sutton Coldfield)

Department of the Environment for Northern Ireland, Stormont, Belfast BT4 3SS (01232 63210)

Department of Environmental Management, Farnborough College of Technology, Boundary Road, Farnborough, Hampshire GV14 6SB (01252 515511)

*Note:* Departments of State will change their names from April 1996.

ECO Environmental Information Trust, 10–12 Picton Street, Montpelier, Bristol BS6 5QA (0117 942 0162)

Institute of Ecology and Environmental Management, 36 Kingfisher Court, Hambridge Road, Newbury, Berks RG14 5UP (01635 37715)

Institute of Environmental Assessment, Gregory Croft House, Fen Road, East Kirkby, Lincs PE23 4DB (01790 763613)

National Environmental Technology Centre, Culham, Abingdon, Oxfordshire, OS14 3DB (01235 463040)

National Environment Research Council, Polaris House, North Star Avenue, Swindon SN2 1EU (01793 411500; Institutes of Terrestrial Ecology at Banchory, Bangor, Edinburgh, Furzebrook, Merlewood and Monks Wood. The Government proposes to transfer Scottish and Welsh activities to the Countryside Commission for Scotland and a new Welsh Countryside Commission.)

Nature Conservancy Council, Northminster House, Peterborough PE1 1UA (01733 340345)

Royal Commission on Environmental Pollution, Church House, Great Smith Street, London SW1P 3BZ (0171 276 20800)

Scottish Development Department Headquarters, New St. Andrew's House, St. James Centre, Edinburgh EH1 3TG (0131 556 8400)

> Environmental Protection, 27 Perth Street, Edinburgh EH3 5RB (0131 244 3060)
> HM Industrial Pollution Inspectorate for Scotland, 27 Perth Street, Edinburgh EH3 5RB (0131 244 3060)

Warren Spring Laboratory, has now merged with AEA Technology to form the National Environmental Technology Centre.

Welsh Office, Water and Environmental Protection Division, Cathays Park, Cardiff CF7 3NQ (01222 823174)

*Non-governmental organizations*

Alternative Technology Council, Machynlleth, Powys, SY20 9AZ (01654 702400)

Association for the Protection of Rural Scotland, Gladstone's Land, 483 Lawnmarket, Edinburgh EH1 2NT (0131 225 7013)

British Society for Social Responsibility in Science, 25 Horsell Road, London NX5 1XL (0171 607 9615)

Chartered Institution of Water and Environmental Management, 15 John Street, London WC1N 2EB (0171 831 3110)

Confederation of British Industry, Centre Point, New Oxford Street, London WC1A 1DU (0171 379 7400)

Conservation Foundation, 1 Kensington Gore, London SW7 2AR (0171 923 8842)

Consumers Association, 2 Marylebone Road, London NW1 4DX (0171 486 5544)

Council for National Parks, 1 Lynmouth Avenue, Abington Vale, Northampton NN3 3LT (01604 B350 48)

Council for the Protection of Rural Wales, Tybwyn, 31 High Street, Welshpool, Powys SY21 7JP (01938 552525) (branch in every county)

CPRE (Council for the Protection of Rural England), Warwick House, 25 Buckingham Palace Road, London SW1W 0PP (0171 239 5959) (branch in every county)

Friends of the Earth, 26–28 Underwood Street, London N1 7JQ (0171 490 1555)

Green Party, 1A Waterloo Road, London N19 5NJ (0171 272 4474)

Greenpeace, Canonbury Villas, London N1 2PN (0171 354 5100)

National Society for Clean Air and Environmental Protection, 136 North Street, Brighton, East Sussex BN1 1RG (01273 326313)

National Trust for Places of Historic Interest or Natural Beauty, 36 Queen Anne's Gate, London SW1H 9AS (0171 222 9251)

Royal Society of Arts, 8 John Adam Street, London WC2N 6EZ (0171 930 5115 (Pollution Abatement Technology Award Scheme for innovation)

Town and Country Planning Association, 17 Carlton House Terrace, London SW1Y 5AS (0171 930 8903)

Ulster Society for the Preservation of the Countryside, Peskett House, 2/2a Windsor Road, Belfast 9 (01232 381304)

The Wildlife Trust, Royal Society for Nature Conservation, The Green, Whitham Park, Waterside South, Lincoln LN57 7JR (01522 544400)

## Air pollution

*Departments of state and other public bodies*

European Ozone Research Co-Ordinating Unit, High Cross, Madingley Road, Cambridge, Cambridgeshire CB3 0ET (01223 251632)

Her Majesty's Industrial Pollution Inspectorate for Scotland, 27 Perth Street, Edinburgh EH3 5RB (0131 244 3060)

Meteorological Office, London Road, Bracknell, Berkshire RG12 2SZ (01344 420242)

*Non-governmental organizations*

Association for the Conservation of Energy, Westgate House, Prebend Street, London N1 8TT (0171 935 1495)

Atmospheric Research and Information Centre, Department of Environmental and Geological Studies, Manchester Metropolitan University, Chester Street, Manchester M1 5GD (0161 247 1593)

Confederation of British Industry, Centre Point, New Oxford Street, London WC1A 1DU (0171 379 7400)

Conservation Foundation, 1 Kensington Gore, London SW7 2AR (0171 923 8842)

Consumer's Association, 2 Marylebone Road, London NW1 4DX (0171 486 5544)

Institute of Refrigeration, 76 Mill Lane, Carshalton, Surrey SM5 2JR (0181 647 7033)

Motor Industry Research Association, Watling Street, Nuneaton, Warwickshire CV10 0TU (01203 348541) (research on technical, social and legal effects, also provides an engineering facility for the industry)

## Countryside, farming, wildlife

*Departments of state and other public bodies*

British Association of Nature Conservationists, Northamptonshire Wildlife Trust, Lings House, Billings Lings, Northamptonshire NN3 4BE (01604 405 285)

Countryside Commission for Wales, Plas Penthos, Ifordd Penthosd, Bangor, Gwynedd LL57 2LQ (01248 370444)

Department of Agriculture and Fisheries for Scotland, Pentland House, 47 Robb's Loan, Edinburgh EH14 1TY (0131 556 8400)

Farming and Wildlife Advisory Group, National Agricultural Centre, Stoneleigh, Kenilworth, Warks CV8 2RX (01203 696699)

Health and Safety Executive, Agricultural Inspectorate Division, Magdalen House, Stanley Precinct, Bootle, Merseyside L20 3QZ (0151 951 4000)

Ministry of Agriculture, Fisheries and Food, Whitehall Place, London SW1A 2HH (0171 270 3000)

> Conservation Policy and Environmental Protection Divisions, Nobel/Ergon House, 17 Smith's Square, London SW1P 3HX (0171 238 6563)

*Non-governmental organizations*

BIPRA (British Industrial Biological Research Association), Woodmansterne Road, Carshalton, Surrey SM5 4DS (0181 652 1000) (toxicology of environmental chemicals, food additives, etc.)

British Trust for Ornithology, The Nunnery, Nunnery Place, Thetford IP24 2PU (01842 750 050)

Chemical Industries Association, King's Building, Smith Square, London SW1 3JJ (0171 834 3399)

Henry Doubleday Research Association, Wolston Lane, Ryton-on-Dunsmore, Coventry CV8 2LG (01203 303 517)

National Farmers' Union, 22 Long Acre, London WC2E 9LY (0171 235 5077)

Royal Society for the Protection of Birds, The Lodge, Sandy, Bedfordshire SG19 2DL (01767 680551)

Soil Association, 86 Colston Street, Bristol BS1 5BB (0117 929 0661)

World Wide Fund for Nature – UK, Panda House, Weyside Park, Catteshall Lane, Godalming, Surrey GU7 1XR (01483 426 444)

## Marine pollution

*Departments of state and other bodies*

Coastguard Control Agency, 105 Commercial Road, Southampton, Hampshire SO15 1EG (01703 329 1000) (deal with cleaning up or incidents of pollution)

International Petroleum Industry Environmental Conservation Association

(IPIECA), 2nd Floor, Monmouth House, 87–93 Westbourne Grove, London W2 4UL (0171 221 2026)

Marine Biological Association of the United Kingdom, The Laboratory, Citadel Hill, Plymouth, Devon PL1 2PB (01752 222772)

Marine Safety Agency, Spring Place, 105 Commercial Road, Southampton, Hampshire SO15 1EG (01703 329 1000) (deal with legislation to prevent pollution)

Ministry of Agriculture, Fisheries and Food, Fisheries Laboratory (and Fisheries Radiobiological Laboratory), Pakefield Road, Lowestoft, Suffolk NR33 0HT (01502 562244)

Scottish Marine Biological Association, Dunstaffnage Marine Research Laboratory, PO Box 3, Oban, Argyll PA34 4AD (01631 562244) (NERC Research Institute)

*Non-governmental organizations*

Advisory Committee on Oil Pollution of the Sea (ACOPS), 11 Dartmouth Street, London SW1H 9BN (0171 779 3033)

British Oil Spill Control Association, 4th Floor, 30 Great Guildford Street, London SE1 0HS (0171 928 9199)

Institute of Petroleum, 61 New Cavendish Street, London W1M 8AR (0171 467 7100)

Marine Conservation Society, 9 Gloucester Road, Ross-on-Wye HR9 5BU (01989 566 017)

**Noise**

*Departments of state and other public bodies*

British Railway Board, Euston House, 24 Eversholt Street, London NW1 1DZ (0171 928 5151)

Building Research Establishment, Building Research Station, Garston, Watford WD2 7JR (01923 894040) (work on traffic noise)

Civil Aviation Authority, 45–49 Kingsway, London WC2B 6TE (0171 379 7311)

Department of the Environment, 2 Marsham Street, London SW1P 3EB (0171 276 3000)

*Non-governmental organizations*

Association of Noise Consultants, 6 Trap Road, Guilden Morden, Nr Royston, Herts S98 0JE (01763 852958)

Chartered Institution of Health, Chadwick House, Hatfield, London SE18 8DJ (0171 928 6006)

Federation of Heathrow Anti-Noise Groups, 95a Walton Road, East Molesey, London KT8 0DR (0181 941 2344)

Institute of Sound and Vibration Research, University of Southampton, Highfield, Southampton SO9 5NH (01703 510 0015)

## Nuclear

*Departments of state and other public bodies*

Department of Energy, Thames House, Millbank, London SW1P 4QJ (0171 215 5000)

Harwell Laboratory, Didcot, Oxford OS11 0RA (01235 433942)

Her Majesty's Nuclear Installations Inspectorate, St. Peter's House, Stanley Precinct, Bootle, Merseyside L20 3QZ (0151 951 3527)

National Radiological Protection Board, Chilton, Didcot OX11 0RQ (01235 831600)

Scottish Power, Cathcart House, Spean Street, Glasgow G44 4BE (0141 637 7177) (nuclear power in Scotland)

UK Nirex Ltd, Curie Avenue, Harwell, Didcot, Oxfordshire OX11 0RH (01235 825500)

*Non-governmental organizations*

British Nuclear Energy Society, 1–7 Great George Street, London SW1P 3AA (0171 222 7722)

Institution of Nuclear Engineers, 1 Penerley Road, London SE6 2LQ (0181 698 1500)

## Occupational health and safety

*Departments of state, research establishments and other public bodies*

Health and Safety Commission, Rose Court, 2 Southwark Bridge, London SE1 9HF (0171 717 6000)

Medical Research Council, Environmental Epidemiology Unit, Southampton General Hospital, Southampton SO16 6YD (01703 777624)

*Non-governmental organizations*

British Occupational Hygiene Society, Suite 2, Georgian House, Great Northern Road, Derby DE1 1LT (01332 298101)

Institution of Occupational Safety and Health, 103 The Grange, Highfield Drive, Wigston, Leicester LE8 1NN (0116 2571399)

Society for the Prevention of Asbestosis and Industrial Diseases, 38 Drapers Road, Enfield, Middlesex EN2 8LU (01707 873025)

Society of Occupational Medicine, 6 St. Andrew's Place, Regents Park, London NW1 4LB (0171 486 2641)

Trades Union Congress, Great Russell Street, London WC1B 3LS (0171 636 4030)

## Public health, toxicology

*Departments of state and other public bodies*

British Toxicology Society, c/o Institute of Biology, 20 Queensberry

Place, London SW7 2DZ (0171 581 8333)

Department of Health, Richmond House, 79 Whitehall, London SW1A 2NS (0171 210 3000)

Medical Research Council, Environmental Epidemiology Unit, Southampton General Hospital, Southampton SO16 6YD (01703 777624)

National Environmental Technology Centre, Culham, Abingdon, Oxfordshire OS14 3DB (01235 463040)

Warren Spring Laboratory, has now merged with AEA Technology to form the National Environmental Technology Centre

*Non-governmental organizations*

Association of County Public Health Officers, c/o N.J. Durnford, Wiltshire County Council, Trowbridge, Wiltshire BA14 8JW (01225 753641, ext. 3543)

BIBRA (British Industrial Biological Research Association), Woodmansterne Road, Carshalton, Surrey SM5 4DS (0181 652 1000) (toxicology of environmental chemicals, food additives, etc.)

Chemical Industries Association, King's Building, Smith Square, London SW1 3JJ (0171 834 3399)

Institution of Environmental Health Officers, Chadwick House, 48 Rushworth Street, London SE1 0RB (0171 928 6006)

Rapra Technology, Shawbury, Shrewsbury SY4 4NR (01939 250383) (rubber and plastics)

Royal Environmental Health Institute of Scotland, Glasgow G1 1TX (0141 227 4397)

Royal Institute of Public Health and Hygiene, 28 Portland Place, London W1N 4DE (0171 580 2731)

Royal Society of Chemistry, Burlington House, Piccadilly, London S1V 0BN (0171 437 8658)

Royal Society of Health, 30a St. George's Drive, London SW1V 4BH (0171 630 0121)

Society of Chemical Industry, 14 Belgrave Square, London SW1X 8PS (0171 235 3681)

## Wastes and waste treatment and disposal

*Departments of state and other public bodies*

Department of the Environment, Directorate Air Noise and Wastes, 43 Marsham Street, London SW1P 3EB (0171 276 3000) (environmental quality)

Hazardous Materials Service, Building 151, Harwell Laboratory, Didcot, Oxford OX11 0RA (01235 433942)

Harwell Laboratory, Didcot, Oxford OX11 0RA (01235 433942)

Local Authority Recycling Advisory Committee, c/o London Waste Regulation Authority, 20 Albert Embankment, London SE1 7TJ (0171

587 3075)

London Waste Regulatory Authority, Hampton House, 20 Albert Embankment, London SE1 7TJ (0171 587 3000)

National Environmental Technology Centre, Culham, Abingdon, Oxfordshire OS14 3DB (01235 463040)

Warren Springs Laboratory has now merged with AEA Technology to form the National Environmental Technology Centre (funded by the Department of Trade and Industry and the Department of the Environment mainly for the benefit of small businesses).

Waste Management Information Bureau, AEA Technology, National Environmental Technology Centre, F6 Culham, Abingdon OX14 3DB (01235 463 162)

*Non-governmental organizations*

Aluminium Can Recycling Association, 5 Gatsby Court, 176 Holliday Street, Birmingham B1 1TJ (0121 633 4656)

British Metals Federation, 16 High Street, Brampton, Huntingdon, Cambridgeshire PE18 8TU (01480 455249)

British Secondary Metals Association, 25 Park Road, Runcorn, Cheshire WA17 4SS (01928 572400)

Chartered Institute of Public Finance and Accountancy, 2 Robert Street, London WC2N 6BH (0171 895 8823)

Chemical Recovery Association,

Industry Council for Packaging and the Environment (INCPEN), Tenterden House, 3 Tenterden Street, London W1R 9AH (0171 409 0949)

Institute of Wastes Management, 9 Saxon Court, St. Peter's Gardens, Northampton NN1 1SX (01604 20426)

National Association of Waste Disposal Contractors, Mountbarrow House, 6–20 Elizabeth Street, London SW1W 9RB (0171 824 8882)

National Association of Waste Disposal Officers, c/o Waste Disposal Authority, Hertfordshire County Council, Goldings, North Road, Hertford SG14 2PY (01992 556160)

PIRA International, Randalls Road, Leatherhead, Surrey KT22 7RU (01372 37616)

Rapra Technology, Shawbury, Shrewsbury SY4 4NR (01939 250383) (rubber and plastics)

Society of Chemical Industry, 14–15 Belgrave Square, London SW1X 8PS (0171 235 3681)

Tidy Britain Group, Premier House, 12–13 Hatton Garden, London EC1N 2NH (0171 831 1462)

Waste Watch, Hobart House, Grosvenor Place, London SW1X 7AW (0171 245 9998)

World Resource Foundation, Bridge House, High Street, Tonbridge, Kent TN9 1DP (01732 368 333)

## Water and water pollution

*Departments of state and other public bodies*

British Waterways, Willow Grange, Church Road, Watford, Hertfordshire WD1 3QA (01923 226 422)

Department of the Environment, Water Directorate, Romney House, 43 Marsham Street, London SW1P 3PY (0171 276 3000)

Her Majesty's Inspectorate of Pollution, Romney House, 43 Marsham Street, London SW1P 3PY (0171 276 8142) (Regional Offices in Bedford, Bristol, Cardiff, Fleet, Lancaster, Leeds and Sutton Coldfield)

National Rivers Authority, 30–34 Albert Embankment, London SE1 7TL (0171 820 0101) (Please note that from 1 April 1996 the National Rivers Authority will be changing its name to The Environment Agency)

National Water Research Centre, Henley Road, Medmenham, PO Box 16, Marlow, Buckinghamshire SL7 2HD (01491 571531)

Scottish River Purification Boards Association, 1 South Street, Perth PH2 8NJ (01738 627 989)

Water Research Centre Scotland, Unit 16, Beta Centre, Stirling University, Innovations Park, Stirling SK9 4NF (01786 471580)

Water Services Association of England and Wales, 1 Queen Anne's Gate, London SW1H 9BT (0171 957 4567)

Welsh Office, Water and Environmental Protection Division, Cathays Park, Cardiff CF7 3NQ (01222 823174)

*Non-governmental organizations*

Association of Water Officers, 15 Market Place, South Shields, Tyne and Wear NE33 1JQ (0191 456 3882)

British Effluent and Water Association, 1 Queen Anne's Gate, London SW1H 9BT (0171 957 4554)

British Water, 1 Queen Anne's Gate, London SW1H 9BT (0171 957 4554)

Freshwater Biological Association, The Ferry House, Far Sawrey, Ambleside, Cumbria LA22 0LP (015394 42468)

Institute of Hydrology, Maclean Building, Crowmarsh, Gifford, Wallingford, Oxfordshire OX10 8BB (01491 838800) (NERC Research Institute)

Institution of Water and Environmental Management, 15 John Street, London WC1 2EB (0171 430 0899)

Institution of Water Officers, 12 Summerhill Terrace, Newcastle upon Tyne NE4 6EB (0191 230 5150)

Water Companies Association, 1 Queen Anne's Gate, London SW1H 9BT (0171 957 4567)

Water Services Association, 1 Queen Anne's Gate, London SW1H 9BT (0171 957 4567)

**International**

For comprehensive sources of information on international organizations, see the Directories Section below.

Council of Europe, BP 431, R6-67006 Strasbourg Cedex (88 61 49 61)

Food and Agriculture Organization, Via delle Terme di Carcalla, 00100 Rome (6 57971)

Institute for European Environmental Policy, 158 Buckingham Palace Road, London WC1H 0DD (0171 824 8787)

International Association on Water Pollution Research and Control, 1 Queen Anne's Gate, London SW1H 9BT (071 222 3848)

International Atomic Energy Agency, POB 100, Wagramerstrasse 5, 1400 Vienna (1 0222 2360)

International Commission on Radiological Protection, POB 35, Didcot OX11 0RJ (0235 833929)

International Maritime Organisation, 4 Albert Embankment, London SE1 7SR (071 735 7611)

International Solid Wastes and Public Cleansing Association, V. Farimagsgade 29, DK-1606 Copenhagen V (01 15 65 65)

International Union for Conservation of Nature and Natural Resources (IUCN), Avenue du Mont-Blanc, CH-1196 Gland, Switzerland (022 64 71 81)

International Union of Air Pollution Prevention Associations Secretariat, 136 North Street, Brighton BN1 1RG (01273 3263§6)

International Water Supply Association, 1 Queen Anne's Gate, London SW1H 9BT (0171 957 4567)

Nuclear Energy Agency of OECD, 38 Boulevard Suchet, 75016 Paris (1 45 24 82 00)

Oslo and Paris Commissions, New Court, 48 Carey Street, London WC2A 2JE (secretariat for the Oslo and Paris Conventions controlling dumping and pollution in the North Sea and north-east Atlantic; 0171 242 9927)

United Nations Environment Programme, POB 30552, Nairobi (2 333930)

World Health Organization, Avenue Appia, 1211 Geneva 27 (022 912 111)

World Meteorological Organization, Case Postale 5, 1211 Geneva 20 (022 346 400)

World Wide Fund for Nature, World Conservation Centre, Avenue du Mont-Blanc, CH-1196 Gland, Switzerland (022 64 71 81)

**United States of America**

This is only a very small selection of national American organizations. In particular, many governmental bodies and the often important State and local organizations have been omitted. The national Wildlife Federation's *Conservation Directory* is a very useful printed source of information on the hundreds of organizations in the field.

*Governmental organizations*

Council on Environmental Quality, 722 Jackson Place, NW, Washington, DC 20503 (202 395 5750)

United States Congress, House of Representatives Committee on Energy and Commerce, 2125 Rayburn House Office Building, Washington, DC 20515 (202 225 2927)

United States Congress, House of Representatives Committee on Interior and Insular Affairs, Room 1324, Longworth House Office Building, Washington, DC 20515 (202 225 2761; various subcommittees on environmental matters)

United States Congress, Senate Committee on Energy and Natural Resources, Room SD-364, Dirksen Building, Washington, DC 20510 (202 224 4971)

United States Congress, Senate Committee on Environment and Public Works, Room SD-458, Dirksen Building, Washington, DC 20510 (202 224 6176)

United States Department of Agriculture, 14th St. and Independence Ave., SW, Washington, DC 20250 (202 447 2791)

United States Department of Energy, Forrestal Building, 1000 Independence Ave., SW, Washington, DC 20585 (202 586 6210)

United States Department of the Interior, Interior Building, C St. between 18th and 19th, NW, Washington, DC 20240 (202 343 1100)

United States Environmental Protection Agency, 401 M St., SW, Washington, DC 20460 (202 382 2090)

United States Nuclear Regulatory Commission, Washington, DC 20555 (301 492 7000)

*Non-governmental organizations*

Air Pollution Control Association, H3W Gateway Building, Pittsburgh, PA 15222 (412 232 3444)

Conservation International, 1015 10th St., NW, Suite 1000, Washington, DC 20036 (202 429 5660)

Earth First!, 1700 Juliet St., Austin, TX 78704 (812 443 8831)

Earth Island Institute, 300 Broadway, Suite 28, San Francisco, CA 94133 (415 788 3666)

Environmental Defense Fund, 1616 P. St., NW, Washington, DC 20036 (202 387 3500)

Environmental Policy Institute, 218 D St., SE, Washington, DC 20003 (202 544 2600)

Friends of the Earth, 218 D St., SE, Washington, DC 20003 (202 544 2600)

Friends of the Earth – Pacific Rim, 4512 University Way, NE, Seattle, WA 98105 (206 789 6139)

Greenpeace, 1436 U. St., NW, Washington, DC 20009 (202 462 1177)

International Rivers Network, 301 Broadway, Suite B, San Francisco, CA 94133 (415 398 4404)

National Audubon Society, 801 Pennsylvania Ave., SE, Suite 301, Washington, DC 20003 (202 547 9009)

National Environmental Health Association, 720 S. Colorado Blvd., South Tower, Denver, CO 80222 (303 756 9090)

National Wildlife Federation, 1400 16th St., NW, Washington, DC 20036 (202 797 6601)

Rainforest Action Network, 301 Broadway, Suite A, San Francisco, CA 94133 (415 398 4404)

Sierra Club, 408 C St., NE, Washington, DC 20002 (202 547 1144)

Water Pollution Control Federation, 601 Wythe St., Alexandria, VA 22314 (703 684 2400)

World Resources Institute, 1709 New York Ave., NW, 7th Floor, Washington, DC 20006 (202 638 6300)

World Wide Fund for Nature – US, 1250 24th St., NW, Washington, DC 20037 (202 293 4800)

## Directories

For further information on organizations in the field of the environment and pollution control, there are many specialized and general directories which are kept more or less up-to-date:

Cairns, T. *et al.* (1988) *Directory of Environmental Journals and Media Contacts*, Reader's Digest (contacts in television, radio and newspapers)

Civic Trust (1988) *Environmental Directory: National and Regional Organisations of Interest to Those Concerned with Amenity and the Environment*, 7th edn., Civic Trust

*Civil Service Year Book*, HMSO (details of departments, offices, personnel)

*Councils, Committees and Boards: a Handbook of Advisory, Consultative, Executive and Similar Bodies in British Public Life* (1989), 7th edn., edited by I. G. Anderson, CBD Research

*Directory for the Environment: Organizations in Britain and Ireland* 1985–86 (1986), 3rd edn., edited by M. J. C. Barker, Routledge & Kegan Paul

*The Green Index: a Directory of Environmental Organisations in Britain and Ireland* (1990), edited by J. E. Milner and others for Environmental Information Bureau, Cassell

*Industrial Research in the United Kingdom: Guide to Organizations and Programmes* (1989) 13th edn., Longman

*Municipal Yearbook and Public Services Directory*, 2 volumes, Municipal Journal (associations, individuals and sources of information; sections include Health: Environmental; Refuse and salvage; Town and country planning; Water; Organizations; plus details of all authorities, with names of members, chief officers, etc.)

*NSCA Members Handbook*, annual, National Society for Clean Air and Environmental Protection (divisional membership, directory of organizations, directory of consultants, trade directory and buyer's guide)

For international organizations, see:

*Encyclopedia of Associations: International Organizations* (1988–1989), 2 volumes with supplement, Gale Research

*Europa World Year Book, volume* 1, Europa Publications (Part 1: International Organizations)

For the USA, see:

*Conservation Directory* 1989 (1989) National Wildlife Federation

*Encyclopedia of Associations, volume* 1: *National Organizations of the U.S.* (1989), 3 parts, 24th edn., Gale Research

*Encyclopedia of Associations: Regional, State, and Local Organizations* (1987–) 7 volumes, Gale

A reliable source of up-to-date information on Parliament and its business is the Public Information Office, House of Commons Library, Norman Shaw (North) Building, Victoria Embankment, London SW1A 2JF (0171 219 4623)

Finally, the information and public relations arm of the government is the Central Office of Information, Hercules Road, Westminster Bridge Road, London SE1 7DV (0171 928 2345)

## The Periodic Table

*giving atomic number and chemical symbol for each element*

| 1 H | | | | | | | | | | | | | | | | | 2 He |
|---|---|---|---|---|---|---|---|---|---|---|---|---|---|---|---|---|---|
| 3 Li | 4 Be | | | | | | | | | | | 5 B | 6 C | 7 N | 8 O | 9 F | 10 Ne |
| 11 Na | 12 Mg | ← | | TRANSITION ELEMENTS | | | | | | | → | 13 Al | 14 Si | 15 P | 16 S | 17 Cl | 18 Ar |
| 19 K | 20 Ca | 21 Sc | 22 Ti | 23 V | 24 Cr | 25 Mn | 26 Fe | 27 Co | 28 Ni | 29 Cu | 30 Zn | 31 Ga | 32 Ge | 33 As | 34 Se | 35 Br | 36 Kr |
| 37 Rb | 38 Sr | 39 Y | 40 Zr | 41 Nb | 42 Mo | 43 Tc | 44 Ru | 45 Rh | 46 Pd | 47 Ag | 48 Cd | 49 In | 50 Sn | 51 Sb | 52 Te | 53 I | 54 Xe |
| 55 Cs | 56 Ba | 57* La | 72 Hf | 73 Ta | 74 W | 75 Re | 76 Os | 77 Ir | 78 Pt | 79 Au | 80 Hg | 81 Tl | 82 Pb | 83 Bi | 84 Po | 85 At | 86 Rn |
| 87 Fr | 88 Ra | 89† Ac | | | | | | | | | | | | | | | |

| *LANTHANONS | 58 Ce | 59 Pr | 60 Nd | 61 Pm | 62 Sm | 63 Eu | 64 Gd | 65 Tb | 66 Dy | 67 Ho | 68 Er | 69 Tm | 70 Yb | 71 Lu |
|---|---|---|---|---|---|---|---|---|---|---|---|---|---|---|

| †ACTINONS | 90 Th | 91 Pa | 92 U | 93 Np | 94 Pu | 95 Am | 96 Cm | 97 Bk | 98 Cf | 99 Es | 100 Fm | 101 Md | 102 No | 103 Lr |
|---|---|---|---|---|---|---|---|---|---|---|---|---|---|---|

 ARTIFICIAL ELEMENTS

## Table of Chemical Elements

Atomic masses are based on the 1969 international agreed values of the International Union of Pure and Applied Chemistry, the basis being the carbon-12 isotope. Values in brackets indicate the mass numbers of the most stable isotopes. See also the Periodic Table.

| Symbol | Name | Atomic | |
| --- | --- | --- | --- |
| | | No. | Mass |
| Ac | Actinium | 89 | (227) |
| Ag | Silver | 47 | 107.868 |
| Al | Aluminium | 13 | 26.9815 |
| Am | Americium | 95 | (243) |
| Ar | Argon | 18 | 39.948 |
| As | Arsenic | 33 | 74.9216 |
| At | Astatine | 85 | (210) |
| Au | Gold | 79 | 196.9665 |
| B | Boron | 5 | 10.81 |
| Ba | Barium | 56 | 137.34 |
| Be | Beryllium | 4 | 9.0122 |
| Bi | Bismuth | 83 | 208.9806 |
| Bk | Berkelium | 97 | (249) |
| Br | Bromine | 35 | 79.904 |
| C | Carbon | 6 | 12.011 |
| Ca | Calcium | 20 | 40.08 |
| Cd | Cadmium | 48 | 112.40 |
| Ce | Cerium | 58 | 140.12 |
| Cf | Californium | 98 | (251) |
| Cl | Chlorine | 17 | 35.453 |
| Cm | Curium | 96 | (247) |
| Co | Cobalt | 27 | 58.9332 |
| Cr | Chromium | 24 | 51.996 |
| Cs | Caesium | 55 | 132.9055 |
| Cu | Copper | 29 | 63.546 |
| Dy | Dysprosium | 66 | 162.50 |
| Er | Erbium | 68 | 167.26 |
| Es | Einsteinium | 99 | (254) |
| Eu | Europium | 63 | 151.96 |
| F | Fluorine | 9 | 18.9984 |
| Fe | Iron (ferrum) | 26 | 55.847 |
| Fm | Fermium | 100 | (253) |
| Fr | Francium | 87 | (223) |
| Ga | Gallium | 31 | 69.72 |
| Gd | Gadolinum | 64 | 157.25 |
| Ge | Germanium | 32 | 72.59 |
| H | Hydrogen | 1 | 1.0080 |
| Ha | Hahnium | 105 | — |
| He | Helium | 2 | 4.0026 |

| Symbol | Name | Atomic | |
|--------|------|--------|--------|
| | | No. | Mass |
| Hf | Hafnium | 72 | 178.49 |
| Hg | Mercury | 80 | 200.59 |
| Ho | Holmium | 67 | 164.9303 |
| I | Iodine | 53 | 126.9045 |
| In | Indium | 49 | 114.82 |
| Ir | Iridium | 77 | 192.22 |
| K | Potassium | 19 | 39.102 |
| Kr | Krypton | 36 | 83.80 |
| La | Lanthanum | 57 | 138.9055 |
| Li | Lithium | 3 | 6.941 |
| Lu | Lutetium | 71 | 174.97 |
| Lr | Lawrencium | 103 | — |
| Md | Mendelevium | 101 | (256) |
| Mg | Magnesium | 12 | 24.305 |
| Mn | Manganese | 25 | 54.9380 |
| Mo | Molybdenum | 42 | 95.94 |
| N | Nitrogen | 7 | 14.0067 |
| Na | Sodium (natrium) | 11 | 22.9898 |
| Nb | Niobrium (columbium) | 41 | 92.9064 |
| Nd | Neodymium | 60 | 144.24 |
| Ne | Neon | 10 | 20.179 |
| Ni | Nickel | 28 | 58.71 |
| No | Nobelium | 102 | (254) |
| Np | Neptunium | 93 | (237) |
| O | Oxygen | 8 | 15.9994 |
| Os | Osmium | 76 | 190.2 |
| P | Phosphorus | 15 | 30.9738 |
| Pa | Protactinium | 91 | 231.0359 |
| Pb | Lead (plumbum) | 82 | 207.2 |
| Pd | Palladium | 46 | 106.4 |
| Pm | Promethium | 61 | (145) |
| Po | Polonium | 84 | (210) |
| Pr | Praseodymium | 59 | 140.9077 |
| Pt | Platinum | 78 | 195.09 |
| Pu | Plutonium | 94 | (242) |
| Ra | Radium | 88 | 226.0254 |
| Rb | Rubidium | 37 | 85.4678 |
| Re | Rhenium | 75 | 186.2 |
| Rf | Rutherfordium | 104 | — |
| Rh | Rhodium | 45 | 102.9055 |
| Rn | Radon (niton) | 86 | (222) |
| Ru | Ruthenium | 44 | 101.07 |
| S | Sulphur | 16 | 32.06 |
| Sb | Antimony | 51 | 121.75 |
| Sc | Scandium | 21 | 44.9559 |
| Se | Selenium | 34 | 78.96 |
| Si | Silicon | 14 | 28.086 |

| Symbol | Name | Atomic | |
|--------|------|--------|--------|
| | | No. | Mass |
| Sm | Samarium | 62 | 150.4 |
| Sn | Tin (stannum) | 50 | 118.69 |
| Sr | Strontium | 38 | 87.62 |
| Ta | Tantalum | 73 | 180.9479 |
| Tb | Terbium | 65 | 158.9254 |
| Tc | Technetium | 43 | (99) |
| Te | Tellurium | 52 | 127.60 |
| Th | Thorium | 90 | 232.0381 |
| Ti | Titanium | 22 | 47.90 |
| Tl | Thallium | 81 | 204.37 |
| Tm | Thulium | 69 | 168.9342 |
| U | Uranium | 92 | 238.029 |
| V | Vanadium | 23 | 50.9414 |
| W | Tungsten (wolfram) | 74 | 183.85 |
| Xe | Xenon | 54 | 131.30 |
| Y | Yttrium | 39 | 88.9059 |
| Yb | Ytterbium | 70 | 173.04 |
| Zn | Zinc | 30 | 65.37 |
| Zr | Zirconium | 40 | 91.22 |

# Table of prefixes for SI units

| Prefix | Symbol | Factor |
|--------|--------|--------|
| tera | T | $10^{12} = 1\,000\,000\,000\,000$ |
| giga | G | $10^{9} = 1\,000\,000\,000$ |
| mega | M | $10^{6} = 1\,000\,000$ |
| kilo | k | $10^{3} = 1000$ |
| hecto | h | $10^{2} = 100$ |
| deca | da | $10^{1} = 10$ |
| deci | d | $10^{-1} = 0.1$ |
| centi | c | $10^{-2} = 0.01$ |
| milli | m | $10^{-3} = 0.001$ |
| micro | $\mu$ | $10^{-6} = 0.000001$ |
| nano | n | $10^{-9} = 0.000000001$ |
| pico | p | $10^{-12} = 0.000\,000\,000\,001$ |
| femto | f | $10^{-15} = 0.000\,000\,000\,000\,001$ |
| atto | a | $10^{-18} = 0.000\,000\,000\,000\,000\,001$ |

APPENDIX IV

# Conversion table for SI and British units

| Physical property | British unit | SI unit | SI unit | British unit | Cgs unit |
|---|---|---|---|---|---|
| length | 1 ft | 0.305 m | 1 m | 3.28 ft | 100 cm |
| | 1 mile | 1.61 km | 1 km | 0.621 mile | $10^5$ cm |
| area | 1 ft$^2$ | 0.0929 m$^2$ | 1 m$^2$ | 10.76 ft$^2$ | $10^4$ cm$^2$ |
| | 1 acre | 4047.0 m$^2$ | 1 km$^2$ | $2.471 \times 10^2$ acres | |
| | 1 acre | 0.405 ha | 1 ha | 2.471 acres | |
| volume | 1 ft$^3$ | 0.0283 m$^3$ | 1 m$^3$ | 35.31 ft$^3$ | $10^8$ cm$^2$ |
| | | | | | $10^6$ cm$^3$ |
| | 1 gall (UK) | 4.55 litres | 1 litre | 0.220 gallon (UK) | $10^3$ cm$^3$ |
| mass | 1 lb | 0.454 kg | 1 kg | 2.204 lb | $10^3$ g |
| density | 1 lb/ft$^3$ | 16.02 kg/m$^3$ | 1 kg/m$^3$ | 0.0624 lb/ft$^3$ | $10^3$ g/cm$^3$ |
| time | 1 sec | 1 s | 1 s | 1 second | 1 s |
| | | | a defined fraction of a solar day 3600 s = 1 h | | |
| velocity | 1 ft/s | 0.305 m/s | 1 m/s | 3.28 ft/s | 100 cm/s |
| | | | 1 km/h | 0.911 ft/s | |
| | 1 mile/h | 1.61 km/h | 1 km/h | 0.621 mile/h | 27.8 cm/s |

| Quantity | | | | |
|---|---|---|---|---|
| force (mass × acceleration) | 1 lbf | 4.45 N | 1 N(kg m/s²) | 0.225 lbf | $10^5$ dyn (dynes) (1 dyn = 1 g cm/s²) |
| pressure (force per unit area) | 1 lbf/ft² | 47.5 N/m² | 1 N n/m² (1 pascal – Pa) | 0.0207 lbf/ft² | 10 dyn/cm² |
| pressure in meteorology and acoustics | | | 1 bar | 29.53 in Hg | 750 mm Hg ($10^6$ dyn/cm²) |
| | | | $10^5$ N/m² | | |
| | | | 1.013 bar | 29.91 in Hg | 760 mm Hg |
| work (force × distance) energy | 1 ft lbf | 1.352 J | 1 J (Nm) (joule) | 0.738 ft lbf | $10^7$ ergs |
| heat equivalent | 1 Btu | 1.055 kJ | 1 kJ | 0.948 Btu | $10^{10}$ ergs |
| power | 1 ft lbf/s | 1.356 W | 1 W(J/s) | 0.738 ft lbf/s | $10^7$ ergs/s |
| (rate of doing work) | 1 hp | 0.746 kW | 1 kW | 1.34 hp | $10^{10}$ ergs/s |
| flow rate | 1 million gall/d | 0.0526 m³/s (Cumecs) | 1 m³ s⁻¹ (Cumecs) | 19.01 million gall/d | $10^6$ cm³/s |
| | | | 1 × $10^6$ m³/d | 220 million gall/d | |
| | | | 1 l/s | 13.20 gall/min | $10^3$ cm³/s |
| | 1 gall/min | 0.0761 s⁻¹ | | | |